Comprehensive Natural Products Chemistry

Comprehensive Natural Products Chemistry

Editors-in-Chief

Sir Derek Barton†
Texas A&M University, USA

Koji Nakanishi
Columbia University, USA

Executive Editor

Otto Meth-Cohn
University of Sunderland, UK

Volume 6
PREBIOTIC CHEMISTRY, MOLECULAR FOSSILS,
NUCLEOSIDES, AND RNA

Volume Editors

Dieter Söll
Yale University, USA

Susumu Nishimura
Banyu Tsukuba Research Institute, Japan

Peter B. Moore
Yale University, USA

1999

ELSEVIER

AMSTERDAM – LAUSANNE – NEW YORK – OXFORD – SHANNON – SINGAPORE – TOKYO

Elsevier Science Ltd., The Boulevard, Langford Lane, Kidlington, Oxford, OX5 1GB, UK

First edition 1999

Library of Congress Cataloging-in-Publication Data
Comprehensive natural products chemistry / editors-in-chief, Sir Derek Barton, Koji Nakanishi ; executive editor, Otto Meth-Cohn. -- 1st ed.
    p. cm.
  Includes index.
  Contents: v. 6. Prebiotic chemistry, molecular fossils, nucleosides, and RNA / volume editors Dieter Söll, Susumu Nishimura, Peter B. Moore
  1. Natural products. I. Barton, Derek, Sir, 1918-1998. II. Nakanishi, Koji, 1925- . III. Meth-Cohn, Otto.
QD415.C63 1999
547.7--dc21   98-15249

British Library Cataloguing in Publication Data
Comprehensive natural products chemistry
  1. Organic compounds
  I. Barton, Sir Derek, 1918-1998 II. Nakanishi Koji III. Meth-Cohn Otto
  572.5

ISBN 0-08-042709-X (set : alk. paper)
ISBN 0-08-043158-5 (Volume 6 : alk. paper)

♾™   The paper used in this publication meets the minimum requirements of the American National Standard for Information Sciences—Permanence of Paper for Printed Library Materials, ANSI Z39.48–1984.

Typeset by BPC Digital Data Ltd., Glasgow, UK.
Printed and bound in Great Britain by BPC Wheatons Ltd., Exeter, UK.

Contents

Introduction

For many decades, Natural Products Chemistry has been the principal driving force for progress in Organic Chemistry.

In the past, the determination of structure was arduous and difficult. As soon as computing became easy, the application of X-ray crystallography to structural determination quickly surpassed all other methods. Supplemented by the equally remarkable progress made more recently by Nuclear Magnetic Resonance techniques, determination of structure has become a routine exercise. This is even true for enzymes and other molecules of a similar size. Not to be forgotten remains the progress in mass spectrometry which permits another approach to structure and, in particular, to the precise determination of molecular weight.

There have not been such revolutionary changes in the partial or total synthesis of Natural Products. This still requires effort, imagination and time. But remarkable syntheses have been accomplished and great progress has been made in stereoselective synthesis. However, the one hundred percent yield problem is only solved in certain steps in certain industrial processes. Thus there remains a great divide between the reactions carried out in living organisms and those that synthetic chemists attain in the laboratory. Of course Nature edits the accuracy of DNA, RNA, and protein synthesis in a way that does not apply to a multi-step Organic Synthesis.

Organic Synthesis has already a significant component that uses enzymes to carry out specific reactions. This applies particularly to lipases and to oxidation enzymes. We have therefore, given serious attention to enzymatic reactions.

No longer standing in the wings, but already on-stage, are the wonderful tools of Molecular Biology. It is now clear that multi-step syntheses can be carried out in one vessel using multiple cloned enzymes. Thus, Molecular Biology and Organic Synthesis will come together to make economically important Natural Products.

From these preliminary comments it is clear that Natural Products Chemistry continues to evolve in different directions interacting with physical methods, Biochemistry, and Molecular Biology all at the same time.

This new Comprehensive Series has been conceived with the common theme of "How does Nature make all these molecules of life?" The principal idea was to organize the multitude of facts in terms of Biosynthesis rather than structure. The work is not intended to be a comprehensive listing of natural products, nor is it intended that there should be any detail about biological activity. These kinds of information can be found elsewhere.

The work has been planned for eight volumes with one more volume for Indexes. As far as we are aware, a broad treatment of the whole of Natural Products Chemistry has never been attempted before. We trust that our efforts will be useful and informative to all scientific disciplines where Natural Products play a role.

D. H. R. Barton†   K. Nakanishi   O. Meth-Cohn

Preface

It is surprising indeed that this work is the first attempt to produce a "comprehensive" overview of Natural Products beyond the student text level. However, the awe-inspiring breadth of the topic, which in many respects is still only developing, is such as to make the job daunting to anyone in the field. Fools rush in where angels fear to tread and the particular fool in this case was myself, a lifelong enthusiast and reader of the subject but with no research base whatever in the field!

Having been involved in several of the *Comprehensive* works produced by Pergamon Press, this omission intrigued me and over a period of gestation I put together a rough outline of how such a work could be written and presented it to Pergamon. To my delight they agreed that the project was worthwhile and in short measure Derek Barton was approached and took on the challenge of fleshing out this framework with alacrity. He also brought his long-standing friend and outstanding contributor to the field, Koji Nakanishi, into the team. With Derek's knowledge of the whole field, the subject was broken down into eight volumes and an outstanding team of internationally recognised Volume Editors was appointed.

We used Derek's 80th birthday as a target for finalising the work. Sadly he died just a few months before reaching this milestone. This work therefore is dedicated to the memory of Sir Derek Barton, Natural Products being the area which he loved best of all.

OTTO METH-COHN
Executive Editor

SIR DEREK BARTON

Sir Derek Barton, who was Distinguished Professor of Chemistry at Texas A&M University and holder of the Dow Chair of Chemical Invention died on March 16, 1998 in College Station, Texas of heart failure. He was 79 years old and had been Chairman of the Executive Board of Editors for Tetrahedron Publications since 1979.

Barton was considered to be one of the greatest organic chemists of the twentieth century whose work continues to have a major influence on contemporary science and will continue to do so for future generations of chemists.

Derek Harold Richard Barton was born on September 8, 1918 in Gravesend, Kent, UK and graduated from Imperial College, London with the degrees of B.Sc. (1940) and Ph.D. (1942). He carried out work on military intelligence during World War II and after a brief period in industry, joined the faculty at Imperial College. It was an early indication of the breadth and depth of his chemical knowledge that his lectureship was in physical chemistry. This research led him into the mechanism of elimination reactions and to the concept of molecular rotation difference to correlate the configurations of steroid isomers. During a sabbatical leave at Harvard in 1949–1950 he published a paper on the "Conformation of the Steroid Nucleus" (*Experientia*, 1950, **6**, 316) which was to bring him the Nobel Prize in Chemistry in 1969, shared with the Norwegian chemist, Odd Hassel. This key paper (only four pages long) altered the way in which chemists thought about the shape and reactivity of molecules, since it showed how the reactivity of functional groups in steroids depends on their axial or equatorial positions in a given conformation. Returning to the UK he held Chairs of Chemistry at Birkbeck College and Glasgow University before returning to Imperial College in 1957, where he developed a remarkable synthesis of the steroid hormone, aldosterone, by a photochemical reaction known as the Barton Reaction (nitrite photolysis). In 1978 he retired from Imperial College and became Director of the Natural Products Institute (CNRS) at Gif-sur-Yvette in France where he studied the invention of new chemical reactions, especially the chemistry of radicals, which opened up a whole new area of organic synthesis involving Gif chemistry. In 1986 he moved to a third career at Texas A&M University as Distinguished Professor of Chemistry and continued to work on novel reactions involving radical chemistry and the oxidation of hydrocarbons, which has become of great industrial importance. In a research career spanning more than five decades, Barton's contributions to organic chemistry included major discoveries which have profoundly altered our way of thinking about chemical structure and reactivity. His chemistry has provided models for the biochemical synthesis of natural products including alkaloids, antibiotics, carbohydrates, and DNA. Most recently his discoveries led to models for enzymes which oxidize hydrocarbons, including methane monooxygenase.

The following are selected highlights from his published work:

The 1950 paper which launched Conformational Analysis was recognized by the Nobel Prize Committee as the key contribution whereby the third dimension was added to chemistry. This work alone transformed our thinking about the connection between stereochemistry and reactivity, and was later adapted from small molecules to macromolecules e.g., DNA, and to inorganic complexes.

Barton's breadth and influence is illustrated in "Biogenetic Aspects of Phenol Oxidation" (*Festschr. Arthur Stoll*, 1957, 117). This theoretical work led to many later experiments on alkaloid biosynthesis and to a set of rules for *ortho-para*-phenolic oxidative coupling which allowed the predication of new natural product systems before they were actually discovered and to the correction of several erroneous structures.

In 1960, his paper on the remarkably short synthesis of the steroid hormone aldosterone (*J. Am. Chem. Soc.*, 1960, **82**, 2641) disclosed the first of many inventions of new reactions—in this case nitrite photolysis—to achieve short, high yielding processes, many of which have been patented and are used worldwide in the pharmaceutical industry.

Moving to 1975, by which time some 500 papers had been published, yet another "Barton reaction" was born—"The Deoxygenation of Secondary Alcohols" (*J. Chem. Soc. Perkin Trans. 1*, 1975, 1574), which has been very widely applied due to its tolerance of quite hostile and complex local environments in carbohydrate and nucleoside chemistry. This reaction is the chemical counterpart to ribonucleotide→ deoxyribonucleotide reductase in biochemistry and, until the arrival of the Barton reaction, was virtually impossible to achieve.

In 1985, "Invention of a New Radical Chain Reaction" involved the generation of carbon radicals from carboxylic acids (*Tetrahedron*, 1985, **41**, 3901). The method is of great synthetic utility and has been used many times by others in the burgeoning area of radicals in organic synthesis.

These recent advances in synthetic methodology were remarkable since his chemistry had virtually no precedent in the work of others. The radical methodology was especially timely in light of the significant recent increase in applications for fine chemical syntheses, and Barton gave the organic community an entrée into what will prove to be one of the most important methods of the twenty-first century. He often said how proud he was, at age 71, to receive the ACS Award for Creativity in Organic Synthesis for work published in the preceding five years.

Much of Barton's more recent work is summarized in the articles "The Invention of Chemical Reactions—The Last 5 Years" (*Tetrahedron*, 1992, **48**, 2529) and "Recent Developments in Gif Chemistry" (*Pure Appl. Chem.*, 1997, **69**, 1941).

Working 12 hours a day, Barton's stamina and creativity remained undiminished to the day of his death. The author of more than 1000 papers in chemical journals, Barton also held many successful patents. In addition to the Nobel Prize he received many honors and awards including the Davy, Copley, and Royal medals of the Royal Society of London, and the Roger Adams and Priestley Medals of the American Chemical Society. He held honorary degrees from 34 universities. He was a Fellow of the Royal Societies of London and Edinburgh, Foreign Associate of the National Academy of Sciences (USA), and Foreign Member of the Russian and Chinese Academies of Sciences. He was knighted by Queen Elizabeth in 1972, received the Légion d'Honneur (Chevalier 1972; Officier 1985) from France, and the Order of the Rising Sun from the Emperor of Japan. In his long career, Sir Derek trained over 300 students and postdoctoral fellows, many of whom now hold major positions throughout the world and include some of today's most distinguished organic chemists.

For those of us who were fortunate to know Sir Derek personally there is no doubt that his genius and work ethic were unique. He gave generously of his time to students and colleagues wherever he traveled and engendered such great respect and loyalty in his students and co-workers, that major symposia accompanied his birthdays every five years beginning with the 60th, ending this year with two celebrations just before his 80th birthday.

With the death of Sir Derek Barton, the world of science has lost a major figure, who together with Sir Robert Robinson and Robert B. Woodward, the cofounders of *Tetrahedron*, changed the face of organic chemistry in the twentieth century.

Professor Barton is survived by his wife, Judy, and by a son, William from his first marriage, and three grandchildren.

A. I. SCOTT
Texas A&M University

Reprinted from *Tetrahedron*, 1998, **54**, 8847
Photograph courtesy of Library and Information Centre, Royal Society of Chemistry. © The Nobel Foundation

Contributors to Volume 6

Dr. S. Altman
Department of Molecular, Cellular and Developmental Biology, Yale University, 266 Whitney Avenue, New Haven, CT 06511, USA

Dr. B. L. Bass
Howard Hughes Medical Institute, University of Utah, 6110a Eccles Institute of Human Genetics, Building 533, Salt Lake City, UT 84112, USA

Dr. J. H. Cate
Department of Biology, Sinsheimer Laboratories, University of California, Santa Cruz, CA 95064, USA

Dr. A. E. Dahlberg
Department of Molecular Biology, Cell Biology and Biochemistry, Brown University, Box G-J4, Providence, RI 02912, USA

Dr. J. A. Doudna
Department of Molecular Biophysics and Biochemistry, Yale University, 266 Whitney Avenue, New Haven, CT 06520-8114, USA

Dr. F. Eckstein
Abteilung Chemie, Max-Planck-Institut für Experimentelle Medizin, Hermann-Rein-Strasse 3, D-37075 Göttingen, Germany

Dr. A. D. Ellington
Institute for Cellular and Molecular Biology, Molecular Biology Building (MBB) #3.424, Ellington Laboratory, University of Texas at Austin, 26th and Speedway, Austin, TX 78712, USA

Dr. C. Florentz
UPR 9002 du CNRS, Institut de Biologie Moléculaire et Cellulaire, 15, rue Rene Descartes, F-67084 Strasbourg-Cédex, France

Dr. R. Giegé
UPR 9002 du CNRS, Institut de Biologie Moléculaire et Cellulaire, 15, rue Rene Descartes, F-67084 Strasbourg-Cédex, France

Dr. S. T. Gregory
Department of Molecular Biology, Cell Biology and Biochemistry, Brown University, Box G-J4, Providence, RI 02912, USA

Dr. M. Helm
UPR 9002 du CNRS, Institut de Biologie Moléculaire et Cellulaire, 15, rue Rene Descartes, F-67084 Strasbourg-Cédex, France

Dr. Y. Komatsu
Graduate School of Pharmaceutical Sciences, Hokkaido University, Sapporo 060-0812, Japan

Dr. T. Kuwabara
Institute of Applied Biochemistry, University of Tsukuba, 1-1-1 Tennoudai, Tsukuba Science City 305, Japan

Dr. D. H. Mathews
Department of Chemistry, University of Rochester, Rochester, NY 14627-0216, USA

Dr. P. B. Moore
Department of Chemistry and Molecular Biophysics and Biochemistry, Yale University, 350 Edwards Street, New Haven, CT 06520-8017, USA

Dr Susumu Nishimura
Director, Banyu Tsukuba Research Institute, Okubo 3, Tsukuba 300-33, Japan

Dr. M. O'Connor
Department of Molecular Biology, Cell Biology and Biochemistry, Brown University, Box G-J4, Providence, RI 02912, USA

Dr. M. Öhman
Department of Molecular Genome Research, Stockholm University, SE-106 91 Stockholm, Sweden

Professor E. Ohtsuka
Graduate School of Pharmaceutical Sciences, Hokkaido University, Sapporo 060-0812, Japan

Dr. R. Parker
Department of Molecular and Cellular Biology, Howard Hughes Medical Institute, University of Arizona, Tucson, AZ 85721, USA

Dr. M. P. Robertson
Institute for Cellular and Molecular Biology, Molecular Biology Building (MBB) #3.424, University of Texas at Austin, 26th and Speedway, Austin, TX 78712, USA

Professor D. Söll
Department of Molecular Biophysics and Biochemistry, Yale University, 266 Whitney Avenue, New Haven, CT 06520-8114, USA

Dr. J. K. Strauss-Soukup
Department of Molecular Biophysics and Biochemistry, Yale University, 260 Whitney Avenue, New Haven, CT 06520, USA

Dr. S. A. Strobel
Department of Molecular Biophysics and Biochemistry, Yale University, 260 Whitney Avenue, New Haven, CT 06520, USA

Professor R. H. Symons
Department of Plant Science, Waite Campus, The University of Adelaide, Glen Osmond, SA 5064, Australia

Dr. K. Taira
Institute of Applied Biochemistry, University of Tsukuba, 1-1-1 Tennoudai, Tsukuba Science City 305, Japan

Dr. S. Tharun
Department of Molecular and Cellular Biology, Howard Hughes Medical Institute, University of Arizona, Tucson, AZ 85721, USA

Dr. D. Turner
Department of Chemistry, University of Rochester, Hutchinson B08, Rochester, NY 14627, USA

Dr. N. K. Vaish
Department of Chemistry, Boston College, Chestnut Hill, MA 02167, USA

Dr. S. Verma
Department of Chemistry, Indian Institute of Technology, Kanpur 208016 (U.P.), India

Dr. M. Warashina
Institute of Applied Biochemistry, University of Tsukuba, 1-1-1 Tennoudai, Tsukuba Science City 305, Japan

Dr. T. Xia
Department of Chemistry, University of Rochester, Rochester, NY 14627-0216, USA

Dr. D.-M. Zhou
Institute of Applied Biochemistry, University of Tsukuba, 1-1-1 Tennoudai, Tsukuba Science City 305, Japan

Abbreviations

The most commonly used abbreviations in *Comprehensive Natural Products Chemistry* are listed below. Please note that in some instances these may differ from those used in other branches of chemistry

A	adenine
ABA	abscisic acid
Ac	acetyl
ACAC	acetylacetonate
ACTH	adrenocorticotropic hormone
ADP	adenosine 5'-diphosphate
AIBN	2,2'-azobisisobutyronitrile
Ala	alanine
AMP	adenosine 5'-monophosphate
APS	adenosine 5'-phosphosulfate
Ar	aryl
Arg	arginine
ATP	adenosine 5'-triphosphate
B	nucleoside base (adenine, cylosine, guanine, thymine or uracil)
9-BBN	9-borabicyclo[3.3.1]nonane
BOC	*t*-butoxycarbonyl (or carbo-*t*-butoxy)
BSA	*N*,*O*-bis(trimethylsilyl)acetamide
BSTFA	*N*,*O*-bis(trimethylsilyl)trifluoroacetamide
Bu	butyl
Bu^n	*n*-butyl
Bu^i	isobutyl
Bu^s	*s*-butyl
Bu^t	*t*-butyl
Bz	benzoyl
CAN	ceric ammonium nitrate
CD	cyclodextrin
CDP	cytidine 5'-diphosphate
CMP	cytidine 5'-monophosphate
CoA	coenzyme A
COD	cyclooctadiene
COT	cyclooctatetraene
Cp	η^5-cyclopentadiene
Cp*	pentamethylcyclopentadiene
12-Crown-4	1,4,7,10-tetraoxacyclododecane
15-Crown-5	1,4,7,10,13-pentaoxacyclopentadecane
18-Crown-6	1,4,7,10,13,16-hexaoxacyclooctadecane
CSA	camphorsulfonic acid
CSI	chlorosulfonyl isocyanate
CTP	cytidine 5'-triphosphate
cyclic AMP	adenosine 3',5'-cyclic monophosphoric acid
CySH	cysteine
DABCO	1,4-diazabicyclo[2.2.2]octane
DBA	dibenz[*a*,*h*]anthracene
DBN	1,5-diazabicyclo[4.3.0]non-5-ene

DBU   1,8-diazabicyclo[5.4.0]undec-7-ene
DCC   dicyclohexylcarbodiimide
DEAC   diethylaluminum chloride
DEAD   diethyl azodicarboxylate
DET   diethyl tartrate (+ or -)
DHET   dihydroergotoxine
DIBAH   diisobutylaluminum hydride
Diglyme   diethylene glycol dimethyl ether (or bis(2-methoxyethyl)ether)
DiHPhe   2,5-dihydroxyphenylalanine
Dimsyl Na   sodium methylsulfinylmethide
DIOP   2,3-*O*-isopropylidene-2,3-dihydroxy-1,4-bis(diphenylphosphino)butane
dipt   diisopropyl tartrate (+ or -)
DMA   dimethylacetamide
DMAD   dimethyl acetylenedicarboxylate
DMAP   4-dimethylaminopyridine
DME   1,2-dimethoxyethane (glyme)
DMF   dimethylformamide
DMF-DMA   dimethylformamide dimethyl acetal
DMI   1,3-dimethyl-2-imidazalidinone
DMSO   dimethyl sulfoxide
DMTSF   dimethyl(methylthio)sulfonium fluoroborate
DNA   deoxyribonucleic acid
DOCA   deoxycorticosterone acetate

EADC   ethylaluminum dichloride
EDTA   ethylenediaminetetraacetic acid
EEDQ   *N*-ethoxycarbonyl-2-ethoxy-1,2-dihydroquinoline
Et   ethyl
EVK   ethyl vinyl ketone

FAD   flavin adenine dinucleotide
Fl   flavin
FMN   flavin mononucleotide

G   guanine
GABA   4-aminobutyric acid
GDP   guanosine 5'-diphosphate
GLDH   glutamate dehydrogenase
gln   glutamine
Glu   glutamic acid
Gly   glycine
GMP   guanosine 5'-monophosphate
GOD   glucose oxidase
G-6-P   glucose-6-phosphate
GTP   guanosine 5'-triphosphate

Hb   hemoglobin
His   histidine
HMPA   hexamethylphosphoramide
                (or hexamethylphosphorous triamide)

Ile   isoleucine
INAH   isonicotinic acid hydrazide
IpcBH   isopinocampheylborane
Ipc$_2$BH   diisopinocampheylborane

KAPA   potassium 3-aminopropylamide
K-Slectride potassium tri-*s*-butylborohydride

LAH	lithium aluminum hydride
LAP	leucine aminopeptidase
LDA	lithium diisopropylamide
LDH	lactic dehydrogenase
Leu	leucine
LICA	lithium isopropylcyclohexylamide
L-Selectride	lithium tri-*s*-butylborohydride
LTA	lead tetraacetate
Lys	lysine
MCPBA	*m*-chloroperoxybenzoic acid
Me	methyl
MEM	methoxyethoxymethyl
MEM-Cl	ß-methoxyethoxymethyl chloride
Met	methionine
MMA	methyl methacrylate
MMC	methyl magnesium carbonate
MOM	methoxymethyl
Ms	mesyl (or methanesulfonyl)
MSA	methanesulfonic acid
MsCl	methanesulfonyl chloride
MVK	methyl vinyl ketone
NAAD	nicotinic acid adenine dinucleotide
NAD	nicotinamide adenine dinucleotide
NADH	nicotinamide adenine dinucleotide phosphate, reduced
NBS	*N*-bromosuccinimider
NMO	*N*-methylmorpholine *N*-oxide monohydrate
NMP	*N*-methylpyrrolidone
PCBA	*p*-chlorobenzoic acid
PCBC	*p*-chlorobenzyl chloride
PCBN	*p*-chlorobenzonitrile
PCBTF	*p*-chlorobenzotrifluoride
PCC	pyridinium chlorochromate
PDC	pyridinium dichromate
PG	prostaglandin
Ph	phenyl
Phe	phenylalanine
Phth	phthaloyl
PPA	polyphosphoric acid
PPE	polyphosphate ester (or ethyl *m*-phosphate)
Pr	propyl
Pr^i	isopropyl
Pro	proline
Py	pyridine
RNA	ribonucleic acid
Rnase	ribonuclease
Ser	serine
Sia$_2$BH	disiamylborane
TAS	tris(diethylamino)sulfonium
TBAF	tetra-*n*-butylammonium fluoroborate
TBDMS	*t*-butyldimethylsilyl
TBDMS-Cl	*t*-butyldimethylsilyl chloride
TBDPS	*t*-butyldiphenylsilyl
TCNE	tetracyanoethene

TES	triethylsilyl
TFA	trifluoracetic acid
TFAA	trifluoroacetic anhydride
THF	tetrahydrofuran
THF	tetrahydrofolic acid
THP	tetrahydropyran (or tetrahydropyranyl)
Thr	threonine
TMEDA	*N,N,N',N'*,tetramethylethylenediamine[1,2-bis(dimethylamino)ethane]
TMS	trimethylsilyl
TMS-Cl	trimethylsilyl chloride
TMS-CN	trimethylsilyl cyanide
Tol	toluene
TosMIC	tosylmethyl isocyanide
TPP	tetraphenylporphyrin
Tr	trityl (or triphenylmethyl)
Trp	tryptophan
Ts	tosyl (or *p*-toluenesulfonyl)
TTFA	thallium trifluoroacetate
TTN	thallium(III) nitrate
Tyr	tyrosine
Tyr-OMe	tyrosine methyl ester
U	uridine
UDP	uridine 5'-diphosphate
UMP	uridine 5'-monophosphate

Contents of All Volumes

An Historical Perspective of Natural Products Chemistry

KOJI NAKANISHI

Columbia University, New York, USA

To give an account of the rich history of natural products chemistry in a short essay is a daunting task. This brief outline begins with a description of ancient folk medicine and continues with an outline of some of the major conceptual and experimental advances that have been made from the early nineteenth century through to about 1960, the start of the modern era of natural products chemistry. Achievements of living chemists are noted only minimally, usually in the context of related topics within the text. More recent developments are reviewed within the individual chapters of the present volumes, written by experts in each field. The subheadings follow, in part, the sequence of topics presented in Volumes 1–8.

1. ETHNOBOTANY AND "NATURAL PRODUCTS CHEMISTRY"

Except for minerals and synthetic materials our surroundings consist entirely of organic natural products, either of prebiotic organic origins or from microbial, plant, or animal sources. These materials include polyketides, terpenoids, amino acids, proteins, carbohydrates, lipids, nucleic acid bases, RNA and DNA, etc. Natural products chemistry can be thought of as originating from mankind's curiosity about odor, taste, color, and cures for diseases. Folk interest in treatments for pain, for food-poisoning and other maladies, and in hallucinogens appears to go back to the dawn of humanity

For centuries China has led the world in the use of natural products for healing. One of the earliest health science anthologies in China is the Nei Ching, whose authorship is attributed to the legendary Yellow Emperor (thirtieth century BC), although it is said that the dates were backdated from the third century by compilers. Excavation of a Han Dynasty (206 BC–AD 220) tomb in Hunan Province in 1974 unearthed decayed books, written on silk, bamboo, and wood, which filled a critical gap between the dawn of medicine up to the classic Nei Ching; Book 5 of these excavated documents lists 151 medical materials of plant origin. Generally regarded as the oldest compilation of Chinese herbs is Shen Nung Pen Ts'ao Ching (Catalog of Herbs by Shen Nung), which is believed to have been revised during the Han Dynasty; it lists 365 materials. Numerous revisions and enlargements of Pen Ts'ao were undertaken by physicians in subsequent dynasties, the ultimate being the Pen Ts'ao Kang Mu (General Catalog of Herbs) written by Li Shih-Chen over a period of 27 years during the Ming Dynasty (1573–1620), which records 1898 herbal drugs and 8160 prescriptions. This was circulated in Japan around 1620 and translated, and has made a huge impact on subsequent herbal studies in Japan; however, it has not been translated into English. The number of medicinal herbs used in 1979 in China numbered 5267. One of the most famous of the Chinese folk herbs is the ginseng root *Panax ginseng*, used for health maintenance and treatment of various diseases. The active principles were thought to be the saponins called ginsenosides but this is now doubtful; the effects could well be synergistic between saponins, flavonoids, etc. Another popular folk drug, the extract of the Ginkgo tree, *Ginkgo biloba* L., the only surviving species of the Paleozoic era (250 million years ago) family which became extinct during the last few million years, is mentioned in the Chinese Materia Medica to have an effect in improving memory and sharpening mental alertness. The main constituents responsible for this are now understood to be ginkgolides and flavonoids, but again not much else is known. Clarifying the active constituents and mode of (synergistic) bioactivity of Chinese herbs is a challenging task that has yet to be fully addressed.

The Assyrians left 660 clay tablets describing 1000 medicinal plants used around 1900–400 BC, but the best insight into ancient pharmacy is provided by the two scripts left by the ancient Egyptians, who were masters of human anatomy and surgery because of their extensive mummification practices. The Edwin Smith Surgical Papyrus purchased by Smith in 1862 in Luxor (now in the New York Academy of Sciences collection), is one of the most important medicinal documents of the ancient Nile Valley, and describes the healer's involvement in surgery, prescription, and healing practices using plants, animals, and minerals. The Ebers Papyrus, also purchased by Edwin Smith in 1862, and then acquired by Egyptologist George Ebers in 1872, describes 800 remedies using plants, animals, minerals, and magic. Indian medicine also has a long history, possibly dating back to the second millennium BC. The Indian materia medica consisted mainly of vegetable drugs prepared from plants but also used animals, bones, and minerals such as sulfur, arsenic, lead, copper sulfate, and gold. Ancient Greece inherited much from Egypt, India, and China, and underwent a gradual transition from magic to science. Pythagoras (580–500 BC) influenced the medical thinkers of his time, including Aristotle (384–322 BC), who in turn affected the medical practices of another influential Greek physician Galen (129–216). The Iranian physician Avicenna (980–1037) is noted for his contributions to Aristotelian philosophy and medicine, while the German-Swiss physician and alchemist Paracelsus (1493–1541) was an early champion who established the role of chemistry in medicine.

The rainforests in Central and South America and Africa are known to be particularly abundant in various organisms of interest to our lives because of their rich biodiversity, intense competition, and the necessity for self-defense. However, since folk-treatments are transmitted verbally to the next generation via shamans who naturally have a tendency to keep their plant and animal sources confidential, the recipes tend to get lost, particularly with destruction of rainforests and the encroachment of "civilization." Studies on folk medicine, hallucinogens, and shamanism of the Central and South American Indians conducted by Richard Schultes (Harvard Botanical Museum, emeritus) have led to renewed activity by ethnobotanists, recording the knowledge of shamans, assembling herbaria, and transmitting the record of learning to the village.

Extracts of toxic plants and animals have been used throughout the world for thousands of years for hunting and murder. These include the various arrow poisons used all over the world. *Strychnos* and *Chondrodendron* (containing strychnine, etc.) were used in South America and called "curare," *Strophanthus* (strophantidine, etc.) was used in Africa, the latex of the upas tree *Antiaris toxicaria* (cardiac glycosides) was used in Java, while *Aconitum napellus*, which appears in Greek mythology (aconitine) was used in medieval Europe and Hokkaido (by the Ainus). The Colombian arrow poison is from frogs (batrachotoxins; 200 toxins have been isolated from frogs by B. Witkop and J. Daly at NIH). Extracts of *Hyoscyamus niger* and *Atropa belladonna* contain the toxic tropane alkaloids, for example hyoscyamine, belladonnine, and atropine. The belladonna berry juice (atropine) which dilates the eye pupils was used during the Renaissance by ladies to produce doe-like eyes (belladona means beautiful woman). The Efik people in Calabar, southeastern Nigeria, used extracts of the calabar bean known as esere (physostigmine) for unmasking witches. The ancient Egyptians and Chinese knew of the toxic effect of the puffer fish, fugu, which contains the neurotoxin tetrodotoxin (Y. Hirata, K. Tsuda, R. B. Woodward).

When rye is infected by the fungus *Claviceps purpurea*, the toxin ergotamine and a number of ergot alkaloids are produced. These cause ergotism or the "devil's curse," "St. Anthony's fire," which leads to convulsions, miscarriages, loss of arms and legs, dry gangrene, and death. Epidemics of ergotism occurred in medieval times in villages throughout Europe, killing tens of thousands of people and livestock; Julius Caesar's legions were destroyed by ergotism during a campaign in Gaul, while in AD 994 an estimated 50,000 people died in an epidemic in France. As recently as 1926, a total of 11,000 cases of ergotism were reported in a region close to the Urals. It has been suggested that the witch hysteria that occurred in Salem, Massachusetts, might have been due to a mild outbreak of ergotism. Lysergic acid diethylamide (LSD) was first prepared by A. Hofmann, Sandoz Laboratories, Basel, in 1943 during efforts to improve the physiological effects of the ergot alkaloids when he accidentally inhaled it. "On Friday afternoon, April 16, 1943," he wrote, "I was seized by a sensation of restlessness... ." He went home from the laboratory and "perceived an uninterrupted stream of fantastic dreams ... ." (*Helvetica Chimica Acta*).

Numerous psychedelic plants have been used since ancient times, producing visions, mystical fantasies (cats and tigers also seem to have fantasies?, see nepetalactone below), sensations of flying, glorious feelings in warriors before battle, etc. The ethnobotanists Wasson and Schultes identified "ololiqui," an important Aztec concoction, as the seeds of the morning glory *Rivea corymbosa* and gave the seeds to Hofmann who found that they contained lysergic acid amides similar to but less potent than LSD. Iboga, a powerful hallucinogen from the root of the African shrub *Tabernanthe iboga*, is used by the Bwiti cult in Central Africa who chew the roots to obtain relief from fatigue and hunger; it contains the alkaloid ibogamine. The powerful hallucinogen used for thousands of years by the American Indians, the peyote cactus, contains mescaline and other alkaloids. The Indian hemp plant, *Cannabis sativa*, has been used for making rope since 3000 BC, but when it is used for its pleasure-giving effects it is called cannabis and has been known in central Asia, China, India, and the Near East since ancient times. Marijuana, hashish (named after the Persian founder of the Assassins of the eleventh century, Hasan-e Sabbah), charas, ghanja, bhang, kef, and dagga are names given to various preparations of the hemp plant. The constituent responsible for the mind-altering effect is 1-tetrahydrocannabinol (also referred to as 9-THC) contained in 1%. R. Mechoulam (1930–, Hebrew University) has been the principal worker in the cannabinoids, including structure determination and synthesis of 9-THC (1964 to present); the Israeli police have also made a contribution by providing Mechoulam with a constant supply of marijuana. Opium (morphine) is another ancient drug used for a variety of pain-relievers and it is documented that the Sumerians used poppy as early as 4000 BC; the narcotic effect is present only in seeds before they are fully formed. The irritating secretion of the blister beetles, for example *Mylabris* and the European species *Lytta vesicatoria*, commonly called Spanish fly, was used medically as a topical skin irritant to remove warts but was also a major ingredient in so-called love potions (constituent is cantharidin, stereospecific synthesis in 1951, G. Stork, 1921–; prep. scale high-pressure Diels–Alder synthesis in 1985, W. G. Dauben, 1919–1996).

Plants have been used for centuries for the treatment of heart problems, the most important being the foxgloves *Digitalis purpurea* and *D. lanata* (digitalin, diginin) and *Strophanthus gratus* (ouabain). The bark of cinchona *Cinchona officinalis* (called quina-quina by the Indians) has been used widely among the Indians in the Andes against malaria, which is still one of the major infectious diseases; its most important alkaloid is quinine. The British protected themselves against malaria during the occupation of India through gin and tonic (quinine!). The stimulant coca, used by the Incas around the tenth century, was introduced into Europe by the conquistadors; coca beans are also commonly chewed in West Africa. Wine making was already practiced in the Middle East 6000–8000 years ago; Moors made date wines, the Japanese rice wine, the Vikings honey mead, the Incas maize chicha. It is said that the Babylonians made beer using yeast 5000–6000 years ago. As shown above in parentheses, alkaloids are the major constituents of the herbal plants and extracts used for centuries, but it was not until the early nineteenth century that the active principles were isolated in pure form, for example morphine (1816), strychnine (1817), atropine (1819), quinine (1820), and colchicine (1820). It was a century later that the structures of these compounds were finally elucidated.

2. DAWN OF ORGANIC CHEMISTRY, EARLY STRUCTURAL STUDIES, MODERN METHODOLOGY

The term "organic compound" to define compounds made by and isolated from living organisms was coined in 1807 by the Swedish chemist Jons Jacob Berzelius (1779–1848), a founder of today's chemistry, who developed the modern system of symbols and formulas in chemistry, made a remarkably accurate table of atomic weights and analyzed many chemicals. At that time it was considered that organic compounds could not be synthesized from inorganic materials *in vitro*. However, Friedrich Wöhler (1800–1882), a medical doctor from Heidelberg who was starting his chemical career at a technical school in Berlin, attempted in 1828 to make "ammonium cyanate," which had been assigned a wrong structure, by heating the two inorganic salts potassium cyanate and ammonium sulfate; this led to the unexpected isolation of white crystals which were identical to the urea from urine, a typical organic compound. This well-known incident marked the beginning of organic chemistry. With the preparation of acetic acid from inorganic material in 1845 by Hermann Kolbe (1818–1884) at Leipzig, the myth surrounding organic compounds, in which they were associated with some vitalism was brought to an end and organic chemistry became the chemistry of carbon compounds. The same Kolbe was involved in the development of aspirin, one of the earliest and most important success stories in natural products chemistry. Salicylic acid from the leaf of the wintergreen plant had long been used as a pain reliever, especially in treating arthritis and gout. The inexpensive synthesis of salicylic acid from sodium phenolate and carbon dioxide by Kolbe in 1859 led to the industrial production in 1893 by the Bayer Company of acetylsalicylic acid "aspirin," still one of the most popular drugs. Aspirin is less acidic than salicylic acid and therefore causes less irritation in the mouth, throat, and stomach. The remarkable mechanism of the anti-inflammatory effect of aspirin was clarified in 1974 by John Vane (1927–) who showed that it inhibits the biosynthesis of prostaglandins by irreversibly acetylating a serine residue in prostaglandin synthase. Vane shared the 1982 Nobel Prize with Bergström and Samuelsson who determined the structure of prostaglandins (see below).

In the early days, natural products chemistry was focused on isolating the more readily available plant and animal constituents and determining their structures. The course of structure determination in the 1940s was a complex, indirect process, combining evidence from many types of experiments. The first effort was to crystallize the unknown compound or make derivatives such as esters or 2,4-dinitrophenylhydrazones, and to repeat recrystallization until the highest and sharp melting point was reached, since prior to the advent of isolation and purification methods now taken for granted, there was no simple criterion for purity. The only chromatography was through special grade alumina (first used by M. Tswett in 1906, then reintroduced by R. Willstätter). Molecular weight estimation by the Rast method which depended on melting point depression of a sample/camphor mixture, coupled with Pregl elemental microanalysis (see below) gave the molecular formula. Functionalities such as hydroxyl, amino, and carbonyl groups were recognized on the basis of specific derivatization and crystallization, followed by redetermination of molecular formula; the change in molecular composition led to identification of the functionality. Thus, sterically hindered carbonyls, for example the 11-keto group of cortisone, or tertiary hydroxyls, were very difficult to pinpoint, and often had to depend on more searching experiments. Therefore, an entire paper describing the recognition of a single hydroxyl group in a complex natural product would occasionally appear in the literature. An oxygen function suggested from the molecular formula but left unaccounted for would usually be assigned to an ether.

Determination of C-methyl groups depended on Kuhn–Roth oxidation which is performed by drastic oxidation with chromic acid/sulfuric acid, reduction of excess oxidant with hydrazine, neutralization with alkali, addition of phosphoric acid, distillation of the acetic acid originating from the C-methyls, and finally its titration with alkali. However, the results were only approximate, since *gem*-dimethyl groups only yield one equivalent of acetic acid, while primary, secondary, and tertiary methyl groups all give different yields of acetic acid. The skeletal structure of polycyclic compounds were frequently deduced on the basis of dehydrogenation reactions. It is therefore not surprising that the original steroid skeleton put forth by Wieland and Windaus in 1928, which depended a great deal on the production of chrysene upon Pd/C dehydrogenation, had to be revised in 1932 after several discrepancies were found (they received the Nobel prizes in 1927 and 1928 for this "extraordinarily difficult structure determination," see below).

In the following are listed some of the Nobel prizes awarded for the development of methodologies which have contributed critically to the progress in isolation protocols and structure determination. The year in which each prize was awarded is preceded by "Np."

Fritz Pregl, 1869–1930, Graz University, Np 1923. Invention of carbon and hydrogen microanalysis. Improvement of Kuhlmann's microbalance enabled weighing at an accuracy of 1 μg over a 20 g range, and refinement of carbon and hydrogen analytical methods made it possible to perform analysis with 3–4 mg of sample. His microbalance and the monograph *Quantitative Organic Microanalysis* (1916) profoundly influenced subsequent developments in practically all fields of chemistry and medicine.

The Svedberg, 1884–1971, Uppsala, Np 1926. Uppsala was a center for quantitative work on colloids for which the prize was awarded. His extensive study on ultracentrifugation, the first paper of which was published in the year of the award, evolved from a spring visit in 1922 to the University of Wisconsin. The ultracentrifuge together with the electrophoresis technique developed by his student Tiselius, have profoundly influenced subsequent progress in molecular biology and biochemistry.

Arne Tiselius, 1902–1971, Ph.D. Uppsala (T. Svedberg), Uppsala, Np 1948. Assisted by a grant from the Rockefeller Foundation, Tiselius was able to use his early electrophoresis instrument to show four bands in horse blood serum, alpha, beta and gamma globulins in addition to albumin; the first paper published in 1937 brought immediate positive responses.

Archer Martin, 1910–, Ph.D. Cambridge; Medical Research Council, Mill Hill, and Richard Synge, 1914–1994, Ph.D. Cambridge; Rowett Research Institute, Food Research Institute, Np 1952. They developed chromatography using two immiscible phases, gas–liquid, liquid–liquid, and paper chromatography, all of which have profoundly influenced all phases of chemistry.

Frederick Sanger, 1918–, Ph.D. Cambridge (A. Neuberger), Medical Research Council, Cambridge, Np 1958 and 1980. His confrontation with challenging structural problems in proteins and nucleic acids led to the development of two general analytical methods, 1,2,4-fluorodinitrobenzene (DNP) for tagging free amino groups (1945) in connection with insulin sequencing studies, and the dideoxynucleotide method for sequencing DNA (1977) in connection with recombinant DNA. For the latter he received his second Np in chemistry in 1980, which was shared with Paul Berg (1926–, Stanford University) and Walter Gilbert (1932–, Harvard University) for their contributions, respectively, in recombinant DNA and chemical sequencing of DNA. The studies of insulin involved usage of DNP for tagging disulfide bonds as cysteic acid residues (1949), and paper chromatography introduced by Martin and Synge 1944. That it was the first elucidation of any protein structure lowered the barrier for future structure studies of proteins.

Stanford Moore, 1913–1982, Ph.D. Wisconsin (K. P. Link), Rockefeller, Np 1972; and William Stein, 1911–1980, Ph.D. Columbia (E. G. Miller); Rockefeller, Np 1972. Moore and Stein cooperatively developed methods for the rapid quantification of protein hydrolysates by combining partition chromatography, ninhydrin coloration, and drop-counting fraction collector, i.e., the basis for commercial amino acid analyzers, and applied them to analysis of the ribonuclease structure.

Bruce Merrifield, 1921–, Ph.D. UCLA (M. Dunn), Rockefeller, Np 1984. The concept of solid-phase peptide synthesis using porous beads, chromatographic columns, and sequential elongation of peptides and other chains revolutionized the synthesis of biopolymers.

High-performance liquid chromatography (HPLC), introduced around the mid-1960s and now coupled on-line to many analytical instruments, for example UV, FTIR, and MS, is an indispensable daily tool found in all natural products chemistry laboratories.

3. STRUCTURES OF ORGANIC COMPOUNDS, NINETEENTH CENTURY

The discoveries made from 1848 to 1874 by Pasteur, Kekulé, van't Hoff, Le Bel, and others led to a revolution in structural organic chemistry. Louis Pasteur (1822–1895) was puzzled about why the potassium salt of tartaric acid (deposited on wine casks during fermentation) was dextrorotatory while the sodium ammonium salt of racemic acid (also deposited on wine casks) was optically inactive although both tartaric acid and "racemic" acid had identical chemical compositions. In 1848, the 25 year old Pasteur examined the racemic acid salt under the microscope and found two kinds of crystals exhibiting a left- and right-hand relation. Upon separation of the left-handed and right-handed crystals, he found that they rotated the plane of polarized light in opposite directions. He had thus performed his famous resolution of a racemic mixture, and had demonstrated the phenomenon of chirality. Pasteur went on to show that the racemic acid formed two kinds of salts with optically active bases such as quinine; this was the first demonstration of diastereomeric resolution. From this work Pasteur concluded that tartaric acid must have an element of asymmetry within the molecule itself. However, a three-dimensional understanding of the enantiomeric pair was only solved 25 years later (see below). Pasteur's own interest shifted to microbiology where he made the crucial discovery of the involvement of "germs" or microorganisms in various processes and proved that yeast induces alcoholic fermentation, while other microorganisms lead to diseases; he thus saved the wine industries of France, originated the process known as "pasteurization," and later developed vaccines for rabies. He was a genius who made many fundamental discoveries in chemistry and in microbiology.

The structures of organic compounds were still totally mysterious. Although Wöhler had synthesized urea, an isomer of ammonium cyanate, in 1828, the structural difference between these isomers was not known. In 1858 August Kekulé (1829–1896; studied with André Dumas and C. A. Wurtz in Paris, taught at Ghent, Heidelberg, and Bonn) published his famous paper in Liebig's *Annalen der Chemie* on the structure of carbon, in which he proposed that carbon atoms could form C–C bonds with hydrogen and other atoms linked to them; his dream on the top deck of a London bus led him to this concept. It was Butlerov who introduced the term "structure theory" in 1861. Further, in 1865 Kekulé conceived the cyclo-hexa-1:3:5-triene structure for benzene (C_6H_6) from a dream of a snake biting its own tail. In 1874, two young chemists, van't Hoff (1852–1911, Np 1901) in Utrecht, and Le Bel (1847–1930) in Paris, who had met in 1874 as students of C. A. Wurtz, published the revolutionary three-dimensional (3D) structure of the tetrahedral carbon Cabcd to explain the enantiomeric behavior of Pasteur's salts. The model was welcomed by J. Wislicenus (1835–1902, Zürich, Würzburg, Leipzig) who in 1863 had demonstrated the enantiomeric nature of the two lactic acids found by Scheele in sour milk (1780) and by Berzelius in muscle tissue (1807). This model, however, was criticized by Hermann Kolbe (1818–1884, Leipzig) as an "ingenious but in reality trivial and senseless natural philosophy." After 10 years of heated controversy, the idea of tetrahedral carbon was fully accepted, Kolbe had died and Wislicenus succeeded him in Leipzig.

Emil Fischer (1852–1919, Np 1902) was the next to make a critical contribution to stereochemistry. From the work of van't Hoff and Le Bel he reasoned that glucose should have 16 stereoisomers. Fischer's doctorate work on hydrazines under Baeyer (1835–1917, Np 1905) at Strasbourg had led to studies of osazones which culminated in the brilliant establishment, including configurations, of the Fischer sugar tree starting from D-(+)-glyceraldehyde all the way up to the aldohexoses, allose, altrose, glucose, mannose, gulose, idose, galactose, and talose (from 1884 to 1890). Unfortunately Fischer suffered from the toxic effects of phenylhydrazine for 12 years. The arbitrarily but luckily chosen absolute configuration of D-(+)-glyceraldehyde was shown to be correct sixty years later in 1951 (Johannes-Martin Bijvoet, 1892–1980). Fischer's brilliant correlation of the sugars comprising the Fischer sugar tree was performed using the Kiliani (1855–1945)–Fischer method via cyanohydrin intermediates for elongating sugars. Fischer also made remarkable contributions to the chemistry of amino acids and to nucleic acid bases (see below).

4. STRUCTURES OF ORGANIC COMPOUNDS, TWENTIETH CENTURY

The early concept of covalent bonds was provided with a sound theoretical basis by Linus Pauling (1901–1994, Np 1954), one of the greatest intellects of the twentieth century. Pauling's totally interdisciplinary research interests, including proteins and DNA is responsible for our present understanding of molecular structures. His books *Introduction to Quantum Mechanics* (with graduate student E. B. Wilson, 1935) and *The Nature of the Chemical Bond* (1939) have had a profound effect on our understanding of all of chemistry.

The actual 3D shapes of organic molecules which were still unclear in the late 1940s were then brilliantly clarified by Odd Hassel (1897–1981, Oslo University, Np 1969) and Derek Barton (1918–1998, Np 1969). Hassel, an X-ray crystallographer and physical chemist, demonstrated by electron diffraction that cyclohexane adopted the chair form in the gas phase and that it had two kinds of bonds, "standing (axial)" and "reclining (equatorial)" (1943). Because of the German occupation of Norway in 1940, instead of publishing the result in German journals, he published it in a Norwegian journal which was not abstracted in English until 1945. During his 1949 stay at Harvard, Barton attended a seminar by Louis Fieser on steric effects in steroids and showed Fieser that interpretations could be simplified if the shapes ("conformations") of cyclohexane rings were taken into consideration; Barton made these comments because he was familiar with Hassel's study on *cis*- and *trans*-decalins. Following Fieser's suggestion Barton published these ideas in a four-page *Experientia* paper (1950). This led to the joint Nobel prize with Hassel (1969), and established the concept of conformational analysis, which has exerted a profound effect in every field involving organic molecules.

Using conformational analysis, Barton determined the structures of many key terpenoids such as ß-amyrin, cycloartenone, and cycloartenol (Birkbeck College). At Glasgow University (from 1955) he collaborated in a number of cases with Monteath Robertson (1900–1989) and established many challenging structures: limonin, glauconic acid, byssochlamic acid, and nonadrides. Barton was also associated with the Research Institute for Medicine and Chemistry (RIMAC), Cambridge, USA founded by the Schering company, where with J. M. Beaton, he produced 60 g of aldosterone at a time when the world supply of this important hormone was in mg quantities. Aldosterone synthesis ("a good problem") was achieved in 1961 by Beaton ("a good experimentalist") through a nitrite photolysis, which came to be known as the Barton reaction ("a good idea") (quotes from his 1991 autobiography published by the American Chemical Society). From Glasgow, Barton went on to Imperial College, and a year before retirement, in 1977 he moved to France to direct the research at ICSN at Gif-sur-Yvette where he explored the oxidation reaction selectivity for unactivated C–H. After retiring from ICSN he made a further move to Texas A&M University in 1986, and continued his energetic activities, including chairman of the *Tetrahedron* publications. He felt weak during work one evening and died soon after, on March 16, 1998. He was fond of the phrase "gap jumping" by which he meant seeking generalizations between facts that do not seem to be related: "In the conformational analysis story, one had to jump the gap between steroids and chemical physics" (from his autobiography). According to Barton, the three most important qualities for a scientist are "intelligence, motivation, and honesty." His routine at Texas A&M was to wake around 4 a.m., read the literature, go to the office at 7 a.m. and stay there until 7 p.m.; when asked in 1997 whether this was still the routine, his response was that he wanted to wake up earlier because sleep was a waste of time—a remark which characterized this active scientist approaching 80!

Robert B. Woodward (1917–1979, Np 1965), who died prematurely, is regarded by many as the preeminent organic chemist of the twentieth century. He made landmark achievements in spectroscopy, synthesis, structure determination, biogenesis, as well as in theory. His solo papers published in 1941–1942 on empirical rules for estimating the absorption maxima of enones and dienes made the general organic chemical community realize that UV could be used for structural studies, thus launching the beginning of the spectroscopic revolution which soon brought on the applications of IR, NMR, MS, etc. He determined the structures of the following compounds: penicillin in 1945 (through joint UK–USA collaboration, see Hodgkin), strychnine in 1948, patulin in 1949, terramycin, aureomycin, and ferrocene (with G. Wilkinson, Np 1973—shared with E. O. Fischer for sandwich compounds) in 1952, cevine in 1954 (with Barton Np 1966, Jeger and Prelog, Np 1975), magnamycin in 1956, gliotoxin in 1958, oleandomycin in 1960, streptonigrin in 1963, and tetrodotoxin in 1964. He synthesized patulin in 1950, cortisone and cholesterol in 1951, lanosterol, lysergic acid (with Eli Lilly), and strychnine in 1954, reserpine in 1956, chlorophyll in 1960, a tetracycline (with Pfizer) in 1962, cephalosporin in 1965, and vitamin B_{12} in 1972 (with A. Eschenmoser, 1925–, ETH Zürich). He derived biogenetic schemes for steroids in 1953 (with K. Bloch, see below), and for macrolides in 1956, while the Woodward–Hoffmann orbital symmetry rules in 1965 brought order to a large class of seemingly random cyclization reactions.

Another central figure in stereochemistry is Vladimir Prelog (1906–1998, Np 1975), who succeeded Leopold Ruzicka at the ETH Zürich, and continued to build this institution into one of the most active and lively research and discussion centers in the world. The core group of intellectual leaders consisted of P. Plattner (1904–1975), O. Jeger, A. Eschenmoser, J. Dunitz, D. Arigoni, and A. Dreiding (from Zürich University). After completing extensive research on alkaloids, Prelog determined the structures of nonactin, boromycin, ferrioxamins, and rifamycins. His seminal studies in the synthesis and properties of 8–12 membered rings led him into unexplored areas of stereochemisty and chirality. Together with Robert Cahn (1899–1981, London Chemical Society) and Christopher Ingold (1893–1970, University College, London; pioneering mechanistic interpretation of organic reactions), he developed the Cahn–Ingold–Prelog (CIP) sequence rules for the unambiguous specification of stereoisomers. Prelog was an excellent story teller, always had jokes to tell, and was respected and loved by all who knew him.

4.1 Polyketides and Fatty Acids

Arthur Birch (1915–1995) from Sydney University, Ph.D. with Robert Robinson (Oxford University), then professor at Manchester University and Australian National University, was one of the earliest chemists to perform biosynthetic studies using radiolabels; starting with polyketides he studied the biosynthesis of a variety of natural products such as the C_6–C_3–C_6 backbone of plant phenolics, polyene macrolides, terpenoids, and alkaloids. He is especially known for the Birch reduction of aromatic rings, metal–ammonia reductions leading to 19-norsteroid hormones and other important products (1942–) which were of industrial importance. Feodor Lynen (1911–1979, Np 1964) performed studies on the intermediary metabolism of the living cell that led him to the demonstration of the first step in a chain of reactions resulting in the biosynthesis of sterols and fatty acids.

Prostaglandins, a family of 20-carbon, lipid-derived acids discovered in seminal fluids and accessory genital glands of man and sheep by von Euler (1934), have attracted great interest because of their extremely diverse biological activities. They were isolated and their structures elucidated from 1963 by S. Bergström (1916–, Np 1982) and B. Samuelsson (1934–, Np 1982) at the Karolinska Institute, Stockholm. Many syntheses of the natural prostaglandins and their nonnatural analogues have been published.

Tetsuo Nozoe (1902–1996) who studied at Tohoku University, Sendai, with Riko Majima (1874–1962, see below) went to Taiwan where he stayed until 1948 before returning to Tohoku University. At National Taiwan University he isolated hinokitiol from the essential oil of *taiwanhinoki*. Remembering the resonance concept put forward by Pauling just before World War II, he arrived at the seven-membered nonbenzenoid aromatic structure for hinokitiol in 1941, the first of the troponoids. This highly original work remained unknown to the rest of the world until 1951. In the meantime, during 1945–1948, nonbenzenoid aromatic structures had been assigned to stipitatic acid (isolated by H. Raistrick) by Michael J. S. Dewar (1918–) and to the thujaplicins by Holger Erdtman (1902–1989); the term tropolones was coined by Dewar in 1945. Nozoe continued to work on and discuss troponoids, up to the night before his death, without knowing that he had cancer. He was a remarkably focused and warm scientist, working unremittingly. Erdtman (Royal Institute of Technology, Stockholm) was the central figure in Swedish natural products chemistry who, with his wife Gunhild Aulin Erdtman (dynamic General Secretary of the Swedish Chemistry Society), worked in the area of plant phenolics.

As mentioned in the following and in the concluding sections, classical biosynthetic studies using radioactive isotopes for determining the distribution of isotopes has now largely been replaced by the use of various stable isotopes coupled with NMR and MS. The main effort has now shifted to the identification and cloning of genes, or where possible the gene clusters, involved in the biosynthesis of the natural product. In the case of polyketides (acyclic, cyclic, and aromatic), the focus is on the polyketide synthases.

4.2 Isoprenoids, Steroids, and Carotenoids

During his time as an assistant to Kekulé at Bonn, Otto Wallach (1847–1931, Np 1910) had to familiarize himself with the essential oils from plants; many of the components of these oils were compounds for which no structure was known. In 1891 he clarified the relations between 12 different monoterpenes related to pinene. This was summarized together with other terpene chemistry in book form in 1909, and led him to propose the "isoprene rule." These achievements laid the foundation for the future development of terpenoid chemistry and brought order from chaos.

The next period up to around 1950 saw phenomenal advances in natural products chemistry centered on isoprenoids. Many of the best natural products chemists in Europe, including Wieland, Windaus, Karrer, Kuhn, Butenandt, and Ruzicka contributed to this breathtaking pace. Heinrich Wieland (1877–1957) worked on the bile acid structure, which had been studied over a period of 100 years and considered to be one of the most difficult to attack; he received the Nobel Prize in 1927 for these studies. His friend Adolph Windaus (1876–1959) worked on the structure of cholesterol for which he also received the Nobel Prize in 1928. Unfortunately, there were chemical discrepancies in the proposed steroidal skeletal structure, which had a five-membered ring B attached to C-7 and C-9. J. D. Bernal, Mineralogical Museums, Cambridge University, who was examining the X-ray patterns of ergosterol (1932) noted that the dimensions were inconsistent with the Wieland–Windaus formula. A reinterpretation of the production of chrysene from sterols by Pd/C dehydrogenation reported by Diels (see below) in 1927 eventually led Rosenheim and King and Wieland and Dane to deduce the correct structure in 1932. Wieland also worked on the structures of morphine/strychnine alkaloids, phalloidin/amanitin cyclopeptides of toxic mushroom *Amanita phalloides*, and pteridines, the important fluorescent pigments of butterfly wings. Windaus determined the structure of ergosterol and continued structural studies of its irradiation product which exhibited antirachitic activity "vitamin D." The mechanistically complex photochemistry of ergosterol leading to the vitamin D group has been investigated in detail by Egbert Havinga (1927–1988, Leiden University), a leading photochemist and excellent tennis player.

Paul Karrer (1889–1971, Np 1937), established the foundations of carotenoid chemistry through structural determinations of lycopene, carotene, vitamin A, etc. and the synthesis of squalene, carotenoids, and others. George Wald (1906–1997, Np 1967) showed that vitamin A was the key compound in vision during his stay in Karrer's laboratory. Vitamin K (K from "Koagulation"), discovered by Henrik Dam (1895–1976, Polytechnic Institute, Copenhagen, Np 1943) and structurally studied by Edward Doisy (1893–1986, St. Louis University, Np 1943), was also synthesized by Karrer. In addition, Karrer synthesized riboflavin (vitamin B_2) and determined the structure and role of nicotinamide adenine dinucleotide phosphate (NADP$^+$) with Otto Warburg. The research on carotenoids and vitamins of Karrer who was at Zürich University overlapped with that of Richard Kuhn (1900–1967, Np 1938) at the ETH Zürich, and the two were frequently rivals. Richard Kuhn, one of the pioneers in using UV-vis spectroscopy for structural studies, introduced the concept of "atropisomerism" in diphenyls, and studied the spectra of a series of diphenyl polyenes. He determined the structures of many natural carotenoids, proved the structure of riboflavin-5-phosphate (flavin-adenine-dinucleotide-5-phosphate) and showed that the combination of NAD-5-phosphate with the carrier protein yielded the yellow oxidation enzyme, thus providing an understanding of the role of a prosthetic group. He also determined the structures of vitamin B complexes, i.e., pyridoxine, *p*-aminobenzoic acid, pantothenic acid. After World War II he went on to structural studies of nitrogen-containing oligosaccharides in human milk that provide immunity for infants, and brain gangliosides. Carotenoid studies in Switzerland were later taken up by Otto Isler (1910–1993), a Ruzicka student at Hoffmann-La Roche, and Conrad Hans Eugster (1921–), a Karrer student at Zürich University.

Adolf Butenandt (1903–1998, Np 1939) initiated and essentially completed isolation and structural studies of the human sex hormones, the insect molting hormone (ecdysone), and the first pheromone, bombykol. With help from industry he was able to obtain large supplies of urine from pregnant women for estrone, sow ovaries for progesterone, and 4,000 gallons of male urine for androsterone (50 mg, crystals). He isolated and determined the structures of two female sex hormones, estrone and progesterone, and the male hormone androsterone all during the period 1934–1939 (!) and was awarded the Nobel prize in 1939. Keen intuition and use of UV data and Pregl's microanalysis all played important roles. He was appointed to a professorship in Danzig at the age of 30. With Peter Karlson he isolated from 500 kg of silkworm larvae 25 mg of α-ecdysone, the prohormone of insect and crustacean molting hormone, and determined its structure as a polyhydroxysteroid (1965); 20-hydroxylation gives the insect and crustacean molting hormone or ß-ecdysone (20-hydroxyecdysteroid). He was also the first to isolate an insect pheromone, bombykol, from female silkworm moths (with E. Hecker). As president of the Max Planck Foundation, he strongly influenced the postwar rebuilding of German science.

The successor to Kuhn, who left ETH Zürich for Heidelberg, was Leopold Ruzicka (1887–1967, Np 1939) who established a close relationship with the Swiss pharmaceutical industry. His synthesis of the 17- and 15-membered macrocyclic ketones, civetone and muscone (the constituents of musk) showed that contrary to Baeyer's prediction, large alicyclic rings could be strainless. He reintroduced and refined the isoprene rule proposed by Wallach (1887) and determined the basic structures of many sesqui-, di-, and triterpenes, as well as the structure of lanosterol, the key intermediate in cholesterol biosynthesis. The "biogenetic isoprene rule" of the ETH group, Albert Eschenmoser, Leopold Ruzicka, Oskar Jeger, and Duilio Arigoni, contributed to a concept of terpenoid cyclization (1955), which was consistent with the mechanistic considerations put forward by Stork as early as 1950. Besides making the ETH group into a center of natural products chemistry, Ruzicka bought many seventeenth century Dutch paintings with royalties accumulated during the war from his Swiss and American patents, and donated them to the Zürich Kunsthaus.

Studies in the isolation, structures, and activities of the antiarthritic hormone, cortisone and related compounds from the adrenal cortex were performed in the mid- to late 1940s during World War II by Edward Kendall (1886–1972, Mayo Clinic, Rochester, Np 1950), Tadeus Reichstein (1897–1996, Basel University, Np 1950), Philip Hench (1896–1965, Mayo Clinic, Rochester, Np 1950), Oskar Wintersteiner (1898–1971, Columbia University, Squibb) and others initiated interest as an adjunct to military medicine as well as to supplement the meager supply from beef adrenal glands by synthesis. Lewis Sarett (1917–, Merck & Co., later president) and co-workers completed the cortisone synthesis in 28 steps, one of the first two totally stereocontrolled syntheses of a natural product; the other was cantharidin (Stork 1951) (see above). The multistep cortisone synthesis was put on the production line by Max Tishler (1906–1989, Merck & Co., later president) who made contributions to the synthesis of a number of drugs, including riboflavin. Besides working on steroid reactions/synthesis and antimalarial agents, Louis F. Fieser (1899–1977) and Mary Fieser (1909–1997) of Harvard University made huge contributions to the chemical community through their outstanding books *Natural Products related to Phenanthrene* (1949), *Steroids* (1959), *Advanced Organic Chemistry* (1961), and *Topics in Organic Chemistry* (1963), as well as their textbooks and an important series of books on Organic Reagents. Carl Djerassi (1923–, Stanford University), a prolific chemist, industrialist, and more recently a novelist, started to work at the Syntex laboratories in Mexico City where he directed the work leading to the first oral contraceptive ("the pill") for women.

Takashi Kubota (1909–, Osaka City University), with Teruo Matsuura (1924–, Kyoto University), determined the structure of the furanoid sesquiterpene, ipomeamarone, from the black rotted portion of spoiled sweet potatoes; this research constitutes the first characterization of a phytoallexin, defense substances produced by plants in response to attack by fungi or physical damage. Damaging a plant and characterizing the defense substances produced may lead to new bioactive compounds. The mechanism of induced biosynthesis of phytoallexins, which is not fully understood, is an interesting biological mechanistic topic that deserves further investigation. Another center of high activity in terpenoids and nucleic acids was headed by Frantisek Sorm (1913–1980, Institute of Organic and Biochemistry, Prague), who determined the structures of many sesquiterpenoids and other natural products; he was not only active scientifically but also was a central figure who helped to guide the careers of many Czech chemists.

The key compound in terpenoid biosynthesis is mevalonic acid (MVA) derived from acetyl-CoA, which was discovered fortuitously in 1957 by the Merck team in Rahway, NJ headed by Karl Folkers (1906–1998). They soon realized and proved that this C_6 acid was the precursor of the C_5 isoprenoid unit isopentenyl diphosphate (IPP) that ultimately leads to the biosynthesis of cholesterol. In 1952 Konrad Bloch (1912–, Harvard, Np 1964) with R. B. Woodward published a paper suggesting a mechanism of the cyclization of squalene to lanosterol and the subsequent steps to cholesterol, which turned out to be essentially correct. This biosynthetic path from MVA to cholesterol was experimentally clarified in stereochemical detail by John Cornforth (1917–, Np 1975) and George Popják. In 1932, Harold Urey (1893–1981, Np 1934) of Columbia University discovered heavy hydrogen. Urey showed, contrary to common expectation, that isotope separation could be achieved with deuterium in the form of deuterium oxide by fractional electrolysis of water. Urey's separation of the stable isotope deuterium led to the isotopic tracer methodology that revolutionized the protocols for elucidating biosynthetic processes and reaction mechanisms, as exemplified beautifully by the cholesterol studies. Using MVA labeled chirally with isotopes, including chiral methyl, i.e., -CHDT, Cornforth and Popják clarified the key steps in the intricate biosynthetic conversion of mevalonate to cholesterol in stereochemical detail. The chiral methyl group was also prepared independently by Duilio Arigoni (1928–, ETH, Zürich). Cornforth has had great difficulty in hearing and speech since childhood but has been helped expertly by his chemist wife Rita; he is an excellent tennis and chess player, and is renowned for his speed in composing occasional witty limericks.

Although MVA has long been assumed to be the only natural precursor for IPP, a non-MVA pathway in which IPP is formed via the glyceraldehyde phosphate-pyruvate pathway has been discovered (1995–1996) in the ancient bacteriohopanoids by Michel Rohmer, who started working on them with Guy Ourisson (1926–, University of Strasbourg, terpenoid studies, including prebiotic), and by Duilio Arigoni in the ginkgolides, which are present in the ancient *Ginkgo biloba* tree. It is possible that many other terpenoids are biosynthesized via the non-MVA route. In classical biosynthetic experiments, ^{14}C-labeled acetic acid was incorporated into the microbial or plant product, and location or distribution of the ^{14}C label was deduced by oxidation or degradation to specific fragments including acetic acid; therefore, it was not possible or extremely difficult to map the distribution of all radioactive carbons. The progress in ^{13}C NMR made it possible to incorporate ^{13}C-labeled acetic acid and locate all labeled carbons. This led to the discovery of the nonmevalonate pathway leading to the IPP units. Similarly, NMR and MS have made it possible to use the stable isotopes, e.g., ^{18}O, ^{2}H, ^{15}N, etc., in biosynthetic studies. The current trend of biosynthesis has now shifted to genomic approaches for cloning the genes of various enzyme synthases involved in the biosynthesis.

4.3 Carbohydrates and Cellulose

The most important advance in carbohydrate structures following those made by Emil Fischer was the change from acyclic to the current cyclic structure introduced by Walter Haworth (1883–1937). He noticed the presence of α- and ß-anomers, and determined the structures of important disaccharides including cellobiose, maltose, and lactose. He also determined the basic structural aspects of starch, cellulose, inulin, and other polysaccharides, and accomplished the structure determination and synthesis of vitamin C, a sample of which he had received from Albert von Szent-Györgyi (1893–1986, Np 1937). This first synthesis of a vitamin was significant since it showed that a vitamin could be synthesized in the same way as any other organic compound. There was strong belief among leading scientists in the 1910s that cellulose, starch, protein, and rubber were colloidal aggregates of small molecules. However, Hermann Staudinger (1881–1965, Np 1953) who succeeded R. Willstätter and H. Wieland at the ETH Zürich and Freiburg, respectively, showed through viscosity measurements and various molecular weight measurements that macromolecules do exist, and developed the principles of macromolecular chemistry.

In more modern times, Raymond Lemieux (1920–, Universities of Ottawa and Alberta) has been a leader in carbohydrate research. He introduced the concept of *endo-* and *exo-*anomeric effects, accomplished the challenging synthesis of sucrose (1953), pioneered in the use of NMR coupling constants in configuration studies, and most importantly, starting with syntheses of oligosaccharides responsible for human blood group determinants, he prepared antibodies and clarified fundamental aspects of the binding of oligosaccharides by lectins and antibodies. The periodate–potassium permanganate cleavage of double bonds at room temperature (1955) is called the Lemieux reaction.

4.4 Amino Acids, Peptides, Porphyrins, and Alkaloids

It is fortunate that we have China's record and practice of herbal medicine over the centuries, which is providing us with an indispensable source of knowledge. China is rapidly catching up in terms of infrastructure and equipment in organic and bioorganic chemistry, and work on isolation, structure determination, and synthesis stemming from these valuable sources has picked up momentum. However, as mentioned above, clarification of the active principles and mode of action of these plant extracts will be quite a challenge since in many cases synergistic action is expected. Wang Yu (1910–1997) who headed the well-equipped Shanghai Institute of Organic Chemistry surprised the world with the total synthesis of bovine insulin performed by his group in 1965; the human insulin was synthesized around the same time by P. G. Katsoyannis, A. Tometsko, and C. Zaut of the Brookhaven National Laboratory (1966).

One of the giants in natural products chemistry during the first half of this century was Robert Robinson (1886–1975, Np 1947) at Oxford University. His synthesis of tropinone, a bicyclic amino ketone related to cocaine, from succindialdehyde, methylamine, and acetone dicarboxylic acid under Mannich reaction conditions was the first biomimetic synthesis (1917). It reduced Willstätter's 1903 13-step synthesis starting with suberone into a single step. This achievement demonstrated Robinson's analytical prowess. He was able to dissect complex molecular structures into simple biosynthetic building blocks, which allowed him to propose the biogenesis of all types of alkaloids and other natural products. His laboratory at Oxford, where he developed the well-known Robinson annulation reaction (1937) in connection with his work on the synthesis of steroids became a world center for natural products study. Robinson was a pioneer in the so-called electronic theory of organic reactions, and introduced the use of curly arrows to show the movements of electrons. His analytical power is exemplified in the structural studies of strychnine and brucine around 1946–1952. Barton clarified the biosynthetic route to the morphine alkaloids, which he saw as an extension of his biomimetic synthesis of usnic acid through a one-electron oxidation; this was later extended to a general phenolate coupling scheme. Morphine total synthesis was brilliantly achieved by Marshall Gates (1915–, University of Rochester) in 1952.

The yield of the Robinson tropinone synthesis was low but Clemens Schöpf (1899–1970) , Ph.D. Munich (Wieland), Universität Darmstadt, improved it to 90% by carrying out the reaction in buffer; he also worked on the stereochemistry of morphine and determined the structure of the steroidal alkaloid salamandarine (1961), the toxin secreted from glands behind the eyes of the salamander.

Roger Adams (1889–1971, University of Illinois), was the central figure in organic chemistry in the USA and is credited with contributing to the rapid development of its chemistry in the late 1930s and 1940s, including training of graduate students for both academe and industry. After earning a Ph.D. in 1912 at Harvard University he did postdoctoral studies with Otto Diels (see below) and Richard Willstätter (see below) in 1913; he once said that around those years in Germany he could cover all *Journal of the American Chemical Society* papers published in a year in a single night. His important work include determination of the structures of tetrahydrocannabinol in marijuana, the toxic gossypol in cottonseed oil, chaulmoogric acid used in treatment of leprosy, and the Senecio alkaloids with Nelson Leonard (1916–, University of Illinois, now at Caltech). He also contributed to many fundamental organic reactions and syntheses. The famous Adams platinum catalyst is not only important for reducing double bonds in industry and in the laboratory, but was central for determining the number of double bonds in a structure. He was also one of the founders of the *Organic Synthesis* (started in 1921) and the *Organic Reactions* series. Nelson Leonard switched interests to bioorganic chemistry and biochemistry, where he has worked with nucleic acid bases and nucleotides, coenzymes, dimensional probes, and fluorescent modifications such as ethenoguanine.

The complicated structures of the medieval plant poisons aconitine (from *Aconitum*) and delphinine (from *Delphinium*) were finally characterized in 1959–1960 by Karel Wiesner (1919–1986, University of New Brunswick), Leo Marion (1899–1979, National Research Council, Ottawa), George Büchi (1921–, mycotoxins, aflatoxin/DNA adduct, synthesis of terpenoids and nitrogen-containing bioactive compounds, photochemistry), and Maria Przybylska (1923–, X-ray).

The complex chlorophyll structure was elucidated by Richard Willstätter (1872–1942, Np 1915). Although he could not join Baeyer's group at Munich because the latter had ceased taking students, a close relation developed between the two. During his chlorophyll studies, Willstätter reintroduced the important technique of column chromatography published in Russian by Michael Tswett (1906). Willstätter further demonstrated that magnesium was an integral part of chlorophyll, clarified the relation between chlorophyll and the blood pigment hemin, and found the wide distribution of carotenoids in tomato, egg yolk, and bovine corpus luteum. Willstätter also synthesized cyclooctatetraene and showed its properties to be wholly unlike benzene but close to those of acyclic polyenes (around 1913). He succeeded Baeyer at Munich in 1915, synthesized the anesthetic cocaine, retired early in protest of anti-Semitism, but remained active until the Hitler era, and in 1938 emigrated to Switzerland.

The hemin structure was determined by another German chemist of the same era, Hans Fischer (1881–1945, Np 1930), who succeeded Windaus at Innsbruck and at Munich. He worked on the structure of hemin from the blood pigment hemoglobin, and completed its synthesis in 1929. He continued Willstätter's structural studies of chlorophyll, and further synthesized bilirubin in 1944. Destruction of his institute at Technische Hochschule München, during World War II led him to take his life in March 1945. The biosynthesis of hemin was elucidated largely by David Shemin (1911–1991).

In the mid 1930s the Department of Biochemistry at Columbia Medical School, which had accepted many refugees from the Third Reich, including Erwin Chargaff, Rudolf Schoenheimer, and others on the faculty, and Konrad Bloch (see above) and David Shemin as graduate students, was a great center of research activity. In 1940, Shemin ingested 66 g of ^{15}N-labeled glycine over a period of 66 hours in order to determine the half-life of erythrocytes. David Rittenberg's analysis of the heme moiety with his home-made mass spectrometer showed all four pyrrole nitrogens came from glycine. Using ^{14}C (that had just become available) as a second isotope (see next paragraph), doubly labeled glycine $^{15}NH_2^{14}CH_2COOH$ and other precursors, Shemin showed that glycine and succinic acid condensed to yield δ-aminolevulinate, thus elegantly demonstrating the novel biosynthesis of the porphyrin ring (around 1950). At this time, Bloch was working on the other side of the bench.

Melvin Calvin (1911–1997, Np 1961) at University of California, Berkeley, elucidated the complex photosynthetic pathway in which plants reduce carbon dioxide to carbohydrates. The critical $^{14}CO_2$ had just been made available at Berkeley Lawrence Radiation Laboratory as a result of the pioneering research of Martin Kamen (1913–), while paper chromatography also played crucial roles. Kamen produced ^{14}C with Sam Ruben (1940), used ^{18}O to show that oxygen in photosynthesis comes from water and not from carbon dioxide, participated in the *Manhattan* project, testified before the House UnAmerican Activities Committee (1947), won compensatory damages from the US Department of State, and helped build the University of California, La Jolla (1957). The entire structure of the photosynthetic reaction center (>10 000 atoms) from the purple bacterium *Rhodopseudomonas viridis* has been established by X-ray crystallography in the landmark studies performed by Johann Deisenhofer (1943–), Robert Huber (1937–), and Hartmut Michel (1948–) in 1989; this was the first membrane protein structure determined by X-ray, for which they shared the 1988 Nobel prize. The information gained from the full structure of this first membrane protein has been especially rewarding.

The studies on vitamin B_{12}, the structure of which was established by crystallographic studies performed by Dorothy Hodgkin (1910–1994, Np 1964), are fascinating. Hodgkin also determined the structure of penicillin (in a joint effort between UK and US scientists during World War II) and insulin. The formidable total synthesis of vitamin B_{12} was completed in 1972 through collaborative efforts between Woodward and Eschenmoser, involving 100 postdoctoral fellows and extending over 10 years. The biosynthesis of fascinating complexity is almost completely solved through studies performed by Alan Battersby (1925–, Cambridge University), Duilio Arigoni, and Ian Scott (1928–, Texas A&M University) and collaborators where advanced NMR techniques and synthesis of labeled precursors is elegantly combined with cloning of enzymes controlling each biosynthetic step. This work provides a beautiful demonstration of the power of the combination of bioorganic chemistry, spectroscopy and molecular biology, a future direction which will become increasingly important for the creation of new "unnatural" natural products.

4.5 Enzymes and Proteins

In the early days of natural products chemistry, enzymes and viruses were very poorly understood. Thus, the 1926 paper by James Sumner (1887–1955) at Cornell University on crystalline urease was received with ignorance or skepticism, especially by Willstätter who believed that enzymes were small molecules and not proteins. John Northrop (1891–1987) and co-workers at the Rockefeller Institute went on to crystallize pepsin, trypsin, chymotrypsin, ribonuclease, deoyribonuclease, carboxypeptidase, and other enzymes between 1930 and 1935. Despite this, for many years biochemists did not recognize the significance of these findings, and considered enzymes as being low molecular weight compounds adsorbed onto proteins or colloids. Using Northrop's method for crystalline enzyme preparations, Wendell Stanley (1904–1971) at Princeton obtained tobacco mosaic virus as needles from one ton of tobacco leaves (1935). Sumner, Northrop, and Stanley shared the 1946 Nobel prize in chemistry. All these studies opened a new era for biochemistry.

Meanwhile, Linus Pauling, who in mid-1930 became interested in the magnetic properties of hemoglobin, investigated the configurations of proteins and the effects of hydrogen bonds. In 1949 he showed that sickle cell anemia was due to a mutation of a single amino acid in the hemoglobin molecule, the first correlation of a change in molecular structure with a genetic disease. Starting in 1951 he and colleagues published a series of papers describing the alpha helix structure of proteins; a paper published in the early 1950s with R. B. Corey on the structure of DNA played an important role in leading Francis Crick and James Watson to the double helix structure (Np 1962).

A further important achievement in the peptide field was that of Vincent Du Vigneaud (1901–1978, Np 1955), Cornell Medical School, who isolated and determined the structure of oxytocin, a posterior pituitary gland hormone, for which a structure involving a disulfide bond was proposed. He synthesized oxytocin in 1953, thereby completing the first synthesis of a natural peptide hormone.

Progress in isolation, purification, crystallization methods, computers, and instrumentation, including cyclotrons, have made X-ray crystallography the major tool in structural. Numerous structures including those of ligand/receptor complexes are being published at an extremely rapid rate. Some of the past major achievements in protein structures are the following. Max Perutz (1914, Np 1962) and John Kendrew (1914–1997, Np 1962), both at the Laboratory of Molecular Biology, Cambridge University, determined the structures of hemoglobin and myoglobin, respectively. William Lipscomb (1919–, Np 1976), Harvard University, who has trained many of the world's leaders in protein X-ray crystallography has been involved in the structure determination of many enzymes including carboxypeptidase A (1967); in 1965 he determined the structure of the anticancer bisindole alkaloid, vinblastine. Folding of proteins, an important but still enigmatic phenomenon, is attracting increasing attention. Christian Anfinsen (1916–1995, Np 1972), NIH, one of the pioneers in this area, showed that the amino acid residues in ribonuclease interact in an energetically most favorable manner to produce the unique 3D structure of the protein.

4.6 Nucleic Acid Bases, RNA, and DNA

The "Fischer indole synthesis" was first performed in 1886 by Emil Fischer. During the period 1881–1914, he determined the structures of and synthesized uric acid, caffeine, theobromine, xanthine, guanine, hypoxanthine, adenine, guanine, and made theophylline-D-glucoside phosphoric acid, the first synthetic nucleotide. In 1903, he made 5,5-diethylbarbituric acid or Barbital, Dorminal, Veronal, etc. (sedative), and in 1912, phenobarbital or Barbipil, Luminal, Phenobal, etc. (sedative). Many of his syntheses formed the basis of German industrial production of purine bases. In 1912 he showed that tannins are gallates of sugars such as maltose and glucose. Starting in 1899, he synthesized many of the 13 α-amino acids known at that time, including the L- and D-forms, which were separated through fractional crystallization of their salts with optically active bases. He also developed a method for synthesizing fragments of proteins, namely peptides, and made an 18-amino acid peptide. He lost his two sons in World War I, lost his wealth due to postwar inflation, believed he had terminal cancer (a misdiagnosis), and killed himself in July 1919. Fischer was a skilled experimentalist, so that even today, many of the reactions performed by him and his students are so delicately controlled that they are not easy to reproduce. As a result of his suffering by inhaling diethylmercury, and of the poisonous effect of phenylhydrazine, he was one of the first to design fume hoods. He was a superb teacher and was also influential in establishing the Kaiser Wilhelm Institute, which later became the Max Planck Institute. The number and quality of his accomplishments and contributions are hard to believe; he was truly a genius.

Alexander Todd (1907–1997, Np 1957) made critical contributions to the basic chemistry and synthesis of nucleotides. His early experience consisted of an extremely fruitful stay at Oxford in the Robinson group, where he completed the syntheses of many representative anthocyanins, and then at Edinburgh where he worked on the synthesis of vitamin B_1. He also prepared the hexacarboxylate of vitamin B_{12} (1954), which was used by D. Hodgkin's group for their X-ray elucidation of this vitamin (1956). M. Wiewiorowski (1918–), Institute for Bioorganic Chemistry, in Poznan, has headed a famous group in nucleic acid chemistry, and his colleagues are now distributed worldwide.

4.7 Antibiotics, Pigments, and Marine Natural Products

The concept of one microorganism killing another was introduced by Pasteur who coined the term antibiosis in 1877, but it was much later that this concept was realized in the form of an actual antibiotic. The bacteriologist Alexander Fleming (1881–1955, University of London, Np 1945) noticed that an airborne mold, a *Penicillium* strain, contaminated cultures of *Staphylococci* left on the open bench and formed a transparent circle around its colony due to lysis of *Staphylococci*. He published these results in 1929. The discovery did not attract much interest but the work was continued by Fleming until it was taken up further at Oxford University by pathologist Howard Florey (1898–1968, Np 1945) and biochemist Ernst Chain (1906–1979, Np 1945). The bioactivities of purified "penicillin," the first antibiotic, attracted serious interest in the early 1940s in the midst of World War II. A UK/USA team was formed during the war between academe and industry with Oxford University, Harvard University, ICI, Glaxo, Burroughs Wellcome, Merck, Shell, Squibb, and Pfizer as members. This project resulted in the large scale production of penicillin and determination of its structure (finally by X-ray, D. Hodgkin). John Sheehan (1915–1992) at MIT synthesized 6-aminopenicillanic acid in 1959, which opened the route for the synthesis of a number of analogues. Besides being the first antibiotic to be discovered, penicillin is also the first member of a large number of important antibiotics containing the ß-lactam ring, for example cephalosporins, carbapenems, monobactams, and nocardicins. The strained ß-lactam ring of these antibiotics inactivates the transpeptidase by acylating its serine residue at the active site, thus preventing the enzyme from forming the link between the pentaglycine chain and the D-Ala-D-Ala peptide, the essential link in bacterial cell walls. The overuse of ß-lactam antibiotics, which has given rise to the disturbing appearance of microbial resistant strains, is leading to active research in the design of synthetic ß-lactam analogues to counteract these strains. The complex nature of the important penicillin biosynthesis is being elucidated through efforts combining genetic engineering, expression of biosynthetic genes as well as feeding of synthetic precursors, etc. by Jack Baldwin (1938–, Oxford University), José Luengo (Universidad de León, Spain) and many other groups from industry and academe.

Shortly after the penicillin discovery, Selman Waksman (1888–1973, Rutgers University, Np 1952) discovered streptomycin, the second antibiotic and the first active against the dreaded disease tuberculosis. The discovery and development of new antibiotics continued throughout the world at pharmaceutical companies in Europe, Japan, and the USA from soil and various odd sources: cephalosporin from sewage in Sardinia, cyclosporin from Wisconsin and Norway soil which was carried back to Switzerland, avermectin from the soil near a golf course in Shizuoka Prefecture. People involved in antibiotic discovery used to collect soil samples from various sources during their trips but this has now become severely restricted to protect a country's right to its soil. M. M. Shemyakin (1908–1970, Institute of Chemistry of Natural Products, Moscow) was a grand master of Russian natural products who worked on antibiotics, especially of the tetracycline class; he also worked on cyclic antibiotics composed of alternating sequences of amides and esters and coined the term depsipeptide for these in 1953. He died in 1970 of a sudden heart attack in the midst of the 7th IUPAC Natural Products

Symposium held in Riga, Latvia, which he had organized. The Institute he headed was renamed the Shemyakin Institute.

Indigo, an important vat dye known in ancient Asia, Egypt, Greece, Rome, Britain, and Peru, is probably the oldest known coloring material of plant origin, Indigofera and Isatis. The structure was determined in 1883 and a commercially feasible synthesis was performed in 1883 by Adolf von Baeyer (see above, 1835–1917, Np 1905), who founded the German Chemical Society in 1867 following the precedent of the Chemistry Society of London. In 1872 Baeyer was appointed a professor at Strasbourg where E. Fischer was his student, and in 1875 he succeeded J. Liebig in Munich. Tyrian (or Phoenician) purple, the dibromo derivative of indigo which is obtained from the purple snail Murex bundaris, was used as a royal emblem in connection with religious ceremonies because of its rarity; because of the availability of other cheaper dyes with similar color, it has no commercial value today. K. Venkataraman (1901–1981, University of Bombay then National Chemical Laboratory) who worked with R. Robinson on the synthesis of chromones in his early career, continued to study natural and synthetic coloring matters, including synthetic anthraquinone vat dyes, natural quinonoid pigments, etc. T. R. Seshadri (1900–1975) is another Indian natural products chemist who worked mainly in natural pigments, dyes, drugs, insecticides, and especially in polyphenols. He also studied with Robinson, and with Pregl at Graz, and taught at Delhi University. Seshadri and Venkataraman had a huge impact on Indian chemistry. After a 40 year involvement, Toshio Goto (1929–1990) finally succeeded in solving the mysterious identity of commelinin, the deep-blue flower petal pigment of the Commelina communis isolated by Kozo Hayashi (1958) and protocyanin, isolated from the blue cornflower Centaurea cyanus by E. Bayer (1957). His group elucidated the remarkable structure in its entirety which consisted of six unstable anthocyanins, six flavones and two metals, the molecular weight approaching 10 000; complex stacking and hydrogen bonds were also involved. Thus the pigmentation of petals turned out to be far more complex than the theories put forth by Willstätter (1913) and Robinson (1931). Goto suffered a fatal heart attack while inspecting the first X-ray structure of commelinin; commelinin represents a pinnacle of current natural products isolation and structure determination in terms of subtlety in isolation and complexity of structure.

The study of marine natural products is understandably far behind that of compounds of terrestrial origin due to the difficulty in collection and identification of marine organisms. However, it is an area which has great potentialities for new discoveries from every conceivable source. One pioneer in modern marine chemistry is Paul Scheuer (1915–, University of Hawaii) who started his work with quinones of marine origin and has since characterized a very large number of bioactive compounds from mollusks and other sources. Luigi Minale (1936–1997, Napoli) started a strong group working on marine natural products, concentrating mainly on complex saponins. He was a leading natural products chemist who died prematurely. A. Gonzalez Gonzalez (1917–) who headed the Organic Natural Products Institute at the University of La Laguna, Tenerife, was the first to isolate and study polyhalogenated sesquiterpenoids from marine sources. His group has also carried out extensive studies on terrestrial terpenoids from the Canary Islands and South America. Carotenoids are widely distributed in nature and are of importance as food coloring material and as antioxidants (the detailed mechanisms of which still have to be worked out); new carotenoids continue to be discovered from marine sources, for example by the group of Synnove Liaaen-Jensen, Norwegian Institute of Technology). Yoshimasa Hirata (1915–), who started research at Nagoya University, is a champion in the isolation of nontrivial natural products. He characterized the bioluminescent luciferin from the marine ostracod *Cypridina hilgendorfii* in 1966 (with his students, Toshio Goto, Yoshito Kishi, and Osamu Shimomura); tetrodotoxin from the fugu fish in 1964 (with Goto and Kishi and co-workers), the structure of which was announced simultaneously by the group of Kyosuke Tsuda (1907–, tetrodotoxin, matrine) and Woodward; and the very complex palytoxin, $C_{129}H_{223}N_3O_{54}$ in 1981–1987 (with Daisuke Uemura and Kishi). Richard E. Moore, University of Hawaii, also announced the structure of palytoxin independently. Jon Clardy (1943–, Cornell University) has determined the X-ray structures of many unique marine natural products, including brevetoxin B (1981), the first of the group of toxins with contiguous *trans*-fused ether rings constituting a stiff ladder-like skeleton. Maitotoxin, $C_{164}H_{256}O_{68}S_2Na_2$, MW 3422, produced by the dinoflagellate *Gambierdiscus toxicus* is the largest and most toxic of the nonbiopolymeric toxins known; it has 32 alicyclic 6- to 8-membered ethereal rings and acyclic chains. Its isolation (1994) and complete structure determination was accomplished jointly by the groups of Takeshi Yasumoto (Tohoku University), Kazuo Tachibana and Michio Murata (Tokyo University) in 1996. Kishi, Harvard University, also deduced the full structure in 1996.

The well-known excitatory agent for the cat family contained in the volatile oil of catnip, *Nepeta cataria*, is the monoterpene nepetalactone, isolated by S. M. McElvain (1943) and structure determined by Jerrold Meinwald (1954); cats, tigers, and lions start purring and roll on their backs in response to this lactone. Takeo Sakan (1912–1993) investigated the series of monoterpenes neomatatabiols, etc.

from Actinidia, some of which are male lacewing attractants. As little as 1 fg of neomatatabiol attracts lacewings.

The first insect pheromone to be isolated and characterized was bombykol, the sex attractant for the male silkworm, *Bombyx mori* (by Butenandt and co-workers, see above). Numerous pheromones have been isolated, characterized, synthesized, and are playing central roles in insect control and in chemical ecology. The group at Cornell University have long been active in this field: Tom Eisner (1929–, behavior), Jerrold Meinwald (1927–, chemistry), Wendell Roeloff (1939–, electrophysiology, chemistry). Since the available sample is usually minuscule, full structure determination of a pheromone often requires total synthesis; Kenji Mori (1935–, Tokyo University) has been particularly active in this field. Progress in the techniques for handling volatile compounds, including collection, isolation, GC/MS, etc., has started to disclose the extreme complexity of chemical ecology which plays an important role in the lives of all living organisms. In this context, natural products chemistry will be play an increasingly important role in our grasp of the significance of biodiversity.

5. SYNTHESIS

Synthesis has been mentioned often in the preceding sections of this essay. In the following, synthetic methods of more general nature are described. The Grignard reaction of Victor Grignard (1871–1935, Np 1912) and then the Diels–Alder reaction by Otto Diels (1876–1954, Np 1950) and Kurt Alder (1902–1956, Np 1950) are extremely versatile reactions. The Diels–Alder reaction can account for the biosynthesis of several natural products with complex structures, and now an enzyme, a Diels–Alderase involved in biosynthesis has been isolated by Akitami Ichihara, Hokkaido University (1997).

The hydroboration reactions of Herbert Brown (1912–, Purdue University, Np 1979) and the Wittig reactions of Georg Wittig (1897–1987, Np 1979) are extremely versatile synthetic reactions. William S. Johnson (1913–1995, University of Wisconsin, Stanford University) developed efficient methods for the cyclization of acyclic polyolefinic compounds for the synthesis of corticoid and other steroids, while Gilbert Stork (1921–, Columbia University) introduced enamine alkylation, regiospecific enolate formation from enones and their kinetic trapping (called "three component coupling" in some cases), and radical cyclization in regio- and stereospecific constructions. Elias J. Corey (1928–, Harvard University, Np 1990) introduced the concept of retrosynthetic analysis and developed many key synthetic reactions and reagents during his synthesis of bioactive compounds, including prostaglandins and gingkolides. A recent development is the ever-expanding supramolecular chemistry stemming from 1967 studies on crown ethers by Charles Pedersen (1904–1989), 1968 studies on cryptates by Jean-Marie Lehn (1939–), and 1973 studies on host–guest chemistry by Donald Cram (1919–); they shared the chemistry Nobel prize in 1987.

6. NATURAL PRODUCTS STUDIES IN JAPAN

Since the background of natural products study in Japan is quite different from that in other countries, a brief history is given here. Natural products is one of the strongest areas of chemical research in Japan with probably the world's largest number of chemists pursuing structural studies; these are joined by a healthy number of synthetic and bioorganic chemists. An important Symposium on Natural Products was held in 1957 in Nagoya as a joint event between the faculties of science, pharmacy, and agriculture. This was the beginning of a series of annual symposia held in various cities, which has grown into a three-day event with about 50 talks and numerous papers; practically all achievements in this area are presented at this symposium. Japan adopted the early twentieth century German or European academic system where continuity of research can be assured through a permanent staff in addition to the professor, a system which is suited for natural products research which involves isolation and assay, as well as structure determination, all steps requiring delicate skills and much expertise.

The history of Japanese chemistry is short because the country was closed to the outside world up to 1868. This is when the Tokugawa shogunate which had ruled Japan for 264 years was overthrown and the Meiji era (1868–1912) began. Two of the first Japanese organic chemists sent abroad were Shokei Shibata and Nagayoshi Nagai, who joined the laboratory of A. W. von Hoffmann in Berlin. Upon return to Japan, Shibata (Chinese herbs) started a line of distinguished chemists, Keita and Yuji Shibata (flavones) and Shoji Shibata (1915–, lichens, fungal bisanthraquinonoid pigments, ginsenosides); Nagai returned to Tokyo Science University in 1884, studied ephedrine, and left a big mark in the embryonic era of organic chemistry. Modern natural products chemistry really began when three extraordinary organic chemists returned from Europe in the 1910s and started teaching and research at their respective faculties:

Riko Majima, 1874–1962, C. D. Harries (Kiel University); R. Willstätter (Zürich): Faculty of Science, Tohoku University; studied urushiol, the catecholic mixture of poison ivy irritant.

Yasuhiko Asahina, 1881–1975, R. Willstätter: Faculty of pharmacy, Tokyo University; lichens and Chinese herb.

Umetaro Suzuki, 1874–1943, E. Fischer: Faculty of agriculture, Tokyo University; vitamin B_1(thiamine).

Because these three pioneers started research in three different faculties (i.e., science, pharmacy, and agriculture), and because little interfaculty personnel exchange occurred in subsequent years, natural products chemistry in Japan was pursued independently within these three academic domains; the situation has changed now. The three pioneers started lines of first-class successors, but the establishment of a strong infrastructure takes many years, and it was only after the mid-1960s that the general level of science became comparable to that in the rest of the world; the 3rd IUPAC Symposium on the Chemistry of Natural Products, presided over by Munio Kotake (1894–1976, bufotoxins, see below), held in 1964 in Kyoto, was a clear turning point in Japan's role in this area.

Some of the outstanding Japanese chemists not already quoted are the following. Shibasaburo Kitazato (1852–1931), worked with Robert Koch (Np 1905, tuberculosis) and von Behring, antitoxins of diphtheria and tetanus which opened the new field of serology, isolation of microorganism causing dysentery, founder of Kitazato Institute; Chika Kuroda (1884–1968), first female Ph.D., structure of the complex carthamin, important dye in safflower (1930) which was revised in 1979 by Obara *et al.*, although the absolute configuration is still unknown (1998); Munio Kotake (1894–1976), bufotoxins, tryptophan metabolites, nupharidine; Harusada Suginome (1892–1972), aconite alkaloids; Teijiro Yabuta (1888–1977), kojic acid, gibberrelins; Eiji Ochiai (1898–1974), aconite alkaloids; Toshio Hoshino (1899–1979), abrine and other alkaloids; Yusuke Sumiki (1901–1974), gibberrelins; Sankichi Takei (1896–1982), rotenone; Shiro Akabori (1900–1992), peptides, C-terminal hydrazinolysis of amino acid ; Hamao Umezawa (1914–1986), kanamycin, bleomycin, numerous antibiotics; Shojiro Uyeo (1909–1988), lycorine; Tsunematsu Takemoto (1913–1989), inokosterone, kainic acid, domoic acid, quisqualic acid; Tomihide Shimizu (1889–1958), bile acids; Kenichi Takeda (1907–1991), Chinese herbs, sesquiterpenes; Yoshio Ban (1921–1994), alkaloid synthesis; Wataru Nagata (1922–1993), stereocontrolled hydrocyanation.

7. CURRENT AND FUTURE TRENDS IN NATURAL PRODUCTS CHEMISTRY

Spectroscopy and X-ray crystallography has totally changed the process of structure determination, which used to generate the excitement of solving a mystery. The first introduction of spectroscopy to the general organic community was Woodward's 1942–1943 empirical rules for estimating the UV maxima of dienes, trienes, and enones, which were extended by Fieser (1959). However, Butenandt had used UV for correctly determining the structures of the sex hormones as early as the early 1930s, while Karrer and Kuhn also used UV very early in their structural studies of the carotenoids. The Beckman DU instruments were an important factor which made UV spectroscopy a common tool for organic chemists and biochemists. With the availability of commercial instruments in 1950, IR spectroscopy became the next physical tool, making the 1950 Colthup IR correlation chart and the 1954 Bellamy monograph indispensable. The IR fingerprint region was analyzed in detail in attempts to gain as much structural information as possible from the molecular stretching and bending vibrations. Introduction of NMR spectroscopy into organic chemistry, first for protons and then for carbons, has totally changed the picture of structure determination, so that now IR is used much less frequently; however, in biopolymer studies, the techniques of difference FTIR and resonance Raman spectroscopy are indispensable.

The dramatic and rapid advancements in mass spectrometry are now drastically changing the protocol of biomacromolecular structural studies performed in biochemistry and molecular biology. Herbert Hauptman (mathematician, 1917–, Medical Foundation, Buffalo, Np 1985) and Jerome Karle (1918–, US Naval Research Laboratory, Washington, DC, Np 1985) developed direct methods for the determination of crystal structures devoid of disproportionately heavy atoms. The direct method together with modern computers revolutionized the X-ray analysis of molecular structures, which has become routine for crystalline compounds, large as well as small. Fred McLafferty (1923–, Cornell University) and Klaus Biemann (1926–, MIT) have made important contributions in the development of organic and bioorganic mass spectrometry. The development of cyclotron-based facilities for crystallographic biology studies has led to further dramatic advances enabling some protein structures to be determined in a single day, while cryoscopic electron micrography developed in 1975 by Richard Henderson and Nigel Unwin has also become a powerful tool for 3D structural determinations of membrane proteins such as bacteriorhodopsin (25 kd) and the nicotinic acetylcholine receptor (270 kd).

Circular dichroism (c.d.), which was used by French scientists Jean B. Biot (1774–1862) and Aimé Cotton during the nineteenth century "deteriorated" into monochromatic measurements at 589 nm after R.W. Bunsen (1811–1899, Heidelberg) introduced the Bunsen burner into the laboratory which readily emitted a 589 nm light characteristic of sodium. The 589 nm $[\alpha]_D$ values, remote from most chromophoric maxima, simply represent the summation of the low-intensity readings of the decreasing end of multiple Cotton effects. It is therefore very difficult or impossible to deduce structural information from $[\alpha]_D$ readings. Chiroptical spectroscopy was reintroduced to organic chemistry in the 1950s by C. Djerassi at Wayne State University (and later at Stanford University) as optical rotatory dispersion (ORD) and by L. Velluz and M. Legrand at Roussel-Uclaf as c.d. Günther Snatzke (1928–1992, Bonn then Ruhr University Bochum) was a major force in developing the theory and application of organic chiroptical spectroscopy. He investigated the chiroptical properties of a wide variety of natural products, including constituents of indigenous plants collected throughout the world, and established semiempirical sector rules for absolute configurational studies. He also established close collaborations with scientists of the former Eastern bloc countries and had a major impact in increasing the interest in c.d. there.

Chiroptical spectroscopy, nevertheless, remains one of the most underutilized physical measurements. Most organic chemists regard c.d. (more popular than ORD because interpretation is usually less ambiguous) simply as a tool for assigning absolute configurations, and since there are only two possibilities in absolute configurations, c.d. is apparently regarded as not as crucial compared to other spectroscopic methods. Moreover, many of the c.d. correlations with absolute configuration are empirical. For such reasons, chiroptical spectroscopy, with its immense potentialities, is grossly underused. However, c.d. curves can now be calculated nonempirically. Moreover, through-space coupling between the electric transition moments of two or more chromophores gives rise to intense Cotton effects split into opposite signs, exciton-coupled c.d.; fluorescence-detected c.d. further enhances the sensitivity by 50- to 100-fold. This leads to a highly versatile nonempirical microscale solution method for determining absolute configurations, etc.

With the rapid advances in spectroscopy and isolation techniques, most structure determinations in natural products chemistry have become quite routine, shifting the trend gradually towards activity-monitored isolation and structural studies of biologically active principles available only in microgram or submicrogram quantities. This in turn has made it possible for organic chemists to direct their attention towards clarifying the mechanistic and structural aspects of the ligand/biopolymeric receptor interactions on a more well-defined molecular structural basis. Until the 1990s, it was inconceivable and impossible to perform such studies.

Why does sugar taste sweet? This is an extremely challenging problem which at present cannot be answered even with major multidisciplinary efforts. Structural characterization of sweet compounds and elucidation of the amino acid sequences in the receptors are only the starting point. We are confronted with a long list of problems such as cloning of the receptors to produce them in sufficient quantities to investigate the physical fit between the active factor (sugar) and receptor by biophysical methods, and the time-resolved change in this physical contact and subsequent activation of G-protein and enzymes. This would then be followed by neurophysiological and ultimately physiological and psychological studies of sensation. How do the hundreds of taste receptors differ in their structures and their physical contact with molecules, and how do we differentiate the various taste sensations? The same applies to vision and to olfactory processes. What are the functions of the numerous glutamate receptor subtypes in our brain? We are at the starting point of a new field which is filled with exciting possibilities.

Familiarity with molecular biology is becoming essential for natural products chemists to plan research directed towards an understanding of natural products biosynthesis, mechanisms of bioactivity triggered by ligand–receptor interactions, etc. Numerous genes encoding enzymes have been cloned and expressed by the cDNA and/or genomic DNA-polymerase chain reaction protocols. This then leads to the possible production of new molecules by gene shuffling and recombinant biosynthetic techniques. Monoclonal catalytic antibodies using haptens possessing a structure similar to a high-energy intermediate of a proposed reaction are also contributing to the elucidation of biochemical mechanisms and the design of efficient syntheses. The technique of photoaffinity labeling, brilliantly invented by Frank Westheimer (1912–, Harvard University), assisted especially by advances in mass spectrometry, will clearly be playing an increasingly important role in studies of ligand–receptor interactions including enzyme–substrate reactions. The combined and sophisticated use of various spectroscopic means, including difference spectroscopy and fast time-resolved spectroscopy, will also become increasingly central in future studies of ligand–receptor studies.

Organic chemists, especially those involved in structural studies have the techniques, imagination, and knowledge to use these approaches. But it is difficult for organic chemists to identify an exciting and worthwhile topic. In contrast, the biochemists, biologists, and medical doctors are daily facing exciting life-related phenomena, frequently without realizing that the phenomena could be understood or at least clarified on a chemical basis. Broad individual expertise and knowledge coupled with multidisciplinary research collaboration thus becomes essential to investigate many of the more important future targets successfully. This approach may be termed "dynamic," as opposed to a "static" approach, exemplified by isolation and structure determination of a single natural product. Fortunately for scientists, nature is extremely complex and hence all the more challenging. Natural products chemistry will be playing an absolutely indispensable role for the future. Conservation of the alarming number of disappearing species, utilization of biodiversity, and understanding of the intricacies of biodiversity are further difficult, but urgent, problems confronting us.

That natural medicines are attracting renewed attention is encouraging from both practical and scientific viewpoints; their efficacy has often been proven over the centuries. However, to understand the mode of action of folk herbs and related products from nature is even more complex than mechanistic clarification of a single bioactive factor. This is because unfractionated or partly fractionated extracts are used, often containing mixtures of materials, and in many cases synergism is most likely playing an important role. Clarification of the active constituents and their modes of action will be difficult. This is nevertheless a worthwhile subject for serious investigations.

Dedicated to Sir Derek Barton whose amazing insight helped tremendously in the planning of this series, but who passed away just before its completion. It is a pity that he was unable to write this introduction as originally envisaged, since he would have had a masterful overview of the content he wanted, based on his vast experience. I have tried to fulfill his task, but this introduction cannot do justice to his original intention.

ACKNOWLEDGMENT

I am grateful to current research group members for letting me take quite a time off in order to undertake this difficult writing assignment with hardly any preparation. I am grateful to Drs. Nina Berova, Reimar Bruening, Jerrold Meinwald, Yoko Naya, and Tetsuo Shiba for their many suggestions.

8. BIBLIOGRAPHY

"A 100 Year History of Japanese Chemistry," Chemical Society of Japan, Tokyo Kagaku Dojin, 1978.
K. Bloch, *FASEB J.*, 1996, **10**, 802.
"Britannica Online," 1994–1998.
Bull. Oriental Healing Arts Inst. USA, 1980, **5**(7).
L. F. Fieser and M. Fieser, "Advanced Organic Chemistry," Reinhold, New York, 1961.
L. F. Fieser and M. Fieser, "Natural Products Related to Phenanthrene," Reinhold, New York, 1949.
M. Goodman and F. Morehouse, "Organic Molecules in Action," Gordon & Breach, New York, 1973.
L. K. James (ed.), "Nobel Laureates in Chemistry," American Chemical Society and Chemistry Heritage Foundation, 1994.
J. Mann, "Murder, Magic and Medicine," Oxford University Press, New York, 1992.
R. M. Roberts, "Serendipity, Accidental Discoveries in Science," Wiley, New York, 1989.
D. S. Tarbell and T. Tarbell, "The History of Organic Chemistry in the United States, 1875–1955," Folio, Nashville, TN, 1986.

6.01
Overview

PETER B. MOORE
Yale University, New Haven, CT, USA

and

SUSUMU NISHIMURA
Banyu Tsukuba Research Institute, Tsukuba, Japan

and

DIETER SÖLL
Yale University, New Haven, CT, USA

The importance of the role RNA plays in all aspects of gene expression has been understood by molecular biologists and biochemists since the late 1950s. Nevertheless, relative to what was going on in the DNA and protein fields, RNA biochemistry remained a backwater for many years primarily because RNA is hard to work with. For example, unless handled carefully, RNAs are rather prone to hydrolytic degradation, and most of the RNAs abundant in nature have molecular weights so large that for a long time it seemed unlikely that anything useful could be learned about them using the physical and chemical techniques of the day. In addition, many biologically important RNAs are so rare that it is difficult to prepare enough of any one of them from natural sources to do all the experiments one would like. Finally, for many years, by comparison with the protein world, the RNA universe appeared to be very small, consisting only of transfer RNAs, ribosomal RNAs, messenger RNAs, all of which are extremely difficult to prepare in quantity, and a few viral RNAs. Why spend one's career struggling to understand the properties of a class of macromolecules so difficult and so limited?

The mind set of those in the RNA field has slowly been transformed from a somewhat pessimistic resignation to near manic optimism by the events of the last twenty years. Powerful methods have been developed for sequencing RNA, and a rich variety of chemical and genetic methods is now available for determining the functional significance of single residues in large RNAs, and even that of individual groups within single residues. On top of that, the supply problem has been solved. Chemical and enzymatic methods now exist that make it possible to synthesize RNAs of any sequence in amounts adequate for even the most material-hungry experimental techniques. In many other respects, RNA is easier to work with today than protein. In addition, the RNA universe has expanded. Scores of new RNAs have been discovered, most of them in eukaryotic organisms, that perform functions of which the biochemical community was entirely ignorant in the 1960s, when the first blossoming of the RNA field occurred. Additional stimulus was provided in the 1980s by the discovery that two different RNAs possess catalytic activity, and several additional catalytic RNAs have since been identified. Their existence has led to renewed interest in the possibility that the first organisms might have used RNA both as genetic material and as catalysts for the reactions required for their survival. Francis Crick's reflection (in 1966) on an RNA molecule's versatility

("It almost appears as if tRNA were Nature's attempt to make an RNA molecule play the role of a protein") can now be extended to many RNA species. One interesting offshoot of these developments has been the invention of a new field of chemistry that is devoted to the production of synthetic RNAs that have novel ligand binding and catalytic activities. Finally, and belatedly, NMR spectroscopists and X-ray crystallographers have begun solving RNA structures.

This volume covers the full range of problems being addressed by workers in the RNA field today. Each chapter has been contributed by a scientist expert in the area it covers, and is thus a reliable guide for those interested in entering the field. The Editors hope that those patient enough to read the entire book will come away with an appreciation of the rapid progress now being made in the RNA field, and will sense the excitement that now pervades it. RNA biochemistry is destined to catch up with DNA and protein biochemistry in the next 10 or 15 years, and it is certain that important new biological insights will emerge in the process.

6.02

A Spectroscopist's View of RNA Conformation: RNA Structural Motifs

PETER B. MOORE

Yale University, New Haven, CT, USA

6.02.1 INTRODUCTION

Biologically, RNA mediates between DNA and protein—DNA makes RNA makes protein—and RNA is also intermediate between DNA and protein chemically. Some RNAs are carriers of genetic information, similar to DNA, and others, like transfer RNAs and ribosomal RNAs, are protein-like. Their functions are determined by their conformations as much as their sequences, and some have enzymatic activity.

Even though RNA biochemists have recognized their need for structures almost as long as protein biochemists, far more is known about proteins than RNAs. Coordinates for more than 5000 proteins have been deposited in the Brookhaven Protein Data Bank, but the number of RNA entries is of the order of 50 and many of them describe RNA fragments, not whole molecules.

All of the RNA structures available before ~1985 were crystal structures and X-ray crystallography remains the dominant, high resolution method for determining RNA conformations. By the late 1980s, nuclear magnetic resonance (NMR) had emerged as a viable alternative, but for many years, only a few structures a year were solved that way. Since 1995, however, the production rate has increased to roughly a structure a month, and because the field is taking off, it is time for biochemists who make use of RNA structures to understand what their spectroscopist colleagues are producing.

This chapter describes how RNA conformations are determined by NMR and summarizes what NMR has taught us about RNA motifs. For this purpose, a motif is any assembly of nucleotides bigger than a base pair or a base triple that has a distinctive conformation and that is common in RNAs. The description of RNA spectroscopy provided is intended to help biochemists understand what NMR structures are, not to indoctrinate the would-be spectroscopist.

6.02.2 THE DETERMINATION OF RNA STRUCTURES BY NMR

The behavior of all atoms that have nuclear spins greater than zero can be studied by NMR and the predominant isotopes of two of the five elements abundant in RNA have nuclear spins of 1/2: ^1H and ^{31}P. NMR signals produced by ^1H atoms can provide information about the conformation and dynamics of the bases and sugars of an RNA and ^{31}P spectra can do the same for its phosphate groups. Those not content with what ^1H and ^{31}P spectra provide, can prepare their RNAs labeled with ^{13}C and/or ^{15}N, which are also spin 1/2 nuclei. Thus NMR spectra can be obtained from all the atoms in a nucleic acid except its oxygens, for which no suitable isotope exists. What can be learned from them?

The answer to this question, of course, is to be found in the primary NMR literature, which is vast and for the most part too technical for nonspecialists. For that reason, rather than fill the text that follows with references that its intended audience is likely to find useless, I direct them here to a few secondary sources. For NMR fundamentals, Slichter's book is recommended.[1] It is complete and its verbal descriptions are good enough that readers need not wade through its (many) derivations. Those interested in multidimensional NMR, about which little is said below, can consult either the short monograph by Goldman,[2] or the treatise of Ernst *et al.*[3] Wüthrich's book on the NMR of proteins and nucleic acids is so useful, the cover has fallen off the local copy.[4] A technically oriented, but more up to date text on protein NMR appeared in 1996, which is also useful.[5]

6.02.2.1 NMR Fundamentals

Nuclei that have spin have magnetic moments and for that reason, they orient like compass needles when placed in magnetic fields. Because nuclei are very small, their response is quantized. Spin 1/2 nuclei orient themselves in magnetic fields in only two ways: parallel to it and antiparallel to it. Because the energy associated with the parallel orientation is slightly lower than that of the antiparallel orientation, in a population of atoms that has come to equilibrium in a magnetic field, the number of nuclei in the parallel orientation is only slightly larger than the number in the antiparallel orientation. In the strongest available magnets, the excess is only a few per million. The tiny net magnetization that results is what NMR spectroscopists study. Sensitivity is not one of NMR's selling points!

An NMR spectrometer consists of a magnet that orients the nuclei in samples, a radio frequency transmitter that perturbs nuclear orientations in controlled ways, and a receiver that detects the electromagnetic signals generated when the orientations of the magnetic moments of aligned populations of nuclei are perturbed. NMR spectrometers produce spectra that display the magnitude of these electromagnetic signals as a function of perturbing frequency. A peak in such a plot is a *resonance*.

6.02.2.2   Chemical Shift

Electromagnetic radiation makes spin 1/2 nuclei reorient in external magnetic fields with maximum efficiency when the product of Planck's constant multiplied by the frequency of the reorienting radiation equals the difference in energy between their two possible orientations, that is $hv = \Delta E$, and that frequency, the *resonant frequency*, is the one at which the intensity of a resonance is maximum. The energy difference that determines a resonant frequency is the product of the strength of the magnetic field experienced by the nuclei responsible for a resonance and their intrinsic magnetic moments. The magnetic moments of all nuclides relevant to biochemists were measured long ago, and they differ so much that there is no overlap whatever in the frequency ranges at which different nuclides resonate; e.g. hydrogen resonances cannot be confused with phosphorus resonances. (Note that the resonant frequency of protons in a 500 MHz NMR spectrometer is about 500 MHz.)

The reason NMR is interesting is that the magnetic fields that nuclei experience, and hence their resonant frequencies, depend on chemical context. Nuclei in molecules are invariably surrounded by electrons, which for these purposes are best thought of as particles in continual motion. Charged particles moving in magnetic fields experience forces that add a circular component to their trajectories, and when charged particles move in circles, magnetic fields result. Thus, in effect, when a molecule is placed in an external magnetic field it becomes a tiny solenoidal magnet whose field (usually) opposes the external field. In solution, where molecular rotations lead to averaging, the atom-to-atom variation in the strength of these induced magnetic fields within a single molecule is seldom more than a few millionths the magnitude of the external, inducing field. Nevertheless, it is easily detected because the receivers in modern NMR spectrometers have frequency resolutions of about 1 part in 10^8. Thus the proton spectrum of a biological macromolecule is a set of resonances differing modestly in frequency, not a single, massive resonance.

The frequency differences that distinguish resonances in a spectrum are called *chemical shifts*, and their importance cannot be overstated. Chemical shift differences are the primary way the resonances of atoms at different positions in a population of identical molecules are distinguished from each other, and if a spectrometer cannot *resolve* a large fraction of the resonances in the proton spectrum of an RNA little progress can be made. (A *resolved* spectrum is one in which each resonance represents an atom in a single position in the molecule of interest.)

Chemical shifts are usually reported using a relative scale the unit of which is the *part per million* (ppm). The chemical shift of a resonances is 10^6 times the difference between its resonant frequency and the resonant frequency of an atom of the same type at a particular position in some standard substance, divided by the frequency of the standard resonance. A virtue of this scale is its independence of spectrometer field strength; if a resonance has a chemical shift of 8 ppm in a 250 MHz spectrometer, its chemical shift will be 8 ppm in a 500 MHz spectrometer also. By convention, if the frequency of a resonance is less than that of the standard, its chemical shift is positive, and it is described as a *down field* resonance. *Up field* resonances have negative chemical shifts. The proton spectrum of an RNA spans about 12 ppm, and spectrometers can measure the chemical shifts of macromolecular resonances to about 0.01 ppm.

6.02.2.3   Couplings and Torsion Angles

The resonant behavior of atoms within molecules in solution is affected by only one other phenomenon we need to worry about; *J*-coupling. If in a solution of identical molecules, a spin 1/2 atom at one position is *J*-coupled to a spin 1/2 atom at another, that atom will contribute two closely spaced resonances to the spectrum of the molecule not the single resonance otherwise expected. More complex splitting patterns arise when an atom is *J*-coupled to several neighbors.

J-coupling occurs because of magnetic interactions that occur when electrons contact nuclei, which they do when they occupy molecular orbitals that have nonzero values at nuclear positions. For example, electrons in σ molecular orbitals contact both of the nuclei they help bond, but electrons in π molecular orbitals do not. Electrons are spin 1/2 particles and have intrinsic magnetic moments, just like spin 1/2 nuclei. If the spin of an electron and the nucleus it contacts have the same orientation, the orbital energy of the electron will be slightly lower than it would be if their spins were antiparallel because of favorable magnetic interactions. If electrons having both spin orientations contact a nucleus equally, the total spin magnetic interaction energy is zero.

Why does contact lead to splitting? Suppose two spin 1/2 nuclei, A and B, are bonded by a molecular orbital that contacts them both and contains two electrons, one with its spin up and the other with its spin down. If the spin of A is parallel to the external magnetic field, electronic configurations that put the electron whose spin is up close to A will be favored because they have lower energies. Since the electrons are paired, if the spin up electron is close to A, the spin down electron must be close to B, and B will experience a slight magnetic field because it is not "seeing" both electrons equally. If the orientation of the spin of nucleus A were reversed, the bias in the spin orientation of the electron contacting nucleus B would also be reversed, as would the magnetic field experienced by B. Thus within a population of identical molecules, nuclei of type B will resonate at two slightly different frequencies, one for each of the two possible orientations of the spin of A. The difference in resonant frequency between the two resonances is called a *splitting* or a *coupling constant*, and splittings are mutual. The splitting of the resonance of B due to A is the same as the splitting of the resonance of A due to B.

Four facts about *J*-coupling are relevant here. First, *J*-coupling effects are transmitted *exclusively* through covalent bonds. Second, *J*-couplings between atoms separated by more than three or four bonds are usually too small to detect. Third, splittings are independent of external magnetic field strength and they vary in magnitude from a few Hz to ~ 100 Hz in biological macromolecules, depending on the species of atoms that are coupled and the way they are bonded together. Fourth, macromolecular torsion angles can be deduced from coupling constants because the magnitudes of three and four bond couplings vary sinusoidally with torsion angle. Thus experiments that explore the couplings in a spectrum of a molecule can identify resonances arising from atoms that are near neighbors in its covalent structure and determine the magnitude of its torsion angles.

6.02.2.4 Spin–Lattice Relaxation: Nuclear Overhauser Effects and Distances

Every resonance in an NMR spectrum has two times associated with it: a *spin–lattice relaxation time*, or T_1, and a *spin–spin relaxation time*, or T_2. Spin–lattice relaxation is important because phenomena that contribute to it are important sources of information about interatomic distances. Spin–spin relaxation is important because it sets limits on the sizes of the RNAs that can be studied by NMR.

It takes time for the magnetic moments of nuclei to become oriented when a sample is placed in a magnetic field, or to return to equilibrium, if their equilibrium orientations have been disturbed. Both processes proceed with first order kinetics and their rate constants are the same. The inverse of a first order rate constant is a time, and in this case, that time is called the spin–lattice relaxation time, or T_1.

Spin–lattice relaxation is caused by magnetic interactions that make pairs of neighboring nuclei change their spin orientations simultaneously. The rate at which these events occur depends on a host of factors, among them the magnitudes of the magnetic moments of the atoms involved, the external magnetic field strength, the distances between atoms, and the speed of any motions that change their relative orientations. Everything else being equal, the slower a macromolecule rotates diffusionally, the longer the T_1 of its atoms. For RNAs of the size that are being characterized by NMR today, proton spin–lattice relaxation times range from 1–10 s.

Transmitters in modern spectrometers can be programmed to irradiate samples with pulses of electromagnetic radiation that under favorable circumstances can more or less instantaneously upset the spin orientation of all the atoms in a molecular population that contribute to a single resonance, without disturbing the orientations of any others. Suppose this is done to the H-1′ resonance of nucleotide (*n*) in some RNA. What happens next? As the disequilibrated H-1′ population returns to equilibrium, exchanges in magnetization between them and the protons adjacent to them, cause the latter to "share" in their disequilibrium. The H-2′ protons of nucleotide (*n*) are certain to be affected, as are nearby protons belonging to nucleotide (*n* + 1). In molecules the size of an RNA, there is a reduction in the magnitude of the resonances of adjacent protons that becomes more pronounced with the passage of time, up to hundreds of milliseconds after the initial disequilibration, and then fades away. These changes in resonance intensity are called nuclear Overhauser effects, or NOEs, for short.

NOEs are transmitted through space, and everything else being equal, their magnitude is inversely proportional to the distance between interacting nuclei raised to the sixth power. In modern spectrometers, proton–proton NOEs, which are the ones usually studied, are undetectably small when distances exceed 5 Å, no matter how long the time waited for them to develop. Thus by studying NOEs, which atom is within 5 Å of which other atom in an RNA can be determined, and their separation can be estimated.

6.02.2.5   Spin–Spin Relaxation: Molecular Weight Limitations

Spin–spin relaxation exists because NMR spectrometers detect signals only when the magnetic moments of entire populations of nuclei are aligned, and moving in synchrony. This condition is met at the outset of the typical NMR experiment, but as time goes by, the motions of individual nuclei vary from the mean, due to random differences in their environments, and as the variation in the population grows, the vector sum of their magnetic moments decays to zero. Since nuclear magnetic signals also lose intensity when individual nuclei return to their equilibrium orientations, all processes that contribute to spin–lattice relaxation contribute to spin–spin relaxation also. T_2 is always shorter than T_1.

NMR signals decay with first order kinetics, and their characteristic times, T_2, can be estimated by measuring the widths of resonances in spectra. If T_2 is short, resonances will be broad. If T_2 is long, resonances will be narrow. For RNAs in the molecular weight range of interest here, T_2 are of the order of 20 ms, and the more slowly a macromolecule tumbles, the shorter its T_2. Thus big molecules have broader resonances than small molecules.

The broadening of resonances that accompanies increased molecular weight makes it even harder to resolve the spectra of large RNAs than it would be otherwise. The chemical shift range over which atoms in RNAs resonate is independent of molecular weight, and large RNAs contain more atoms in chemically distinct environments than small RNAs. Hence the larger an RNA, the more crowded its spectra and the more difficult it is to resolve its resonances. T_2 effects add insult to injury because broad resonances are intrinsically harder to resolve than narrow resonances. Since resolution of spectra is a *sine qua non* for spectroscopic analysis, these two effects, spectral crowding and resonance broadening, combine to set an upper bound to the molecular weights of the RNAs that can be studied by NMR. The molecular weight frontier stands today (1998) at about 45 nucleotides.

There is nothing permanent about this molecular weight frontier. For example, the higher the field strength of a spectrometer, the better resolved the spectra it produces. Thus as long as the field strengths of the spectrometer magnets available continue to increase, as they have in recent years, the frontier will continue to move forward, albeit at enormous expense. The sensitivity improvement that accompany increases in field strength is an important added benefit of this brute force approach to improving spectral resolution.

Isotopic labeling can also contribute. When multidimensional experiments are done on samples labeled with ^{13}C and ^{15}N, spectra result in which proton resonances that have similar chemical shifts are distinguished by the chemical shifts of the ^{13}C or ^{15}N atoms to which the protons in question are bonded. Surprisingly, these techniques have had a much bigger impact on NMR size limits for protein than they have for RNA. Proton T_2s in macromolecules labeled with ^{13}C and ^{15}N are always shorter than those in unlabeled macromolecules because of ^1H (^{13}C, ^{15}N) interactions, and the sensitivity of heteronuclear experiments degrades as T_2 decreases. For reasons that have yet to be understood, this isotope-T_2 effect seems to be greater in RNAs than it is in proteins, and so in contrast to the experience of protein spectroscopists, only modest increases in the molecular weights of the RNAs that can be studied by NMR have resulted from the application of heteronuclear strategies. What these approaches have done is increase the reliability and completeness of the assignments that are obtained for the spectra of RNAs of "ordinary" size.

RNA T_2 can be reduced by selective deuteration because the relaxation rates of protons tends to be dominated by their interactions with neighboring protons. Thus when some protons in a molecule are replaced with deuterons (^2H), which have much lower magnetic moments, the relaxation rates of the remaining protons decrease. Note that because deuterium resonates at frequencies well outside the proton range, deuterium labeling can be used to remove specific resonances from the proton spectra of macromolecules, which can also help solve assignment problems.

6.02.2.6   Samples

A single sample consisting of 0.2 ml of a 2 mM solution of an RNA can suffice for its structural analysis. Not all RNAs can be investigated under all possible solvent conditions, however. A structure will not emerge from a spectroscopic investigation unless the RNA of interest is monomeric under the conditions chosen and has a single conformation. Sometimes, RNA samples are required that are labeled with ^{13}C, ^{15}N, and ^2H, either generally or site specifically. The technology for producing such samples is constantly improving, and the cost of making them continues to fall.[6–9]

6.02.2.7 Multidimensional NMR

The modern era of macromolecular NMR began in the late 1970s, when the two-dimensional spectra first began being collected on proteins.[4] Among the first experiments done were the COSY (or Correlation Spectroscopy) and NOESY (or Nuclear Overhauser Spectroscopy) experiments. The former generates a two-dimensional spectrum in which resonances that are *J*-coupled are displayed, and the latter does the same for resonances that give NOEs. The more complicated multidimensional experiments introduced subsequently accomplish similar ends by different means. Happily, there is not the slightest reason for the consumer of NMR structures to worry about the details.

6.02.2.8 Assignments

Ribonucleotides contain 8–10 protons of which 7–8 are bonded directly to carbon atoms, and hence do not exchange rapidly with water protons. The remainder are bonded to nitrogens and oxygens, and exchange rapidly. The resonances of the nonexchangeable and more slowly exchanging protons of an RNA can be observed in spectra taken from samples dissolved in H_2O, and the resonances of its nonexchangeable protons can be studied selectively using samples dissolved in D_2O. As Figure 1 shows, RNA resonances cluster in four groups, depending on chemical type. Note, however, that the chemical shift separations between groups of resonances are about the same size as environmental chemical shift effects that disperse resonances within groups, and hence resonances can appear between clusters or even in the "wrong" cluster. The ^{13}C and ^{15}N spectra of RNAs are similarly complex, but since the chemical shift separation between groups is significantly larger (see Varani and Tinoco[10]), "misplacement" of resonances is less likely (but still not impossible).[11] The ^{31}P spectrum of an RNA is always its worst dispersed because all its phosphorus atoms appear in a single chemical context. Fortunately, there is only one phosphorus resonance per residue.

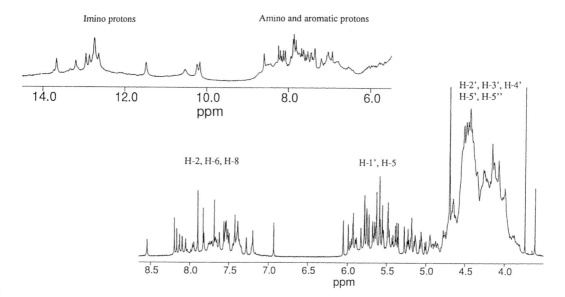

Figure 1 Proton spectrum of a typical RNA. The lower spectrum shows resonances that can be observed in D_2O and the upper spectrum shows the additional resonances observed when an RNA is dissolved in H_2O. The types of protons that contribute to each region of the spectrum are indicated. This figure is copied, with permission, from the Ph.D. Thesis of A. Szewczak, Yale University, 1994.

The first order of business for the RNA spectroscopist is assignment of spectra, and this is invariably the most time-consuming phase of any NMR project. A resonance is assigned when the atom (or atoms) responsible for it have been identified. Assignments are vital because until they are obtained, nothing can be inferred about molecular conformation from NOESY and COSY crosspeaks. A number of strategies for assigning RNA spectra are available, all derived from techniques pioneered by protein spectroscopists (see Varani and co-workers,[9,10] Nikonowicz and Pardi,[13] and Moore[14]). As is the case with multidimensional spectroscopy, there is no need for the nonspecialists to worry about the details.

6.02.2.9 Helices and Torsion Angles

By the time NMR spectroscopists get involved, the A-form helices of an RNA have usually been identified by other means and their existence is easy to confirm spectroscopically. The imino proton resonances of AU, GC, and GU base pairs are easily distinguished on the basis of their chemical shifts and the NOEs they give to other kinds of protons. Furthermore, imino–imino NOEs, which are characteristic of double helices, can be used to determine the order of base pairs in helices, and a distinctive pattern of NOEs involving nonexchangeable proton resonances is observed in double-helical RNAs.[4]

In principle, the conformation of the nonhelical parts of an RNA could be determined by measuring glycosidic torsion angles, and the backbone torsion angles of each nucleotide (see Figure 2).[15] As a practical matter, however, it is hard to measure coupling constants that speak to many of these torsion angles, and difficult to measure any of them with sufficient accuracy. Nevertheless, data that define the rotamer ranges of torsion angles are relatively easy to obtain from NMR spectra and that information is immensely helpful. The two torsion angles that are easiest to access spectroscopically are δ and χ.

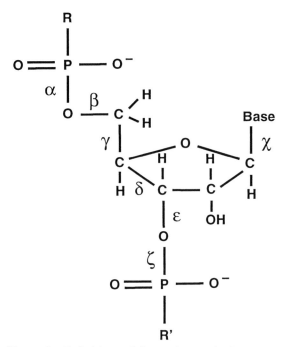

Figure 2 Definitions of the torsion angles in RNAs.

The glycosidic torsion angles of nucleotides (χ) fall into two nonoverlapping ranges, *syn* and *anti*, which are easily distinguished. The intranucleotide distance between pyrimidine H-6 or purine H-8 protons and H-1′ ribose protons is short if nucleotides are *syn* and long if they are *anti*, and since NOE intensities are proportional to r^{-6}, the difference in ((H-6 or H-8) to H-1′) NOE intensity is huge. The only way to get χ wrong is by misassigning resonances.

Sugar pucker, which corresponds to δ, is also easy to determine. The riboses of most nucleotides in RNA have a C-3′-*endo* pucker, but some are found in the DNA-like C-2′-*endo* configuration. Sugar puckers can be deduced from H-1′–H-2′ coupling constants, which are large for C-2′-*endo* riboses, small for C-3′-*endo* riboses, and intermediate if a ribose is exchanging rapidly between the two alternatives. H-1′–H-2′ cross-peaks in COSY-like spectra fall in a distinctive chemical shift range and because their appearances are determined by the magnitude of couplings they represent, coupling constants can be estimated by measuring their substructures. H-1′–H-2′ coupling constants also have a bearing on ε. For steric reasons, ε is never found in the $+$gauche range, and if a ribose is C-3′-endo, the trans rotamer is also impossible[16] (See Saenger[15] for rotamer definitions.)

Soft information about α and ζ can be gleaned from the ^{31}P spectrum of an RNA because ^{31}P chemical shifts are sensitive to both.[17] [31]P chemical shifts fall in a narrow range in A-form RNA, and thus α and ζ are likely to have A-form values when ^{31}P shifts are within that range (see Varani et al.[9]). If an unusual ^{31}P chemical shift is observed, neither angle can be constrained.

6.02.2.10   Distance Estimation

In the simple situations, the initial rates at which cross-peaks increase in intensity in proton–proton NOESY spectra are proportional to the distances between the protons they relate between, raised to negative sixth power. RNA NOESY spectra are internally calibrated because every pyrimidine contributes an intranucleotide H-5/H-6 cross-peak to it and the separation between those protons is fixed covalently. Thus, if the intensity of each cross-peak in the NOESY spectrum of an RNA is evaluated relative to the intensities of its H-5–H-6 cross-peaks, estimates of proton–proton distances can be obtained:

$$NOE_{i,j} \cdot d_{i,j}{}^6 = NOE_{H5,H6} \cdot d_{H5,H6}{}^6$$

where $NOE_{i,j}$ is the intensity of the cross-peak assigned to protons i and j, $d_{i,j}$ is the distance between them, and $NOE_{H5,H6}$ and $d_{H5,H6}$ are the corresponding quantities for pyrimidine H-5 and H-6 protons. Unfortunately, distance estimates obtained this way are quite crude.

First, it is usually impractical to collect RNA NOESY spectra under conditions that prevent the alteration of cross-peak intensities by transfers of magnetization to protons other than the two each cross-peak represents. When "third party" protons are involved, the conversion of cross-peak intensities into interatomic distances outlined above is invalid. Techniques exist to take these effects into account during the computation of NMR structures (for an application see White et al.[18]), but they are imperfect. The relative motions of the protons in a molecule have to be understood in detail if the rates at which magnetization transfer between them are to be estimated accurately. Because the detailed information required is invariably lacking, *faut de mieux*, it is assumed that the dynamics of all the protons in an RNA can be characterized by a single correlation time, which is a gross oversimplification.

Second, the distances between nonbonded protons in a molecule fluctuate all the time due to thermal motions. If the fluctuations are fast, a molecule will look as though it has a unique conformation spectroscopically and average NMR data will be measured. Unfortunately, because NOE intensities depend on distance raised to the negative sixth power, the average NOE intensity observed for a pair of protons whose separation fluctuates will always be greater than the intensity that would be observed if their separation was fixed at the average value. (Averaging is also a problem when torsion angles are estimated quantitatively using coupling constants (see Varani et al.[9]). The reason is fundamentally the same as for NOEs. The conformational parameter sought is not linearly related to the data used to estimate it.

Third, NOE intensities relate to distances in a simple way only if NOESY spectra are acquired under conditions that allow sample magnetization to equilibrate completely between each iteration of the experiment that is averaged to produce them. It takes about five times the T_1 for this to occur and for many RNAs, this implies the need for (at least) a 50 s (!) wait between iterations. Since multidimensional experiments commonly consume days of spectrometer time when 5 s cycle times are used, fully relaxed spectra are seldom accumulated. If all the protons in a molecule have the same T_1, the effect of hyperfast data collection is the same for all NOEs, but this is not the case for RNA. For all of these reasons, most RNA spectroscopists are content to classify their NOE cross-peaks as being "weak," "medium," and "strong" and to assign broad, overlapping distance ranges to them on that basis.

6.02.2.11   Structure Calculations

Once distances and torsion angles have been estimated, structure computations can begin. There are several algorithms for extracting conformations from NMR information. Debate about their relative merits remains lively, but there is no need for the nonspecialist to be concerned about the details. However, it is important for the nonspecialist to realize that even though the objective is to find the single structure that best accounts for the data, unique structures never emerge. What is produced instead are families of structures that, within error, are consistent with all, or almost all the information available. If the data are sufficiently constraining, the members of the family will be closely similar, and the spectroscopist responsible will claim that the structure is solved.

6.02.3 SOLUTION STRUCTURES AND CRYSTAL STRUCTURES COMPARED

Both crystallographers and spectroscopists file lists of atomic coordinates in data banks, and produce molecular images that look exactly alike. Thus biochemists can be forgiven for believing that the information in an NMR structure is equivalent to that in a crystal structure. It is not and it is important to understand why.

6.02.3.1 On the Properties of Crystallographic Structures

Crystallographic analyses produce molecular images equivalent to those that an X-ray microscope of large numerical aperture would produce, if such a thing existed. Almost no assumptions are made in generating these images, which are called electron density maps, and the ones that are published are seldom wrong. One reason is that atomic resolution electron density maps are easy to verify. A map of a nucleic acid, for example, had better contain blobs of density that look like nucleotides, and if the number of such blobs present is not the same as the number of nucleotides in the sequence crystallized, something is wrong.

Macromolecular electron density maps are interpreted by fitting into them representations of the biopolymer of interest that have appropriate bond lengths and bond angles. (Note that the bond lengths and angles used derive from small molecule crystallography!) The lower the resolution of an electron density map, i.e., the longer the wavelength of the shortest wavelength Fourier components included in its computation, the less detail it contains, and the harder it is to fit models into it unambiguously. Electron density contributed by bound ions and water molecules may lead to confusion, for example. Once the initial fitting process is finished, the conformation of the model is adjusted to optimize the correspondence between the diffraction pattern it implies and that observed. The refined product is the structure that gets published. When published structures are wrong, which they sometimes are, model building errors are invariably to blame.

Not surprisingly, the quality of the product depends on the resolution of the electron density map interpreted. A 4 Å map of an RNA is likely to be difficult to interpret unambiguously, but may be useful. A 3 Å RNA map should lead to a structural model that accurately depicts the overall shape of the molecule and reliably reports the general placement of its bases and the trajectory of the backbone. Some bound waters and metals may be evident. A map in the low 2 Å range will provide additional information about waters and metals ions and will accurately define all torsion angles. An RNA map that has a resolution in the low 1 Å range should be totally unambiguous and should specify atomic positions with an accuracy of a few tenths of an Å.

6.02.3.2 Solution Structures

Like the crystallographer, the spectroscopist interprets the data assuming the molecule of interest is a biopolymer of the appropriate type and sequence and that it has standard bond lengths and bond angles. Nevertheless, as the preceding discussion demonstrates, spectroscopic structures are not interpretations of molecular images.

Because spectroscopic structures are not based on images, they do not have resolutions. The experimental constraints that determine a solution structure are all local: estimates of individual torsion angles, and interatomic distances less than 5 Å. Thus NMR structures tend to depict small scale detail more accurately than they depict overall molecular shape, which is exactly the reverse of crystallography. Spectroscopic models do have precisions, which can be estimated by computing the pairwise root-mean-squared deviations (r.m.s.d.) between atomic positions within families of structures. Well-determined families of RNA structures have an average pairwise r.m.s.d around 1.0 Å and computational studies suggest that structures determined using large numbers of crudely specified constraints are likely to be more precise and more accurate than structures determined using small numbers of precisely specified constraints; quantity is more important than quality.[19]

Spectroscopic constraints are seldom distributed evenly throughout the volume of a molecule and hence some parts of a spectroscopically derived model will be more precisely determined than others. In regions where the data are highly constraining, the members of a structure family will be closely superimposable. Where the data are sparse, the scatter between independently computed structures will be large. The reader is warned that there is an alarming tendency of authors to describe the poorly determined regions of their solution structures as "flexible". Information about molecular dynamics can be obtained by NMR, but it is extracted from measurements of relaxation times, not from COSY and NOESY spectra. Regions of structures where r.m.s.d are large may be flexible, but then again, they may not. Crystal structures often suffer from a similar problem. Because of local, static, crystal disorder or dynamic disorder, the data obtained from a crystal may determine the conformation of one part of a structure less well than it determines others. Here too there is no simple way to determine the degree to which the local lack of structural definition is due to dynamics or not.

6.02.3.3 Constraints and Computations

By protein NMR standards, the number of constraints per unit molecular weight that can be extracted from RNA spectra is small because the number of protons per unit molecular weight of RNA is (relatively) small. Furthermore, they are not evenly distributed. Many of the easiest intranucleotide NOEs to observe are determined by χ, and a large fraction of the easily observed internucleotide NOEs are determined primarily by the distance between the H-2′ of nucleotide (n) and the H-6 or H-8 proton of nucleotide ($n+1$). For both reasons, RNA solution structures tend to be less accurate than protein solution structures. In fact, most NMR-derived RNA structures would be of poor quality indeed if the only information used in their computation was their covalent structures and the spectroscopic data.

The reason RNA solution structures having respectable precisions emerge is that lots of additional information is fed into the computation that produces them. The lengths of hydrogen bonds in standard base pairs are often specified exactly, for example, and most structure-producing programs minimize the conformational energies of the structures they produce. In fact, some of them can fold nucleic acid sequences into compact conformations in the total absence of experimental information. The contributions made by these programs would not be objectionable if it were certain that they are capable of evaluating conformational energies accurately, but they are not, and it would take an entire treatise to explain why. Thus in addition to helping these programs select the right conformation from the set of "low energy" alternatives, the experimental data also have to keep them honest.

The interpretation of NMR structures is further vexed by the fact that no two laboratories compute structures the same way, and each structure a laboratory publishes is likely to have been computed differently from its predecessors. Thus two laboratories can easily produce models for the same RNA that differ by far more than the precision claimed by each for its model. The only reliable way to decide whether differences between competing NMR models are real is to compare spectra. If the spectra are different, their conformations differ. Otherwise, differences in data treatment must be looked into.

Finally, the unfavorable ratio of experimental observations to coordinates characteristic of RNA spectroscopy makes NMR-derived RNA models hypersensitive to assignments. A single, mis-assigned NOE cross-peak can have a devastating impact on the conformation proposed for an RNA because important qualitative features of structures are often supported by single NOE cross-peaks. (For a modest example of this effect, compare Cheong *et al*[20] with Allain and Varani[21]).

Most of the shortcomings of RNA spectroscopy are characteristic of a physical technique still in its infancy. There is reason to hope that many of them will be ironed out in time and that standards of practice will develop that reduce the impact of the rest. Until that day arrives, however, *caveat emptor*.

6.02.3.4 Experimental Comparisons of Solution and Crystal Structures

Until recently, there were no RNAs whose structures had been determined by both NMR and X-ray crystallography, and hence no way to assess the accuracy of NMR structures. It was not for lack of trying. Several oligonucleotides that had been characterized in solution were crystallized so that comparisons could be made, but, frustratingly, their conformations changed radically during crystallization (e.g. Cheong *et al.*,[20] Holbrook and co-workers,[22,23] and Heus and Pardi[24]). Fortunately, there are now four systems where comparisons can be made: (i) the anticodon stem-loops of tRNAs,[25–29] (ii) fragment 1 from *Escherichia coli* 5S rRNA,[30,31] (iii) a cobalt hexamine-binding stem-loop from the group I intron,[32,33] and (iv) the sarcin/ricin loop from 28S rRNA.[11,34] The news they convey is that spectroscopists have been doing quite well.

The tRNA study cited was motivated by the absence of unambiguous information about anticodon loop conformation in the two initiator tRNA crystal structures published previously[35,36] and concern that initiator anticodons might differ conformationally from elongator anticodons. The structure of the anticodon loop of yeast initiator methionyl tRNA was compared spectroscopically with that of *E. coli* elongator methionyl tRNA, which has the same sequence. In solution, both anticodon loops have conformations resembling that seen crystallographically in the anticodon loop in yeast phenylalanyl tRNA, which is an elongator tRNA; the r.m.s.d. between the anticodon backbone atoms of yeast phenylalanyl tRNA and the yeast initiator tRNA NMR model was 1.2 Å. The bases on the 3′ side of the initiator loop do not stack as neatly as those in phenylalanyl tRNA, but the difference is real. All the anticodon riboses in the yeast phenylalanyl tRNA crystal structure are C-3′-*endo*, but several in the solution structure are C-2′-*endo*.

In 1996–1997, both crystal and solution structures were obtained for several molecules containing the helix IV, helix V, loop E region from *E. coli* 5S rRNA. The 18-nucleotide, loop E regions of both structures superimpose with an all-atom r.m.s.d. of about 1.0 Å, and the irregular non-Watson–Crick pairing in the middle of loop E seen in the crystal structure is faithfully represented in the NMR structure. When longer segments of the two models are compared, the superposition degrades because the relative orientations of distant segments of the 42-base RNA studied by NMR are not well determined. This is bound to be a problem in any elongated structure that is determined using a method that measures only short distances.

The third comparison is provided by a small stem-loop from the P4–P6 domain of the group I intron from *Tetrahymena*. Crystallographic studies have shown that this loop binds cobalt hexamine when it is part of the larger RNA and it binds cobalt hexamine in isolation also. The conformation of the loop in solution closely resembles that seen in the P4–P6 crystal structure and cobalt hexamine binds to both molecules in the same position.

The sarcin/ricin loop (SRL) from rat 28S rRNA provides the last comparison. It is the only example so far of an RNA where the oligonucleotide crystallized is identical to the one characterized spectroscopically. The molecule is organized in the same way in both structures. The same base pairs are seen in both, but the relationship between its loop and its stem is not well determined spectroscopically. Even though the r.m.s.d. difference between the loops of the two models is only about 1.5 Å, the solution structure of SRL was not close enough to the crystal structure so that the structure of the crystal could be solved by molecular replacement using the solution structure as the starting model.[37]

These comparisons demonstrate that solution structures are capable of describing the topology of an RNA correctly, i.e. specifying its base pairs and the approximate trajectory of its backbone. At least locally, solution structures are likely to superimpose on the corresponding X-ray structure with r.m.s.d. less than 2 Å. For many purposes, this level of accuracy is good enough for biochemists and molecular biologists, and it is not clear that the differences between solution structures and crystal structures should all be attributed to error in solution structures.

6.02.4 LESSONS LEARNED ABOUT MOTIFS BY NMR

For reasons already elucidated, the RNA spectroscopist cannot determine the conformations of entire naturally occurring RNAs. Consequently, RNA spectroscopists have concentrated on three classes of RNAs: (i) small, synthetic oligonucleotides that contain interesting base-pairing irregularities, (ii) RNA aptamers, and (iii) domains excised from large, natural RNAs.

The work done on synthetic oligonucleotide has been motivated by the belief that RNA structures are modular, which is to say that the conformations of motifs in small oligonucleotides of otherwise arbitrary sequence are identical to the conformations of the same motifs in all other RNAs. Aptamers are RNA sequences selected from random populations *in vitro* on the basis of their capacity to bind specific ligands or to perform other selectable functions (see Gold *et al.*[38]). In order for sequence space to be sampled thoroughly, the lengths of oligonucleotides in the RNA populations from which aptamers are selected must be quite small, and consequently, most aptamers are small enough for spectroscopists to study intact (see Cech and Szewczak,[39] Marshall *et al.*[40]). The objective of most aptamer spectroscopists is to understand how aptamers bind their ligands. Those who concentrate on domains do not need to invoke modularity to justify their activities. By definition, a domain is a portion of a macromolecule that is conformationally autonomous; the conformation determined for a domain in isolation has to be the same as that in the larger RNA from which it derives. The only problem students of domain structure confront, therefore, is proving their oligonucleotides are domains in the first place.

6.02.4.1   RNA Organization in General

Qualitatively, the way single stranded RNAs organize themselves was understood in 1960.[41,42] They fold so that the short sequences they contain that are "accidentally" complementary form short double helices to (approximately) the maximum extent possible. The dominant structural element that results is the *hairpin loop*, or *stem-loop*, which is produced when an RNA chain folds back on itself so that complementary sequences close to each other in its sequence can pair. Thus most RNAs have secondary structures that consist of a series of stem-loops separated by sequences of less certain conformation that are usually represented as being single stranded.

Inevitably, in RNA stems where strands of "random" sequence are aligned to maximize Watson–Crick pairing, bases are juxtaposed that cannot form canonical pairs and because stems are stabilized if hydrogen bonds form and bases stack, they pair anyway. GU pairs within otherwise regular helical stems are a case in point. They are so common that wobble GUs, which fit easily into helices, are considered "honorary" Watson–Crick pairs. In addition to occasional noncanonical base pairs, helical stems are often interrupted by *bulged bases*, which is to say bases on one strand that have no partner to pair with on the other, and by *internal loops*, in which longer sequences on both strands that are juxtaposed cannot obviously be paired. Some internal loops have sequences long enough to include stem-loops of their own; they are called *junctions*. Whether the stem of a stem-loop contains irregularities or not, it must have a *terminal loop*, i.e., a sequence that links the 5′ to the 3′ strand of its stem, and their conformations cannot be predicted *a priori* either. The terminal loops of some stem-loops are big enough to contain stem-loops of their own.

The evidence available suggests that most stem-loops are domains, and since many of them contain less than 45 nucleotides, and those that do not can often be "trimmed", stem-loops derived from natural RNAs are favorite targets for spectroscopic investigation. By characterizing them, the conformations of important elements of RNA secondary structure are being investigated.

Large RNAs have tertiary structures, of course; some of them are as compactly folded as globular proteins. The interactions that stabilize RNA tertiary structure involve both stem-loops and the "unstructured" sequences that link them together, but they are unusual in RNAs of the sizes that RNA spectroscopists can study. For that reason, NMR has provided little insight into this aspect of RNA conformation.

6.02.4.2   Terminal Loops

A great deal has been learned about terminal loop structure by NMR, particularly about the conformations of terminal loops that have short sequences. Short terminal loop sequences play the same role in RNA as β-turns in proteins. They are concise structures that stabilize 180° changes in backbone direction.

6.02.4.2.1   *U-turns*

The U-turn is a four-base terminal loop motif, the consensus sequence of which is UNRN (note "N" stands for any nucleotides and "R" means any purine.) They were first characterized in the mid-1970s by crystallographers working on transfer RNAs,[26–29] and their existence in tRNAs in solution has been confirmed.[25] Spectroscopic studies have demonstrated that they occur in other contexts. The L11 binding region of 23S rRNA includes a U-turn[43,44] as does loop IIa in yeast U2snRNA.[45]

Figure 3 shows a typical U-turn. Like all other U-turns, it is stabilized by a hydrogen bond between the imino proton of U-1 and an oxygen belonging to the phosphate group of R-3, and the 2'OH of U-1 and N-7 of R-3.[46] All of the U-turns characterized so far are components of larger terminal loops.

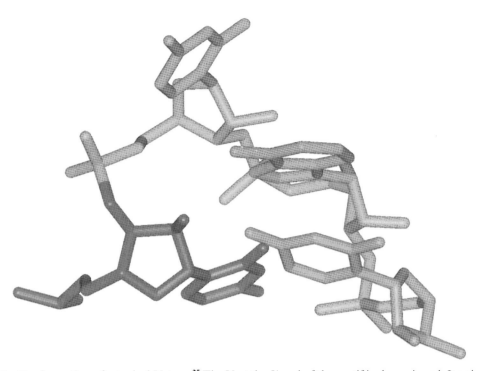

Figure 3   Conformation of a typical U-turn.[25] The U at the 5'-end of the motif is shown in red. It points away from the viewer. The three bases that follow (blue) form a stack the bases of which point out towards the viewer.

6.02.4.2.2   Tetraloops

In the late 1980s, it was noticed that helical stems terminated by 4-nucleotide loops, or *tetraloops*, having the sequence UNCG are abundant in rRNAs, and it was demonstrated that they are unusually stable.[47] Further analysis revealed the existence of two other "special" tetraloops sequences: GNRA and CUNG.[48] Spectroscopic studies done subsequently have demonstrated that these three kinds of tetraloops have distinctive conformations, as expected, and those who work with short RNA oligonucleotides now routinely include them in sequences intended to form stem-loops.

The conformation of the UNCG motif was analyzed initially in Tinoco's laboratory in 1990,[20] and five years later, their structural proposal was revised using a larger set of NMR-derived restraints.[21] The most striking feature of the UNCG turn is the unusual *syn–anti* pair that forms between U-1 and G-4, which has a phosphate–phosphate distance so small it can be spanned by the middle two residues, N-2 and C-3.

GNRA tetraloops have also received a great deal of attention, and, as expected, they all have similar conformations.[24,49] As is the case with UNCG tetraloops, the "secret" of these structures is the slipped, or side-by-side pair that forms between G-1 and A-4, which greatly reduces the distance between the backbones of the two strands of the loop being capped. Interestingly, the trajectory of the backbone in GNRA tetraloops is so similar to that in U-turns that some now refer to GNRA tetraloops as U-turns. It would be wiser to apply that phrase only to turns whose sequence is UNRN.

A GNRA tetraloop has been observed in an entirely unexpected context; that provided by an aptamer which binds AMP.[50,51] In the presence of AMP, an otherwise unstructured internal loop in this RNA folds so that the AMP can interact with the RNA as though it were A-4 in a GNRA tetraloop. The similarity between the conformation of the resulting loop and that of a normal GNRA tetraloop is striking.

The structure of the last member of the set, CUNG, is quite different from that of the other standard tetraloops.[52] C-1 and G-4 form a Watson–Crick base pair, and U-2 reaches down into the minor groove of the helical stem being capped and interacts with its last base pair. This interaction appears to require that the last base pair be a GC, an inference strongly supported by phylogenetic data. Thus conformationally, CUNG tetraloops are really UN diloops, but they have a consensus sequences that is six bases long: G(CUNG)C.

6.02.4.2.3   *Other terminal loops*

Many terminal loops are not, or do not appear to be, motifs. It would be a mistake to assume they lack structure, however. Conformations have just been obtained for two such loops: the conserved UGAA loop found at the 3′ end of all 18S rRNAs[53] and the UGGGGCG loop that is a universal component of the peptidyl transferase region of 23S-like rRNAs.[54] Their conformations will not be discussed here because they are not motifs, but the reader should examine them anyway. Both are highly structured and inspection of them should induce a sense of humility. No one could possibly have predicted their conformations in advance.

6.02.4.3   Internal Loops

The dominant motif in RNA stem-loops is the A-form helix, the conformation of which was well understood long before NMR spectroscopy was mature enough to contribute in any way. It is a two-stranded, antiparallel double helix of indefinite length having geometry so well known it does not need to be described here.[15] There is no restriction on the nucleotide sequence in either of the two strands of an A-form helix, provided the sequence of the other strand is its Watson–Crick complement. If GU wobble pairs are accepted as equivalent to Watson–Crick GCs and AUs, roughly two-thirds of the bases in an RNA like a ribosomal RNAs are involved in A-form helix.

As pointed out earlier, the helical continuity of many stem-loops is interrupted by internal loops, only a small number of which have been characterized spectroscopically (or crystallographically, for that matter). They come in two varieties: symmetric and asymmetric. In a symmetric internal loop, the number of loop nucleotides is the same in the two strands and in an asymmetric loop, it is not. Only a small number on internal loop motifs have been identified so far; there are bound to be more.

6.02.4.3.1   *Symmetric internal loop motifs*

NMR and crystal structures provide numerous examples of an internal loop motifs called "cross-strand purine stacks". In A-form helix, the bases in each strand form a continuous stack that runs the length of the helix. In cross-strand purine stacks, a purine in one strand stacks on a purine from an adjacent base pair that belongs to the other strand. This alters the relative sizes of the major and minor grooves.

The first cross-strand purine stacks observed spectroscopically were the cross-strand A stacks found in loop E from eukaryotic 5S rRNA[55] and the sarcin ricin loop from rat 28S rRNA.[56] The consensus sequence for this kind of stack is 5′ (G or C)GA paired with 5′UA(G or C), and the pairing is a Watson–Crick GC, in either orientation, followed by a slipped GA and a reverse Hoogsteen AU. The six-membered ring of the A in the GA stacks on the six membered ring of the A in the AU (Figure 4). Two more examples have been found in loop E from prokaryotic 5S rRNA.[31]

Loop E also contains a cross-strand G stack that is composed of two wobble GU pairs sandwiched between two Watson–Crick GCs. In this motif, 5′UG is paired with 5′UG and the six-membered rings of its Gs are stacked (Figure 5). Note that since GUs embedded in helices are thought of as equivalent to GCs and AUs, it may be somewhat surprising that this motif has a distinctive conformation. It is clear from crystallographic studies that the sequences other than those mentioned here also cause cross-strand purine stacks (e.g., Cate *et al.*[33]).

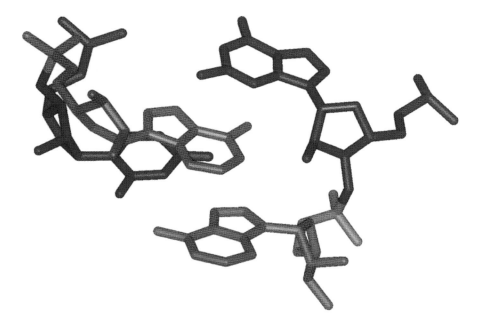

Figure 4   Cross-strand A stack.[31] The reverse-Hoogsteen AU belonging to this stack lies below its side-by-side AG in this diagram. The two As are red and the G and U with which they pair are blue. The stacking of the six-membered rings of the As is obvious.

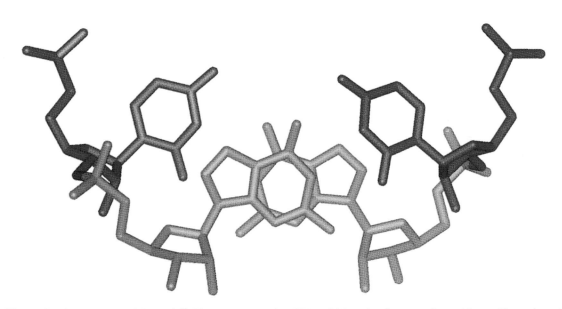

Figure 5   A cross-strand G stack.[31] The two successive GU wobble pairs that constitute this motif are viewed down the axis of the double helix to which they belong. The six-membered rings of the two Gs (red) stack almost perfectly. There is an approximate two-fold axis in this motif running between the planes of the two Gs, perpendicular to the helix axis.

As it happens, loop E from *E. coli* 5S rRNA is one of the only symmetric internal loops whose conformation is known.[31] Thus even though the conformation adopted of the six bases in the middle of this loop are not a motif, its conformation is worthy of contemplation.

6.02.4.3.2   *Asymmetric internal loop motifs*

Both prokaryotic loop E and the sarcin ricin loop include a three-base structure called a "bulged G motif".[11,55,56] The sequence is 5'(G or C)GAA paired with 5'AGUG(G or C). G-2 of the second strand reaches across the minor groove of the motif so that its imino proton can hydrogen bond to the phosphate group that links G-2 and A-3 in the first strand. The remaining bases (5′(G or C)GA and 5′...UG(G or C)) form a cross-strand A stack, and A-4 from the first stand forms a symmetric, parallel, *anti–anti* pair with A-1 of the second strand. It is not clear what nucleotides can follow the AA pair, but so far, only antiparallel, *anti-*, *anti-*, all pyrimidine pairs have been found at that position. Because the AA pair is symmetric, the backbone of this motif has a distinctive, S-shaped trajectory on its bulged G side (Figure 6).

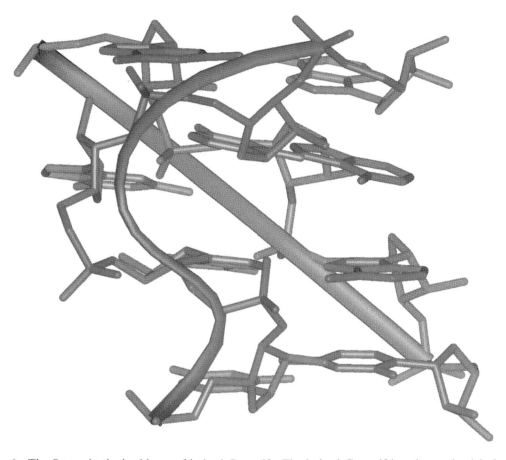

Figure 6   The S-turn in the backbone of bulged-G motifs. The bulged-G motif ion the sarcin ricin loop is shown.[55] The 5′ strand of the motif, which contains the bulged G, is shown in red, and the 3′ strand is blue. The backbone trajectories of both strands are indicated by continuous oval lines.

This motif is just one example of how "extra" bases in asymmetric internal loops get "taken care of." A rich variety of alternatives is on display in the many structures of aptamers and ligand-binding natural RNAs that have been published, none of which are motifs.[57–68] Examination of these structures leaves a single strong impression. A remarkable fraction of these loops is grossly similar to normal double helix, even though, on a local level, many of them are remarkably nonhelical. Most of them are distorted double helices that interrupt the regular helices they separate without breaking their continuities.

6.02.4.4   Pseudoknots

Many RNAs contain pseudoknots, which are structures in which the loop of a stem-loop forms a double helix by pairing with other nucleotides from some other part of the same molecule. When the sequence that base-pairs with the loop starts immediately after the stem of the stem-loop, the object that results is two stem-loops joined side-by-side, like Siamese twins, because the loop bases of both stem-loops are one strand of the stem of their partners (for details see Wyatt and Tinoco[69]).

The structures that result are motifs topologically, even though their sequences vary a lot. In the late 1980s, a series of synthetic pseudoknots were studied by NMR,[70] and a natural pseudoknot has since been characterized.[71]

6.02.5 REFERENCES

1. C. P. Slichter, "Principles of Magnetic Resonance," 3rd edn., Springer-Verlag, New York, 1989.
2. M. Goldman, "Quantum Description of High-Resolution NMR in Liquids," Oxford University Press, Oxford, 1988.
3. R. R. Ernst, G. Bodenhausen, and A. Wokaun, "Principles of Nuclear Magnetic Resonance in One and Two Dimensions," Oxford University Press, Oxford, 1987.
4. K. Wuthrich, "NMR of Proteins and Nucleic Acids," Wiley, New York, 1986.
5. J. Cavanagh, W. J. Fairbrother, A. G. Palmer, III, and N. J. Skelton, "Protein NMR Spectroscopy. Principles and Practice," Academic Press, San Diego, CA, 1996.
6. E. P. Nikonowicz, A. Sirr, P. Legault, F. M. Jucker, L. M. Baer, and A. Pardi, *Nucleic Acids Res.*, 1992, **20**, 4507.
7. R. T. Batey, M. Inada, E. Kujawinski, J. D. Puglisi, and J. R. Williamson, *Nucleic Acids Res.*, 1992, **20**, 4515.
8. R. T. Batey, J. L. Battiste, and J. R. Williamson, *Methods Enzymol.*, 1995, **261**, 300.
9. G. Varani, F. Aboul-ela, and F. H.-T. Allain, *Progr. Nucl. Magn. Reson. Spectr.*, 1996, **29**, 51.
10. G. Varani and I. Tinoco, Jr., *Quart. Rev. Biophys.*, 1991, **24**, 479.
11. A. A. Szewczak and P. B. Moore, *J. Mol. Biol.*, 1995, **247**, 81.
12. A. Szewczak, Ph.D. Thesis, Yale University, 1994.
13. E. P. Nikonowicz and A. Pardi, *J. Mol. Biol.*, 1993, **232**, 1141.
14. P. B. Moore, *Acc. Chem. Res.*, 1995, **28**, 251.
15. W. Saenger, "Principles of Nucleic Acid Structure," Springer-Verlag, New York, 1984.
16. C. Altona, *Recueil. J. Roy. Netherlands Chem. Soc.*, 1982, **101**, 413.
17. D. G. Gorenstein, "Phosphorus -31 NMR. Principles and Applications," Academic Press, Orlando, FL, 1984.
18. S. A. White, M. Nilges, A. Huang, A. T. Brunger, and P. B. Moore, *Biochemistry*, 1992, **31**, 1610.
19. F. H.-T. Allain and G. Varani, *J. Mol. Biol.*, 1997, **267**, 338.
20. C. J. Cheong, G. Varani, and I. Tinoco, *Nature*, 1990, **346**, 680.
21. F. H.-T. Allain and G. Varani, *J. Mol. Biol.*, 1995, **250**, 333.
22. S. R. Holbrook, C. Cheong, I. Tinoco, and S.-H. Kim, *Nature*, 1991, **353**, 579.
23. K. J. Baeyens, H. L. De Bondt, A. Pardi, and S. R. Holbrook, *Proc. Natl. Acad. Sci. USA*, 1996, **93**, 12851.
24. H. A. Heus and A. Pardi, *Science*, 1991, **253**, 191.
25. D. C. Schweisguth and P. B. Moore, *J. Mol. Biol.*, 1997, **267**, 505.
26. B. Hingerty, R. S. Brown, and A. Jack, *J. Mol. Biol.*, 1979, **124**, 523.
27. S. R. Holbrook, J. L. Sussman, R. W. Warrant, and S.-H. Kim, *J. Mol. Biol.*, 1978, **123**, 631.
28. E. Westhof and M. Sundaralingam, *Biochemistry*, 1986, **25**, 4868.
29. E. Westhof, P. Dumas, and D. Moras, *Acta Crystallogr A*, 1988, **44**, 112.
30. C. C. Correll, B. Freeborn, P. B. Moore, and T. A. Steitz, *Cell*, 1997, **91**, 705.
31. A. Dallas and P. B. Moore, *Structure*, 1997, **5**, 1639.
32. J. S. Kieft and I. Tinoco, Jr., *Structure*, 1997, **5**, 713.
33. J. H. Cate, A. R. Gooding, E. Podell, K. H. Zhou, B. L. Golden, C. E. Kundrot, T. R. Cech, and J. A. Doudna, *Science*, 1996, **273**, 1678.
34. C. C. Correll, personal communication.
35. N. H. Woo, B. A. Roe, and A. Rich, *Nature*, 1980, **286**, 346.
36. R. Basavappa and P. B. Sigler, *EMBO J.*, 1991, **10**, 3105.
37. C. C. Correll, personal communication.
38. L. Gold, B. Polisky, O. Uhlenbeck, and M. Yarus, *Ann. Rev. Biochem.*, 1995, **64**, 763.
39. T. R. Cech and A. A. Szewczak, *RNA*, 1996, **2**, 625.
40. K. A. Marshall, M. P. Robertson, and A. D. Ellington, *Structure*, 1997, **5**, 729.
41. J. R. Fresco and B. M. Alberts, *Proc. Natl. Acad. Sci. USA*, 1960, **46**, 311.
42. J. R. Fresco, B. M. Alberts, and P. Doty, *Nature*, 1960, **188**, 98.
43. M. A. Fountain, M. J. Serra, T. R. Krugh, and D. H. Turner, *Biochemistry*, 1996, **35**, 6539.
44. S. G. Huang, Y. X. Wang, and D. E. Draper, *J. Mol. Biol.*, 1996, **258**, 308.
45. S. C. Stallings and P. B. Moore, *Structure*, 1997, **5**, 1173.
46. G. J. Quigley and A. Rich, *Science*, 1976, **194**, 796.
47. C. Tuerk, P. Gauss, C. Thermes, D. R. Groebe, M. Gayle, N. Guild, G. Stormo, Y. d'Aubenton-Carafa, O. C. Uhlenbeck, I. Tinoco, E. N. Brody, and L. Gold, *Proc. Natl. Acad. Sci. USA*, 1988, **85**, 1364.
48. C. R. Woese, S. Winker, and R. R. Gutell, *Proc. Natl. Acad. Sci. USA*, 1990, **87**, 8467.
49. F. M. Jucker, H. A. Heus, P. F. Yip, E. H. M. Moors, and A. Pardi, *J. Mol. Biol.*, 1996, **264**, 968.
50. F. Jiang, R. A. Kumar, R. A. Jones, and D. J. Patel, *Nature*, 1996, **382**, 183.
51. T. Dieckmann, E. Suzuki, G. K. Nakamura, and J. Feigon, *RNA*, 1996, **2**, 628.
52. F. M. Jucker and A. Pardi, *Biochemistry*, 1995, **34**, 14416.
53. S. E. Butcher, T. Dieckmann, and J. Feigon, *J. Mol. Biol.*, 1997, **268**, 348.
54. E. V. Puglisi, R. Green, H. F. Noller, and J. D. Puglisi, *Nat. Struct. Biol.*, 1997, **4**, 775.
55. B. Wimberly, G. Varani, and I. Tinoco, Jr., *Biochemistry*, 1993, **32**, 1078.
56. A. A. Szewczak, P. B. Moore, Y.-L. Chan, and I. G. Wool, *Proc. Natl. Acad. Sci. USA*, 1993, **90**, 9581.
57. J. D. Puglisi, R. Y. Tan, B. J. Calnan, A. D. Frankel, and J. R. Williamson, *Science*, 1992, **257**, 76.
58. F. Aboul-ela, J. Karn, and G. Varani, *J. Mol. Biol.*, 1995, **253**, 313.
59. J. D. Puglisi, L. Chen, S. Blanchard, and A. D. Frankel, *Science*, 1995, **270**, 1200.
60. X. M. Ye, R. A. Kumar, and D. J. Patel, *Chem. Biol.*, 1995, **2**, 827.

61. J. L. Battiste, R. Tan, A. Fraenkel, and J. R. Williamson, *Biochemistry*, 1994, **33**, 2741.
62. J. L. Battiste, H. Y. Mao, N. S. Rao, R. Y. Tan, D. R. Muhandiram, L. E., Kay, A. D. Frankel, and J. R. Williamson, *Science*, 1996, **273**, 1547.
63. K. Kalurachchi, K. Uma, R. A. Zimmermann, and E. P. Nikonowicz, *Proc. Nat. Acad. Sci. USA*, 1997, **94**, 2139.
64. D. Fourmy, M. I. Recht, S. C. Blanchard, and J. D. Puglisi, *Science*, 1996, **274**, 1367.
65. Y. S. Yang, M. Kochoyan, P. Burgstaller, E. Westhof, and M. Famulok, *Science*, 1996, **272**, 1343.
66. P. Fan, A. K. Suri, R. Fiala, D. Live, and D. J. Patel, *J. Mol. Biol.*, 1996, **258**, 480.
67. G. R. Zimmerman, R. D. Jenison, C. L. Wick, J.-P. Simorre, and A. Pardi, *Nature, Structural Biology*, 1997, **4**, 644.
68. L. C. Jiang, A. K. Suri, R. Fiala, and D. J. Patel, *Chem. Biol.*, 1997, **4**, 35.
69. J. R. Wyatt and I. Tinoco, Jr., in "The RNA World," eds. R. F. Gesteland and J. F. Atkins, Cold Spring Harbor Laboratory, Cold Spring Harbor, NY, 1993, p. 465.
70. J. D. Puglisi, J. R. Wyatt, and I. Tinoco, Jr., *Nature*, 1988, **331**, 283.
71. Z. H. Du, D. P. Giedroc, and D. W. Hoffman, *Biochemistry*, 1996, **35**, 4187.

6.03
Thermodynamics of RNA Secondary Structure Formation

TIANBING XIA, DAVID H. MATHEWS, and
DOUGLAS H. TURNER
University of Rochester, NY, USA

6.03.1   INTRODUCTION

RNA is an active component in many cellular processes.[1] For example, RNA alone can act as an enzyme to catalyze RNA transformations.[2–4] It is also possible that the RNA in ribosomes[5,6] and signal recognition particles[7] is actively involved in protein synthesis and protein translocation across membranes, respectively. Retroviruses, including HIV, are RNA–protein complexes.

Nucleic acids are now being sequenced at a rate of more than one million nucleotides per day,[8,9] and the entire three billion bases in the human genome are expected to be sequenced by 2005.[10] This will provide sequences for many important RNA molecules. While such sequence information facilitates investigations of RNA, in-depth understanding of structure–function relationships requires knowledge of three-dimensional structure, energetics, and dynamics.

Due to their complexity and dynamic behavior, it is difficult and time-consuming to determine three-dimensional structures for natural RNA molecules. Thus from 1973 to 1996 the only three-dimensional structures determined by X-ray crystallography (see Chapter 6.04) for natural RNAs longer than 30 nucleotides were tRNAs,[11–13] hammerhead ribozymes,[14–17] and one domain of a group I intron.[18] Structures of some natural fragments of RNA have also been determined by NMR.[19–32] These methods cannot keep pace with the rate of discovery and sequencing of interesting new RNA molecules. Thus there is a need for other reliable methods of determining RNA structure. If the energetics of RNA were completely understood, it would be possible to predict their folding, reactivity, and functional properties directly from their sequences.

The first stage in predicting RNA structure is determination of secondary structure, essentially a listing of base pairs contained in its folded structure. Determination of secondary structure also defines the various loops present in a given RNA. Figure 1 shows a secondary structure illustrating most of the loop motifs. Often, these non-Watson–Crick regions of an RNA are particularly important for function since unusual arrays of functional groups are available there for tertiary interactions[33] or recognition of other cellular components.[34] Thus determination of secondary structure helps identify nucleotides that may be important for function.

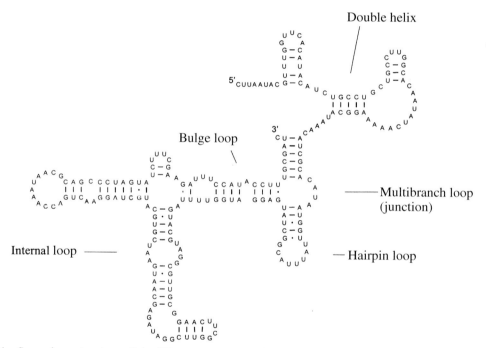

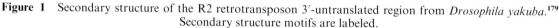

Figure 1 Secondary structure of the R2 retrotransposon 3′-untranslated region from *Drosophila yakuba*.[179] Secondary structure motifs are labeled.

Phylogenetic sequence comparison is one way to determine RNA secondary structure, provided large numbers of homologous sequences from different organisms are available.[35,36] When not enough related sequences are known, however, alternative methods must be used, and the most popular is free energy minimization. It is based on two assumptions: (i) the dominant interactions responsible for RNA structures are local,[37–41] presumably hydrogen bonding between bases and stacking between adjacent base pairs;[42–44] (ii) the conformations RNA adopts are equilibrium, lowest free energy conformations.[42,45]

At least two factors limit the success of secondary structure prediction by free energy minimization. First, algorithms do not exist that include all possible folding motifs and deal efficiently with the enormous numbers of possible secondary structures for a long sequence.[42,46–48] Second, our knowledge of the contributions of various RNA motifs to the total free energy of RNA structures is still incomplete. Rapid methods for the synthesis of oligonucleotides[49,50] (see Chapter 6.06) make it possible to study the sequence dependence of RNA secondary structure thermodynamics in a systematic way. Accumulation of this knowledge has steadily improved predictions[51,52] and incorrect predictions often occur at motifs for which little experimental data are available. Thus, understanding of the thermodynamics of RNA secondary structure is crucial for successful structure prediction. This chapter reviews the methods available for measuring the thermodynamics of RNA motifs, the known sequence dependence of these thermodynamics, and applications to predicting RNA secondary structure, modeling tertiary structure, and designing therapeutics.

6.03.2 THERMODYNAMIC ANALYSIS OF RNA STRUCTURAL TRANSITIONS

6.03.2.1 Hypochromism: Basis of Transition Analysis

Many techniques for investigating order–disorder structural transitions follow changes that occur in a spectroscopic property when the transition is induced thermally. A convenient property to follow for nucleic acids is UV absorption, which results from complex $n \rightarrow \pi^*$ and $\pi \rightarrow \pi^*$ transitions of the bases.[53–55] A decrease in UV absorption is observed in nucleic acids upon duplex formation. This decrease is called "hypochromism".[55] For short oligonucleotides, 30–40% hypochromicity at 260 or 280 nm is typical.

Hypochromism is largely due to interactions between electrons in different bases.[56–61] In particular, the transition dipole moment of the absorbing base interacts with the light-induced dipoles of neighboring bases. For a polymeric array of chromophore residues, such as bases in a nucleic acid, this interaction depends on the relative orientation and separation of bases. If the bases are stacked parallel so that the transition dipole moments of adjacent bases are oriented more or less head-to-head (helical form), the probability of photon absorbance by a base is reduced due to light-induced dipoles in neighboring bases. Because the shape of the UV absorption of nucleotide bases is not significantly affected by these interactions,[56] the order–disorder transition of RNA can be followed by monitoring the UV absorption at a single wavelength, typically at 260 nm for AU-rich or 280 nm for GC-rich sequences.[62–64]

6.03.2.2 Equilibrium Transition: Two-state Model

A UV absorption vs. temperature profile is called a "UV melting curve" by analogy with true phase changes. A typical experimental curve for a duplex to random coil transition for a short oligonucleotide is shown in Figure 2. Often, short duplexes melt in a two-state, all-or-none, manner, i.e., an RNA strand is either in a completely double helical or in a completely random conformation state; no partially ordered states are significantly populated. This is because the initiation step in helix formation is unfavorable compared to helix growth steps.[54]

General formulas have been presented for analyzing melting curves.[65] The majority of equilibria of interest to molecular biologists are bimolecular or unimolecular in nature:

$$A \Leftrightarrow B \quad \text{(unimolecular)} \tag{1}$$

$$2A \Leftrightarrow A_2 \quad \text{(bimolecular, self-complementary)} \tag{2}$$

$$C + D \Leftrightarrow E \quad \text{(bimolecular, non-self-complementary)} \tag{3}$$

The equilibrium constant for duplex formation is

$$K = \frac{\alpha/2}{(C_T/a)(1-\alpha)^2} \quad \text{(bimolecular)} \tag{4}$$

where C_T is the total single strand concentration, a is 1 for self-complementary duplexes and 4 for non-self-complementary duplexes; α is the mole fraction of single strand in duplex form. The equilibrium constant is related to the free energy change at temperature T, $\Delta G°(T)$, of the transition by

$$K = e^{-\Delta G°(T)/RT} \tag{5}$$

Here, T is temperature in kelvins, $T = 273.15 + t$, where t is temperature in °C. The free energy change is related to the enthalpy and entropy changes, $\Delta H°$ and $\Delta S°$, by:

UV melting curve

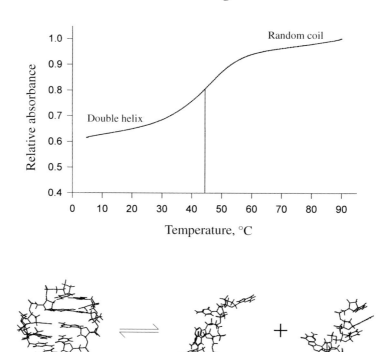

Figure 2 Typical melting curve for a double helix to random coil transition. The rate of heating must be much slower than the rate of conformational relaxation of the RNA, i.e., equilibrium is established at each temperature during measurement of the melting curve. The vertical line indicates the melting temperature, T_M, where half the strands are in double helix and half in random coil conformations.

$$\Delta G^{\circ}(T) = \Delta H^{\circ} - T\Delta S^{\circ} \tag{6}$$

The melting temperature, T_M, is defined as the temperature at which $\alpha = 1/2$. At the T_M, equilibrium constants are given by:

$$K_{T_M} = \frac{a}{C_T} \quad \text{(bimolecular)} \tag{7}$$

Equations (5)–(7) predict a concentration dependence of T_M for duplexes:

$$\frac{1}{T_M} = \frac{R}{\Delta H^{\circ}} \ln{(C_T/a)} + \frac{\Delta S^{\circ}}{\Delta H^{\circ}} \quad \text{(bimolecular)} \tag{8}$$

For unimolecular transitions, the corresponding equations are

$$K = \frac{\alpha}{1-\alpha}, \; K_{T_M} = 1, \text{ and } T_M = \frac{\Delta H^{\circ}}{\Delta S^{\circ}} \quad \text{(unimolecular)} \tag{9}$$

Note that the T_M is concentration independent for unimolecular transitions.

6.03.2.3 Data Analysis

Experimental techniques for measuring melting curves have been described in detail.[64,66] Given an absorbance vs. temperature melting curve of a duplex to single strand transition like the one in Figure 2, thermodynamic information can be derived by analyzing α as a function of temperature. Under the two-state transition assumption, the measured extinction coefficient, $\varepsilon(T)$, at any temperature can be expressed as a mole-fraction weighted linear combination of two components, ε_{ss} and ε_{ds}:[66]

$$\varepsilon(T) = \varepsilon_{ss}(1-\alpha) + \varepsilon_{ds}\alpha \qquad (10)$$

where ε_{ss} is the average extinction coefficient for the single-stranded states and ε_{ds} is the extinction coefficient per strand for the double-stranded state. Since lower and upper baselines for a typical melting curve are relatively straight, ε_{ss} and ε_{ds} are usually approximated as linear functions of temperature,[66]

$$\varepsilon_{ss} = m_{ss}T + b_{ss}, \quad \text{and} \quad \varepsilon_{ds} = m_{ds}T + b_{ds} \qquad (11)$$

When melting temperatures are low, base stacking in the single strands can sometimes produce nonlinear upper baselines. For non-self-complementary duplexes, it is sometimes possible to separately measure the temperature dependence of ε_{ss} instead of using the linear approximation.[67,68] The fraction of strands in duplex, α, can be expressed as follows:

$$\alpha = \frac{\varepsilon_{ss}(T) - \varepsilon(T)}{\varepsilon_{ss}(T) - \varepsilon_{ds}(T)} \qquad (12)$$

The parameter α as a function of temperature is related to ΔH° and ΔS° through the equilibrium constant K by Equation (4).

At high temperature, each strand can exist only in the random coil single-stranded state. Thus, the total single-strand concentration can be estimated from the absorbance at high temperature (normally $>80\,^\circ C$), and extinction coefficients of single strands calculated from published dimer and monomer extinction coefficients[69] using the nearest-neighbor approximation.[70] Thermodynamic parameters can be derived by using a nonlinear least-squares routine[71] to fit experimental curves to the two-state model (Equation (10)) with m_{ss}, m_{ds}, b_{ss}, b_{ds}, ΔH°, and ΔS° being the adjustable parameters.[29,66]

Thermodynamic parameters of duplex formation can be averaged over melting curves measured at different concentrations or obtained from plots of the reciprocal of the melting temperatures, T_M^{-1} vs. $\ln(C_T/a)$, (Equation (8)).[72] The data are normally taken as being consistent with a two-state transition if the ΔH° values calculated by the two analysis methods agree within 15%.[42,66,73,74] A 100-fold range in strand concentration is normally explored. Typical discrepancies in ΔH°, ΔS°, and ΔG°_{37} obtained from the two analysis methods are 5.8%, 6.5%, and 1.8%, respectively.[75] Note that the 15% criterion is a necessary but not sufficient condition for proving two-state behavior. The derived parameters are indirect and model dependent.

Methods for estimating errors in thermodynamic parameters have been described in detail.[75,76] Instrumental fluctuations contribute negligibly to the uncertainties. Standard deviations of parameters for single measurements are typically about 6.5%, 7.3%, and 2.4% for ΔH°, ΔS°, and ΔG°_{37}, respectively. Standard deviations for parameters calculated from Equation (8) based on 7–10 measurements are typically 2.9%, 3.3%, and 1.0% for ΔH°, ΔS°, and ΔG°_{37}, respectively. The relative uncertainty in ΔS° is usually about 13% larger than that in ΔH°, because ΔS° depends on more experimental parameters than ΔH°. Uncertainty in T_M is normally about 1.6 °C.[75] The errors in ΔG° and T_M are less than those in ΔH° and ΔS° because errors in ΔH° and ΔS° are highly correlated, with average observed correlation coefficients being greater than 0.999.[75] Thus, ΔG° and T_M are more accurate parameters than either ΔH° or ΔS° individually.[66,75–78]

6.03.2.4 Complications and Caveats

The above treatment assumes that ΔH° and ΔS° are temperature independent, which need not be the case. ΔH° and ΔS° will be temperature dependent if ΔC_P°, the difference in heat capacity (where $C_P^\circ = (\partial H^\circ/\partial T)_P$) between single- and double-stranded states, is not zero. A simulation of how a temperature-independent, nonzero ΔC_P° affects van't Hoff analyses[79] showed that a small ΔC_P° can make a hidden contribution to data analysis that biases the slope of van't Hoff plots. Since the curvature that should result from the small ΔC_P° is likely to be lost within the noise, this may lead to systematic errors in ΔH°_{vH}. In principle, one could explicitly include a nonzero ΔC_P° in the fitting function.[66,79] The error associated with the ΔC_P°, however, is likely to be as large as the parameter itself.[79]

While many assumptions and simplifications have been used in the analysis of RNA optical melting data, the results obtained have proven useful to predict the stabilities of many new sequences. Evidently, totally accurate values for the thermodynamic parameters are not required.

6.03.2.5 Calorimetry

Calorimetry is another technique for investigating the energetics of biomolecules. Experimental techniques for differential scanning calorimetry (DSC) and isothermal titration calorimetry (ITC) have been described in great detail.[80–83] Compared to the model-dependent ΔH°_{vH} values indirectly derived from the measurement of temperature-dependent spectroscopic properties, transition enthalpies determined calorimetrically do not depend on the nature of the transition. DSC measures the excess heat capacity, ΔC^{ex}_P. From the ΔC^{ex}_P vs. temperature profile, one can obtain ΔH° and ΔS° directly, after subtracting baselines appropriately:

$$\Delta H^\circ = \int \Delta C^{ex}_P \, dT \tag{13}$$

$$\Delta S^\circ = \int \frac{\Delta C^{ex}_P}{T} \, dT \tag{14}$$

The shapes of such curves depend on the nature of the transitions they represent, but it is the area underneath them (ΔC^{ex}_P vs. T or $\Delta C^{ex}_P/T$ vs. T) that gives ΔH° and ΔS°.

If a transition actually proceeds in a two-state manner, ΔH° values determined from optical melting curves and calorimetry will agree. If intermediate states are significantly populated, a transition will be broadened and this will make the apparent ΔH°_{vH} smaller than the true ΔH° as determined calorimetrically. The ratio $\Delta H^\circ_{vH}/\Delta H^\circ_{DSC}$ provides a measure of the size of the cooperative unit, i.e., the fraction of the structure that melts cooperatively.[65] If the ratio is 1, then the transition is two-state; if the ratio is less than 1, then the transition involves intermediate states. There are, however, exceptions.[84] Calorimetry has not been as widely used as optical methods in studies of RNA because it requires more material. In DSC, moreover, errors in ΔH° and ΔS° appear to be uncorrelated. Thus, errors in ΔG° are larger for DSC than for optical experiments. For example, in one study, ΔG° values reported from calorimetric and optical melting of duplexes differ by 5 kcal mol^{-1}, which corresponds to more than a 1000-fold difference in equilibrium constant.[85] Methods have been developed, however, for using calorimetric and optical data simultaneously to determine thermodynamic parameters.[86,87]

6.03.2.6 Statistical Treatment of Transitions

The two-state assumption is normally only applicable to relatively short oligonucleotides (less than 20 base pairs). In long oligomers or polymers, the helix growth steps dominate the initiation step, therefore intermediate states are significantly populated, and the two-state model is no longer valid. Since the helix growth steps are unimolecular, the concentration dependence of T_M, which is characteristic of the multimolecular initiation step, is not observed with polymers. Even some short oligomer sequences do not have two-state transitions. For example, base pairs at the end of a double helix may open before central base pairs.[88] Statistical models must be used to analyze transitions that are not two-state.[53,54,88–95]

The general procedure for a statistical treatment is first to write the partition function q for the molecule's conformations, which by definition, contains a complete description of the thermo-dynamics of its transitions. From the partition function, the expected average properties of the system can be expressed as a function of relevant parameters like equilibrium constants. These parameters can then be extracted by fitting predictions derived from the partition function to the experimentally accessible data.

Assuming that an RNA sequence can adopt random coil and n different duplex conformations, the molecular partition function is

$$q = \sum_{i=0}^{n} \exp(-G_i^\circ/RT) \tag{15}$$

where G_i is the free energy of the ith conformation and the summation is over all possible conformations including all the different duplex conformations and the random coil conformation. If we set the free energy for random coil, G_0, at 0, and remove the contribution to the partition function from the random coil state, we are left with the conformational partition function, q_c,

$$q_c = q - 1 = \sum_{i=1}^{n} \exp\left(-\frac{\Delta G_i^\circ}{RT}\right) = \sum_{i=1}^{n} K_i \tag{16}$$

where ΔG_i is the free energy difference between the ith duplex conformation and random coil and K_i is the corresponding equilibrium constant.

A simple statistical model, the zipper model, is probably adequate for most transitions of small RNAs.[53,54,94,96] This model assumes that each residue exists in either a double helical or coil state, that initiation of a base-paired region can occur at any residue in the sequence, and that all of the base-paired residues occur contiguously in a single region, i.e., only one double helical region is allowed. To calculate K_i, we further assume that only perfectly aligned duplexes make significant contributions to q_c (perfectly matching zipper model); the equilibrium constant for initiating the duplex is $\kappa = \sigma \cdot s$, where s is the equilibrium constant for adding one base pair to an existing double helical region. To simplify the presentation, we assume that s is independent of sequence, although this is not generally true.

With these assumptions, the equilibrium constant for forming a duplex with j base pairs is κs^{j-1}. If we ignore the symmetry number for simplicity, the degeneracy of a duplex with j base pairs is $g_j(L) = L - j + 1$, where L is the length of the polymer. If the summation is taken over the energy levels instead of over the individual conformational states, then the conformational partition function becomes:

$$q_c = \sum_{j=1}^{L} g_j(L)\kappa s^{j-1} = \kappa\left[(L+1)\sum_{j=1}^{L} s^{j-1} - \sum_{j=1}^{L} js^{j-1}\right] \tag{17}$$

The first term is just a simple geometric series,

$$\sum_{j=1}^{L} s^{j-1} = \frac{s^L - 1}{s - 1},$$

and the second term can be converted into one since it is equal to $\partial(\Sigma s^j)/\partial s$.[54] Thus the equation reduces to:

$$q_c = \kappa\frac{s^{L+1} - (L+1)s + L}{(s-1)^2} \tag{18}$$

The two-state assumption, where the only duplex conformation that contributes to q_c has L base pairs, corresponds to the condition of large s and finite L, i.e., short oligomers with favorable helix growth steps. In this case, the conformational partition function is just the equilibrium constant, $K_{eq} = q_c = \kappa s^{L-1}$.[54]

When analyzing UV melting data, the absorbance A is assumed to be linear in the fraction of total bases paired, X_b:

$$X_b = \frac{A - A_s}{A_d - A_s} \tag{19}$$

where A_s and A_d are the absorbances of the mixture of single strands and the pure fully-formed duplex, respectively. Since q_c is a summation over equilibrium constants, it can be written as

$$q_c = \frac{\sum_i [H_i]}{[SS]^2} = \frac{X}{2(1-X)^2 C_T} \quad \text{(self-complementary)} \tag{20}$$

$$q_c = \frac{\sum_i [H_i]}{[SS_1][SS_2]} = \frac{2X}{(1-X)^2 C_T} \quad \text{(non-self-complementary)} \tag{21}$$

where H_i are the possible helical conformations, SS are single stranded states, C_T is the total strand concentration, and X is the fraction of strands in all possible duplex states. The average number of base pairs per duplex, $\langle j \rangle$, can be obtained by calculating the expectation value of j,

$$\langle j \rangle = \frac{1}{q_c} \sum_{j=1}^{L} j\kappa g_j(L) s^{j-1} = \frac{1}{q_c} \frac{d}{ds}\left(\sum_{j=1}^{L} \kappa g_j(L) s^j \right) = \frac{1}{q_c} \frac{d(sq_c)}{ds} \tag{22}$$

This allows X_b to be expressed in terms of X and q_c:

$$X_b = \frac{X\langle j \rangle}{L} = \frac{X}{Lq_c} \frac{d(sq_c)}{ds} \tag{23}$$

We can substitute X in terms of q_c and C_T from Equations (20) and (21) and obtain

$$X_b = \frac{1 + 4q_c C_T - (1 + 8q_c C_T)^{1/2}}{4Lq_c^2 C_T} \frac{d(sq_c)}{ds} \quad \text{(self-complementary)} \tag{24}$$

$$X_b = \frac{1 + q_c C_T - (1 + 2q_c C_T)^{1/2}}{Lq_c^2 C_T} \frac{d(sq_c)}{ds} \quad \text{(non-self-complementary)} \tag{25}$$

Thus, κ and s can be determined by fitting to X_b, which is experimentally accessible by Equation (19).

Real transitions in RNA are usually more complicated than the zipper model proposes, e.g., more than one helical region may be present at the same time, and s is sequence dependent.[54] In that case, additional κ and s values have to be added. The number of parameters increases rapidly for realistic models. On the other hand, it is sometimes possible to reasonably approximate a transition with a single or a few intermediate states.[41,97]

6.03.3 THERMODYNAMICS OF RNA SECONDARY STRUCTURE MOTIFS

As discussed above, thermodynamic parameters for structural transitions can be determined by analyzing the melting behavior of RNA molecules. This information can be used to predict the thermodynamic properties of other RNA molecules of known sequence, if the measured energetics can be partitioned to separate contributions for various structural motifs. For example, the free energy change and therefore equilibrium constant (Equation (5)) for forming an ordered structure from a random coil can be predicted if it is assumed that the total free energy change is simply the sum of the free energy changes for forming the separate motifs.

6.03.3.1 Watson–Crick Helical Regions

Watson–Crick base pairs are the most abundant motif in RNA structures, typically accounting for about 50% of nucleotides in a sequence. They form the helical framework for RNA and are also central to RNA–RNA recognition processes.[1,98,99] Therefore, they are the most extensively studied motif. Measurements on the sequence dependence of energetics of Watson–Crick duplex formation by RNA oligonucleotides and polynucleotides show that base pair composition (mole fraction of AU and GC base pairs) cannot alone explain the thermodynamics.[72,100,101] Sequences with identical composition but different permutations of base pairs may have different free energy changes for duplex formation. For example, the duplexes (5′GUUCGAAC3′)$_2$ and (5′UUGGCCAA3′)$_2$ both have four AU and four GC base pairs, but have free energy changes of duplex formation at 37 °C of −8.8 and −11.0 kcal mol^{-1}, respectively.[49,75] This difference of 2.2 kcal mol^{-1} translates into a 35-fold difference in equilibrium constants for duplex formation at 37 °C (Equation (5)). Thus, stability depends on more than the number of hydrogen bonds formed.

Presumably, vertical stacking interactions between neighboring base pairs are the sequence-dependent variable.

The sequence dependence of duplex formation involving only Watson–Crick base pairs is approximated well by a nearest-neighbor model.[42,72,74,75,100,101] Several such models have been proposed.[72,75,102] The INN–HB (Individual Nearest Neighbor-Hydrogen Bonding) model of Xia *et al.*[75] assumes that the stability of a duplex depends on an initiation term for formation of the first pair, 10 helix propagation terms for the 10 possible nearest-neighbor combinations of adjacent Watson–Crick pairs, a base pair composition term that accounts for the number of hydrogen bonds in the duplex, and a symmetry term that is zero except when the duplex is self-complementary. The base composition term in this model is simply a term associated with each terminal AU pair, since duplexes with identical nearest neighbors but different base pair composition must have different numbers of terminal AU pairs.[75]

For example, the duplexes $\frac{5'\text{ACGCA3}'}{3'\text{UGCGU5}'}$ and $\frac{5'\text{GCACG3}'}{3'\text{CGUGC5}'}$ have the same nearest neighbors, but different base compositions and therefore different terminal base pairs. The measured free energies of duplex formation at 37 °C are −4.97 and −6.17 kcal mol^{-1}, respectively. Duplexes with terminal GC pairs are consistently more stable than duplexes with the same nearest neighbors but with terminal AU pairs. Thus, the free energy term for terminal AU pairs is unfavorable.[75] This INN–HB model is statistically identical to the ISS (Independent Short Sequence) nearest-neighbor model of Gray,[102] although the physical models differ.

In the INN–HB model, the free energy change for formation of a duplex with only Watson–Crick pairs is given by:

$$\Delta G^\circ(\text{duplex}) = \Delta G^\circ_{\text{init}} + \sum_j n_j \Delta G^\circ_j(\text{NN}) + m_{\text{term}-\text{AU}}\Delta G^\circ_{\text{term}-\text{AU}} + \Delta G_{\text{sym}} \qquad (26)$$

Each $\Delta G^\circ_j(\text{NN})$ term is the free energy contribution of the *j*th nearest neighbor with n_j occurrences in the sequence. The $\Delta G^\circ_{\text{init}}$ term is the free energy of initiation. Among other factors, initiation includes translational and rotational entropy loss for converting two particles into one. This free energy of initiation is assumed to be independent of length, although this assumption is not rigorously correct. The parameters $m_{\text{term}-\text{AU}}$ and $\Delta G^\circ_{\text{term}-\text{AU}}$ are number of terminal AU pairs and the free energy parameter per terminal AU pair, respectively. The ΔG_{sym} term is required because there is twofold rotational symmetry in self-complementary duplexes, but not in single-stranded states or non-self-complementary duplexes. The maintenance of this symmetry makes self-complementary duplexes less favorable by a factor of two in conformational space,[54,103] i.e., there is an extra entropy loss of $R\ln 2 = 1.4$ eu upon duplex formation. At 37 °C, $\Delta G_{\text{sym}} = -T\Delta S = -(310.15)(-1.4) = 0.43$ kcal mol^{-1} for self-complementary duplexes and zero for non-self-complementary duplexes.[74,75] Equations similar to Equation (26) can be written for ΔH° and ΔS°, except there is no symmetry term for ΔH°.

Parameters for the INN–HB model are obtained by applying multiple linear regression analysis of experimental data to the regression function of Equation (26).[75] Parameters based on a set of 90 RNA duplexes with only Watson–Crick base pairs are listed in Table 1. On average, these parameters predict ΔG°_{37}, ΔH°, ΔS°, and T_M of duplex formation within 3.2%, 6.0%, 6.8%, and 1.3 °C. This accuracy is compatible with the limits of a nearest-neighbor model as determined from measurements on different sequences predicted to have identical thermodynamic parameters.[49,75,104,105] Although these parameters are derived from duplexes with two-state melting behavior, they can also provide useful approximations for RNA duplexes that are not two-state.[75]

From Table 1, the ΔG°_{37} values of AU-only, one-AU/one-GC, and GC-only nearest-neighbors have average values of 1.1, 2.2, and 3.0 kcal mol^{-1}. Evidently, base composition is an important factor. The ΔG°_{37} for $\frac{5'\text{GC3}'}{3'\text{CG5}'}$ propagation (−3.42 kcal mol^{-1}) is more favorable than for $\frac{5'\text{CG3}'}{3'\text{GC5}'}$ propagation (−2.36 kcal mol^{-1}) by more than 1 kcal mol^{-1}, suggesting that orientations of base pairs and therefore stacking patterns are also important. In the INN–HB model, each nearest-neighbor parameter includes contributions from stacking between the two base pairs of the doublet, and half of the hydrogen bonding interaction from each base pair.

6.03.3.2 GU Pairs

GU base pairs are the most common motif besides Watson–Crick base pairs. GU pairs are often conserved in RNA secondary structures, suggesting that they may play either a structural or

Table 1   Thermodynamic parameters for Watson–Crick[75] and GU[113] base pairs in 1 M NaCl, pH 7.

Parameters	ΔG°_{37} (kcal mol^{-1})	ΔH° (kcal mol^{-1})	$\Delta S^{\circ a}$ (eu)
Propagation of Watson–Crick nearest-neighbors			
5'AA3' 3'UU5'	−0.93	−6.82	−19.0
5'AU3' 3'UA5'	−1.10	−9.38	−26.7
5'UA3' 3'AU5'	−1.33	−7.69	−20.5
5'CU3' 3'GA5'	−2.08	−10.48	−27.1
5'CA3' 3'GU5'	−2.11	−10.44	−26.9
5'GU3' 3'CA5'	−2.24	−11.40	−29.5
5'GA3' 3'CU5'	−2.35	−12.44	−32.5
5'CG3' 3'GC5'	−2.36	−10.64	−26.7
5'GG3' 3'CC5'	−3.26	−13.39	−32.7
5'GC3' 3'CG5'	−3.42	−14.88	−36.9
Propagation of GU-containing nearest-neighbors			
5'GU3' 3'UG5'	1.29 (−1.06)^b	−14.59 (−14.14)^b	−51.2 (−42.2)^b
5'GG3' 3'UU5'	0.47	−13.47	−44.9
5'UG3' 3'GU5'	0.30	−9.26	−30.8
5'AG3' 3'UU5'	−0.55	−3.21	−8.6
5'UG3' 3'AU5'	−1.00	−6.99	−19.3
5'GA3' 3'UU5'	−1.27	−12.83	−37.3
5'GU3' 3'UA5'	−1.36	−8.81	−24.0
5'CG3' 3'GU5'	−1.41	−5.61	−13.5
5'GG3' 3'CU5'	−1.53	−8.33	−21.9
5'GG3' 3'UC5'	−2.11	−12.11	−32.2
5'GC3' 3'UG5'	−2.51	−12.59	−32.5
Duplex parameters			
Initiation^c	4.09	3.61	−1.5
Each terminal-AU or GU^d	0.45	3.72	10.5
Symmetry correction^e	0.43	0	−1.4

^{a} Calculated from parameters for ΔG°_{37} and ΔH°. ^{b} Parameters in parentheses are for propagation of $^{5'GU3'}_{3'UG5'}$ in context of $^{5'GGUC3'}_{3'CUGG5'}$ only. ^{c} Includes potential GC end effects. ^{d} Parameter per terminal AU or GU pair. ^{e} Only for self-complementary duplexes.

functional role[106,107] rather than simply being replacements of Watson–Crick base pairs. For example, GU pairs are conserved in the P1 helix of group I introns[33] because they allow the formation of tertiary interactions with the intron's catalytic core.[108]

The thermodynamics of GU pairs have been systematically studied.[29,109–112] The results show that GU pairs are often thermodynamically similar to AU pairs. The data can be reasonably fit to a nearest-neighbor model except that the tandem $^{5'GU3'}_{3'UG5'}$ motif has to be treated differently depending on context.[112] The $^{5'GGUC3'}_{3'CUGG5'}$ motif is found to have remarkable stability compared to $^{5'GU3'}_{3'UG5'}$ in other contexts. Parameters of GU-containing nearest-neighbors have been evaluated from a database of duplexes containing single or tandem GU pairs,[113] and are listed in Table 1. Note that it is assumed terminal GU pairs should be penalized the same as terminal AU pairs (0.45 kcal mol^{-1} per terminal GU pair) because both have only two hydrogen bonds.

The parameters in Table 1 allow reasonable predictions of the thermodynamic properties of RNA duplexes with Watson–Crick and GU pairs. Examples of applications of these parameters in predicting ΔG°, ΔH°, ΔS°, T_M, and dissociation constants, K_D, are shown in Figure 3 for a self-complementary and a non-self-complementary duplex.

6.03.3.3 Dangling Ends and Terminal Mismatches

The average length of an uninterrupted helix of Watson–Crick and GU base pairs in RNA secondary structures is about seven base pairs.[42] Therefore there are many helix ends in RNA secondary structures. Thus it is important to understand the effects of terminal unpaired nucleotides on the stability of a helix. A 5′ or 3′ single unpaired nucleotide is called a dangling end, and a pair of 5′ and 3′ unpaired nucleotides at the same helix end is called a terminal mismatch.

Effects of dangling ends and terminal mismatches can be directly determined by comparing stabilities of helices with and without them in model systems, i.e., they can be treated as additional nearest neighbors. For example, the effect of a 3′ dangling A is taken as half of the difference in free energy change of duplex formation between a duplex with two 3′ dangling As and the core duplex without a 3′ dangling A:[66]

$$\Delta G^\circ(3' \text{ dangling A}) = \frac{1}{2}\left[\Delta G^\circ\left(\frac{5'\text{GCGCA}}{\text{ACGCG5}'}\right) - \Delta G^\circ\left(\frac{5'\text{GCGC3}'}{3'\text{CGCG5}'}\right)\right] \tag{27}$$

The effect of 5′ dangling ends and terminal mismatches can be obtained similarly. Many of these parameters have been measured[66,110,113–119] and the values are summarized in Tables 2 and 3.

The parameters in Table 2 show that a 5′ dangling end contributes little to the stability of a duplex, averaging $\Delta G^\circ_{37} = -0.2$ kcal mol^{-1}. This is similar to the effect of a 5′ phosphate alone, probably because a 5′ phosphate restricts the conformation available to the following sugar. Evidently, 5′ dangling nucleotides do not interact with the adjacent helix.[116] In contrast, 3′ dangling end effects are very sequence dependent, ranging from -0.1 to -1.7 kcal mol^{-1}, with dangling purines more favorable than pyrimidines. Certain 3′ dangling ends can stabilize a helix almost as much as a base pair. For example, a 3′ dangling A adjacent to a terminal CG pair, $\frac{5'\text{CA3}'}{3'\text{G}}$, stabilizes a duplex by 1.7 kcal mol^{-1}, compared to the nearest-neighbor parameter of -2.1 kcal mol^{-1} for $\frac{5'\text{CA3}'}{3'\text{GU5}'}$.

The difference between stabilizing increments of 5′ and 3′ dangling ends can be rationalized by structural considerations.[44,116] In A-form geometry, a 5′ dangling base is not close to a base on the opposite strand, while a 3′ dangling base can stack directly on the base of the opposite strand in the terminal pair, thus helping to hold the duplex together. Dangling end effects can help rationalize and predict local three-dimensional structure.[42,117,120] Comparisons of free energy increments measured for dangling nucleotides in oligonucleotides with X-ray structures of large RNAs show that dangling ends which favor oligonucleotide duplex formation by more than 1 kcal mol^{-1} in short RNA duplexes are stacked in structures of large RNAs. Dangling ends that favor duplex formation by less than 0.4 kcal mol^{-1} in short RNA duplexes are usually unstacked in large RNAs unless they are in GA pairs. Unstacked dangling ends can also allow the RNA backbone to make a turn.

Stability increments for terminal mismatches are essentially the sum of increments for the constituent dangling ends (Table 3) when they are pyrimidine–pyrimidine and CA or AC mismatches. For purine–purine mismatches, the increment is less than the sum of constituent dangling ends, indicating that they do not interact synergistically to increase helix stability.

To use the parameters in Tables 2 and 3 to calculate the thermodynamic properties of duplexes with dangling ends or terminal mismatches, simply add the terms for dangling ends or terminal mismatches to the duplex parameters predicted by the INN–HB model using Table 1.[75] Note that when dangling ends or terminal mismatches follow terminal AU or GU pairs, the penalty for a terminal AU is still applied.

(a)   Non-self-complementary duplex

$$\Delta G^{\circ}_{37}\left(\begin{array}{l}5'\text{AUGACU3}'\\3'\text{UACUGG5}'\end{array}\right) = \Delta G^{\circ}_{\text{init}} + \Delta G^{\circ}_{37}\left(\begin{array}{l}5'\text{AU3}'\\3'\text{UA5}'\end{array}\right) + \Delta G^{\circ}_{37}\left(\begin{array}{l}5'\text{UG3}'\\3'\text{AC5}'\end{array}\right) + \Delta G^{\circ}_{37}\left(\begin{array}{l}5'\text{GA3}'\\3'\text{CU5}'\end{array}\right)$$

$$+ \Delta G^{\circ}_{37}\left(\begin{array}{l}5'\text{AC3}'\\3'\text{UG5}'\end{array}\right) + \Delta G^{\circ}_{37}\left(\begin{array}{l}5'\text{CU3}'\\3'\text{GG5}'\end{array}\right) + 2\times \Delta G^{\circ}_{37}\left(\begin{array}{l}5'\text{A}\\3'\text{U}\end{array}\Big/\begin{array}{l}\text{G}\\\text{U}\end{array}\right) + \Delta G_{\text{sym}}$$

$$= 4.09 + (-1.10) + (-2.11) + (-2.35) + (-2.24) + (-2.11) + 2\times 0.45 + 0$$

$$= -4.92 \text{ kcal mol}^{-1}$$

$$\Delta H^{\circ}\left(\begin{array}{l}5'\text{AUGACU3}'\\3'\text{UACUGG5}'\end{array}\right) = \Delta H^{\circ}_{\text{init}} + \Delta H^{\circ}\left(\begin{array}{l}5'\text{AU3}'\\3'\text{UA5}'\end{array}\right) + \Delta H^{\circ}\left(\begin{array}{l}5'\text{UG3}'\\3'\text{AC5}'\end{array}\right) + \Delta H^{\circ}\left(\begin{array}{l}5'\text{GA3}'\\3'\text{CU5}'\end{array}\right)$$

$$+ \Delta H^{\circ}\left(\begin{array}{l}5'\text{AC3}'\\3'\text{UG5}'\end{array}\right) + \Delta H^{\circ}\left(\begin{array}{l}5'\text{CU3}'\\3'\text{GG5}'\end{array}\right) + 2\times \Delta H^{\circ}\left(\begin{array}{l}5'\text{A}\\3'\text{U}\end{array}\Big/\begin{array}{l}\text{G}\\\text{U}\end{array}\right)$$

$$= 3.61 + (-9.38) + (-10.44) + (-12.44) + (-11.40) + (-12.11) + 2\times 3.72$$

$$= -44.72 \text{ kcal mol}^{-1}$$

$$\Delta S^{\circ}\left(\begin{array}{l}5'\text{AUGACU3}'\\3'\text{UACUGG5}'\end{array}\right) = \Delta S^{\circ}_{\text{init}} + \Delta S^{\circ}\left(\begin{array}{l}5'\text{AU3}'\\3'\text{UA5}'\end{array}\right) + \Delta S^{\circ}\left(\begin{array}{l}5'\text{UG3}'\\3'\text{AC5}'\end{array}\right) + \Delta S^{\circ}\left(\begin{array}{l}5'\text{GA3}'\\3'\text{CU5}'\end{array}\right)$$

$$+ \Delta S^{\circ}\left(\begin{array}{l}5'\text{AC3}'\\3'\text{UG5}'\end{array}\right) + \Delta S^{\circ}\left(\begin{array}{l}5'\text{CU3}'\\3'\text{GG5}'\end{array}\right) + 2\times \Delta S^{\circ}\left(\begin{array}{l}5'\text{A}\\3'\text{U}\end{array}\Big/\begin{array}{l}\text{G}\\\text{U}\end{array}\right) + \Delta S_{\text{sym}}$$

$$= -1.5 + (-26.7) + (-26.9) + (-32.5) + (-29.5) + (-32.2) + 2\times 10.5 + 0$$

$$= -128.3 \text{ eu}$$

$$T_{\text{M}} = \frac{\Delta H^{\circ}}{\Delta S^{\circ} + R\ln(C_{\text{T}}/4)} = \frac{-44.72\times 1000}{-128.3 + 1.987\times \ln\left(2\times 10^{-4}\big/4\right)} = 302.2 \text{ K} = 29.1 \text{ }^{\circ}\text{C}$$

$$K_{\text{D}} = \exp(\Delta G^{\circ}_{37}/RT) = 0.34 \text{ mM}$$

Figure 3   Calculation of predicted thermodynamic properties for duplexes with Watson–Crick and GU base pairs. (a) Non-self-complementary duplex from binding of a short oligonucleotide (5′AUGACU3′) to the internal guide sequence (5′GGUCAU3′) from a group I intron of *Pneumocystis carinii*,[177] T_{M} is calculated for total strand concentration of 2×10^{-4} M. (b) Self-complementary duplex, T_{M} is calculated for total strand concentration of 1×10^{-4} M. Note that thermodynamic parameters are given for association, and the dissociation constant $K_{\text{D}} = K_{\text{A}}^{-1}$, the inverse of the association constant.

6.03.3.4   Loops

RNA loops are regions of sequence not involved in canonical pairs that are flanked by one or more canonical paired regions. Here a canonical pair is defined as a Watson–Crick or wobble GU pair. The types of loops common in RNA are illustrated in Figure 1. The thermodynamics of RNA

(b) Self-complementary duplex

$$\Delta G_{37}^{\circ}\left(\dfrac{5'\text{UGGCCA3'}}{3'\text{ACCGGU5'}}\right) = \Delta G_{\text{init}}^{\circ} + 2\times\Delta G_{37}^{\circ}\left(\dfrac{5'\text{CA3'}}{3'\text{GU5'}}\right) + 2\times\Delta G_{37}^{\circ}\left(\dfrac{5'\text{GG3'}}{3'\text{CC5'}}\right)$$

$$+\,\Delta G_{37}^{\circ}\left(\dfrac{5'\text{GC3'}}{3'\text{CG5'}}\right) + 2\times\Delta G_{37}^{\circ}\left(\dfrac{\text{A3'}}{\text{U5'}}\right) + \Delta G_{\text{sym}}$$

$$= 4.09 + 2\times(-2.11) + 2\times(-3.26) + (-3.42) + 2\times0.45 + 0.43$$

$$= -8.74 \;\text{kcal}\;\text{mol}^{-1}$$

$$\Delta H^{\circ}\left(\dfrac{5'\text{UGGCCA3'}}{3'\text{ACCGGU5'}}\right) = \Delta H_{\text{init}}^{\circ} + 2\times\Delta H^{\circ}\left(\dfrac{5'\text{CA3'}}{3'\text{GU5'}}\right) + 2\times\Delta H^{\circ}\left(\dfrac{5'\text{GG3'}}{3'\text{CC5'}}\right)$$

$$+\,\Delta H^{\circ}\left(\dfrac{5'\text{GC3'}}{3'\text{CG5'}}\right) + 2\times\Delta H^{\circ}\left(\dfrac{\text{A3'}}{\text{U5'}}\right)$$

$$= 3.61 + 2\times(-10.44) + 2\times(-13.39) + (-14.88) + 2\times3.72$$

$$= -51.49 \;\text{kcal}\;\text{mol}^{-1}$$

$$\Delta S^{\circ}\left(\dfrac{5'\text{UGGCCA3'}}{3'\text{ACCGGU5'}}\right) = \Delta S_{\text{init}}^{\circ} + 2\times\Delta S^{\circ}\left(\dfrac{5'\text{CA3'}}{3'\text{GU5'}}\right) + 2\times\Delta S^{\circ}\left(\dfrac{5'\text{GG3'}}{3'\text{CC5'}}\right)$$

$$+\,\Delta S^{\circ}\left(\dfrac{5'\text{GC3'}}{3'\text{CG5'}}\right) + 2\times\Delta S^{\circ}\left(\dfrac{\text{A3'}}{\text{U5'}}\right) + \Delta S_{\text{sym}}$$

$$= -1.5 + 2\times(-26.9) + 2\times(-32.7) + (-36.9) + 2\times10.5 + (-1.4)$$

$$= -138.0 \;\text{eu}$$

$$T_{\text{M}} = \dfrac{\Delta H^{\circ}}{\Delta S^{\circ} + R\ln(C_T)} = \dfrac{-51.49\times1000}{-138.0 + 1.987\times\ln\left(1\times10^{-4}\right)} = 329.4\;\text{K} = 56.3\,°\text{C}$$

$$K_{\text{D}} = \exp(\Delta G_{37}^{\circ}/RT) = 0.69\;\mu\text{M}$$

Figure 3 (continued)

loops have not been fully investigated due to their enormous sequence diversity. With more and more experimental data available, however, our understanding of their contributions increases and models for approximating their stabilities are becoming more and more realistic.

6.03.3.4.1 *Hairpin loops*

Hairpin loops occur when nucleic acid strands fold back on themselves to make base pairs. Hairpin loops are a large part of RNA secondary structure. For example, nearly 70% of the small subunit rRNA of *Escherichia coli* is found in small stem-hairpin loop structures. Hairpins can provide nucleation sites for overall three-dimensional folding, and be involved in tertiary interactions.[18,30,33,121,122]

Hairpin loops in RNA can be very large, or, in at least one case, as small as two nucleotides.[122]

Table 2 Thermodynamic parameters for unpaired dangling nucleotides in 1 M NaCl, pH 7.[a,b,c]

	$X = A$			$X = C$			$X = G$			$X = U$		
	$\Delta H°$	$\Delta S°$	$\Delta G°_{37}$	$\Delta H°$	$\Delta S°$	$\Delta G°_{37}$	$\Delta H°$	$\Delta S°$	$\Delta G°_{37}$	$\Delta H°$	$\Delta S°$	$\Delta G°_{37}$
3'-Dangling nucleotides												
CX̌ / G	−9.0	−23.4	−1.7	−4.1	−10.7	−0.8	−8.6	−22.2	−1.7	−7.5	−20.4	−1.2
GX̌ / C	−7.4	−20.0	−1.1	−2.8	−7.9	−0.4	−6.4	−16.6	−1.3	−3.6	−9.7	−0.6
RX̌ / U	−4.9	−13.2	−0.8	−0.9	−1.2	−0.5	−5.5	−15.0	−0.8	−2.3	−5.4	−0.6
UX̌ / R	−5.7	−16.4	−0.7	−0.7	−1.8	−0.1	−5.8	−16.4	−0.7	−2.2	−6.8	−0.1
5'-Dangling nucleotides												
XČ / G	−2.4	−6.0	−0.5	3.3	11.8	−0.3	0.8	3.4	−0.2	−1.4	−4.3	−0.1
XǦ / C	−1.6	−4.5	−0.2	0.7	3.1	−0.3	−4.6	−14.8	0.0	−0.4	−1.2	0.0
XŘ / U	−1.6	6.1	−0.3	2.2	7.9	−0.3	0.7	3.4	−0.4	3.1	10.6	−0.2
XǓ / R	−0.5	−0.7	−0.3	6.9	22.8	−0.1	(0.6)[c]	(2.7)[c]	(−0.2)[c]	(0.6)[c]	(2.7)[c]	(−0.2)[c]

[a] Parameters are in units of kcal mol^{-1} for $\Delta H°$ and $\Delta G°_{37}$ and eu for $\Delta S°$. [b] R is A or G. Dangling ends on AU have been measured and those on GU are estimated as the same. [c] Parameters in parentheses are estimated.[203]

Tetraloops (hairpins with four nucleotides) are the predominant hairpin loops in ribosomal RNAs, with GNRA and UNCG being the most common sequences,[123] where N is any nucleotide and R is purine. In general, stabilities of hairpins depend on the stem, the first mismatch on the closing base pair, the size of the loop, and occasionally intraloop interactions dependent on sequence. These factors have been extensively studied.[21,119,124–128] The free energy of loop formation is obtained by measuring the free energy of forming the entire hairpin with the stem, and subtracting out the contribution from the stem, as approximated by the parameters of the INN–HB model, including GU parameters (Table 1), but without the duplex initiation or symmetry term. Note that hairpin stems with an AU or GU base pair at either end of the stem are penalized by the terminal AU term. The free energy calculated for the hairpin loop $\Delta G°_{HL}$ is assumed to be the sum of various interactions:

$$\Delta G°_{HL} = \Delta G°_{hairpin} - \Delta G°_{stem} = \Delta G°_{init}(n) + \Delta G°(\text{first mismatch}) + \Delta G°(\text{bonus/penalty}) \qquad (28)$$

The contributions of first mismatches are approximated by the values for terminal mismatches (Table 3), except for loops smaller than four nucleotides which are too constrained to allow the same stacking possible for terminal mismatches at the end of a duplex. The relative sequence independence of stability for loops of three is consistent with this model.[125] First mismatches of GA or UU stabilize more than terminal mismatches, so a bonus of −0.8 kcal mol^{-1} is added.[113] The free energy for hairpin loop initiation is unfavorable, due primarily to the unfavorable entropy associated with constraining the nucleotides in the loop. The initiation values depend on loop length n and are listed in Table 4.[113]

Several effects can be included in the $\Delta G°$(bonus/penalty) term. For example, Giese *et al.*[119] found that hairpins closed by GU have an enhanced stability of 2.1 kcal mol^{-1} if the G is directly preceded by two Gs.[113] Interestingly, hairpins closed by GU are often preceded by two Gs in known secondary structures. Some tetraloops also have enhanced stability that can be added as part of the $\Delta G°$(bonus/penalty) term. The UNCG tetraloops are the most stable, but GNRA loops are also somewhat more stable than random sequence tetraloops.[126–129] The enhanced stability can be attributed to hydrogen bonding in the loop.[19–21,25,26,128,130–133] Tetraloops are sometimes involved in tertiary interactions.[14,16,18,33,134–136] Bonuses for specific tetraloops based on their phylogenetic occurrences are sometimes applied in RNA secondary structure prediction to mimic the effects of such tertiary interactions.[113] Poly-C hairpins are less stable than other hairpin loops of the same size.[137] When $n > 3$, the penalty can be fit to a linear equation: $\Delta G°(n) = An + B$, where A and B are found to be 0.3 and 1.8 kcal mol^{-1}, respectively.[113] When $n = 3$, this penalty is 1.4 kcal mol^{-1}.

With the above parameters, the thermodynamic properties of hairpins can be approximated. Calculations of $\Delta G°_{37}$, $\Delta H°$, $\Delta S°$, K, and T_M for a hairpin are illustrated in Figure 4. The hairpin transition is unimolecular, therefore the T_M is independent of concentration.

Table 3 Thermodynamic parameters for terminal mismatches in 1 M NaCl, pH 7.[a,b]

Base pair	$X\downarrow$	$Y\rightarrow$	A	C	G	U
5'AX3'		$\Delta H°$	−3.9	(2.0)	(−3.5)	
3'UY5'	A	$\Delta S°$	−10.2	(9.6)	(−8.7)	—
		$\Delta G°_{37}$	−0.8	(−1.0)	(−0.8)	
		$\Delta H°$	−2.3	(6.0)		(−0.3)
	C	$\Delta S°$	−5.3	(21.6)	—	(1.5)
		$\Delta G°_{37}$	−0.6	(−0.7)		(−0.7)
		$\Delta H°$	−3.1		(−3.5)	
	G	$\Delta S°$	−7.3	—	(−8.7)	—
		$\Delta G°_{37}$	−0.8		(−0.8)	
		$\Delta H°$		(4.6)		(−1.7)
	U	$\Delta S°$	—	(17.4)	—	(−2.7)
		$\Delta G°_{37}$		(−0.8)		(−0.8)
5'CX3'		$\Delta H°$	−9.1	−5.6	−5.6	
3'GY5'	A	$\Delta S°$	−24.5	−13.5	−13.4	—
		$\Delta G°_{37}$	−1.5	−1.5	−1.4	
		$\Delta H°$	(−5.7)	(−3.4)		−2.7
	C	$\Delta S°$	(−15.2)	(−7.6)	—	−6.3
		$\Delta G°_{37}$	(−1.0)	(−1.1)		−0.8
		$\Delta H°$	−8.2		−9.2	
	G	$\Delta S°$	−21.8	—	−24.6	—
		$\Delta G°_{37}$	−1.4		−1.6	
		$\Delta H°$		−5.3		−8.6
	U	$\Delta S°$	—	−12.6	—	−23.9
		$\Delta G°_{37}$		−1.4		−1.2
5'GX3'		$\Delta H°$	−5.2	(−4.0)	−5.6	
3'CY5'	A	$\Delta S°$	−13.2	(−8.2)	−13.9	
		$\Delta G°_{37}$	−1.1	(−1.5)	−1.3	
		$\Delta H°$	−7.2	(0.5)		(−4.2)
	C	$\Delta S°$	−19.6	(3.9)	—	(−12.2)
		$\Delta G°_{37}$	−1.1	(−0.7)		(−0.5)
		$\Delta H°$	−7.1		−6.2	
	G	$\Delta S°$	−17.8	—	−15.1	—
		$\Delta G°_{37}$	−1.6		−1.4	
		$\Delta H°$		(−0.3)		(−5.0)
	U	$\Delta S°$	—	(−2.1)	—	(−14.0)
		$\Delta G°_{37}$		(−1.0)		(−0.7)
5'GX3'		$\Delta H°$	3.4	(2.0)	(−3.5)	
3'UY5'	A	$\Delta S°$	10	(9.6)	(−8.7)	—
		$\Delta G°_{37}$	0.3	(−1.0)	(−0.8)	
		$\Delta H°$	(−2.3)	(6.0)		(−0.3)
	C	$\Delta S°$	(−5.3)	(21.6)	—	(1.5)
		$\Delta G°_{37}$	(−0.6)	(−0.7)		(−0.7)
		$\Delta H°$	0.6		(−3.5)	
	G	$\Delta S°$	0.0	—	(−8.7)	—
		$\Delta G°_{37}$	0.6		(−0.8)	
		$\Delta H°$		(4.6)		(−1.7)
	U	$\Delta S°$	—	(17.4)	—	(−2.7)
		$\Delta G°_{37}$		(−0.8)		(−0.8)
5'UX3'		$\Delta H°$	−4.0	−6.3	−8.9	
3'AY5'	A	$\Delta S°$	−9.7	−17.7	−25.2	—
		$\Delta G°_{37}$	−1.0	−0.8	−1.1	
		$\Delta H°$	−4.3	−5.1		−1.8
	C	$\Delta S°$	−11.6	−14.6	—	−4.2
		$\Delta G°_{37}$	−0.7	−0.6		−0.5
		$\Delta H°$	−3.8		−8.9	
	G	$\Delta S°$	−8.5	—	−25.0	—
		$\Delta G°_{37}$	−1.1		−1.2	
		$\Delta H°$		−1.4		1.4
	U	$\Delta S°$	—	−2.5	—	6.0
		$\Delta G°_{37}$		−0.6		−0.5

continued

Table 3 (continued)

Base pair	$X\downarrow$	$Y\rightarrow$	A	C	G	U
5'UX3'		ΔH°	−4.8	(−6.3)	(−8.9)	
3'GY5'	A	ΔS°	−12.8	(−17.7)	(−25.2)	—
		ΔG°_{37}	−0.8	(−0.8)	(−1.1)	
		ΔH°	(−4.3)	(−5.1)		(−1.8)
	C	ΔS°	(−11.6)	(−14.6)	—	(−4.2)
		ΔG°_{37}	(−0.7)	(−0.6)		(−0.5)
		ΔH°	−3.1		1.5	
	G	ΔS°	−11.2	—	2.1	—
		ΔG°_{37}	0.5		0.8	
		ΔH°		(−1.4)		(1.4)
	U	ΔS°	—	(−2.5)	—	(6.0)
		ΔG°_{37}		(−0.6)		(−0.5)

[a] Parameters are listed in units of kcal mol^{-1}, eu, and kcal mol^{-1} for ΔH°, ΔS°, and ΔG°_{37}, respectively. Blanks are Watson–Crick or GU nearest-neighbor parameters. [b] Parameters in parentheses are estimated from measured numbers.[113,203]

Table 4   Free energy changes at 37 °C (kcal mol^{-1}) for initiation of various loops in 1 M NaCl, pH 7.

Types of loop	Number of nucleotides in loop								
	1	*2*	*3*	*4*	*5*	*6*	*7*	*8*	*9*
Hairpin loops[a]	—	—	5.6	5.5	5.6	5.3	5.8	5.4	6.4
Bulge loops[b]	3.8 [b]	2.8	3.2	(3.6)[b]	(4.0) [b]	(4.4)[b]	d	d	d
Internal loops	—	—	—	1.7 [c]	1.8	2.0	d	d	d

[a] Hairpin loops with less than three nucleotides are prohibited. For hairpin loops larger than nine nucleotides, the initiation can be approximated by $\Delta G^\circ(n > 9) = \Delta G^\circ_{\text{init}H}(9) + 1.75RT\ln(n/9)$.[204]   [b] For a bulge loop of 1, stacking of adjacent Watson–Crick or GU pairs has to be included. Based on limited experimental data, the free energies are made 0.4 kcal mol^{-1} less favorable for the next larger size from 4 to 6. Larger bulge and internal loops can be approximated by $\Delta G^\circ(n > 6) = \Delta G^\circ_{\text{init}B}(6) + 1.75RT\ln(n/6)$.   [c] Value for loops of 4 are for 3×1 internal loops only, 2×2 internal loops are treated differently (Tables 5 and 6). [d] Larger internal loops can be approximated by $\Delta G^\circ(n > 6) = \Delta G^\circ_{\text{init}I}(6) + 1.75RT\ln(n/6)$.

6.03.3.4.2   Bulge loops

Bulge loops have unpaired nucleotides on only one strand of a double helix (Figure 1). Bulge loops can be either extrahelical or intrahelical. In natural RNA, one-nucleotide bulges are the most common bulge loops, and single purine bulges tend to stack in the helix and bend the helix, whereas single pyrimidine bulges are extra helical.[138–140] Single bulged As are known to be important for protein binding and perhaps for tertiary folding.[141–143] It is thought that single nucleotide bulges do not interrupt nearest-neighbor stacking, while larger bulges do interrupt stacking.[51,144] Bulges become more destabilizing as the number of nucleotides in the bulge increases.[68,144] Due to lack of data, the destabilizing effects of bulge loops are taken as sequence independent.[68,145,146] Free energy parameters for bulge loops of different size are given in Table 4.[113]

6.03.3.4.3   Internal loops

Internal loops are flanked by two helices with canonical pairs, and contain nucleotides on both strands that are not in canonical pairs (Figure 1). RNA internal loops may play important roles in tertiary interactions[18,33,134] and in protein recognition.[34] Stabilities of internal loops are very dependent on the identity and orientation of closing base pairs, on the sequence in the loop, and on the size and symmetry of the loop. The contributions of internal loops can be estimated by comparing stabilities of duplexes with and without the loop, corrected for the Watson–Crick nearest-neighbor interaction absent in the duplex with the loop and present in the duplex without the loop:[88]

$$\Delta G^\circ_{37}(\text{internal loop}) = \Delta G^\circ_{37}(\text{duplex w/ loop}) - \Delta G^\circ_{37}(\text{duplex w/o loop}) + \Delta G^\circ_{37}(\text{NN}) \qquad (29)$$

where $\Delta G^\circ_{37}(\text{NN})$ is the relevant nearest-neighbor parameter of the INN–HB model (Table 1). This

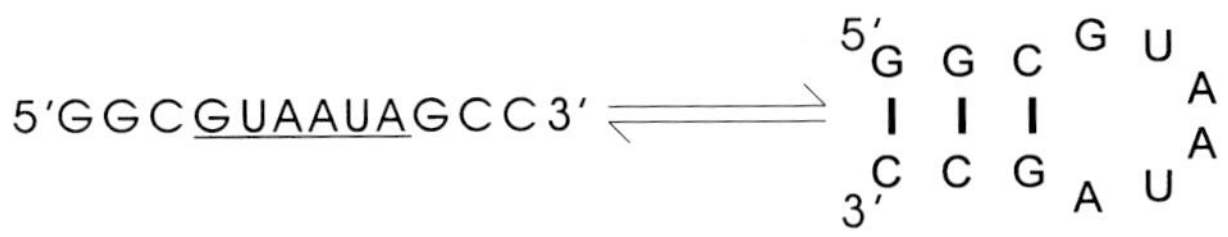

5'G G C G U A A U A G C C 3' $\rightleftharpoons$

$$
\begin{array}{cccccc}
5' & & & & G & U \\
G & G & C & & & A \\
| & | & | & & & A \\
C & C & G & & & A \\
3' & & & A & U & \\
\end{array}
$$

$$\Delta G^{\circ}_{37}(\text{Hairpin}) = \Delta G^{\circ}_{\text{init}}(n=6) + \Delta G^{\circ}_{37}(\text{First Mismatch}) + \Delta G^{\circ}_{37}(\text{Stem})$$

$$= 5.3 + (-1.4) + (-0.8) + (-3.26) + (-3.42) = -3.58 \ \text{kcal mol}^{-1}$$

$$\Delta H^{\circ}(\text{Hairpin}) = \Delta H^{\circ}_{\text{init}}(n=6) + \Delta H^{\circ}(\text{First Mismatch}) + \Delta H^{\circ}(\text{Stem})$$

$$= 0.0 + (-8.2) + (-0.8) + (-13.39) + (-14.88) = -37.27 \ \text{kcal mol}^{-1}$$

$$\Delta S^{\circ}(\text{Hairpin}) = \Delta S^{\circ}_{\text{init}}(n=6) + \Delta S^{\circ}(\text{First Mismatch}) + \Delta S^{\circ}(\text{Stem})$$

$$= -17.1 + (-21.8) + (-32.7) + (-36.9) = -108.5 \ \text{eu}$$

$$T_{\text{M}} = \frac{\Delta H^{\circ}}{\Delta S^{\circ}} = \frac{(-37.27) \times 1000}{(-108.5)} = 343.5 \ K = 70.4 \ ^{\circ}\text{C}$$

$$K = \exp(-\Delta G^{\circ}_{37}/RT) = 333.3$$

Figure 4 Calculation of predicted thermodynamic properties for a hairpin that is highly conserved in the L11 protein binding region of the large subunit ribosomal RNA.[27] Nucleotides in the loop are underlined. Note that the bonus for the GA mismatch is approximated as completely due to a favorable ΔH° and that ΔH° for hairpin initiation is assumed to be zero. The measured values of ΔG°_{37}, ΔH°, ΔS°, and T_{M} are -3.42 kcal mol^{-1}, -36.9 kcal mol^{-1}, 107.8 eu, and 68.7 $^{\circ}$C, respectively.[124]

assumes that the loop does not affect the regions beyond the closing base pairs. Compared to the enormous possible sequence dependence of internal loops, available experimental data are only sufficient for making rough approximations, and future improvements are therefore expected.

(i) Single mismatches (1 × 1 internal loops)

Single mismatches in a helix are the smallest internal loops. A few single mismatches have been studied.[67,88,147] Most single mismatches that have been studied destabilize duplexes, but some are stabilizing. A reasonable approximation for single mismatches with two adjacent GC pairs is that $\Delta G^{\circ}_{37} = 0.4$ kcal mol^{-1} when the mismatch is not GG and $\Delta G^{\circ}_{37} = -1.7$ kcal mol^{-1} when the mismatch is GG.[148] The stability is less favorable by roughly 0.65 kcal mol^{-1} per adjacent AU or GU pair, largely due to the 0.45 kcal mol^{-1} term assigned to such "terminal" pairs of a helix.

(ii) Tandem mismatches (2 × 2 internal loops)

Tandem mismatches are formed when there are two opposing nucleotides on each strand that are not involved in Watson–Crick or GU pairs. These 2×2 internal loops can be symmetric or nonsymmetric in terms of loop sequence and closing base pairs. Tandem mismatches are the most extensively studied internal loops,[76,77,149–151] and the results show that their stabilities are very sequence dependent.

Symmetric tandem mismatches contain adjacent identical mismatches closed by the same base pair, e.g., $\frac{5'GGAC3'}{3'CAGG5'}$, $\frac{5'CUUG3'}{3'GUUC5'}$, etc. The dependence of stabilities of symmetric tandem mismatches and tandem GU pairs on closing base pairs and loop sequences fits a pattern,[151] as shown in Table 5, which lists values recalculated[113] with parameters of the INN–HB model. Loops closed by GC base pairs are more stable than those closed by AU base pairs. The decreasing order of stabilities of mismatches is

$$\frac{UG}{GU} > \frac{GU}{UG} > \frac{GA}{AG} \geqslant \frac{AG}{GA} > \frac{UU}{UU} > \frac{GG}{GG} \approx \frac{CA}{AC} \approx \frac{CU}{UC} \approx \frac{UC}{CU} \approx \frac{CC}{CC} \approx \frac{AC}{CA} \approx \frac{AA}{AA}.$$

Including tandem UG sequences, the range in free energy changes is −4.9 to 2.8 kcal mol^{-1}, corresponding to about a 270 000-fold range in equilibrium constants for folding at 37 °C. GC closing base pairs give enhanced stability over CG closing pairs, especially with GA mismatches. For example, $\frac{5'GGAC3'}{3'CAGG5'}$ is almost 2 kcal mol^{-1} more stable than $\frac{5'CGAG3'}{3'GAGC5'}$. Tandem UG, GA, and UU mismatches can be stabilizing, partly due to hydrogen bonding between mismatched bases as evidenced by NMR spectra.[23,29,32,149–153] Other mismatches, CA, CU, CC, and AA are destabilizing and there is no NMR evidence for hydrogen bonding interactions between mismatched bases, suggesting they are more flexible. Formation of these destabilizing tandem mismatches is also associated with a more favorable entropy change than for GU, GA, or UU, consistent with greater flexibility.[151]

Table 5 Free energy changes for symmetric tandem mismatches $\frac{5'PXYQ3'}{3'QYXP5'}$ in 1 M NaCl, pH 7.[a,b]

Mismatches XY / YX→ Closing base pairs P↓ Q	UG GU	GU UG	GA AG	AG GA	UU UU	GG GG	CA AC	CU UC	UC CU	CC CC	AC CA	AA AA
G C	−4.9	−4.1	−2.6	−1.3	−0.5	(0.8)	1.0	1.1	(1.0)	(1.0)	0.9	1.5
C G	−4.2	−1.1	−0.7	−1.0	−0.4[c]	0.8[c]	1.1[c]	1.4[c]	1.4[c]	1.7[c]	2.0[c]	1.3
Ū A	−2.9	−0.3	0.7	(0.7)	1.1	2.2[d]	1.9	2.2	2.8	(2.8)	(2.8)	2.8
Ā U	−2.1	−0.1	0.3	(0.3)	0.6	1.9[d]	2.3	(2.2)	(2.2)	(2.2)	2.5	2.8

[a] Parameters are based on results of Wu *et al.*[151] and references cited therein except for $\frac{GG}{GG}$. Free energy changes are calculated by Equation (29). [b] Parameters in parentheses are estimated. [c] Ref. 149. [d] M. E. Burkard and D. H. Turner, unpublished results.

There are over 2000 possible tandem mismatch sequences, most of which are nonsymmetric, where two different mismatches are adjacent or the two closing base pairs are different.[77] At present, it is impossible to study exhaustively all the sequence dependence of tandem mismatch stability. Thus it is necessary to identify the most important factors that determine the stability of nonsymmetric tandem mismatches and develop a model that can reasonably approximate their stabilities. Xia *et al.*[77] studied a series of nonsymmetric tandem mismatches in the context of $\frac{5'GXYG3'}{3'CWZC5'}$, where XW and YZ can be either the same or different mismatches. The results are more complicated than for symmetric tandem mismatches.

The stabilities of nonsymmetric tandem mismatches cannot be predicted by simply averaging values for symmetric tandem mismatches.[77] The contribution of one mismatch to the free energy increment for nonsymmetric tandem mismatch formation depends on the identity of the other mismatch. This is partly because the structure of a mismatch is dependent on the structure of the adjacent mismatch and partly because mismatches can have different sizes, e.g., purine–purine, purine–pyrimidine, pyrimidine–pyrimidine. Comparison of stabilities of nonsymmetric tandem mismatches to the average values for symmetric tandem mismatches reveals a pattern. In general, it is more favorable to have two stabilizing mismatches of the same size adjacent to each other than to have two stabilizing mismatches of different sizes. For example, two adjacent GA mismatches stabilize the helix regardless of their orientations. In contrast, GA and UU are two stabilizing mismatches of very different sizes, and any combination of them destabilizes the helix.[77]

From these results, a model has been developed that can approximate the stabilities of any tandem mismatch. The general equation is:[77]

$$\Delta G^{\circ}\left(\begin{array}{l}5'PXYS3'\\3'QWZT5'\end{array}\right)=\frac{1}{2}\left[\Delta G^{\circ}\left(\begin{array}{l}5'PXWQ3'\\3'QWXP5'\end{array}\right)+\Delta G^{\circ}\left(\begin{array}{l}5'TZYS3'\\3'SYZT5'\end{array}\right)\right]+\Delta_{P} \qquad (30)$$

where PQ and ST are the closing base pairs, and XW and YZ are any mismatch combinations. The free energy changes in the bracket are values for symmetric tandem mismatches (Table 5). Δ_P is a penalty term that depends on the identity of the mismatches. In general, combinations of two different stabilizing mismatches have larger penalties, and combinations of destabilizing mismatches have smaller penalties. These penalties are collected in Table 6.[113]

Table 6 Penalty terms for nonsymmetric tandem mismatches.[a]

Category of combinations[b]	Penalty (Δ_P) (kcal mol^{-1})
Two stabilizing mismatches of different size	1.8
One stabilizing and one destabilizing mismatch (not AC) of different size	1.0
Other combinations	0.0

[a] To be used in Equation (30). Penalties are the same for GC and AU closing pairs.   [b] See text.

(iii) 2 × 1 internal loops

Internal loops of three contain two unpaired nucleotides opposing one nucleotide. Schroeder *et al.*[154] investigated internal loops of three closed by GC base pairs, and found a considerable sequence dependence to the stabilities. Loops with potential for forming GA and UU mismatches generally have more favorable free energies of formation than loops without such potential. For example, $\frac{CAC}{GAAG}$ is 1.2 kcal mol^{-1} less stable at 37 °C than $\frac{CAC}{GGAG}$. This enhancement, however, depends on the orientation of the closing base pairs and of the potential GA mismatch in the loop. Loops without potential GA or UU mismatches fall in a relatively narrow range with an average free energy of loop formation of 2.2 kcal mol^{-1}. This trend is similar to the results for symmetric tandem mismatches. The results allow approximations for loops that have not been measured.[113,154] These approximations are listed along with measured values in Table 7.

Table 7 Free energy increments (kcal mol^{-1}) at 37 °C for 2 × 1 internal loops, $\frac{5'P\ X\ P3'}{3'QZYQ5'}$ in 1 M NaCl, pH 7.[a,b]

3'ZY5'	X = A	X = C	X = G	X = U
AA	2.3 2.5	2.3	1.7	
AC	2.1	(2.2)		
AG	0.8		0.8	
AU		2.5		
CA	(2.2)	(2.2)		
CC	1.7	2.5		2.2
CG	(0.6)			
CU		1.9		
GA	1.1 2.1		0.8	
GC	(1.6)			
GG	0.4		(2.2)	
UA		(2.2)		
UC		(2.2)		1.5
UU		(2.2)		1.2

[a] Parameters are for two GC closing pairs. Each AU closing pair is penalized by $0.45 + 0.2 = 0.65$ kcal mol^{-1}, where the 0.45 is the INN–HB model penalty for terminal AU pairs in a helix (see text). When there is only one number, it is for $\frac{5'C\ X\ C3'}{3'GZYG5'}$; when there are two numbers, the top one is for $\frac{5'C\ X\ C3'}{3'GZYG5'}$, and the bottom one is for $\frac{5'G\ X\ G3'}{3'CZYC5'}$.   [b] Parameters in parentheses are estimated.[113,154]

(iv) Other asymmetric internal loops

Asymmetric internal loops containing only As[67] or Cs[144] have been studied. They are found to be more destabilizing than symmetric loops with the same total number of nucleotides. This trend was predicted by Papanicolaou *et al.*[155] based on comparisons of predicted and known RNA secondary structures. From the experimental results, a loop asymmetry penalty can be defined as:

$$\Delta G^\circ(\text{asymmetry penalty}) = \Delta G^\circ(\text{asymmetric loop}, n) - \Delta G^\circ(\text{symmetric loop}, n) \tag{31}$$

Experimental data[67] for internal loops with only As is fit well by a model where $\Delta G^\circ_{37}(\text{asymmetry penalty}) = 0.45|n_1 - n_2|$ kcal mol^{-1},[113] where n_1 and n_2 are the number of unpaired nucleotides on each strand. It is possible, however, that the asymmetry penalty depends on the sequence as well as the size.

(v) Larger internal loops

Measurements cannot currently be made on all possible internal loop sequences. A simple model has been developed to approximate the free energy changes of internal loops with more than four nucleotides.[113] This model includes the loop size, closing base pairs, first mismatches, and loop asymmetry. The free energy change is approximated as:

$$\Delta G^\circ_{\text{loop}} = \Delta G^\circ_{\text{init}}(n_1 + n_2) + \Delta G^\circ_{\text{asymm}}(|n_1 - n_2|) + \Delta G^\circ_{\text{AU/GUclosure}} + m\Delta G^\circ_{\text{GA/UU}} \tag{32}$$

where n_1 and n_2 are the numbers of nucleotides on each side of the loop, $\Delta G^\circ_{\text{init}}(n_1 + n_2)$ is a penalty term for closing the loop (Table 4), $\Delta G^\circ_{\text{asymm}}(|n_1 - n_2|)$ is the asymmetry penalty term discussed above, $\Delta G^\circ_{\text{AU/GUclosure}}$ is a penalty of 0.2 kcal mol^{-1} for loops closed by either AU or GU base pairs. This penalty is applied in addition to the terminal AU/GU term of the INN–HB model. $\Delta G^\circ_{\text{GA/UU}}$ is a favorable bonus for each GA or AG (-1.1 kcal mol^{-1}), or UU (-0.7 kcal mol^{-1}) first mismatch, and m is the number of GA, AG, or UU first mismatches. For nucleotides in the loop other than in the two first mismatches, the sequence dependence is neglected. This is likely an incomplete model for some sequences. An example of using Equation (32) to estimate the free energy increment for an internal loop is shown in Figure 5.

Internal Loop

$$\Delta G^\circ_{\text{loop}}\left(\begin{matrix} 5'U^{GA}U3' \\ 3'G_{AAG}G \end{matrix}\right) = \Delta G^\circ_{\text{init}}(n_1 + n_2) + \Delta G^\circ_{\text{asymm}}(|n_1 - n_2|) + \Delta G^\circ_{\text{AU/GUclosure}} + m\Delta G^\circ_{\text{GA/UU}}$$

$$= 1.8 + 0.45 \times (3-2) + 2 \times 0.2 + 2 \times (-0.8)$$

$$= 1.05 \ \text{kcal} \ \text{mol}^{-1}$$

Figure 5 Calculation of predicted free energy increment for an 2×3 internal loop in J4/5 of group I intron of *Pneumocystis carinii*.[177] Note that the calculation of free energies for the double helical regions flanking this internal loop would include a penalty of 0.45 kcal mol^{-1} for each of the GU pairs closing the loop.

6.03.3.5 Coaxial Stacks and Multibranch Loops (or Junctions)

Coaxial stacking can occur when two helices are directly adjacent or separated by a mismatch. Coaxial stacks have been seen in crystal structures of tRNAs[11–13,156,157] and are essential for the three-dimensional shape of tRNA. Stability increments from coaxial stacking of Watson–Crick pairs and of GA or CC mismatches have been investigated in model systems where a short oligomer binds to a four- or five-nucleotide overhang at the base of a hairpin stem.[52,158,159] This binding creates an interface between two helical regions joined by one continuous backbone and a break in the backbone on the other strand (Figure 6). The contribution of coaxial stacking is calculated as the difference in the free energy change between forming the complex and the isolated short-stem duplex:

$$\Delta G^{\circ}_{\text{coax stack}} = \Delta G^{\circ}(\text{hairpin/oligomer}) - \Delta G^{\circ}(\text{oligomer duplex}) \qquad (33)$$

Since the isolated short oligomer duplexes are usually too short to be experimentally measured, their free energy changes of duplex formation are calculated with nearest-neighbor parameters of the INN–HB model.[75] Note that for interfaces with AU base pairs, the terminal AU term is applied.

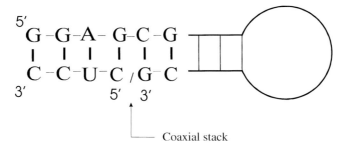

Figure 6 Coaxial stack formed by binding of a short oligonucleotide to the overhang of a hairpin stem. For this case, the free energy of coaxial stacking is determined by $\Delta G^{\circ}_{\text{coax stack}} = \Delta G^{\circ}(\text{hairpin/oligomer complex shown}) - \Delta G^{\circ}\left(\dfrac{5'\text{GGAG}3'}{3'\text{CCUC}5'}\right)$. Here the free energy changes are for formation of the complex between hairpin and short oligomer, and of the isolated short duplex.

The enhanced binding by coaxial stacking of Watson–Crick pairs is found to be independent of the position of the break in the sugar–phosphate backbone, but is dependent on the sequence at the interface, suggesting that the interaction of nearest-neighbor base pairs is the primary determinant of the enhancement.[52,158] These values are collected in Table 8.[113] Compared with the corresponding regular Watson–Crick nearest-neighbor interactions (Table 1), coaxial stacking is 0.8–1.5 kcal mol^{-1} more favorable. Presumably this is because the break in the backbone on one strand offers more flexibility to the base pairs at the interface. Thus, stacking and hydrogen bonding interactions can be optimized. Because the free energy increments for coaxial stacking do not include additional hydrogen bonding interactions, while regular Watson–Crick nearest neighbors do include hydrogen bonding interactions,[75] the real enhancements are even larger. When the chains are extended with an unpaired nucleotide beyond the break, the free energy increments for coaxial stacking are approximately equal to that of a regular Watson–Crick nearest-neighbor interaction.[52,113] This could be because the backbone has to make an unfavorable turn to allow the coaxial stack.

Table 8 Free energy increments of coaxial stacking interaction at 37 °C (kcal mol^{-1}) in 1 M NaCl, pH 7.

Coaxial stacking interface ^{a}	G-C C/G	G-G C/C	C-C G/G	C-G G/C	C/G G-C	G-A C/U	G-U C/A	A-G U/C	U-G A/C	U-A A/U
Stacking interaction (kcal mol^{-1})	−4.20	−4.01	−4.23	−3.16	−3.46	−3.84	−3.34	2.95	−2.92	−2.74

^{a} A hyphen "-" represents continuation of backbone, while a back slash "/" represents a break in the backbone.

GA mismatches between helices are common in known RNA secondary structures. For example, in yeast phenylalanine tRNA, the D and anticodon arms are separated by a GA mismatch, but coaxially stack with each other. GA mismatches between helices are also found in self-alkylating,[160] and self-ligating[161] ribozymes. Thermodynamic effects of GA mismatches and the much less frequently occurring CC mismatch at the interface have been studied.[159] Despite the difference in abundance, both GA and CC mismatches at the interface contribute roughly 2 kcal mol^{-1}, and are relatively insensitive to the identity of the adjacent base pairs, and to extension beyond the interface.[113]

Multibranch loops or junctions are loops where more than two helices intersect. They usually also contain unpaired nucleotides, and are a major determinant of RNA three-dimensional structure.[120] Coaxial stacking is a major contributor to stabilization of multibranch loops. The thermodynamics of other factors that determine stabilities of multibranch loops have not been studied in detail. In general, the stability of a multibranch loop can depend on the number of helices involved, the number of unpaired nucleotides in the loop, and interactions such as base triples involving nucleotides in and near the loop. The INN–HB model for Watson–Crick nearest neighbors suggests that each helix that ends with an AU pair should be made less favorable by about 0.5 kcal mol^{-1}.[75] A crude model has been proposed[52,113] that predicts the free energies of multibranch loops based on an unfavorable initiation term and favorable coaxial stacking. These parameters can be estimated by optimizing the accuracy of the folding algorithm for predicting known secondary structures.[113]

6.03.3.6   Environmental Effects on RNA Secondary Structure Thermodynamics

RNA is a polyanion. In order to form stable structures, it requires counterions to neutralize the repulsions of the negatively charged backbone phosphates. Manning's polyelectrolyte theory predicts that, due to counterion condensation, the local concentration of monovalent cation near a polynucleotide duplex is around 1 M, regardless of the bulk salt concentration if no multivalent cations are present.[162] In solutions containing only monovalent cations, e.g., Na$^+$ and K$^+$, the stability of a long duplex increases with salt concentration up to 1 M, with the increase in T_M being linear with log[Na$^+$] up to about 0.2 M Na$^+$. At salt concentrations above 1 M, addition of more salt decreases duplex stability. The destabilizing effect at high salt concentrations depends on the anion with $CCl_3COO^- > SCN^- > ClO_4^- > Me_2CO_2^- > Br^-,Cl^-$.[163] The effects depend on RNA length. Less charge will be neutralized in oligonucleotides than in polynucleotides because of the reduced charge density at the ends. The dependence of salt effects on length in DNA oligonucleotides has been analyzed.[164] The length effect is negligible at high salt concentrations.[165,166]

In the presence of multivalent ions such as Mg^{2+}, the monovalent ions around the RNA backbone will be essentially replaced by multivalent ions.[162,167] If saturated with Mg^{2+}, increasing concentrations of Na$^+$ decrease duplex stability, because saturation with Mg^{2+} leaves the random coil with a higher density of negative charge than the duplex. Thus, a higher monovalent cation concentration favors a random coil resulting in a lower T_M.

The Mg^{2+} ion may provide specific binding to many potential sites in various RNA structural motifs, especially in non-Watson–Crick regions.[18,168] Experimental results show that duplex stability in 1 M NaCl is similar to that observed in the presence of a few millimolar Mg^{2+} plus 0.1 M Na$^+$ or K$^+$, which is similar to physiological conditions.[29,32,77,169] Therefore, thermodynamic information derived at 1 M NaCl is useful for predicting RNA properties in biologically relevant environments.

Addition of cosolvents to aqueous solutions of RNA usually destabilizes RNA structures.[170,171] Typically, the T_M of a duplex is a linear function of cosolvent concentration.[171]

The stability of most duplexes is relatively insensitive to pH between 5 and 9.[44] At low or high pH, bases are protonated or deprotonated precluding normal hydrogen bonding and the formation of duplexes. In RNA, CC$^+$ mismatches have been observed in the middle of the duplex (CGC̲C̲C̲GCG)$_2$ at pH 5.5.[149]

6.03.4   APPLICATIONS

Our knowledge of the thermodynamics of RNA secondary structure is steadily increasing. Its applications to calculating the thermodynamic properties of RNA molecules containing specific structural motifs have been shown above. Calculations for RNA molecules with combinations of different motifs can be done in a similar fashion. Other important applications include estimation of tertiary interactions, prediction of RNA structure from sequence, and designing ribozymes to target mRNA.

6.03.4.1   Estimation of Tertiary Interactions

In addition to the formation of secondary structures, RNA folding usually involves tertiary interactions, about which little is known. One way to detect the effect of tertiary interactions is to compare dissociation constants, K_D, for RNA–RNA association with those predicted based on the thermodynamics for formation of RNA secondary structure.[172] When tertiary interactions appear to be important, their net effect can be determined by comparing K_D values for the complete system with those measured for a model system that can only form secondary structure.[173] The contribution of various groups to tertiary interactions can be determined by measuring the K_D for complexes with functional group substitutions.[108,173–176] Such comparisons provide insight into structure–function relationships, and can suggest ways to target RNA by exploiting tertiary interactions.[177,178]

For example, the predicted K_D for the duplex, $\begin{smallmatrix} 5'\text{AUGACU3}' \\ 3'\text{UACUGG5}' \end{smallmatrix}$ is 0.34 mM in 1 M NaCl (Figure (3a)), close to the K_D of 0.32 mM measured in 135 mM KCl, 15 mM $MgCl_2$.[177] Binding of 5'AUGACU3' to a group I ribozyme from *Pneumocystis carinii* involves the same pairing interactions, but the K_D is 5.2 nM. Thus, tertiary interactions must strengthen this binding 60 000-fold.[177] Such binding enhancement by tertiary interactions (BETI) can be used to increase the specificity and binding of short antisense agents to RNA.[178]

6.03.4.2 RNA Secondary Structure Prediction and Modeling of Three Dimensional Structure

Thermodynamic parameters are crucial for RNA secondary structure predictions because of the large number of possible pairings. The parameters form the basis of algorithms that can predict the most likely foldings.[46,47,113,179–181] Recursive (dynamic programming) algorithms[46,47] use predicted free energies of smaller fragments to predict free energies for larger fragments until the free energy for the whole structure is calculated. Knotted structures are not allowed in this recursive algorithm, but could be allowed in genetic algorithms.[180,181] At this time, little is known about the thermodynamics of knotted structures, and many other motifs require gross approximations, as discussed above. For this reason, folding algorithms have also been designed to calculate suboptimal structures, and to incorporate experimental data such as enzymatic mapping.[47,113,182] These programs facilitate the design of experimental approaches to single out and test the real structure.

Reliable secondary structure is also the basis for modeling three-dimensional structure, and the thermodynamics of RNA secondary structure can help predict three-dimensional structure. For example, as discussed above, sequences with 3' dangling end stacking more favorable than 1 kcal mol^{-1} in model systems are usually stacked in three-dimensional structures.[42,120] Moreover, conserved features in secondary structures often suggest small model systems whose three-dimensional structures can be solved by NMR or X-ray crystallography. Two such examples are illustrated in Figure 7.

6.03.4.3 Targeting RNA with Ribozymes

Ribozymes have therapeutic potential for targeting RNA either for cleavage[4,183–186] or repair.[187,188] There are several obstacles to realizing this potential. For example, ribozymes may fold into inactive conformations,[189,190] targets may have secondary and tertiary structures that prevent ribozyme binding, and it may be difficult to achieve target specificity.[191–194] In principle, knowledge of the thermodynamics of secondary structure formation can help overcome these obstacles. Secondary structure prediction algorithms can help design ribozymes that have an optimal folding much favored over other foldings.[189] With constraints from chemical and enzymatic mapping, they may also make it possible to predict suitable sites on RNA targets. Specificity may be difficult to achieve because it requires that ribozymes dissociate from mismatched substrate much faster than the rate of the chemical reaction they catalyze.[191] The strength of Watson–Crick base pairing makes this difficult except for short helixes, but sequences that can form short helixes are likely to bind many perfect matches in cellular RNA. One potential solution to this problem is to design ribozymes with enhanced dissociation rates. LeCuyer and Crothers[195] have shown that intramolecular helixes may rearrange much faster than expected, suggesting that the rate-limiting step does not necessarily require disruption of all base pairs in a helix. Another potential solution is to design or evolve a ribozyme that recognizes a three-dimensional shape in the target RNA, rather than simply base sequence. Knowledge of the thermodynamics of secondary structure formation should be useful in implementing these and other strategies.

6.03.5 FUTURE PERSPECTIVES

Our understanding of the thermodynamics of RNA secondary structure is a powerful tool. The current thermodynamic database provides reasonable prediction of RNA secondary structures. It can also be used to aid the design and interpretation of many experiments. Our knowledge is still

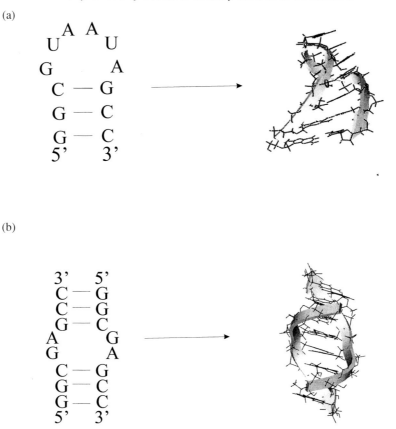

Figure 7　Secondary and three-dimensional structures of two fragments of natural RNAs. Arrows point from 5′ to 3′ along the backbones. (a) Conserved hairpin in the L11 protein binding region of large subunit ribosomal RNA.[27] (b) A tandem mismatch commonly found in RNA.[23]

crude, however; the stabilities of most motifs have not been experimentally determined. Therefore many approximations have to be made and need to be tested, especially for loop regions. High-density oligonucleotide array technologies,[196–201] may make it possible to speed up the accumulation of this kind of experimental data. Random selection methods will provide a way to identify motifs with unusual properties.[202] Moreover, better understanding of intermolecular interactions should improve approximations for motifs that have not been studied. Thus, secondary structure thermodynamics should become an even more powerful and reliable tool in the future.

6.03.6　REFERENCES

1. J. D. Watson, N. H. Hopkins, J. W. Roberts, J. A. Steitz, and A. M. Weiner, "Molecular Biology of the Gene," Benjamin Cummings Inc., Menlo Park, CA, 1987.
2. T. R. Cech and B. L. Bass, *Annu. Rev. Biochem*, 1986, **55**, 599.
3. T. R. Cech, *Annu. Rev. Biochem.*, 1990, **59**, 543.
4. S. Altman, *Adv. Enzymol.*, 1989, **62**, 1.
5. F. Noller, V. Hoffarth, and L. Zimniak, *Science*, 1992, **256**, 1416.
6. Nitta, Y. Kamada, H. Noda, T. Ueda, and K. Watanabe, *Science*, 1998, **281**, 666.
7. P. Walter and G. Blobel, *Nature*, 1982, **299**, 691.
8. N. Williams, *Science*, 1997, **275**, 301.
9. D. Gershon, *Nature*, 1997, **389**, 417.
10. E. Marshall, *Science*, 1995, **267**, 783.
11. S. H. Kim, G. J. Quigley, F. L. Suddath, A. McPherson, D. Sneden, J. J. Kim, J. Weinzierl, and A. Rich, *Science*, 1973, **179**, 285.
12. S. H. Kim, F. L. Suddath, G. J. Quigley, A. McPherson, J. L. Sussman, A. H. J. Wang, N. C. Seeman, and A. Rich, *Science*, 1974, **185**, 435.
13. J. D. Robertus, J. E. Ladner, J. T. Finch, D. Rhodes, R. S. Brown, B. F. C. Clark, and A. Klug, *Nature*, 1974, **250**, 546.
14. H. W. Pley, K. M. Flaherty, and D. B. McKay, *Nature*, 1994, **372**, 68.
15. W. G. Scott, J. T. Finch, and A. Klug, *Cell*, 1995, **81**, 991.

16. W. G. Scott, J. B. Murray, J. R. P. Arnold, B. L. Stoddard, and A. Klug, *Science*, 1996, **274**, 2065.
17. J. B. Murray, D. P. Terwey, L. Maloney, A. Karpeisky, N. Usman, L. Beigelman, and W. G. Scott, *Cell*, 1998, **92**, 665.
18. J. H. Cate, A. R. Gooding, E. Podell, K. Zhou, B. L. Golden, C. E. Kundrot, T. R. Cech, and J. A. Doudna, *Science*, 1996, **273**, 1678.
19. C. Cheong, G. Varani, and I. Tinoco, Jr., *Nature*, 1990, **346**, 680.
20. H. A. Heus and A. Pardi, *Science*, 1991, **253**, 191.
21. G. Varani, C. Cheong, and I. Tinoco, Jr., *Biochemistry*, 1991, **30**, 3280.
22. J. D. Puglisi, R. Tan, B. J. Calnan, A. D. Frankel, and J. R. Williamson, *Science*, 1992, **257**, 76.
23. J. SantaLucia, Jr. and D. H. Turner, *Biochemistry*, 1993, **32**, 12612.
24. B. Wimberly, G. Varani, and I. Tinoco, Jr., *Biochemistry*, 1993, **32**, 1078.
25. F. H.-T. Allain and G. Varani, *J. Mol. Biol.*, 1995, **250**, 333.
26. A. A. Szewczak and P. B. Moore, *J. Mol. Biol.*, 1995, **247**, 81.
27. M. A. Fountain, M. J. Serra, T. R. Krugh, and D. H. Turner, *Biochemistry*, 1996, **35**, 6539.
28. S. Huang, Y. X. Wang, and D. E. Draper, *J. Mol. Biol.*, 1996, **258**, 308.
29. J. A. McDowell and D. H. Turner, *Biochemistry*, 1996, **35**, 14077.
30. S. E. Butcher, T. Dieckmann, and J. Feigon, *J. Mol. Biol.*, 1997, **268**, 348.
31. A. Dallas and P. B. Moore, *Structure*, 1997, **5**, 1639.
32. J. A. McDowell, L. Y. He, X. Y. Chen, and D. H. Turner, *Biochemistry*, 1997, **36**, 8030.
33. F. Michel and E. Westhof, *J. Mol. Biol.*, 1990, **216**, 585.
34. C. Zwieb, *J. Biol. Chem.*, 1992, **267**, 15650.
35. B. D. James, G. J. Olsen, and N. R. Pace, *Methods Enzymol.*, 1989, **180**, 227.
36. R. R. Gutell, N. Larsen, and C. R. Woese, *Microbiol. Rev.*, 1994, **58**, 10.
37. D. M. Crothers, P. E. Cole, C. W. Hilbers, and R. G. Shulman, *J. Mol. Biol.*, 1974, **87**, 63.
38. C. W. Hilbers, G. T. Robillard, R. G. Shulman, R. D. Blake, P. K. Webb, R. Fresco, and D. Riesner, *Biochemistry*, 1976, **15**, 1874.
39. A. R. Banerjee, J. A. Jaeger, and D. H. Turner, *Biochemistry*, 1993, **32**, 153.
40. L. Jaeger, E. Westhof, and F. Michel, *J. Mol. Biol.*, 1993, **234**, 331.
41. L. G. Laing and D. E. Draper, *J. Mol. Biol.*, 1994, **237**, 560.
42. D. H. Turner, N. Sugimoto, and S. M. Freier, *Annu. Rev. Biophys. Chem.*, 1988, **17**, 167.
43. D. H. Turner and P. C. Bevilacqua, in "The RNA World," eds. R. F. Gesteland, and J. F. Atkins, Cold Spring Harbor Laboratory Press, Cold Spring Harbor, NY, 1993.
44. D. H. Turner, in "Nucleic Acids: Structures, Properties, and Functions," eds. V. A. Bloomfield, D. M. Crothers, and I. Tinoco, Jr., University Science Books, Mill Valley, CA, 1999.
45. I. Tinoco, Jr., O. C. Uhlenbeck, and M. D. Levine, *Nature*, 1971, **230**, 362.
46. M. Zuker and P. Stiegler, *Nucleic Acids Res.*, 1981, **9**, 133.
47. M. Zuker, *Science*, 1989, **244**, 48.
48. D. H. Mathews, T. C. Andre, J. Kim, D. H. Turner, and M. Zuker, in "Molecular Modeling of Nucleic Acids," ACS Symposium Series 682, eds. N. B. Leontis and J. SantaLucia, Jr., American Chemical Society, Washington, DC, 1998, p. 246.
49. R. Kierzek, M. H. Caruthers, C. E. Longfellow, D. Swinton, D. H. Turner, and S. M. Freier, *Biochemistry*, 1986, **25**, 7840.
50. N. Usman, K. K. Ogilvie, M.-Y. Jiang, and R. J. Cedergren, *J. Am. Chem. Soc.*, 1987, **109**, 7845.
51. J. A. Jaeger, D. H. Turner, and M. Zuker, *Proc. Natl. Acad. Sci. USA*, 1989, **86**, 7706.
52. A. E. Walter, D. H. Turner, J. Kim, M. H. Lyttle, P. Müller, D. H. Mathews, and M. Zuker, *Proc. Natl. Acad. Sci. USA*, 1994, **91**, 9218.
53. V. A. Bloomfield, D. M. Crothers, and I. Tinoco, Jr., "Physical Chemistry of Nucleic Acids," Harper and Row, New York, 1974.
54. C. R. Cantor and P. R. Schimmel, "Biophysical Chemistry," Part III, Freeman, San Francisco, CA, 1980.
55. C. R. Cantor and P. R. Schimmel, "Biophysical Chemistry," Part II, Freeman, San Francisco, CA, 1980.
56. I. Tinoco, Jr., *J. Am. Chem. Soc.*, 1960, **82**, 4785.
57. I. Tinoco, Jr., *J. Chem. Phys.*, 1960, **33**, 1332.
58. I. Tinoco, Jr., *J. Chem. Phys.*, 1961, **34**, 1067.
59. W. Rhodes, *J. Chem. Phys.*, 1961, **83**, 3609.
60. H. DeVoe and I. Tinoco, Jr., *J. Mol. Biol.*, 1962, **4**, 518.
61. D. F. Bradley, I. Tinoco, Jr., and R. W. Woody, *Biopolymers*, 1963, **1**, 239.
62. J. R. Fresco, L. C. Klotz, and E. D. Richards, *Cold Spring Harbor Symp. Quant. Biol.*, 1963, **28**, 83.
63. G. Felsenfeld and S. Z. Hirschman, *J. Mol. Biol.*, 1965, **13**, 407.
64. J. D. Puglisi and I. Tinoco, Jr., *Methods Enzymol.*, 1989, **180**, 304.
65. L. A. Marky and K. J. Breslauer, *Biopolymers*, 1987, **26**, 1601.
66. M. Petersheim and D. H. Turner, *Biochemistry*, 1983, **22**, 256.
67. A. E. Peritz, R. Kierzek, N. Sugimoto, and D. H. Turner, *Biochemistry*, 1991, **30**, 6428.
68. C. E. Longfellow, R. Kierzek, and D. H. Turner, *Biochemistry*, 1990, **29**, 278.
69. E. G. Richards, in "Handbook of Biochemistry and Molecular Biology: Nucleic Acids," ed. G. D. Fasman, 3rd edn., CRC Press, Cleveland, OH, 1975, vol. I, p. 57.
70. P. N. Borer, in "Handbook of Biochemistry and Molecular Biology: Nucleic Acids," ed. G. D. Fasman, 3rd edn., CRC Press, Cleveland, OH, 1975, vol. I, p. 589.
71. P. R. Bevington and D. K. Robinson, "Data Reduction and Error Analysis for the Physical Sciences," 2nd edn., McGraw-Hill, Boston, 1992.
72. P. N. Borer, B. Dengler, I. Tinoco, Jr., and O. C. Uhlenbeck, *J. Mol. Biol.*, 1974, **86**, 843.
73. D. De P. Albergo, L. A. Marky, K. J. Breslauer, and D. H. Turner, *Biochemistry*, 1981, **20**, 1409.
74. S. M. Freier, R. Kierzek, J. A. Jaeger, N. Sugimoto, M. H. Caruthers, T. Neilson, and D. H. Turner, *Proc. Natl. Acad. Sci. USA*, 1986, **83**, 9373.
75. T. Xia, J. SantaLucia, Jr., M. E. Burkard, R. Kierzek, S. J. Schroeder, C. Cox, and D. H. Turner, *Biochemistry*, 1998, **37**, 14719.

76. J. SantaLucia, Jr., R. Kierzek, and D. H. Turner, *J. Am. Chem. Soc.*, 1991, **113**, 4313.
77. T. B. Xia, J. A. McDowell, and D. H. Turner, *Biochemistry*, 1997, **36**, 12 486.
78. H. T. Allawi and J. SantaLucia, Jr., *Biochemistry*, 1997, **36**, 10 581.
79. J. B. Chaires, *Biophys. Chem.*, 1997, **64**, 15.
80. J. M. Sturtevant, *Annu. Rev. Phys. Chem.*, 1987, **38**, 463.
81. P. L. Privalov and S. A. Potekhin, *Methods Enzymol.*, 1986, **131**, 4.
82. K. J. Breslauer, E. Freire, and M. Straume, *Methods Enzymol.*, 1992, **211**, 533.
83. H. F. Fisher and N. Singh, *Methods Enzymol.*, 1995, **259**, 194.
84. J. SantaLucia, Jr. and D. H. Turner, *Biopolymers*, 1997, **44**, 309.
85. S. M. Law, R. Eritja, M. F. Goodman, and K. J. Breslauer, *Biochemistry*, 1996, **35**, 12 329.
86. S. M. Freier, K. O. Hill, T. G. Dewey, L. A. Marky, K. J. Breslauer, and D. H. Turner, *Biochemistry*, 1981, **20**, 1419.
87. D. E. Draper and T. C. Gluick, *Methods Enzymol.*, 1995, **259**, 281.
88. J. Gralla and D. M. Crothers, *J. Mol. Biol.*, 1973, **78**, 301.
89. A. Wada, S. Yubuki, and Y. Husimi, *Crit. Rev. Biochem.*, 1980, **9**, 87.
90. O. Gotoh, *Adv. Biophys.*, 1983, **16**, 1.
91. R. M. Wartell and A. S. Benight, *Phys. Rep.*, 1985, **126**, 67.
92. M. Schmitz and G. Steger, *CABIOS*, 1992, **8**, 389.
93. D. Poland and H. Scheraga, "Theory of Helix–Coil Transitions in Biopolymers," Academic Press, New York, 1970.
94. D. Poland, "Cooperative Equilibria in Physical Biochemistry, " Clarendon Press, Oxford, 1978.
95. G. Steger, *Nucleic Acids Res.*, 1994, **22**, 2760.
96. D. Pörschke and M. Eigen, *J. Mol. Biol.*, 1971, **62**, 361.
97. T. C. Gluick and D. E. Draper, *J. Mol. Biol.*, 1994, **241**, 246.
98. B. Lewin, "Genes," Oxford University Press, Oxford, 1997.
99. R. F. Gesteland and J. F. Atkins (eds.), "The RNA World: The Nature of Modern RNA Suggests a Prebiotic RNA World, " Cold Spring Harbor Laboratory Press, New York, 1993.
100. I. Tinoco, Jr., P. N. Borer, B. Dengler, M. D. Levine, O. C. Uhlenbeck, D. M Crothers, and J. Gralla, *Nature New Biol.*, 1973, **246**, 40.
101. J. Gralla and D. M. Crothers, *J. Mol. Biol.*, 1973, **73**, 497.
102. D. Gray, *Biopolymers*, 1997, **42**, 783.
103. W. F. Bailey and A. S. Monahan, *J. Chem. Ed.*, 1978, **55**, 489.
104. N. Sugimoto, K. Honda, and M. Sasaki, *Nucleosides Nucleotides*, 1994, **13**, 1311.
105. N. Sugimoto, S. Nakano, M. Katoh, A. Matsumura, H. Nakamuta, T. Ohmichi, M. Yoneyama, and M. Sasaki, *Biochemistry*, 1995, **34**, 11 211.
106. P. H. van Knippenberg, L. J. Formenoy, and H. A. Heus, *Biochem. Biophys. Acta*, 1990, **1050**, 14.
107. D. Gautheret, D. Konings, and R. R. Gutell, *RNA*, 1995, **1**, 807.
108. S. A. Strobel and T. R. Cech, *Science*, 1995, **267**, 675.
109. O. C. Uhlenbeck, F. H. Martin, and P. Doty, *J. Mol. Biol.*, 1971, **57**, 217.
110. S. M. Freier, R. Kierzek, M. H. Caruthers, T. Neilson, and D. H. Turner, *Biochemistry*, 1986, **25**, 3209.
111. N. Sugimoto, R. Kierzek, S. M. Freier, and D. H. Turner, *Biochemistry*, 1986, **25**, 5755.
112. L. He, R. Kierzek, J. SantaLucia, Jr., A. E. Walter, and D. H. Turner, *Biochemistry*, 1991, **30**, 11 124.
113. D. H. Mathews, J. Sabina, M. Zuker, and D. H. Turner, submitted.
114. S. M. Freier, B. J. Burger, D. Alkema, T. Neilson, and D. H. Turner, *Biochemistry*, 1983, **22**, 6198.
115. D. R. Hickey and D. H. Turner, *Biochemistry*, 1985, **24**, 3987.
116. S. M. Freier, D. Alkema, A. Sinclair, T. Neilson, and D. H. Turner, *Biochemistry*, 1985, **24**, 4533.
117. N. Sugimoto, R. Kierzek, and D. H. Turner, *Biochemistry*, 1987, **26**, 4554.
118. N. Sugimoto, R. Kierzek, and D. H. Turner, *Biochemistry*, 1987, **26**, 4559.
119. M. R. Giese, K. Betschart, T. Dale, C. K. Riley, C. Rowan, K. J. Sprouse, and M. J. Serra, *Biochemistry*, 1998, **37**, 1094.
120. M. E. Burkard, D. H. Turner, and I. Tinoco, Jr., in "The RNA World II," eds. T. R. Cech, R. F. Gesteland, and J. F. Atkins, Cold Spring Harbor Laboratory Press, New York, 1998, p. 233.
121. V. Lehnert, L. Jaeger, F. Michel, and E. Westhof, *Chem. Biol.*, 1996, **3**, 993.
122. F. M. Jucker and A. Pardi, *Biochemistry*, 1995, **34**, 14 416.
123. C. R. Woese, S. Winker, and R. R. Gutell, *Proc. Natl. Acad. Sci. USA*, 1990, **87**, 8467.
124. M. J. Serrra, T. J. Axenson, and D. H. Turner, *Biochemistry*, 1994, **33**, 14 289.
125. M. J. Serra, T. W. Barnes, K. Betschart, M. J. Gutierrez, K. J. Sprouse, C. K. Riley, L. Stewart, and R. E. Temel, *Biochemistry*, 1997, **36**, 4844.
126. V. P. Antao and I. Tinoco, Jr., *Nucleic Acids Res.*, 1992, **20**, 819.
127. V. P. Antao, S. Y. Lai, and I. Tinoco, Jr., *Nucleic Acids Res.*, 1991, **19**, 5901.
128. J. SantaLucia, Jr., R. Kierzek, and D. H. Turner, *Science*, 1992, **256**, 217.
129. C. Tuerk, P. Gauss, C. Thermes, D. R. Groebe, M. Gayle, N. Guild, G. Stormo, Y. D'Aubenton-Carafa, O. C. Uhlenbeck, I. Tinoco, Jr., E. N. Brody, and L. Gold, *Proc. Natl. Acad. Sci. USA*, 1988, **85**, 1364.
130. M. Orita, F. Nishikawa, T. Shimayama, K. Taira, Y. Endo, and S. Nishikawa, *Nucleic Acids Res.*, 1993, **21**, 5670.
131. L. Mueller, P. Legault, and A. Pardi, *J. Am. Chem. Soc.*, 1995, **117**, 11 043.
132. G. Varani, *Annu. Rev. Biophys. Biomol. Struct.*, 1995, **24**, 379.
133. F. M. Jucker, H. A. Heus, P. F. Yip, E. H. M. Moors, and A. Pardi, *J. Mol. Biol.*, 1996, **264**, 968.
134. M. Costa and F. Michel, *EMBO J.*, 1997, **16**, 3289.
135. Y. Endo, Y. L. Chan, A. Lin, K. Tsurugi, and I. G. Wool, *J. Biol. Chem.*, 1988, **263**, 7917.
136. A. Glück, Y. Endo, and I. G. Wool, *J. Mol. Biol.*, 1992, **226**, 411.
137. D. R. Groebe and O. C. Uhlenbeck, *Nucleic Acids Res.*, 1988, **16**, 11 725.
138. P. N. Borer, Y. Lin, S. Wang, M. W. Roggenbuck, J. M. Gott, O. C. Uhlenbeck, and I. Pelczer, *Biochemistry*, 1995, **34**, 6488.
139. Y. T. van der Hoogen, A. A. van Beuzekom, E. de Vroom, G. A. van der Marel, J. H. van Boom, and C. Altona, *Nucleic Acids Res.*, 1988, **16**, 5013.

140. R. S. Tang and D. E. Draper, *Nucleic Acids Res.*, 1994, **22**, 835.
141. D. A. Peattie, S. Southwaite, R. A. Garett, and H. F. Noller, *Proc. Natl. Acad. Sci. USA*, 1981, **78**, 7331.
142. P. J. Romaniuk, P. Lowary, H-N. Wu, G. Stormo, and O. C. Uhlenbeck, *Biochemistry*, 1987, **26**, 1563.
143. P. J. Flor, J. B. Flanegan, and T. R. Cech, *EMBO J.*, 1989, **8**, 3391.
144. K. M. Weeks and D. M. Crothers, *Science*, 1993, **261**, 1574.
145. D. R. Groebe and O. C. Uhlenbeck, *Biochemistry*, 1989, **28**, 742.
146. T. R. Fink and D. M. Crothers, *J. Mol. Biol.*, 1972, **66**, 1.
147. S. E. Morse and D. E. Draper, *Nucleic Acids Res.*, 1995, **23**, 302.
148. R. Kierzek, M. E. Burkard, and H. Turner, unpublished results.
149. J. SantaLucia, Jr., R. Kierzek, and D. H. Turner, *Biochemistry*, 1991, **30**, 8242.
150. A. E. Walter, M. Wu, and D. H. Turner, *Biochemistry*, 1994, **33**, 11 349.
151. M. Wu, J. A. McDowell, and D. H. Turner, *Biochemistry*, 1995, **34**, 3204.
152. M. Wu and D. H. Turner, *Biochemistry*, 1996, **35**, 9677.
153. M. Wu, J. SantaLucia, Jr., and D. H. Turner, *Biochemistry*, 1997, **36**, 4449.
154. S. J. Schroeder, J. Kim, and D. H. Turner, *Biochemistry*, 1996, **35**, 16 105.
155. C. Papanicolaou, M. Gouy, and J. Ninio, *Nucleic Acids Res.*, 1984, **13**, 1717.
156. E. Westhof, P. Dumas, and D. Moras, *J. Mol. Biol.*, 1985, **184**, 119.
157. V. Biou, A. Yaremchuk, M. Tukalo, and S. Cusack, *Science*, 1994, **263**, 1404.
158. A. E. Walter and D. H. Turner, *Biochemistry*, 1994, **33**, 12 715.
159. J. Kim, A. E. Walter, and D. H. Turner, *Biochemistry*, 1996, **35**, 13 753.
160. C. Wilson and J. W. Szostak, *Nature*, 1995, **374**, 777.
161. E. H. Ekland, J. W. Szostak, and D. P. Bartel, *Science*, 1995, **269**, 364.
162. G. Manning, *Q. Rev. Biophys.*, 1978, **11**, 179.
163. K. Hamaguchi and E. P. Geiduschek, *J. Am. Chem. Soc.*, 1962, **84**, 1329.
164. J. SantaLucia, Jr., *Proc. Natl. Acad. Sci. USA*, 1998, **95**, 1460.
165. M. T. Record, Jr. and T. M. Lohman, *Biopolymers*, 1978, **17**, 159.
166. M. C. Olmsted, C. F. Anderson, and M. T. Record, Jr., *Proc. Natl. Acad. Sci. USA*, 1989, **86**, 7766.
167. M. T. Record, Jr., C. F. Anderson, and T. M. Lohman, *Q. Rev. Biophys.*, 1978, **2**, 103.
168. L. G. Laing, T. C. Gluick, and D. E. Draper, *J. Mol. Biol.*, 1994, **237**, 577.
169. A. P. Williams, C. E. Longfellow, S. M. Freier, R. Kierzek, and D. H. Turner, *Biochemistry*, 1989, **28**, 4283.
170. D. D. Albergo and D. H. Turner, *Biochemistry*, 1981, **20**, 1413.
171. D. R. Hickey and D. H. Turner, *Biochemistry*, 1985, **24**, 2086.
172. N. Sugimoto, R. Kierzek, and D. H. Turner, *Biochemistry*, 1988, **27**, 6384.
173. P. C. Bevilacqua and D. H. Turner, *Biochemistry*, 1991, **30**, 10 632.
174. N. Sugimoto, M. Tomka, R. Kierzek, P. C. Bevilacqua, and D. H. Turner, *Nucleic Acids Res.*, 1989, **17**, 355.
175. A. M. Pyle and T. R. Cech, *Nature*, 1991, **350**, 628.
176. S. A. Strobel and T. R. Cech, *Biochemistry*, 1993, **32**, 13 593.
177. S. M. Testa, C. G. Haidaris, F. Gigliotti, and D. H. Turner, *Biochemistry*, 1997, **36**, 15 303.
178. S. M. Testa, S. M. Gryaznov, and D. H. Turner, *Biochemistry*, 1998, **37**, 9379.
179. D. H. Mathews, A. R. Banerjee, D. D. Luan, T. H. Eickbush, and D. H. Turner, *RNA*, 1997, **3**, 1.
180. A. P. Gultyaev, F. H. G. van Batenburg, and C. W. A. Pleij, *J. Mol. Biol.*, 1995, **250**, 37.
181. F. H. D. van Batenburg, A. P. Gultyaev, and C. W. A. Pleij, *J. Theor. Biol.*, 1995, **174**, 269.
182. J. A. Jaeger, D. H. Turner, and M. Zuker, *Methods Enzymol.*, 1990, **183**, 281.
183. O. C. Uhlenbeck, *Nature*, 1987, **328**, 596.
184. J. M. Burke, *Nucleic Acids. Mol. Biol.*, 1994, **8**, 105.
185. J. J. Rossi, E. M. Cantin, N. Sarver, and P. F. Chang, *Pharmacol. Ther.*, 1991, **50**, 245.
186. D. Castanotto, H. T. Li, W. Chow, J. J. Rossi, and J. O. Deshler, *Antisense Nucleic Acid Drug Dev.*, 1998, **8**, 1.
187. B. A. Sullenger and T. R. Cech, *Nature*, 1994, **371**, 619.
188. N. Lan, R. P. Howrey, S.-W. Lee, C. A. Smith, and B. A. Sullenger, *Science*, 1998, **280**, 1593.
189. N. Usman and J. A. McSwiggen, *Annu. Rep. Med. Chem.*, 1992, **30**, 285.
190. O. C. Uhlenbeck, *RNA*, 1995, **1**, 4.
191. D. Herschlag, *Proc. Natl. Acad. Sci. USA*, 1991, **88**, 6921.
192. R. W. Roberts and D. M. Crothers, *Proc. Natl. Acad. Sci. USA*, 1991, **88**, 9397.
193. T. M. Woolf, D. A. Melton, and C. G. B. Jennings, *Proc. Natl. Acad. Sci. USA*, 1992, **89**, 7305.
194. K. J. Hertel, D. Herschlag, and O. C. Uhlenbeck, *EMBO J.*, 1996, **15**, 3751.
195. K. A. LeCuyer and D. M. Crothers, *Proc. Natl. Acad. Sci. USA*, 1994, **91**, 3373.
196. W. Bains and G. C. Smith, *J. Theor. Biol.*, 1988, **135**, 303.
197. S. P. A. Fodor, J. L. Read, M. C. Pirrung, L. Stryer, A. T. Lu, and D. Solas, *Science*, 1991, **270**, 467.
198. U. Maskos and E. M. Southern, *Nucleic Acids Res.*, 1992, **20**, 1679.
199. M. J. O'Donnell-Maloney, C. L. Smith, and C. R. Cantor, *TIBTECH*, 1996, **14**, 401.
200. M. Schena, R. A. Heller, T. P. Theriault, K. Konrad, E. Lachenmeier, and R. W. Davis, *TIBTECH*, 1998, **16**, 301.
201. A. V. Fotin, A. L. Drobyshev, D. Y. Proudnikov, A. N. Perov, and A. D. Mirzabekov, *Nucleic Acids Res.*, 1998, **26**, 1515.
202. J. M. Bevilacqua and P. C. Bevilacqua, *Biochemistry*, 1998, **37**, 15 877.
203. M. J. Serra and D. H. Turner, *Methods Enzymol.*, 1995, **259**, 242.
204. H. Jacobson and W. H. Stockmayer, *J. Chem. Phys.*, 1950, **18**, 1600.

6.04
RNA Structures Determined by X-ray Crystallography

JENNIFER A. DOUDNA

Yale University, New Haven, CT, USA

and

JAMIE H. CATE

University of California, Santa Cruz, CA, USA

6.04.1 INTRODUCTION

Many RNA molecules have complex three-dimensional structures under physiological conditions, and the chemical basis for their functional properties cannot be understood unless these structures are known. In recent years it has become practical to determine RNA structures by X-ray crystallography, which can provide high-resolution information not only about RNA conformation but also about RNA interactions with ligands such as metal ions. The study of RNA by X-ray crystallography has become technically feasible due to the development of methods for producing milligram quantities of virtually any RNA molecule, and for crystallizing RNA and producing heavy atom derivatives of RNA crystals. This chapter discusses these methods, and reviews the RNA crystal structures that are currently known.

6.04.2   CRYSTALLIZATION OF RNA

6.04.2.1   Production of Homogeneous RNA by *In Vitro* Transcription

In vitro transcription with bacteriophage T7 RNA polymerase is the method of choice for obtaining milligram quantities of RNA for crystallization. Its only drawback is that transcription with this enzyme results in molecules that are heterogeneous at their 3′-termini and, depending on template sequence, may also be heterogeneous at their 5′-termini.[1,2] Transcripts containing these extra residues cannot be removed from RNA preparations by preparative purification techniques when chain lengths exceed ~50 nucleotides.

Terminal heterogeneity can be removed from transcripts using ribozymes in *cis* and *trans* geometries. When included as part of an RNA transcript, hammerhead, hairpin, and hepatitis delta virus ribozyme sequences will self-cleave during or after transcription to produce RNA with defined termini. This method has been used by several groups to obtain RNA samples suitable for structure determination.[3,4]

6.04.2.2   Chemical Synthesis

Short oligoribonucleotides are often conveniently prepared using automated, solid-phase DNA synthesis machines. Chemically protected ribonucleoside phosphoramidites are sequentially coupled to a protected nucleoside attached at its 3′ end to a solid support such as controlled-pore glass or polystyrene. When synthesis of the sequence is complete, base hydrolysis is used to cleave its linkage to the solid support, releasing a 2′-*O*-silyl protected oligomer. Silyl protecting groups are removed using tetrabutylammonium fluoride (TBAF) or similar chemical reagents. Improvements in the 2′-OH protecting groups and deprotection methods, as well as development of effective oligonucleotide purification methods, have made the chemical synthesis of RNA oligonucleotides of up to 40–50 nucleotides routine (for more information, see Chapter 6.06). With extreme care, RNA oligomers of up to 80 nucleotides in length can be produced in milligram quantities by solid-phase synthesis.[5,6] In practice, however, chemical synthesis of RNA for crystallization is practical for oligonucleotides 30 nucleotides or less in length.

One advantage of this approach is that nucleotide analogues are readily incorporated into synthetic oligoribonucleotides at specific sites. This is useful for the production of heavy atom derivatives (see below) and for investigating ligand–RNA interactions. Enzymatic ligation of short synthetic RNAs and longer RNA molecules prepared by *in vitro* transcription can be used to produce chimeric molecules that contain modified bases at specific locations[7] (see Chapter 6.14).

6.04.2.3   Purification of RNA for Crystallization

Prior to crystallization experiments, contaminating salts and chemical reagents must be removed from RNA samples. This is usually accomplished using ion exchange or reverse phase chromatography, for RNA molecules up to ~40 nucleotides long, and by denaturing polyacrylamide gel electrophoresis for larger RNAs. Following purification, the RNA is dialyzed extensively into a low-salt buffer, and often it is then annealed by heating to 60–90 °C and slow cooling in the presence of 1–10 mM magnesium ion.

6.04.2.4   Establishing the Suitability of RNA Preparations for Crystallization

Whenever possible, purified RNA samples are tested for biological activity prior to crystallization. In the case of a tRNA, this might involve assaying for charging by tRNA synthetase, or, in the case of ribozymes, measuring catalytic activity. Once the activity of a sample is confirmed, it is tested for conformational homogeneity (polydispersity) using native polyacrylamide gel electrophoresis, size exclusion chromatography, or dynamic light scattering.[8] The first two techniques can evaluate polydispersity only under low ionic strength conditions, while light scattering allows the determination of conformational homogeneity of RNA in solutions containing a variety of electrolytes and additives.

6.04.2.5 Sparse Matrix Approaches to RNA Crystallization

The crystallization of macromolecules is a trial and error process, and it is usually necessary to screen a wide range of conditions to find any that are conducive to crystal nucleation and growth. In the case of RNA, additional factors may complicate crystallization, such as the source and purity of material and the inherent instability of RNA. Furthermore, since some RNA molecules adopt several different conformations in solution, conditions that favor a single conformer must be found and used for crystallization.

To facilitate the search for crystallization conditions, sets of precipitating solutions have been developed that are biased towards conditions that have generated RNA crystals in the past.[9–11] These sets are applied to RNA using approaches based on the incomplete factorial and sparse matrix methods developed for protein crystallization. Satisfactory crystals of RNA duplexes, the hammerhead ribozyme, the P4–P6 domain of the *Tetrahymena* group I intron, and a fragment of 5S ribosomal RNA have all been obtained this way.

6.04.3 HEAVY ATOM DERIVATIVES OF RNA CRYSTALS

Once satisfactory crystals of a macromolecule are obtained, the phases for structure factors must be determined so that an electron density map can be calculated. For new structures this is usually achieved by making heavy atom derivatives of crystals, measuring diffraction intensities, and calculating phases based on the positions of the heavy atom(s).[12,13] Heavy atom derivatives of tRNA crystals were produced by soaking lanthanides into crystals, or by reacting crystals with osmium pyridine.[14] For the hammerhead ribozyme, crystal derivatives were prepared by lanthanide soaks and by covalent modification of the RNA with bromine. Covalent modification with bromine or iodine has also been used to solve the structures of short RNA duplexes and the loop E fragment of 5S ribosomal RNA. The crystal structure of the P4–P6 domain of the *Tetrahymena* ribozyme was solved by osmium hexammine substitution of magnesium binding sites in the major groove of the RNA.

6.04.4 DUPLEX STRUCTURES

The A-form helix is the structural unit from which complex, three-dimensional RNA structures are built. Isolated RNA helices often crystallize readily, and their structures can be solved using molecular replacement or covalent modification of the RNA. The high resolution (>2 Å) of some of the structures that have resulted has allowed a detailed look at metal ion binding sites, non-Watson–Crick base pairings, base bulging, helix packing in crystal lattices, and hydration.

6.04.4.1 Metal Ion Interactions in RNA Duplexes

Most structured RNAs require divalent metal ions for folding, and ribozymes generally need them for catalysis.[15] Several divalent ions have been located in the hammerhead ribozyme crystal structures (see below), but their functional significance remains unclear. One divalent metal ion binding site seen in hammerhead structures has also been found in a duplex containing sheared G·A and asymmetric A·A base pairs.[16] The site occurs at a C·G pair followed by the sheared G·A pair. Interestingly, tandem G·A mismatches have been found near the active sites of a lead-dependent ribozyme and an RNA ligase ribozyme, and they occur frequently in ribosomal RNA.[17–19] Thus, this motif may turn out to be a common way to position divalent metal ions within an RNA structure.

6.04.4.2 Noncanonical Base Pairs

Noncanonical or mismatch base pairs are common in RNA, and internal loops in rRNA often contain a high proportion of adenosines.[20] How are these nucleotides arranged, and how do they alter helical geometry? One example has been seen in a symmetric duplex, which includes a 5′-GAAA-3′ bulge surrounded by Watson–Crick pairs.[16] In this structure, tandem asymmetric A · A base pairs are sandwiched between sheared G · A pairs. Another common motif in rRNA involves tandem U · U pairs.[20] The three known duplex structures that contain this motif demonstrate that its structure varies depending on flanking sequences.[21–23] Two of these duplexes contain U · U wobble pairs,[21,23] but interestingly, the number of hydrogen bonds between the U · U pairs depends on the flanking base pairs. This result is consistent with effects seen in thermodynamic studies in solution,[24] but crystal packing forces may also affect the base pair geometry. In the third example, the U · U tandems form at the end of a duplex in an intermolecular contact.[22] These U · U tandems form unusual Hoogsteen pairs in which the N-3—H and O-4 of one uridine hydrogen bond to the O-4 and C-5—H of the other. It still is not clear whether tandem Hoogsteen U · U pairs like this can form in the middle of a duplex region, but they certainly might occur at the end of a helix. More importantly, the structure provides clear examples (at 1.4 Å resolution) of CH—O hydrogen bonds in base pairs and provides a model for U · Ψ base pairs in RNA.

6.04.4.3 RNA Packing and Hydration

RNA packing and hydration play important roles in RNA function, as highlighted in experiments involving large entropic contributions to ΔG.[25,26] While deceptively simple in form, the A-form helix can be greatly distorted, as seen in a structure of an RN–DNA chimeric duplex with a single looped-out adenosine.[27] In addition, the conformation of the extruded adenosine sheds some light on why the backbone of looped-out bases is often susceptible to magnesium-induced hydrolysis. Two high-resolution structures reveal in detail the pattern of hydration of G–C base pairs,[22,28,29] while a lower resolution structure sheds new light on the hydration of A–U pairs.[30,31] In these structures, the backbone plays key roles in the observed hydration patterns: the 2′-OH and the *pro*-R_p phosphate oxygen. As 2′-OH groups play important roles in RNA packing, exemplified by the ribose zipper (see Section 6.04.7), the heavy involvement of the 2′-OH group in hydration is a major factor to consider in thermodynamic studies of RNA–RNA interactions.

6.04.5 TRANSFER RNA

The first RNA molecule to be solved by X-ray crystallography that is large enough to have a tertiary structure was transfer RNA (tRNA). They were first because tRNAs are quite small and so abundant that they are readily purified from cells in adequate quantities. Modern methods for RNA production were not required. The crystal structure of tRNA[Phe], which was determined independently by three groups, became the basis for much of the RNA structural and functional biology that was done for the next 20 years.

tRNA plays a crucial role in protein biosynthesis. It is an adaptor molecule, one end of which interacts with amino acids and the other of which interacts with messenger RNA. Unlike normal double-stranded DNA, tRNA contains short helical elements interspersed with loops and its secondary structure is often drawn as a "cloverleaf" (Figure 1(a)). On the acceptor end, it carries an amino acid that corresponds to the genetic code triplet in its anticodon loop. The anticodon loop forms base pairs with messenger RNA on the ribosome, which then catalyzes peptide bond formation between the amino acid covalently bonded to one tRNA and the growing peptide chain covalently attached to a second one. Our understanding of tRNA structure and function has been reviewed in far more detail than is appropriate here;[32] however, some experiments regarding nucleotide modifications in tRNA deserve mention.

In the crystal structures of tRNA[Phe], the anticodon loop is ~70 Å away from the acceptor end of the molecule where the amino acid is attached. The four helical stems in that tRNA form an L-shaped molecule, each arm of which consists of a stack of two helices. A network of tertiary interactions between the D and T loops stabilize the assembly (Figure 1(b)),[33,34] a pattern that is conserved in other tRNA crystal structures.[35,36] While some tRNAs fold properly in the absence of divalent ions, Mg^{2+} stabilizes the tertiary structure of all of them, and the binding sites of some of the Mg^{2+} ions involved have been inferred from tRNA crystal structures.[37,38]

Although tRNAs synthesized *in vitro* from the four naturally occurring nucleotides are active,

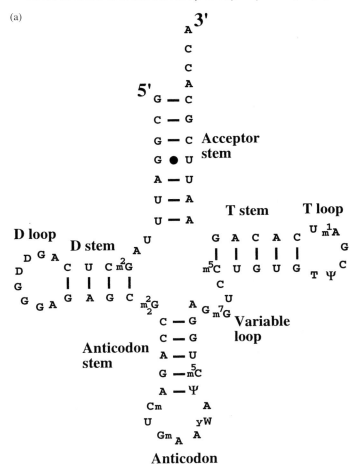

Anticodon

Figure 1 Structure of yeast phenylalanyl tRNA. (a) The secondary structure shows numerous naturally occurring modified nucleosides within the conserved tRNA fold. (b) Ribbon diagram of the crystal structure. Two sets of stacked helices form the L-shaped structure: the T-stem and the acceptor stem are in purple, and the D-stem and anticodon stem are in green.

tRNAs purified from cells always contain at least a few modified nucleotides. Modifications have been found in all of the stems and loops of tRNA.[32] One class of modifications, in the D and T loops, for example, optimizes the tertiary folding of tRNA.[33,34] Other modifications in the acceptor arm (acceptor stem and T stem) play a role in helping the protein-synthesizing machinery distinguish between initiator tRNAs and elongator tRNAs.[36] In the anticodon, modifications often play key roles in specifying codon recognition.[39]

Two advances in the chemical analysis of tRNAs have accelerated our understanding of these modifications. First, mass spectroscopic analysis of nucleotides in natural tRNAs has greatly expanded our knowledge of the kinds of modifications that occur in tRNA. In addition, the extent of modification in different types of organisms can be quickly assessed in the same way. For example, tRNAs from bacteria that grow in cold habitats have a higher abundance of dihydrouridine, which may increase conformational flexibility,[40] while thermophiles have modifications that may stabilize conformation.[41] Second, the chemical synthesis of tRNA-length RNAs may allow for milligram quantities of tRNA to be made that contain single-site modifications.[6] For example, both intra- and interhelical disulfide cross-links have already been incorporated into tRNA for the purpose of biophysical studies.[42] Combined with our increased knowledge of how tRNA interacts with amino-acyl tRNA synthetases and elongation factors, the ability to make designed modifications in tRNA, natural or otherwise, opens many new areas for biochemical and biophysical study.

6.04.6 THE HAMMERHEAD RIBOZYME

The hammerhead motif is a self-cleaving RNA sequence found in small RNAs that are plant pathogens. They make it possible for the multimeric genomes produced by rolling circle replication

(b)

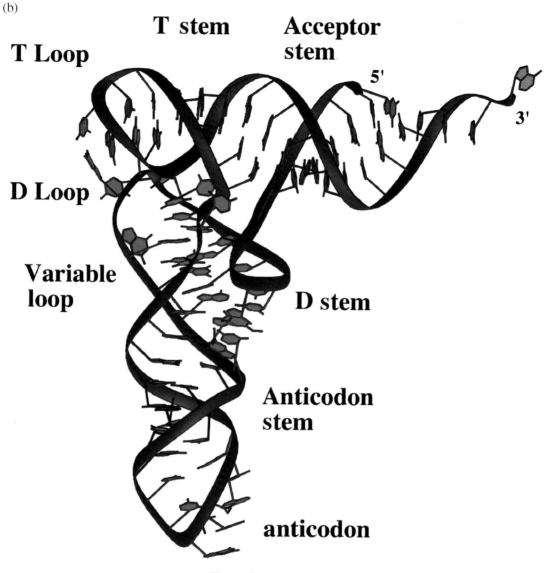

Figure 1　(continued)

to cleave into unit length molecules.[43] Unlike ribozymes such as self-splicing introns and the catalytic RNA subunit of ribonuclease P, the hammerhead domain is small, consisting of three helices that adjoin a core of phylogenetically conserved nucleotides (Figure 2). Cleavage occurs via nucleophilic attack of the 2′-hydroxyl of a specific nucleotide within the core on its adjacent phosphodiester bond to produce a 2′,3′-cyclic phosphate and a 5′-hydroxyl terminus.[44] Normally a single-turnover catalyst, the hammerhead is readily made into a multiple-turnover enzyme by separating the strand containing the cleavage site from the rest of the core.[44,45] Divided molecules like this have proven useful for crystallization because they allow replacement of the substrate strand with an all-DNA strand, or with an RNA strand modified at the cleavage site by a 2′-O-methyl group, neither of which can be cleaved. The crystal structures of these hammerhead–inhibitor complexes have revealed the overall geometry of the ribozyme but have raised almost as many new questions concerning the catalytic mechanism as they have answered.

In three dimensions, the hammerhead is shaped like a wishbone or γ, with stems I and II forming the arms, and stem III and the core forming the base.[46,47] This fold is seen in both of the inhibitor complexes solved so far despite differences in RNA backbone connectivities, substrate strand identities, crystallization conditions, and crystal packing. Whereas the three stems are all A-form helices, the structure of the central core is created, in part, by noncanonical pairings of the phylo-

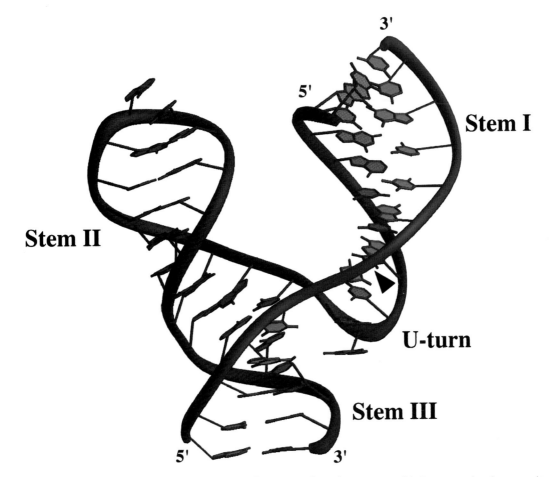

Figure 2 Crystal structure of the hammerhead ribozyme. The substrate strand is in green; the cleavage site is indicated by a black arrow within the U-turn.

genetically conserved nucleotides. Stems II and III sandwich two sheared G·A base pairs and an A–U base pair to form one long pseudocontinuous helix from which stem I and the catalytic site emanate. The highly conserved sequence CUGA between stems I and II forms a tight turn nearly identical in conformation to the uridine turn previously seen in the X-ray crystal structure of yeast phenylalanine transfer RNA.[34,48] The cytosine at the cleavage site between stems I and III is positioned near the CUGA cleft by interactions with the C and A of that sequence. This proximity led Klug and co-workers[47] to propose that the uridine turn, called domain I by McKay's group,[46] constitutes the catalytic pocket of the ribozyme.

Since the hammerhead-inhibitor complexes do not position the scissile bond correctly for the in-line nucleophilic attack that is believed to be part of the catalytic mechanism, these crystal structures probably represent the ground state of the ribozyme.[46,47] This has led to speculation that the hammerhead ribozyme may have to undergo a conformational change in order for cleavage to occur. Eckstein and co-workers[49] have used fluorescence resonance energy transfer (FRET) data to build a three-dimensional model of the hammerhead ribozyme that is similar to the X-ray models, except that the helical groove of stem I facing stem II differs. To distinguish between the solution and X-ray models, an elegant set of disulfide cross-linking experiments was carried out.[50] When stems I and II are cross-linked in conformations that exclude either the FRET or X-ray models, only the ribozyme cross-linked in a manner consistent with the X-ray structures is active. In addition, gel electrophoresis and transient electric birefringence have shown that the three stems are roughly co-planar and do not rearrange significantly after cleavage.[51,52] On the basis of these data, it is unlikely that the cleavage reaction requires a large change in conformation of the ribozyme.

More recent crystal structures of a hammerhead ribozyme complexed with a cleavable substrate have provided new insight into the rearrangements that occur in the catalytic pocket. Unlike the previous hammerhead ribozymes, the construct examined is active in the crystal lattice, allowing the experimenters to "trap" an intermediate in the reaction pathway.[53,54] The major difference between this structure and its two predecessors is a repositioning of the substrate nucleotides in the catalytic pocket. In the trapped structures, a divalent metal ion is bound to the *pro*-Rp oxygen of the phosphate involved in the cleavage reaction, as previously proposed. The scissile bond, however, while still not positioned for in-line attack, is rotated closer to the required orientation.

Although Scott *et al.* propose a new model for the transition state based on this structure,[54] some key questions remain unanswered. First, what are the actual positions of the metal ions involved in catalysis? Second, what is the role of G5 in the CUGA U-turn in the catalytic pocket? Biochemical studies have clearly shown that all of the Watson–Crick base functionalities of G5 and its 2'-hydroxyl are critical for catalysis, yet none of the crystal structures reveals a clear role for this nucleotide. In later structures,[55] weak density interpreted as a divalent metal ion appears next to this guanosine, which is consistent with uranium-induced cleavage at that site, but no function for this metal ion has been shown.

Finally, the tandem sheared G · A base pairs seen in the crystal structures seem to be incompatible with the available biochemical data. Functional group modification studies have suggested that the G · A base pairs could not be in the sheared conformation.[56] Besides the clear need to distinguish between ground state and transition state structure stabilities, there are other factors that should be considered when attempting to relate biochemical data to crystal structures. For example, the thermodynamic stability and even the base pairing conformations of tandem G · A pairs are affected dramatically by their context.[57–59] In addition, the chemically modified RNAs used to "trap" the ribozyme in mid-reaction may have many alternate conformations that are not easily detected in the biochemical experiments.[60]

6.04.7 THE P4–P6 DOMAIN OF THE *TETRAHYMENA* GROUP I SELF-SPLICING INTRON

Group I introns, which are defined by a conserved catalytic core and reaction pathway, splice precursor RNAs so that mature ribosomal, transfer, or messenger RNAs can be formed.[61] Half of the conserved core in the *Tetrahymena thermophila* intron is found in an independently folding domain consisting of the base-paired (P) regions P4 through P6 (P4–P6).[62] By itself, the P4–P6 domain folds into a structure whose chemical protection pattern is very similar to that seen for the P4–P6 region of the intact intron.[62,63] The crystal structure of this domain, a 160-nucleotide RNA, has revealed several new aspects of RNA secondary and tertiary folding, and provides the first example of a kind of helical packing that is thought to occur in large ribozymes and RNA–protein complexes.[64]

In the 2.8 Å crystal structure of the P4–P6 domain, a sharp bend allows stacked helices of the conserved core to pack alongside helices of an extension (helices P5a, P5b, and P5c, or P5abc) that is important for folding and catalytic efficiency (Figure 3).[64,65] Two specific sets of tertiary interactions clamp the two halves of the domain together: an adenosine-rich corkscrew plugs into the minor groove of helix P4, and a GAAA tetraloop binds to a conserved 11-nucleotide internal loop,[66] termed the tetraloop receptor. The A-rich bulge coordinates two magnesium ions via its phosphate oxygens, allowing the backbone to invert and the bases to flip out. The adenosines make numerous tertiary contacts that connect the core helices to the helices in the P5abc extension. From biochemical evidence, these interactions are crucial to the stability of the entire domain.[62–64] The other half of the clamp, equally important to the packing of helices P5abc against the core (although not to the folding of the P5abc region itself), involves a GAAA tetraloop in the same conformation as seen previously.[67,68] The tetraloop receptor, a motif seen in many RNAs, has a widened minor groove that enables it to dock with the tetraloop in a highly specific manner.

The ribose 2'-hydroxyl group is involved in a common motif that occurs in both clamp interactions between the helical stacks. Pairs of riboses form an interhelical "ribose zipper"—a major component of the packing interactions (Figure 4). McKay and co-workers[68] also observed packing that involves pairs of 2'-hydroxyl contacts between a GAAA loop and the stem II minor groove of another hammerhead molecule in the crystal lattice. In a group II intron, riboses likely to be involved in a ribose zipper each contribute 2 kcal mol^{-1} of binding energy via their 2'-OH groups.[69] The number of ribose zippers seen so far suggests that this is a common way to pack RNA helices together.

One unexpected motif seen in the P4–P6 domain structure mediates both intramolecular and intermolecular interactions. At three separate locations in the 160-nucleotide domain, adjacent adenosines in the sequence lie side by side and form a pseudo-base pair within a helix (Figure 5).[70]

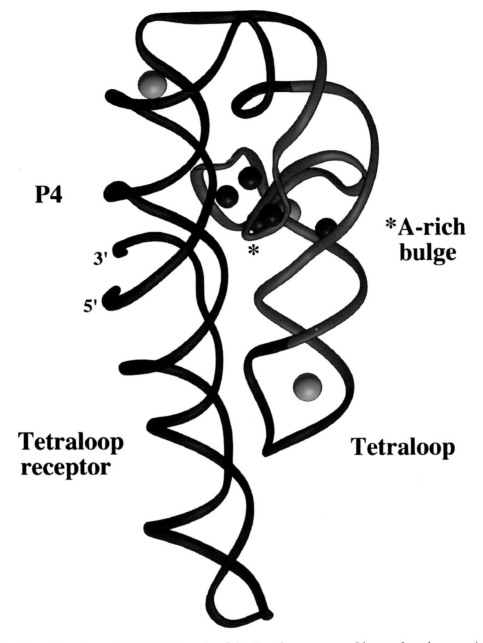

Figure 3 Crystal structure of the P4–P6 domain of the *Tetrahymena* group I intron. In red, magnesium ions in the metal ion core; in gold, osmium hexammine binding sites in the major groove; in light blue, the P5abc helices. The tertiary contacts between the helical stacks are indicated: the A-rich bulge docks into the minor groove of P4, and the tetraloop docks into the minor groove of the tetraloop receptor.

This AA platform opens the minor groove for base stacking or base pairing with nucleotides from a noncontiguous RNA strand.[70] The platform motif has a distinctive chemical modification signature which may make it possible to detect it in other RNAs chemically.[62,70] The ability of this motif to facilitate higher order folding provides at least one explanation for the abundance of adenosine residues in internal loops of many RNAs.

Many of the contacts that stabilize the P4–P6 domain structure, as well as its packing in the crystal lattice, involve the wide and shallow minor groove as opposed to the deep and narrow major groove. However, osmium hexammine, the compound used to determine the RNA structure by multiwavelength anomalous diffraction, binds at three locations in the major groove where non-standard base pairs create pockets of negative electrostatic potential.[71] In two cases, the heavy atoms

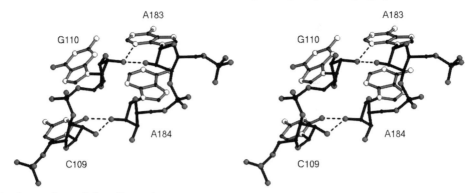

Figure 4 Stereoview of the ribose zipper. Pairs of nucleotides from different strands form a network of directed hydrogen bonds via 6 2′OH groups and the minor-groove functionalities of the bases.

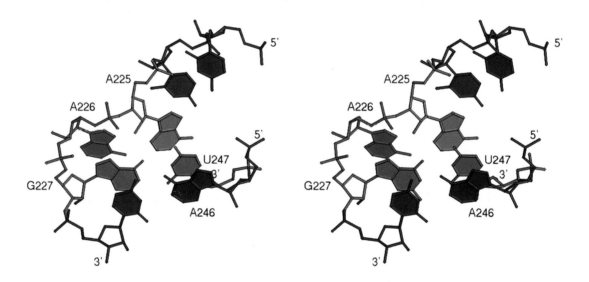

Figure 5 Stereoview of an AA platform. Viewed from the major groove, the AA platform allows cross-strand stacking, opening the minor groove for long-range tertiary contacts. The tertiary contact would be to the upper left of the AA platform as shown.

occupy sites normally bound by hydrated magnesium ions in the native RNA. One of the motifs involved, tandem G–U wobble pairs, occurs frequently in ribosomal RNAs,[72] suggesting a mechanism for metal binding in the ribosome.

6.04.8 5S RIBOSOMAL RNA FRAGMENT

E. coli ribosomal 5S RNA (5S rRNA), which contains 120 nucleotides, forms part of the 50S ribosomal subunit and binds three proteins—L25, L18, and L5. Like all structured RNAs, it has internal non-Watson–Crick base paired regions of loops. One of these, loop E, adopts its biologically functional structure only in the presence of millimolar magnesium ion concentrations. Mild nuclease digestion of 5S rRNA yields a 62 nucleotide fragment I that includes helices I and IV and loop E. The ribosomal protein L25 binds to both 5S rRNA and fragment I and protects helix IV and loop E from chemical modification.

Crystals of fragment I were originally obtained in 1983, but the structure proved difficult to solve due to limited diffraction resolution (~4 Å) and the lack of suitable heavy atom derivatives. Heavy atom derivatives were ultimately obtained by incorporation of chemically modified nucleotides into the RNA.[73] The structures of both fragment I and a smaller dodecamer duplex RNA containing the sequence of loop E have been determined at 3 Å and 1.5 Å resolution, respectively.[74]

Together these two structures reveal an interesting helical molecule in which the loop E region is distorted by three "cross-strand purine stacks" (Figure 6).[74] In this motif, the helical backbones are pinched together by stacking of A bases from opposite sides of the helix that are part of a sheared A · G and a Hoogsteen A · U, respectively. Furthermore, four magnesium ions bind in the narrowed major groove of the helix, creating a unique binding surface for the cognate ribosomal protein L25.

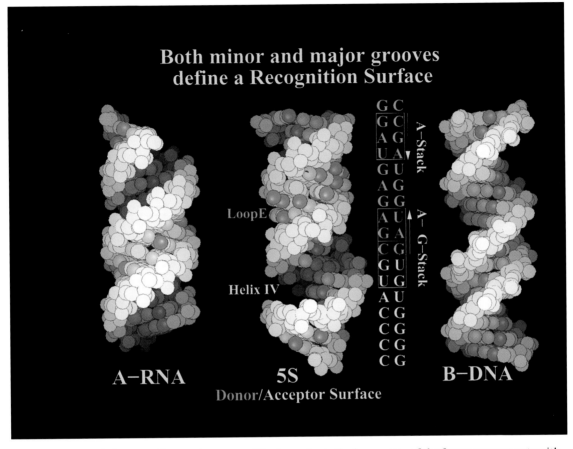

Figure 6 5S rRNA fragment I crystal structure. The irregular helical geometry of the fragment contrasts with A-form RNA (left) or B-form DNA (right). The two cross-strand purine stack motifs in the structure are indicated on the sequence by yellow boxes and arrows.

6.04.9 FUTURE DIRECTIONS

While the crystal structures of several RNAs provide new insights into RNA folding and catalysis, exciting challenges lie ahead. The hammerhead catalytic center is now the best understood of numerous ribozyme active sites; the others remain mysteries. The structures and roles of RNA in ribonucleoprotein particles including telomerase, signal recognition particle, the spliceosome, and the ribosome remain to be tackled by crystallographers. Chemical and biochemical experiments, critical for solving and understanding the hammerhead and P4–P6 domain structures, will be key to structural studies of these other RNAs also.

6.04.10 REFERENCES

1. J. F. Milligan, D. R. Groebe, G. W. Witherell, and O. C. Uhlenbeck, *Nucleic Acids Res.*, 1987, **15**, 8783.
2. J. F. Milligan and O. C. Uhlenbeck, *Methods Enzymol.*, 1989, **180**, 51.
3. S. R. Price, N. Ito, C. Oubridge, J. M. Avis, K. Nagai, *J. Mol. Biol.*, 1995, **249**, 398.

4. A. R. Ferre-D'Amare and J. A. Doudna, *Nucleic Acids Res.*, 1996, **24**, 977.
5. K. K. Ogilvie, N. Usman, K. Nicoghosian, and R. J. Cedergren, *Proc. Natl. Acad. Sci. USA*, 1988, **85**, 5764.
6. J. T. Goodwin, W. A. Stanick, and G. D. Glick, *J. Org. Chem.*, 1994, **59**, 7941.
7. M. J. Moore and P. A. Sharp, *Science*, 1992, **256**, 992.
8. A. R. Ferre-d'Amare and J. A. Doudna, *Methods Mol. Biol.*, 1997, **74**, 371.
9. J. A. Doudna, C. Grosshans, A. Gooding, and C. E. Kundrot, *Proc. Nat. Acad. Sci. USA*, 1993, **90**, 7829.
10. W. G. Scott, J. T. Finch, R. Grenfell, J. Fogg, T. Smith, M. J. Gait, and A. Klug, *J. Mol. Biol.*, 1995, **250**, 327.
11. B. L. Golden, E. R. Podell, A. R. Gooding, and T. R. Cech, *J. Mol. Biol.*, 1997, **270**, 711.
12. C. W. J. Carter and R. M. Sweet (eds.) "Macromolecular Crystallography Part A. Methods in Enzymology." Academic Press, New York, Vol. 276, 1997.
13. C. W. J. Carter and R. M. Sweet (eds.), "Macromolecular Crystallography Part B. Methods in Enzymology." Academic Press, New York, Vol. 277, 1997.
14. W. Saenger, "Principles of Nucleic Acid Structure," Springer, New York, 1984.
15. A. M. Pyle, *Science*, 1993, **261**, 709.
16. K. J. Baeyens, H. C. De Bondt, A. Pardi, and S. R. Holbrook, *Proc. Natl. Acad. Sci. USA*, 1996, **93**, 12 851.
17. E. H. Ekland, J. W. Szostak, and D. P. Bartel, *Science*, 1995, **269**, 364.
18. T. Pan and O. C. Uhlenbeck, *Nature*, 1992, **358**, 560.
19. D. Gautheret, D. Konings, and R. R. Gutell, *J. Mol. Biol.*, 1994, **242**, 1.
20. R. R. Gutell, N. Larsen, and C. R. Woese, *Microbiol. Rev.*, 1994, **58**, 10.
21. K. J. Baeyens, H. L. De Bondt, and S. R. Holbrook, *Nature Struct. Biol.*, 1995, **2**, 56.
22. M. C. Wahl, S. T. Rao, and M. Sundaralingam, *Nature Struct. Biol.*, 1996, **3**, 24.
23. S. E. Lietzke, C. L. Barnes, J. A. Berglund, and C. E. Kundrot, *Structure*, 1996, **4**, 917.
24. M. Wu, J. A. McDowell, and D. H. Turner, *Biochemistry*, 1995, **34**, 3204.
25. Y. Li, P. C. Bevilacqua, D. Mathews, and D. H. Turner, *Biochemistry*, 1995, **34**, 14 394.
26. T. S. McConnell and T. R. Cech, *Biochemistry*, 1995, **34**, 4056.
27. S. Portmann, S. Grimm, C. Workman, N. Usman, and M. Egli, *Chem. Biol.*, 1996, **3**, 173.
28. S. Portmann, N. Usman, and M. Egli, *Biochemistry*, 1995, **34**, 7569.
29. M. Egli, S. Portmann, and N. Usman, *Biochemistry*, 1996, **35**, 8489.
30. M. C. Wahl, C. Ban, S. Sekharudu, B. Ramakrishnan, and M. Sundaralingam, *Acta Crystallogr.*, Ser. D, 1996, **52**, 655.
31. A. C. Dock-Bregeon, B. Chevrier, A. Podjarny, J. Johnson, J. S. Debear, G. R. Gough, P. T. Gilham, and D. Moras, *J. Mol. Biol.*, 1989, **209**, 459.
32. D. Söll and U. L. RajBhandary, (eds.), "tRNA: Structure, Biosynthesis and Function." ASM Press; Washington, DC, 1995.
33. F. L. Suddath, G. J. Quigley, A. McPherson, D. Sneden, J. J. Kim, S. H. Kim, and A. Rich, *Nature*, 1974, **248**, 20.
34. J. D. Robertus, J. E. Ladner, J. T. Finch, D. Rhodes, R. S. Brown, B. F. Clark, and A. Klug, *Nature*, 1974, **250**, 546.
35. D. Moras, M. B. Comarand, J. Fischer, R. Weiss, J. C. Thierry, J. P. Ebel, and R. Giege, *Nature*, 1980, **288**, 669.
36. R. Basavappa and P. B. Sigler, *EMBO J.*, 1991, **10**, 3105.
37. A. Jack, J. E. Ladner, D. Rhodes, R. S. Brown, and A. Klug, *J. Mol. Biol.*, 1977, **111**, 315.
38. S. R. Holbrook, J. L. Sussman, R. W. Warrant, G. H. Church, and S. H. Kim, *Nucleic Acids Res.*, 1977, **4**, 2811.
39. S. Yokoyama and S. Nishimura, in "tRNA: Structure, Function and Recognition," eds. D. Soll and U.L. RajBhandary, ASM Press, Washington, DC, 1995, p. 207.
40. J. J. Dalluge, T. Hamamoto, K. Horikoshi, R. Y. Morita, K. O. Steller, and J. A. McCloskey, *J. Bacteriol.*, 1997, **179**, 1918.
41. J. A. Kowalak, J. J. Dalluge, J. A. McCloskey, and K. O. Steller, *Biochemistry*, 1994, **33**, 7869.
42. J. T. Goodwin, S. E. Osborne, E. J. Scholle, and G. D. Glick, *J. Am. Chem. Soc.*, 1996, **118**, 5207.
43. G. A. Prody, J. T. Bakos, J. M. Buzayan, I. R. Schneider, and G. Bruening, *Science*, 1986, **231**, 1577.
44. O. C. Uhlenbeck, *Nature*, 1987, **328**, 596.
45. J. Haseloff and W. L. Gerlach, *Nature*, 1988, **334**, 585.
46. H. W. Pley, K. M. Flaherty, and D. B. McKay, *Nature*, 1994, **372**, 68.
47. W. G. Scott, J. T. Finch, and A. Klug, *Cell*, 1995, **81**, 991.
48. S. H. Kim, G. J. Quigley, F. L. Suddath, A. McPherson, D. Sneden, J. J. Kim, J. Weinzierl, and A. Rich, *Science*, 1973, **179**, 285.
49. T. Tuschl, C. Gohlke, T. M. Jovin, E. Westhof, and F. Eckstein, *Science*, 1994, **266**, 785.
50. S. T. Sigurdsson, T. Tuschl, and F. Eckstein, *RNA*, 1995, **1**, 575.
51. K. M. Amiri and P. J. Hagerman, *Biochemistry*, 1994, **33**, 13 172.
52. K. M. Amiri and P. J. Hagerman, *J. Mol. Biol.*, 1996, **261**, 125.
53. J. B. Murray, D. P. Terwey, L. Maloney, A. Karpeisky, N. Usman, L. Beigelman, and W. G. Scott, *Cell*, 1998, **92**, 665.
54. W. G. Scott, J. B. Murray, J. R. P. Arnold, B. L. Stoddard, and A. Klug, *Science*, 1996, **274**, 2065.
55. A. L. Feig, W. G. Scott, and O. C. Uhlenbeck, *Science*, 1998, **279**, 81.
56. T. Tuschl, M. M. P. Ng, W. Pieken, F. Benseler, and F. Eckstein, *Biochemistry*, 1993, **32**, 11 658.
57. J. SantaLucia, Jr. and D. H. Turner, *Biochemistry*, 1993, **32**, 12 612.
58. A. E. Walter, M. Wu, and D. H. Turner, *Biochemistry*, 1994, **33**, 11 349.
59. M. Wu and D. H. Turner, *Biochemistry*, 1996, **35**, 9677.
60. O. C. Uhlenbeck, *RNA*, 1995, **1**, 4.
61. T. R. Cech, *Annu. Rev. of Biochem.*, 1990, **59**, 543.
62. F. L. Murphy and T. R. Cech, *Biochemistry*, 1993, **32**, 5291.
63. F. L. Murphy and T. R. Cech, *J. Mol. Biol.*, 1994, **236**, 49.
64. J. H. Cate, A. R. Gooding, E. Podell, K. Zhou, B. L. Golden, G. E. Kundrot, T. R. Cech, and J. A. Doudna, *Science*, 1996, **273**, 1678.
65. G. Van der Horst, A. Christian, and T. Inoue, *Proc. Natl. Acad. Sci. USA*, 1991, **88**, 184.
66. M. Costa and F. Michel, *EMBO J.*, 1995, **14**, 1276.
67. H. A. Heus and A. Pardi, *Science*, 1991, **253**, 191.
68. H. W. Pley, K. M. Flaherty, and D. B. McKay, *Nature*, 1994, **372**, 111.

69. D. L. Abramovitz, R. A. Friedman, and A. M. Pyle, *Science*, 1996, **271**, 1410.
70. J. H. Cate, A. R. Gooding, E. Podell, K. Zhou, B. L. Golden, A. A. Szewczak, C. E. Kundrot, T. R. Cech, and J. A. Doudna, *Science*, 1996, **273**, 1696.
71. J. H. Cate and J. A. Doudna, *Structure*, 1996, **4**, 1221.
72. D. Gautheret, D. Konings, and R. R. Gutell, *RNA*, 1995, **1**, 807.
73. C. C. Correll, B. Freeborn, P. B. Moore, and T. A. Steitz, *J. Biomol. Struct. Dyn.*, 1997, **15**, 165.
74. C. C. Correll, B. Freeborn, P. B. Moore, and T. A. Steitz, *Cell*, 1997, **91**, 705.

6.05
Chemical and Enzymatic Probing of RNA Structure

RICHARD GIEGÉ, MARK HELM, and CATHERINE FLORENTZ

Institut de Biologie Moléculaire et Cellulaire, Strasbourg, France

6.05.1 INTRODUCTION

6.05.1.1 Historical and Theoretical Background

The chemistry of nucleobases, nucleosides, and nucleotides is a large and well-established field of organic chemistry. The study of their reactivities to alkylating agents has contributed to the understanding of mutagenic and carcinogenic mechanisms of the these agents.[1] In addition, today's widely used techniques of sequencing and structural probing of nucleic acids, and related techniques like footprinting and chemical interference techniques, have their origins in organic chemistry. In particular, alkylating agents are now used in an analytical fashion to deduce structural features of nucleic acids. The basic idea is that RNA structure influences the extent to which parts of it, i.e., nucleotides as a convenient unit, can be modified by a given chemical compound or attacked by a nuclease. If the specificity of the reagent is known, structural features of the RNA can be deduced, given the existence of an efficient detection method for the modification. The chemically modified nucleotides are judged noninvolved in the structure, while nonmodified nucleotides would be protected from the reagent by hydrogen-bonding or other steric encumbrances. Similarly, preferential cuts by nucleases provide information about accessible regions in RNA.

The history of structural probing has always paralleled that of nucleic acid sequencing. Table 1 shows historically important events that mark progress in the field. In a typical early structural probing experiment, a relatively large quantity of unlabeled RNA (classically tRNA) was quantitatively reacted with a probe, then digested with an RNase, like RNase T1 and/or RNase A. The resulting fragments were separated by chromatographic techniques like those used for sequencing of the first tRNAs, and then analyzed for their content of nucleotides modified in the experiment. This early probing approach reflected the then current state of the art of RNA sequencing. With the development of fingerprint techniques it became possible to use internally labeled RNA that had to be purified from radioactive cell cultures. However, the analytical procedure was labor intensive and the approach suffered from the intrinsic problem that a large proportion of the target residues had to be modified in order to be detectable. The large body of early work was efficiently validated when the first crystal structure of an RNA became known and, satisfyingly, computed theoretical accessibilities displayed a high degree of correlation with the experimental reactivities.[7,8]

Table 1 Detection methods for modifications of RNA by structural probes.

Detection/ revelation method	RNA	Extent of modification	Fragmentation	Fractionation	Corresponding sequencing method	Refs.
A	unlabeled	complete	various RNases	various chromatographic methods	first-generation biochemistry	2
B	internally labeled	complete	RNases A and T1	electrophoresis/ TLC	fingerprint	3
C	end labeled	statistical	chemical or enzymatic scission at modified residues	PAGE	Maxam and Gilbert/Donis-Keller	4, 5
D	unlabeled	statistical	reverse transcription	PAGE	primer elongation	6

The invention of end-labeling sequencing techniques by Maxam and Gilbert[4,5,9] greatly reformed the technique of structural probing. End-labeling enables the use of only small amounts of material and permits sensitive detection.[10] Because of the augmented sensitivity in signal detection, the principle of statistic modification can be used, i.e., only a few percent of the molecules have to be modified to yield a signal. Thus, the number of molecules with multiple modifications is negligible, and a signal results from a molecule that has not been structurally altered by previous interaction with other probe molecules. Intrinsic parts of the technique of Maxam and Gilbert are chemical treatments that induce chain scission of the nucleic acid at the modified sites. In the case of alkylated bases in RNA, the aromaticity of the purine or pyrimidine rings, which contain nitrogen atoms involved in hemiaminal bonding of the base to the sugar, is usually affected. Subsequent treatment with buffered aniline provokes chain scission by β-elimination.[9,11,12] By this mechanism, three nitrogens in RNA (N-3C, N-7G, and N-7A; see Table 2 for atom abbreviations) can be targeted with DMS (dimethyl sulfate) and DEPC (diethyl pyrocarbonate).[10] Soon after, ENU (ethylnitrosourea) was introduced[13] as a useful probe for mapping phosphates, regardless of the corresponding nucleobases and secondary structure.

Paralleling this evolution, the same reagents were used for footprinting studies of RNAs in interactions with proteins or other macromolecules. Here, a crystal structure of an RNA–protein complex allowed an assessment of the usefulness of this technique. Thus, in the case of yeast tRNA[Asp] complexed to its cognate synthetase, contact points determined in solution[14,15] could be compared to those on the X-ray structure[16] (see Section 6.05.5.1). Even though it had been shown early on that most RNases cut preferably in the anticodon loops of tRNAs, RNases like RNase

V1[17] only became popular in structural probing and footprinting with the development of end-labeling techniques.[18]

As a result of this progress, many RNAs in addition to tRNAs became appropriate for structural probing. However, the resolution of polyacrylamide gel electrophoresis (PAGE) limits the application of the "direct" detection method to roughly the first 150 residues from either extremity of an RNA molecule. Again, adaptation of state-of-the-art sequencing techniques[6] to RNA structural probing permitted important methodological advances. The use of primer elongation with reverse transcriptase[19] allows the exploration of larger molecules such as ribosomal RNAs. A further advantage of the primer elongation technique is that probes, which cannot provoke strand scission, but modify bases to efficiently stall the enzyme, can be utilized on a statistical basis (e.g., CMCT [1-cyclohexyl-3-(2-morpholino-4-ethyl)carbodiimide methotosylate] kethoxal, see Table 2 and Section 6.05.3.1). This considerably enriches the arsenal of small probes and permits most of the atoms important for the architecture of RNA to be probed.

The growing body of structure-function data indicates that the biological activity of many RNAs is intimately related to the recognition of its tertiary structure and irregular structural features, like base-triples and mismatches (this distinguishes RNA from DNA which, in general, displays fewer three-dimensional peculiarities). As a consequence, this has in recent times shifted attention to the development of a second generation of structural probes. These more complex molecules often contain parts specially designed for the recognition of such structures (see Section 6.05.3.2 and Table 2) and, being of intermediate size, are large enough to possess significant conformational selectivity.

6.05.1.2 Fields of Application

The techniques outlined above have led to many important biochemical results. Together with computer algorithms for secondary structure prediction[63,64] and sequence alignments, they allow establishment of Watson–Crick interactions and therefore secondary structure models as a basis for further work. In combination with computer modeling and phylogenetic comparisons,[65] structural probing has led to the construction of valuable three-dimensional models of many RNAs, among which are tRNAs with unusual structures,[66–68] tRNA-like viral RNAs,[69,70] tmRNA,[71,72] gRNA,[73] recognition elements in mRNAs,[74] dimerization initiation site of HIV-1 genomic RNA,[75] RNase P,[76,77] and *Tetrahymena* ribozyme.[78,79] These models are of great heuristic value and permit better understanding of the molecular functioning of the RNAs and definition of the domains of importance so that further (e.g., mutational) analyses can be well aimed. Definition of functional domains in this way is also important for other studies, like NMR or crystallogenesis, which cannot (yet) be conducted on large or flexible multidomain molecules.

6.05.2 VALIDATION OF RNA STRUCTURAL PROBING APPROACHES

6.05.2.1 Probing with Chemical Reagents

When nucleic acids are mapped with chemical reagents, it is often implicitly assumed that the probes map simple steric accessibilities. Strictly speaking, the experimental results, referred to as reactivities, must theoretically comprise steric and electronic parameters of the targeted residues. For most of the known probes, the correlation of reactivity (experimental value) and accessibility (theoretical value) was first qualitatively validated by comparison with the crystallographic structure of tRNA[Phe] from yeast. The targeted residues were, by visual inspection of the crystal structure, classified accessible or not, and compared to the reactivities. A more quantitative procedure used computed accessibilities as a product of the "contact surface area" for residues in tRNA[Phe] and empirically determined the effective radii of probes.[8] It accounted for the reactivities of methoxyamine, TiCl$_3$/KI, semicarbazide-bisulfite, monoperphthalic acid, kethoxal, DMS, and DEPC. A more refined concept included electrostatic potentials calculated with an *ab initio* procedure.[7,80] The resulting theoretical values for expected reactivities were termed ASIF values (accessible surface

Table 2   Structural probes for RNAs

Probe	Size	Target	Modification extent	Strand scission	Detection method	Refs.
(a) Nucleophiles						
NaBH$_4$	+	D, m^{7}G, m^{1}A	C	–	B	20–22
Methoxyamine and hydroxylamines	+	C-4C, C-5C, C-6C	C	–	B	23
Semicarbazide and bisulfite	+	C-4C, C-5C	C,S	–	B,C	24
KI/TiCl$_3$	+	C-5C	C	–	B	25
(b) Electrophiles and alkylating reagents						
Acrylnitrile		s^{4}U, I, Ψ	C	–		20
Chloroacetaldehyde	+	N-1A, N-3C	C	–	B	26
DMS (DES) #	+	N-7G, N-3C, N-1A	C,S	–*	B,C,D	10,27
DEPC #	+	N-7A	S	–*	B,C,D	10,27
ENU (MNU) #	+	phosphate,	S	–*	C	13
Ketoxal, glyoxal #	+	N-1G, N-2G	C,S	–	B,D	11,20
CMCT #	+	N-3U, N-1G, N-3D	C,S	–	B,D	23,28
Tritiated water	+	HC-8Py	C	–	B	29
1-Fluoro-2,4-dinitrobenzene	+	2'OH ss A	C	–	B	30
T4MPyP, TMAP, T2MyP/*hv*	+ +	helix hinges	S	–*	C,D	31
BenzN$_2$$^+$ and derivatives/*hv*	+ +	phosphate	S	–*	C	32
(c) Radical generators/oxidants						
Op–Cu	+ +	binding pocket	S	+	C	33
Rh(phen)$_2$phi^{3+}/*hv*	+ +	tertiary features	S	+	C	34
Rh(DIP)$_3$$^{3+}$	+ +	G–U bp	S	+	C	35
Fe^{2+}/EDTA/H$_2$O$_2$ #	+	HC-2/4/5Rib	S	+	C,D	36
Fe^{2+}/MPE/H$_2$O$_2$	+ +	ds RNA	S	+	C,D	37
KONOO	+	HC-2/4/5Rib	S	+	C	38
X rays	+	HC-2/4/5Rib	S	+	D	39
Fe-bleomycin	+ +	specific	S	+	C	40
NiCR and derivatives/KHSO$_5$	+ +	exposed N-7G	S	–*	C,D	41
Ozone	+	exposed G			B	42
Monoperphthalic acid	+	N-1A, N-3C	C	–	B	20
Isoalloxazine derivatives/*hv*	+ +	G–U bp	S		C	43
(d) Hydrolytic cleavages and nuclease mimics						
Water (OH$^-$)	+	phosphates	S	+	C	44
Metal ion mediated						45
Mg^{2+}	+	specific	S	+	C	46,47
Ca^{2+}, Sr^{2+}	+	specific	S	+	D	48
Pb^{2+}	+	specific	C,S	+	C,D	49,50
Lanthanides (Ln^{3+})	+	specific	S	+	C	47,51
Zn^{2+}	+	specific		+		52
Imidazole	+	ss RNA	S	+	C	53
Spermine-imidazole	+ +	ss Py-A	S	+	C	52–54
(e) Biological nucleases						
Nuclease S1 #	+ + +	ss RNA	S	+	C,D	55–57
Nuclease *N.crassa*	+ + +	ss RNA	S	+	C,D	58
RNase U2	+ + +	ss A	S	+	C,D	18
RNase T1 #	+ + +	ss G	S	+	C,D	56
RNase T2 #	+ + +	ss RNA	S	+	C,D	59
RNase CL3	+ + +	ss C	S	+	C,D	60
RNase V1 #	+ + +	ds RNA	S	+	C,D	17,57
Rn nuclease I	+ + +	ss RNA	S	+	C	61
RNase P1	+ + +	ss RNA	S	+	C	62

Probes: CMCT, 1-cyclohexyl-3-(2-morpholino-4-ethyl)carbodiimide methoptosylate; DMS, dimethyl sulfate; DES, diethylsulfate; DEPC, diethyl pyrocarbonate; ENU or MNU, ethyl- or methylnitrosourea; kethoxal, β-ethoxy-α-ketobutyraldehyde; KONOO, potassium peroxonitrite; Fe^{2+}/MPE, methidiumpropyl-EDTA-iron(II); NiCR, 2,12-dimethyl-3,7,11,17-tetraazabicyclo[11.3.1]heptadeca-1(17)-2,11,13,15-pentaenato nickel(II) perchlorate; Op–Cu, 1,10-phenanthroline–copper; Rh(phen)$_2$phi^{3+}, bis(phenanthroline)(phenanthrene quinone diimine)rhodium(III); Rh(DIP)$_3$$^{3+}$, tris(4,7-diphenyl-1,10-phenanthroline)rhodium(III); T4MPyP, *meso*-tetrakis-(4-*N*-methylpyridyl)porphine; TMAP, *meso*-tetrakis-(*para*-*N*-trimethyl anilinium)porphine; T2MyP, *meso*-tetrakis-(2-*N*-methylpyridyl)porphine; BenzN$_2$$^+$, *p*-*N*-dimethylaminobenzenediazonium cation. Commonly used probes are highlighted by #. **Size:** + small 1–10 Å; + + intermediate 10–100 Å; + + + large over 100 Å. **Target:** N-1A, nitrogen 1 in adenine; N-7A, nitrogen 7 in adenine; N-3C, nitrogen 3 in cytidine; C-4C, C-5C, C-6C, carbons 4, 5, and 6 in cytidine; N-3D, nitrogen 3 in dihydrouracil; N-1G, nitrogen 1 in guanine; N-2G, nitrogen 2 in guanine; N-7G, nitrogen 7 in guanine; N-3U, nitrogen 3 in uracil; m^{1}A, 1-methyladenine; m^{7}G, 7-methylguanine; HC-2/4/5 Rib, hydrogen at carbons 2, 4, and 5 of ribose; HC-8R, hydrogen on carbon 8 of purines; ss, single-stranded; ds, double-stranded; specific, binding pocket particular for a certain probe. **Chemistry:** *modification extent:* C, complete; S, statistical; *strand scission:* –, no strand scission; –*, scission by additional treatment; +, strand scission by probe (nuclease activity). **Detection method:** B, fingerprinting; C, end-labeling; D, primer extension (as in Table 1). **Other abbreviation:** *hv*: requires irradiation.

integrated field).[81] Using yeast tRNA[Phe] and tRNA[Asp] as model RNAs, the probes DMS, DEPC,[27,80] and ENU[13,82] were validated this way in a thorough comparison of reactivities and ASIF values. It was found that the addition of an electrostatic field to the geometric accessibility leads to a good model in that it yields good predictions of chemical reactivities. This correlation is illustrated in Figure 1. In other words chemical reactivities represent predominantly architectural features of the RNA. Interestingly, the probes reflect even subtle conformational differences found in the crystal structures of the two tRNAs.[14,27] They are also capable of monitoring tertiary interactions and Mg^{2+} coordination and thus they represent an analytical tool of high precision and fidelity.

(a) (b)

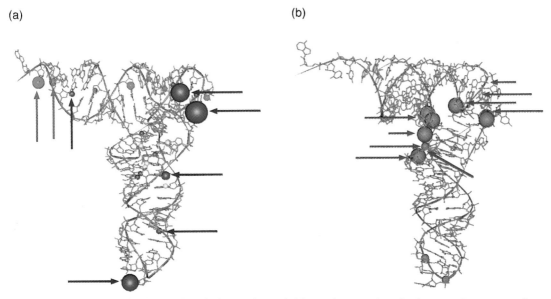

Figure 1 Comparison of calculated and observed reactivities and protections in the crystal structure of yeast tRNA[Phe] [83,84] as model RNA. (a) Nitrogens are indicated by spheres of sizes proportional to computed reactivities corresponding to their ASIF values for N-3C (green), N-7G (blue), and N-7A (yellow).[80] Experimental results for DMS (N-3C and N-7G) and DEPC (N-7A)[27] are indicated by arrows of corresponding colors. (b) Phosphates protected from alkylation by ENU are in purple, their sizes proportional to calculated protections according to their ASIF values.[80,81] Arrows indicate observed protections.[13,85] Images were made with DRAWNA.[86]

6.05.2.2 Probing with Nucleases

Nucleases have become popular probes because of their easy use and fast results. As for the chemical probes, most of the commonly used nucleases have been validated by studying their specificity on tRNAs of known structure.[55–57,59] However, nucleases are much larger than the commonly used chemical probes and they do not require, by definition of their enzymatic activity, additional treatment to induce chain scission. The large size of nucleases brings some pitfalls, and thus, some care has to be taken in interpreting the results. Indeed, a nuclease might establish in the process of protein–RNA recognition several interactions at a time with the RNA, and therefore there is a possibility that these interactions induce conformational changes in the RNA and yield artifacts. Also, because the enzyme is much larger in size than the target, the probe can no longer be assumed to be spherical and therefore a validation approach as used for chemical probes is not possible, hence their frequent use in mapping global domains rather than details in RNA structure.

For all probes that directly cleave RNA, the problem of secondary cleavages can in principle occur. Primary cleavage of the RNA by the nuclease generates two fragments, which might adopt a structure different from the intact RNA. Upon a second cleavage, such a fragment will yield a signal that does not reflect the intact RNA structure. Because secondary fragments are always shorter than the primary fragments, they can be detected by comparing fragmentation patterns of both 3′- and 5′-labeled RNA.[11,12,17,60] Only cleavages that are found in both experiments are of primary origin.

6.05.3 PROBES, TARGETS, AND METHODOLOGY

6.05.3.1 Probes and Targets

Of all the probes described in the literature, only a handful are commonly used (highlighted in Table 2). Among the most popular chemical probes are DEPC and DMS. They were the first to be used with the direct revelation method and their specificity for purine and pyrimidine rings has been well characterized by a large body of data. As with the equally well-characterized ENU (mapping phosphates), chemical chain scission reactions are specific to one type of modification at a time. Reactive N-7A, acylated by DEPC, can be directly revealed by chain scission with aniline. DMS preferentially alkylates N-7G, followed by N-3C and N-1A. Minor methylation sites are N-3A, N-7A, and even N-3G and O-6G. Interestingly, N-7G modifications can be exclusively revealed by treatment of the RNA with sodium borohydride prior to chain scission with aniline, while treatment with hydrazine reveals methylations at N-3C, the other methylations remaining undetected. This specificity in cleavage is possible because sodium borohydride reduces the N-7-methylated imidazole ring of guanine residues, thereby creating a site for aniline scission. On the other hand, N-3-methylated cytidine residues are rendered susceptible to aniline scission by a nucleophilic attack of their pyrimidine ring by hydrazine.[9–12] Thus, breaking down the different modification sites yields a favorable signal-to-noise ratio for each type of modification. In a similar way, mild alkaline treatment reveals exclusively the phosphates alkylated by ENU, even though in principle ENU alkylates all oxygens and some nitrogens in RNA.[1] It is due to the above mentioned properties that these probes work with high precision. Commonly used with the primer extension revelation method are DMS (mapping N-1A and N-3C, but not N-7G), kethoxal (N-1G and N-2G), and CMCT (N-3U and N-1G). Because these probes map, for each nucleobase, at least one nitrogen atom involved in Watson–Crick interactions, they are an excellent means for determining the secondary structure of a given RNA.

As mentioned above, the use of these probes as revealed by the direct method has been validated with the crystallographic structures of yeast tRNA^{Phe} and tRNA^{Asp} as references. This is, however, not the case for the same probes when detected with the primer extension method, and this creates a certain drawback. While the direct method, with end-labeling and chain scission specific for one type of modified nucleotide at a time, guarantees reliability, primer extension produces a relatively high amount of background signal, because all modifications at different sites capable of stopping reverse transcription give signals at the same time.[19] The efficiency of stalling the enzyme differs among the different modifications. Usually the reaction rates of one probe with several different target sites are not identical, e.g., the methylation of N-1A by DMS cannot directly be compared to that of N-3C. Also, strongly structured regions sometimes evade mapping because they hinder the enzyme and create pauses in reverse transcription. These pauses cause regions with a high background in the control experiments. Therefore, the precision of these probes must be judged as slightly lower than those mentioned above. However, their reliability is illustrated by their wide-spread utilization especially for mapping large RNAs (see Section 6.05.1.2) but also tRNAs.[68]

6.05.3.2 Enzyme Mimics, Chemical Nucleases, and Tethered Probes

The phosphates in RNA are several orders of magnitude less resistant to hydrolytic cleavage than those in DNA. Studies on model substrates are consistent with the hydrolysis being subject to general base as well as general acid catalysis.[87] General base catalysis proceeds by deprotonation of the 2′OH, promoting a nucleophilic attack of the 2′-oxygen at the phosphorus to form the often cited 2′–3′ cyclic intermediate. The phosphate can be activated for nucleophilic attack by coordination of an electrophile, e.g., a proton (hence the general acid catalysis[87]) or a divalent metal ion. Therefore, even highly purified RNA in water is subject to some spontaneous hydrolysis, due to trace amounts of hydroxide ions and protons (by definition 10^{-7} M at pH 7). Metal ions can also coordinate water molecules that are then more easily deprotonated. Consequently, divalent metal ions can promote RNA hydrolysis with high efficiency. Because they also favor structuring, especially tertiary interactions, they can be seen as additives that help RNA to adopt its conformation and thus promote cleavage in an indirect manner. Such a role is especially crucial in catalytic RNAs.[88]

In 1976 Werner *et al.* discovered that tRNA^{Phe} from yeast is site-specifically cleaved by Pb^{2+} ions at a rate that assigns catalytic activity to the plombus ions.[49] Lead ions preferably map single-stranded regions unless they coordinate to elements of tertiary interaction, for example in tRNA between the D- and T-loops. They can thus be used as specific probes to verify the structural integrity of a canonical tRNA. Crystallography permitted localization of the coordination pocket[50] and aided the mechanistic studies, which were further supplemented with detailed mutational analyses.[89,90] Other "leadzymes" have been selected, and their properties studied.[45,91] Many other di- and trivalent cations were also found to induce RNA cleavage (see Table 2). As several tRNAs are cleaved by Mg^{2+} ions coordinated near the classical coordination site,[47] such ions, besides their structural role, seem to actively participate in RNA hydrolysis as well (note that above 60 °C and/or at slightly alkaline pH, Mg^{2+} ions exhibit strong RNase activity).[92] A proposed model of the active center of a group I intron suggests coordination of different divalent cations to the same binding pocket in which Mg^{2+} is required for cleavage of the natural substrate. Interestingly Ca^{2+}, Sr^{2+}, and Pb^{2+} induce cleavage of the ribozyme itself, rather than its natural substrate.[48] Not only do these findings render the borders between leadzymes, other metalloenzymes, and ribozymes even more fluent, but they also illustrate the possible use of metal ions to monitor structure in such binding pockets. In view of all the above, plus the fact that some nucleases require divalent cations for functionality (Zn^{2+} for S1 nuclease[93]), metal ions can be regarded as the stripped active center of an RNase or as an "RNase mimic."

With a more detailed knowledge of RNA structures and cleavage mechanisms, it became possible to engineer chemical compounds for structure recognition. Inferred from the crystal structures of RNase A is the knowledge of how imidazole residues, in the form of two histidines, activate phosphates and 2′OH groups of RNA for hydrolysis.[94,95] Imidazole is especially suited because its pK_b is 7.0 and thus it is capable of both acid and base catalysis at neutral pH (see above). This situation has been successfully mimicked by supplying one or two imidazole groups tethered to different ligands with affinity for RNA structural features.[53,54] Cleavage occurs readily with mimics containing two imidazoles; if only one imidazole group is present, as in the spermine–imidazole conjugate, cleavage is triggered when the reaction medium is supplied by imidazole buffer.[53] These probes exhibit site-specificity and, like RNase A, show preference for pyrimidine/A sequences.

A group of transition metal complexes is capable of cleaving RNA upon irradiation or addition of redox-active co-reagents (e.g., H_2O_2) by mechanisms involving free or metal-coordinated radicals.[96] These compounds, termed "chemical nucleases",[96,97] occupy an intermediate position in size between the small electrophiles and the large nucleases, and are therefore potentially valuable for the recognition and mapping of structural features of intermediate size. Indeed, $Rh(DIP)_3^{3+}$ (tris(4,7-diphenyl-1,10-phenathroline)rhodium(III)) selectively cleaves at the 3′-end of the G of G–U mismatches[35] and Fe^{2+}/MPE (methidiumpropyl EDTA-iron(II)) was shown to specifically map double-stranded regions[37] due to the intercalating nature of the methidium moiety. In the latter case, as in the case of RNase A mimics (see above), the cleavage-active center has been tethered to a recognition-active ligand to obtain structural specificity for the cleavage. In a larger view, using a biological macromolecule as the recognition-active ligand is the next logical step. Indeed, Fe^{2+}/EDTA has been tethered to a tRNA,[98] as well as to a protein (EF-G),[99] to probe the corresponding binding sites on the ribosome.

6.05.3.3 Methodology

There are a number of research papers and reviews[10–13,27,100] covering the chemistry and theoretical and experimental aspects of the different probing methods. We will restrict our remarks in this section to a number of technical details and pitfalls concerning the practical aspects of structural probing that demand special attention. Among these are the choice of temperature, buffer, and salt conditions because of their possible influence on RNA structure. Important for the choice of buffer is its inertness towards the probe. For example, Tris, as a popular buffer in the pH range 7.5–8.5, is suitable for probing with nucleases. However, being a nucleophile, and its capacity being strongly temperature dependent, it is not recommended for use with electrophiles and in temperature-dependent probing. For these applications it is commonly replaced by cacodylate. Tris is also considered a radical quencher and, although used with chemical nucleases, should be replaced by a phosphate buffer.[40] Even though probing is in most cases performed at 20–25 °C or 37 °C, it is recommended that UV melting experiments are conducted[101] to facilitate the proper choice of parameters for the denaturing/renaturing procedure that usually preceeds probing.

A very important requirement for structural probing is a homogeneous population of labeled molecules, be it RNA for the direct method or DNA for primer extension.[18] 5′-Labeling is usually performed using T4 polynucleotide kinase and [γ-^{32}P]ATP. Unless exchange reactions are used, an existing nonradioactive 5′-phosphate has to be removed by bacterial alkaline phosphatase or calf intestine phosphatase. In the case of RNAs with a stable structure near the 5′-extremity which hampers dephosphorylation (typically in native tRNAs), very stringent conditions have to be chosen. 3′-Labeling is usually done by ligation of [5′-^{32}P]pCp (5′,3′ cytidine diphosphate) to the 3′OH of the RNA by T4-RNA ligase. Purification to the nucleotide level can be a problem when *in vitro* transcribed RNAs are used, because RNA polymerases do not yield homogeneous populations on the 3′-end upon run-off transcription. RNA labeled this way and not properly purified will yield multiple signals for each nucleotide in probing experiments. A different approach can be used for 3′-labeling of tRNAs and tRNA-like molecules. After removal of the 3′-CCA terminal sequence by endonuclease digestion, reconstitution by ATP/CTP-tRNA nucleotidyltransferase (ATP and CTP are adenosine and cytidine triphosphates) is performed in the presence of [α-^{32}P]ATP. Labeled RNA is usually purified by denaturing PAGE, subsequent excision, passive elution, and ethanol precipitation in the presence of carrier RNA. It is recommended to submit the redissolved RNA to gel filtration to exclude contamination by acrylamide residues, residual salt, and ethanol before structure probing.

For direct detection and assignment of cleavage sites, the radioactive RNA is electrophoresed in parallel with (i) an untreated sample to detect spontaneous and unspecific cleavages in the experimental procedure, (ii) a ladder created by degradation of the sample in the denatured state by heating with alkali or imidazole[53] or simply in water or formamide[100] or by P1 digestion, (iii) the products of a partial digestion with RNase T1 (specific for G) or U2 (specific for A). These allow the correct assignment of bands observed in the sample lane to the primary sequence. Note that aniline, radicals, ENU, S1 type nucleases, and T1 type nucleases all leave chemically different chain scission products. This fact is reflected in the different migration behaviors of fragments corresponding to approximately the first 20 nucleotides on the labeled extremity. Signal assignment in these regions can therefore be difficult or ambiguous. In fragments longer than 30 nucleotides, the chemical differences become irrelevant in comparison with other parameters that determine migration behavior, i.e., sterical encumbrances and charge density. Consequently, assignment of these fragments is reliable. For correct assignment with the primer extension method, sequencing reactions with dideoxy chain terminators are run in parallel. When blocked by chemical modification, the reverse transcriptase stops elongation one nucleotide before the modified residue. Thus, its signal is one nucleotide shorter than the corresponding sequencing signal.

6.05.4 APPROACHING ARCHITECTURAL FEATURES OF RNAs BY STRUCTURAL PROBING

6.05.4.1 Two- and Three-dimensional Structures

tRNA mimics are nontransfer RNAs that are recognized by tRNA-specific enzymes.[102] They have been among the most prominent models to demonstrate the usefulness of enzymatic and chemical probes in unraveling intricate structural features in the RNA world. This relies on the plausible assumption that similar functions are sustained by similar structures. Thus, efforts were made to find in tRNA mimics structural elements recapitulating features of tRNA two-dimensional cloverleaves and/or of their L-shaped three-dimensional architecture. As an example, we take the tRNA-like domain present at the 3′-extremity of turnip yellow mosaic virus (TYMV) RNA. This domain is found within the last 159 nucleotides of the viral RNA and is charged by valyl-tRNA synthetase. Thus it should share structural similarities with tRNA^{Val}. Extensive nuclease probing on 5′- and 3′-end-labeled fragments, up to 159 nt long, revealed a folding that partly differs from a cloverleaf[60,103] suggesting that the mimicry with tRNA occurs at the three-dimensional level (Figure 2). Indeed, the longest fragment is organized into six domains, with the 5′-domain loosely structured, and likely not involved in the tRNA-like fold. Domains I–V are well-organized stem and loop regions and it is easy to recognize in domains II, III, and IV, the mimics of the T-, anticodon-, and D-arms of canonical tRNAs. In contrast, domains I and V, as well as the joining oligonucleotides between domains I and II, do not show any resemblance to tRNA features, except for the 3′-terminal

-ACCA$_{OH}$ sequence characteristic of valine-accepting tRNAs. The understanding of the role of domain I emerged from enzymatic and chemical probing as well as from the observation of compensatory mutations in several strains of TYMV that maintained base pairing between three residues of loop I and the three most 5′-residues in the single-stranded stretch joining stem I to stem II. This was the clue for the proposal of a new RNA folding principle, the pseudoknot[103,104] that allowed construction of a model of the amino acid accepting stem of the tRNA-like molecule involving the sole 3′-end sequence of the viral RNA. Further probing with DMS and DEPC confirmed the pseudoknot fold and ENU probing revealed T-loop features in loop II. A more accurate three-dimensional structure, with correct nucleotide geometries, was then computer modeled, utilizing all the available crystallographic knowledge on tRNAs and the probing data on the isolated RNA fragments. The model showed that the tRNA-like structure is 83 nt long and is thus restricted to domains I–IV. This model, and especially the pseudoknot fold, was of great heuristic value, since it rapidly gave the answer to the folding of other tRNA-like structures, and later allowed prediction of the histidine identity of all plant viral tRNA-like structures since a residue from the pseudoknot loop L1 (residues 21–24, see Figure 2(c)) is stacked 5′ on top of the accepting helix and mimics the histidine major identity residue N-1.[105] An NMR structure of a pseudoknot derived from the TYMV tRNA-like structure[106] has confirmed many of the conclusions from the modeling, in particular its overall fold and the conformation of the histidine identity element (U21) in loop L1. It is important to mention, however, that there were fine structural details revealed by NMR that could not be predicted by structural probing.

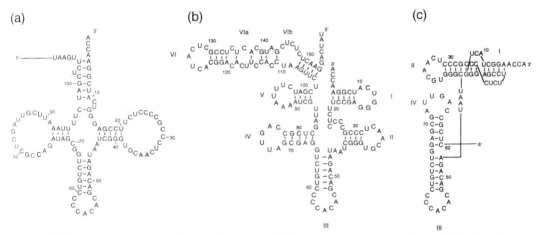

Figure 2 Different steps in the understanding of an RNA structure: the case of the tRNA-like domain of TYMV RNA.[60,103,104] All three displayed models are based on the presence of a CAC valine anticodon located in a loop mimicking a tRNA anticodon loop (bottom of the models). (a) Two-dimensional folding mimicking a cloverleaf established on the sole sequence without additional experimental data (nt about 109), (b) experimentally established two-dimensional folding (nt = 159); (c) pseudoknotted fold mimicking three-dimensional features of canonical tRNA, in particular its architectural organization in amino acid accepting (domains I and II) and anticodon (domains III and IV) branches (nt = 82). Noteworthy, the NMR model[106] showed the stacking of the two helical regions in the pseudoknotted fold (domain I) and a conformation of loop L1 (residues 21–24) compatible with functional data.[105]

In general, the problem of how three-dimensional features can be extracted from probing data is a complex procedure always utilizing computer modeling and if possible additional information from sequence comparisons of phylogenetically related molecules and detection of long-range conservations of base pairings.

6.05.4.2 Comparison of Mutants

Construction and structural analysis of RNA, mutated at strategically important positions, render the technique even more powerful. Thus, it is possible to detect whether a certain mutated position in an RNA is directly responsible for function or rather for structural features. Drastically altered probing patterns indicate an essential role of the mutated position(s) for the RNA architecture. Three-dimensional features are commonly found and verified this way. Typical examples include studies on human mitochondrial tRNA[Lys] [107] and the RNA binding site of the *E. coli* ribosomal protein S8.[108]

Definition of the precise structural features in the binding site of protein S8 on rRNA from *E. coli* was based on an *in vitro* selection procedure. This powerful mutagenesis method generated RNA aptamers able to bind to the ribosomal protein. Establishment of the consensus sequence of the aptamers and hydroxyl radical probing studies with Fe^{2+}/EDTA revealed a number of noncanonical features in the stem of the large hairpin in 16S rRNA, known to contain the binding site. Thus the core structure of the binding site comprises three interdependent bases (nt 597–641–643) forming a base triple, a conserved A residue at position 642, and above these elements a base pair (598–640) and a bulged residue at position 595.[108] In the case of human mitochondrial tRNA[Lys], a combined mutational and probing analysis showed the necessity of a methylation at N-1 of A9 for the correct cloverleaf folding of this molecule. In unmodified tRNA transcripts, lack of this methylation that occurs at a Watson–Crick position allows three additional pairings (A8-U65, A9-U64, G10-C63), with the consequence that the transcript folds into an extended hairpin, as deduced from extensive structural probing. In contrast, transcripts with mutations that prevent Watson–Crick pairing between positions 9 and 64 (e.g., replacing the A–U pair by A/A, A/C, or C/U combinations) showed the cloverleaf fold of these molecules, as did the native tRNA with the post-transcriptional modifications.[107]

6.05.4.3　Defining Domains for Biochemical and Biophysical Studies

Biophysical methods, like NMR and crystallography, yield high-resolution data, but require large amounts of material. If a large RNA molecule is the target of such a study, it has proven useful to first study smaller domains of it because of experimental and financial limitations. To define such domains, and to verify that smaller molecules corresponding to these domains display the same or at least similar structural features, structural probing is an extremely useful if not essential tool. The case of 5S RNA illustrates well how structural probing in combination with phylogenetic studies provided the basis for two- and three-dimensional models that in return allowed selection of subdomains for further refining studies with different methods. This ubiquitous RNA, discovered in 1963, is part of the large ribosomal subunit but, despite numerous studies, its function could not be unambiguously determined. The first secondary structure models to emerge were based on phylogenetic comparisons and thermodynamic calculations.[109,110] The secondary structure of 5S RNA is highly conserved in evolution and was soundly confirmed in a number of papers featuring thorough structural probing work and mutational analysis.[111–113] Based on these data, models of 5S RNAs from different origins were constructed. Also resulting from this body of work was the discovery that 5S RNA of different origins can exist in several metastable isoforms, a fact that had all the while hampered the studies. The E-loop emerged as a structurally highly interesting region, because it was shown to be highly structured, despite not being a classical Watson–Crick helix. The structure was also shown to be strongly dependent on the Mg^{2+} concentration. Crystals of entire 5S RNA from *Thermus flavus* had been obtained but reportedly diffracted only to 8 Å resolution.[114] Thus, the structure was broken down into domains previously defined by chemical probing and these were studied separately by NMR and crystallographic techniques.[113,115–117]

6.05.4.4　Detection of Structural Plasticity of Free RNA

Structural probing permits insight into conformational changes of RNA that can be provoked by a variety of parameters including pH, temperature, salt conditions, and even interaction with proteins (see Section 6.05.5.1). Not being prone to effects that take place in crystals, probing reflects the status of a given RNA in solution. Thus, given the choice of an adequate probe, the conditions can be chosen freely and are not unique as for crystallization. Under crystallization conditions, yeast tRNA[Asp] forms anticodon–anticodon interactions with a second tRNA[Asp] molecule, thus representing a model of a tRNA in interaction with an mRNA. The tertiary interaction G18–C56 is disrupted in both molecules. Under milder structural probing conditions, on the other hand, this tertiary interaction is present. However, it is also absent in transcripts devoid of modified bases.[118] Other effects might be induced by agents added for crystallization that would not usually be chosen when an RNA structure is to be monitored.

Temperature-dependent chemical probing can be an alternative to UV melting curves. In addition to information concerning the melting temperature, it gives an indication of the structural domains that melt in a transition. Probing of tRNA[Phe 85] and tRNA[Asp 14] with ENU allowed the direct demonstration of thermal denaturation beginning in the D-stem and continuing in the T-stem. From an opposite point of view, the structuring process of RNA can be monitored. Starting from denatured RNA molecules at high temperatures, the successive formation of structural domains can be followed by disappearing reactivities at lowered temperatures. This allows a more definite assignment of interactions between different regions of the molecule, because residues involved in the same interaction simultaneously become protected below the melting temperature. One of the first three-dimensional RNA models to be constructed according to probing data was that of bovine mt tRNA[Ser]. This model is based on sequence comparison and temperature-dependent structural probing of N-7G, N-7A, and N-3C.[66,119] As is now typical, experiments were carried out in a comparative fashion in the absence and in the presence of Mg^{2+}. The results indicated a stabilization of the tertiary structure by the divalent cations, an effect that had already been observed on canonical tRNAs[120] and through further experiments on other RNAs has proven to be of general importance.[45,88] For ribozymes in particular, divalent cations and salt conditions have proven to be essential for proper structure and function (see Section 6.05.3.2), as evidenced for the *sun*Y group I ribozyme by temperature-dependent structural probing.[121,122]

6.05.5   PROBING RNAs IN COMPLEXES

6.05.5.1   RNA–Protein Complexes

Since structure formation can protect a nucleic acid from being attacked by enzymatic or chemical probes, it is logical that any other molecule that binds to the nucleic acid can also give protection. These protected regions on the nucleic acid then indicate the binding site. RNA–protein complexes were investigated early on by digestion with nucleases (such as nuclease S1 or RNase V1).[17] This approach, however, suffers from the size of the probe (see Section 6.05.2.2) which is often close to the size of the investigated protein and thus might suffer from sterical hindrances.

In vitro footprinting with chemical probes is restricted to complexes where the protein withstands treatment with the probe without substantial damage affecting its binding or functional properties. The chemical probes listed below have been shown to be harmless to various proteins under the experimental conditions used for footprinting. Artifactual data can be minimized if experimental conditions are optimized so that a maximum of RNA molecules are complexed to the protein, and if protein-induced cleavages of the RNA have been excluded in control experiments. The minimal amount of protein can be estimated if K_D or K_m values are available. If not, footprinting should be performed in the presence of increasing amounts of protein. The smaller chemical probes offer a more detailed insight into the interaction pattern between RNA and protein. Thus DMS, ENU, hydroxyl radicals, and iodine are frequently employed probes. The specificities of ENU (phosphates), hydroxyl radicals (riboses), and iodine (thiophosphates) allow probing of all nucleotides of an RNA chain. DMS has slightly less specificity since it probes the base components of A, C, and G residues. These probes have been applied for the investigation of most if not all RNA families and their interacting proteins. Selected examples include tRNAs interacting with aminoacyl-tRNA synthetases,[123] elongation factor,[124] RNase P,[125] modification enzyme,[126] and ribosome.[127] Other examples include rRNAs interacting with ribosomal proteins,[128] mRNAs binding to the ribosome[129] or to translation regulator factors,[130] snRNP RNAs and proteins,[131] as well as viral RNAs interacting with coat proteins[132] or with other viral proteins.[133]

In a few instances, the interaction domains of a protein on an RNA have not only been determined by chemical probing, but also by X-ray crystallography of the complex. Thus, a comparative study of both approaches is feasible. The structure of yeast tRNA[Asp] complexed to its cognate aspartyl-tRNA synthetase has been established at high resolution.[16,134] It has the great advantage of defining at once hydrogen bonds between any type of atom of the RNA and, in addition, to reveal the partners in the amino acids of the protein. This complex has also been investigated by several probes including the very sensitive and phosphate-specific probes ENU[14] and iodine.[135] All three approaches lead to very similar data, namely that the enzyme binds to the tRNA along the side containing the variable region and the T-loop (Figure 3). A detailed analysis reveals several fine differences between the two chemical probes as well as between solution and crystallographic data. Indeed, some phosphates appear either protected by the enzyme or not, depending on the probe. Also, some moderately protected phosphates appear far from the protein in the crystal structure (these protections likely reflect nonspecific interactions between the tRNA and the synthetase).[135] Again, as already pointed out, reactivities do not reflect simple geometrical accessibilities, and the intervention of proteins potentially adds more significance to electrostatic and quantum chemical effects in the modification. Apart from differences arising from the sizes of the probes, the differential protection patterns would also be linked to the fact that ENU can attack either of the two nonesterified oxygens of the phosphate, while iodine only attacks the sulfur that substitutes one of the two oxygens. Whereas the resolution of chemical probes to determine points of contact is clearly higher than that of nucleases, interpretation of footprinting data must be done with some care.

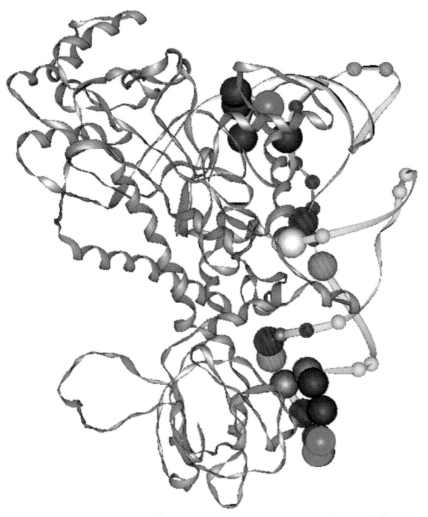

Figure 3 Crystal structure of yeast tRNA^Asp complexed to its cognate synthetase.[16,134] For simplicity, the figure displays only one monomer of the α2 dimeric complex. Spheres indicate contacts of nucleotides with the protein found either by crystallography (blue) or by phosphate mapping as monitored with ENU[14] (red) and iodine[135] (yellow). Coincidences of ENU and iodine protections are marked in orange. Coincidences of crystal structure contacts with probing data are indicated in purple (ENU), green (iodine), and white (ENU and iodine).

Whereas most footprinting experiments are performed *in vitro* with purified couples of macromolecules, investigations *in vivo* have become popular. They consist of a short treatment of growing cells with the probe, followed by quenching, RNA extraction, and searching for modified positions in target RNAs by reverse transcription.[136] Classically, DMS is used because it can readily pass cell membranes and can be quenched by mercaptoethanol. Other probes such as CMCT are also used. Protections of nucleotides are deduced by comparison with *in vitro* modification of free target RNA extracted from untreated cells. Examples of *in vivo* footprinting concern the *E. coli* rRNA[137] or U3A snoRNP from *S. cerevisiae*.[131]

6.05.5.2 RNA–RNA Complexes

Whereas proteins recognize practically all functional groups of RNA with comparable frequency, RNA–RNA interactions are strongly biased towards Watson–Crick interactions of complementary strands, although tertiary interactions are also frequently encountered. Intermolecular RNA–RNA association can be studied in much detail by chemical probing. A recent example is the reverse transcription initiation complex of HIV-1. Reverse transcription of the genomic viral RNA into cDNA requires a short RNA primer to initiate transcription. HIV reverse transcriptase uses tRNA-$^{Lys}_3$ (anticodon s^{4}UUU) which hybridizes to the primer binding site (PBS) of the viral RNA. The formation of this complex can be effected *in vitro* and has been monitored by structural probing with chemical (DMS, DEPC, kethoxal, CMCT) and enzymatic probes (S1, V1, T$_2$). Both RNAs have been probed in their free and complexed state.[138] In the process of primer annealing, the tRNA structure melts and completely rearranges upon hybridization of an 18 nt sequence on the 3'-end to a complementary stretch of the PBS loop. A second domain of functionally essential interaction is the s^{4}UUU tRNA anticodon hybridized to an A-rich loop of the PBS. Interestingly, this second interaction requires thiolation at position 34 for an appreciable efficiency.[138,139] A second interesting example of an RNA–RNA complex is an antisense–sense RNA association in *E. coli*. The secondary structures of the antisense RNA (micF), its binding site on the mRNA (ompF mRNA), and of both molecules when complexed, have been established with both chemical and enzymatic probes.[140] A third example concerns RNA editing of pre-mRNAs in kinetoplastid organisms, which is specified by guide RNAs. Editing involves a covalent binding between both RNAs converting intermolecular to intramolecular interactions. Advantage has been taken of this fact to investigate the secondary structures of the interacting guide complexed to the mRNA as one single molecule with kethoxal, CMCT, DEPC, and DMS.[141]

6.05.5.3 RNA–Antibiotic Complexes

RNAs are primary targets for interaction with many antibiotics. Sensitivity, and thus localization of binding sites for these molecules on various RNAs, has been widely investigated by mutagenic approaches of the RNA but also by footprinting with chemical probes. DMS, kethoxal, and lead have been preferentially used for investigation of the interaction between antibiotics and either rRNAs (both 16S and 23S), ribozymes, or a catalytic group I intron. To be noted is the pioneering work on the interaction of several families of antibiotics with 16S rRNA that allowed definition of concise sets of highly conserved nucleotides as strategic binding domains of the antibiotics.[142]

6.05.6 OTHER APPLICATIONS OF NUCLEOTIDE MODIFICATIONS

Alternative strategies to those described above, that also make use of base modifications in RNA, have been successfully applied to investigate both structural and functional aspects of RNA. These include interference experiments, site-selected insertion of defined modified residues, and exploiting the presence of enzyme-catalyzed modified nucleotides in natural RNAs.

6.05.6.1 Interference Experiments

Interference experiments are aimed to identify at once those elements within an RNA that are involved in a functional process (e.g., binding to a protein or another RNA) and those which are not. Conceptually these experiments resemble classical probing as described above in that they are based on a statistical modification of the RNA in its native conformation. However, the resulting mixed population of molecules is then complexed with the partner molecule and submitted to a selection procedure (e.g., band shift on gel). RNA molecules modified at residues important for interaction with the target molecule are discriminated since they are no longer able to bind. Depending on the experimental strategy and circumstances, either the negatively selected molecules are analyzed for their modified nucleotides or the positively selected molecules are analyzed for all nucleotides that were not modified, i.e., those important for the interaction. All the formerly discussed detection methods can in principle be used. RNA can be modified by external chemicals or during *in vitro* transcription by statistical insertion of internal probes such as phosphorothioates or deoxyribonucleotides. A major innovation in the field is the use of base analogues in combination with thiophosphates. After the interference procedure, iodine cleavage of the thiophosphates provides a means to easily identify the interference site.[143]

A straightforward example was the determination of nucleotides in tRNA important for interaction with nucleotidyl-transferases from different organisms.[144] Different tRNAs from *E. coli* were first modified by DEPC, then end-labeled by 3′-terminal incorporation of [α-^{32}P]ATP by the transferase, then submitted to chain scission. Note the elegance of performing labeling and selection procedures in a single step. An example of application to RNA–RNA interaction refers to the dimerization initiation site of HIV-1 genomic RNA as identified by means of treatment with CMCT, DMS, DEPC, and kethoxal prior to dimerization/hybridization.[145] Monomeric RNA was separated from dimeric RNA by agarose gel electrophoresis under nondenaturing conditions, and the modified residues determined by primer extension. A three-dimensional model of this part of the HIV genomic RNA based on probing data has been published.[75]

6.05.6.2 Site-selected Modification of RNA

Detailed understanding of RNA structure and function may require analysis exceeding simple base mutagenesis. Base analogues may differ from the natural bases by as little as one atom, which is why the technique of chemical and enzymatic incorporation of these modified bases can be seen as atomic mutagenesis. Incorporation by RNA polymerases requires triphosphate of modified nucleotides compatible with enzymatic machinery.[146] 2′-Deoxynucleotides were incorporated into tRNAs to study the impact of 2′OH-groups in structure and function[147] and replacement of U by 4-thiouridines in tRNA^{Phe} during *in vitro* transcription was used to define conjunction sites for attachment of the cleavage reagent 5-iodoacetamido-1,10-*o*-phenanthroline (IoP). This reagent will bind to other RNA such as ribosomal 16S RNA in a reducing environment.[148]

Progress in chemical RNA synthesis, combined with enzymatic cut-and-paste techniques, allows site-selected insertion of modified nucleotides at virtually any desired position of an RNA, thus presenting a powerful tool allowing one to address specific questions on structure–function relationships of these macromolecules. A large number of phosphoramidite derivatives of modified nucleotides, required for chemical synthesis of RNA, have been created and allow insertion of minimally modified heterocyclic bases, ribose rings, and phosphates.[146,149,150] To be mentioned are inosine, isoguanosine, deazaguanosine, deazaadenosines, 2-aminopurines, 2′-fluororibose, 2′-aminoribose, phorphorothioates, and methylphosphates. Much larger modifications can also be introduced, as is the case, for example, with uridine derivatives including photoreactive components.[151]

Utilization of such derivatives has contributed to the understanding of tRNAs,[152,153] ribozymes,[154,155] as well as anti-HIV[156,157] and antisense strategies.[158] Internal insertion of modified elements have found applications in cross-linking experiments, e.g., insertion of disulfide derivatives allows intrahelical cross-linking of RNA chains.[159,160]

Introduction of fluorescent markers to RNA, either internally or at terminal positions, has been a main issue over the years.[161] Chemical insertion of fluorophores (a donor and acceptor, respectively) at two distant positions within the RNA lead to the most interesting fluorescence resonance energy transfer (FRET) experiments which allow the relative spatial orientation of RNA domains to be established. Donor and acceptor fluorophores are classically attached to the extremities of domains and the energy transfer resulting from a dipolar coupling between the transition moments of the two fluorophores measured. This energy transfer is inversely dependent on the distance between the fluorophores. As an example, a three-dimensional model for the hammerhead ribozyme has been proposed on the basis of such an analysis where 5-carboxyfluorescein was the donor and 5-carboxytetramethylrhodamine the acceptor fluorophore.[162] More recently the ion-induced folding of this ribozyme has been evaluated[163] using fluorescein and cyanine-3 as donor and acceptor fluorophores, respectively.

6.05.6.3 Targeting Modified Residues

Natural RNA modification is an enzyme-catalyzed process that takes place in all organisms. Interest in these modified bases steadily increases as they are discovered in more and more different types of RNAs. Also, the number of different modifications of ribonucleotides is on a steady climb.[164] Modified nucleotides stand out among the four major nucleotides thanks to their peculiar properties which make them preferential targets for investigators. Many examples in the older literature within the tRNA field, the RNA family where most modifications are found, concern bases such as wybutosine or m^7G[165,166] for specific strand scission, or modified riboses which hinder enzymatic cleavage. Alternatively, the photoreactive nucleotide s^4U and its derivative s^4T have been used for cross-linking studies. Many natural modifications are also used as insertion points for additional modifications. Thus, for example, the undermodified *E. coli* tRNA^{Tyr} possessing a 7-(aminomethyl)-7-deazaguanosine instead of Q can be modified chemically at this specific position by dansylchloride, a fluorescent compound.[167] *E. coli* tRNA^{fMet} has been derivatized with side chains containing a disulfide bond capable of reaction with cysteine residues of a protein and with *N*-hydroxysuccinimide ester groups capable of coupling to lysine. This tRNA has been used for cross-linking to methionyl-tRNA synthetase.[168] Along the same view, t^6A and acp^3U are target sites for introduction of photoreactive azidointrophenyl probes which allow cross-linking at 20 Å distance.[169]

6.05.7 PERSPECTIVES

The field of structural probing is in constant motion and the new trends that are emerging concern both novel fields of application and development of new methodological tools. Trends towards developing a deeper understanding of RNA structure in solution point to finding novel nonclassical structural features, like noncanonical base pairs and tertiary interactions, to approaching the dynamics of RNA as well as its folding and unfolding behavior, and to searching for conformational changes and alternate structures. The development of new RNA cleaving reagents will continue to be a challenge for chemists. Even though research will take manifold directions, the authors believe that two principal directions will be favored. First, for the design of more efficient cleavers and, second, for that of structure-specific ligands on which they will be tethered. In other words, chemical constructs will be engineered that like enzymes comprise an active site and a ligand binding domain. Because of their size, which can be small, the artificial RNases will have recognition specificities very different from those of the protein RNases. An interesting related issue is the use of antisense oligonucleotides as ligands to confer sequence specificity to the artificial RNases. Another innovation is the development of time-resolved probing methods.[39]

6.05.8 REFERENCES

1. B. Singer, *Nature*, 1976, **264**, 333.
2. R. W. Holley, J. Apgar, G. A. Everett, J. T. Madison, M. Marquisee, S. H. Merrill, J. R. Penswick, and A. Zamir, *Science*, 1965, **147**, 1462.
3. A. D. Branch, B. J. Benenfeld, and H. D. Robertson, *Methods Enzymol.*, 1989, **180**, 130.
4. H. Donis-Keller, A. M. Maxam, and W. Gilbert, *Nucleic Acids Res.*, 1977, **4**, 2527.
5. A. M. Maxam and W. Gilbert, *Proc. Natl. Acad. Sci. USA*, 1977, **74**, 560.
6. F. Sanger, S. Nicklen, and A. R. Coulson, *Proc. Natl. Acad. Sci. USA*, 1977, **74**, 5463.
7. R. Lavery, A. Pullman, B. Pullman, and M. d. Oliveira, *Nucleic Acids Res.*, 1980, **8**, 5095.
8. S. R. Holbrook and S. H. Kim. *Biopolymers*, 1983, **22**, 1145.
9. D. A. Peattie, *Proc. Natl. Acad. Sci. USA*, 1979, **76**, 1760.
10. D. A. Peattie and W. Gilbert, *Proc. Natl. Acad. Sci. USA*, 1980, **77**, 4679.
11. C. Ehresmann, F. Baudin, M. Mougel, P. Romby, J.-P. Ebel, and B. Ehresmann, *Nucleic Acids Res.*, 1987, **15**, 9109.
12. N. A. Kolchanov, I. I. Titov, I. E. Vlassova, and V. V. Vlassov, *Prog. Nucleic Acid Res. Mol. Biol.*, 1996, **53**, 131.
13. V. V. Vlassov, R. Giegé, and J.-P. Ebel, *FEBS Lett.*, 1980, **120**, 12.
14. P. Romby, D. Moras, M. Bergdoll, P. Dumas, V. V. Vlassov, E. Westhof, J.-P. Ebel, and R. Giegé, *J. Mol. Biol.*, 1985, **184**, 455.
15. A. Garcia and R. Giegé, *Biochem. Biophys. Res. Commun.*, 1992, **186**, 956.
16. M. Ruff, S. Krishnaswamy, M. Boeglin, A. Poterszman, A. Mitschler, A. Podjarny, B. Rees, J.-C. Thierry, and D. Moras, *Science*, 1991, **252**, 1682.
17. O. O. Favorova, F. Fasiolo, G. Keith, S. K. Vassilenko, and J.-P. Ebel, *Biochemistry*, 1981, **20**, 1006.
18. G. Knapp, *Methods Enzymol.*, 1989, **180**, 192.
19. L. Lempereur, M. Nicoloso, N. Riehl, C. Ehresmann, B. Ehresmann, and J.-P. Bachcllerie, *Nucleic Acids Res.*, 1985, **13**, 8339.
20. F. Cramer, *Prog. Nucleic Acid Res. Mol. Biol.*, 1971, **11**, 391.

21. W. Wintermeyer and H. G. Zachau, *FEBS Lett.*, 1970, **11**, 160.
22. W. Wintermeyer and H. G. Zachau, *FEBS Lett.*, 1975, **58**, 306.
23. D. Rhodes, *J. Mol. Biol.*, 1975, **94**, 449.
24. H. Hayatsu, *J. Biochem.* (*Tokyo*), 1996, **119**, 391.
25. I. L. Batey and D. M. Brown, *Biochim. Biophys. Acta*, 1977, **474**, 378.
26. W. J. Krzyosiak and J. Ciesolka, *Nucleic Acids Res.*, 1983, **11**, 6913.
27. P. Romby, D. Moras, P. Dumas, J.-P. Ebel, and R. Giegé, *J. Mol. Biol.*, 1987, **195**, 193.
28. M. Litt, *Biochemistry*, 1969, **8**, 3249.
29. R. C. Gamble and P. R. Schimmel, *Proc. Natl. Acad. Sci. USA*, 1974, **71**, 1356.
30. K. Watanabe and F. Cramer, *Eur. J. Biochem.*, 1978, **89**, 425.
31. D. W. Celander and J. M. Nussbaum, *Biochemistry*, 1996, **35**, 12 061.
32. A. Garcia, R. Giegé, and J.-P. Behr, *Nucleic Acids Res.*, 1990, **18**, 89.
33. T. Hermann and H. Heumann, *RNA*, 1995, **1**, 1009.
34. C. S. Chow, L. S. Behlen, O. C. Uhlenbeck, and J. K. Barton, *Biochemistry*, 1992, **31**, 972.
35. C. S. Chow and J. K. Barton, *Biochemistry*, 1992, **31**, 5423.
36. J. A. Latham and T. R. Cech, *Science*, 1989, **245**, 276.
37. C. P. H. Vary and J. N. Vournakis, *Proc. Natl. Acad. Sci. USA*, 1984, **81**, 6978.
38. M. Götte, R. Marquet, C. Isel, V. E. Anderson, G. Keith, H. J. Gross, C. Ehresmann, B. Ehresmann, and H. Heumann, *FEBS Lett.*, 1996, **390**, 226.
39. B. Sclavi, S. Woodson, M. Sullivan, M. R. Chance, and M. Brenowitz, *J. Mol. Biol.*, 1997, **266**, 144.
40. C. E. Holmes, A. T. Abraham, S. M. Hecht, C. Florentz, and R. Giegé, *Nucleic Acids Res.*, 1996, **24**, 3399.
41. C. J. Burrows and S. E. Rokita, *Acc. Chem. Res.*, 1994, **27**, 295.
42. N. Shinriki, K. Ishizaki, A. Ikehata, K. Miura, T. Ueda, N. Kato, and F. Harada, *Symposium Series Nucleic Acids Res.*, 1981, **10**, 211.
43. P. Burgstaller, T. Hermann, C. Huber, E. Westhof, and M. Famulok, *Nucleic Acids Res.*, 1997, **25**, 4018.
44. A.-C. Dock-Bregeon and D. Moras, *Cold Spring Harbor Symp. Quant. Biol.*, 1987, **52**, 113.
45. T. Pan, D. M. Long, and O. C. Uhlenbeck, in "The RNA World," eds. R. F. Gesteland and J. F. Atkins, Cold Spring Harbor Laboratory Press, Cold Spring Harbor, NY, 1993.
46. M. Matsuo, T. Yokogawa, K. Nishikawa, K. Watanabe, and N. Okada, *J. Biol. Chem.*, 1995, **270**, 10 097.
47. T. Marciniec, J. Ciesiolka, J. Wrzesinski, and W. J. Krzyosiak, *FEBS Lett.*, 1989, **243**, 293.
48. B. Streicher, E. Westhof, and R. Schroeder, *EMBO J.*, 1996, **15**, 2556.
49. C. Werner, B. Krebs, G. Keith, and G. Dirheimer, *Biochim. Biophys. Acta*, 1976, **432**, 161.
50. R. S. Brown, J. C. Dewan, and A. Klug, *Biochemistry*, 1985, **24**, 4785.
51. M. Komiyama, *J. Biochem.* (*Tokyo*), 1995, **118**, 665.
52. R. Breslow, D.-L. Huang, and E. Anslyn, *Proc. Natl. Acad. Sci. USA*, 1989, **86**, 1746.
53. V. V. Vlassov, G. Zuber, B. Felden, J.-P. Behr, and R. Giegé, *Nucleic Acids Res.*, 1995, **23**, 3161.
54. M. A. Podyminogin, V. V. Vlassov, and R. Giegé, *Nucleic Acids Res.*, 1993, **21**, 5950.
55. R. M. Wurst, J. N. Vournakis, and A. M. Maxam, *Biochemistry*, 1978, **17**, 4493.
56. P. Wrede, R. Wurst, J. Vournakis, and A. Rich, *J. Biol. Chem.*, 1979, **254**, 9608.
57. P. E. Auron, L. D. Weber, and A. Rich, *Biochemistry*, 1982, **21**, 4700.
58. M. Garret, P. Romby, R. Giegé, and S. Litvak, *Nucleic Acids Res.*, 1984, **12**, 2259.
59. C. P. H. Vary and J. N. Vournakis, *Nucleic Acids Res.*, 1984, **12**, 6763.
60. C. Florentz, J.-P. Briand, P. Romby, L. Hirth, J.-P. Ebel, and R. Giegé, *EMBO J.*, 1982, **1**, 269.
61. A. Przykorska, C. E. Adlouni, G. Keith, J. W. Szarkowski, and G. Dirheimer, *Nucleic Acids Res.*, 1992, **20**, 659.
62. K. S. Aultmann and S. H. Chang, *Eur. J. Biochem.*, 1982, **124**, 471.
63. M. Zucker, *Science*, 1989, **244**, 48.
64. C. Gaspin and E. Westhof, *J. Mol. Biol.*, 1995, **254**, 163.
65. D. Gautheret and R. R. Gutell, *Nucleic Acids Res.*, 1997, **25**, 1559.
66. M. H. L. de Bruijn and A. Klug, *EMBO J.*, 1983, **2**, 1309.
67. A. C. Dock-Bregeon, E. Westhof, R. Giegé, and D. Moras, *J. Mol. Biol.*, 1989, **206**, 707.
68. C. Baron, E. Westhof, A. Böck, and R. Giegé, *J. Mol. Biol.*, 1993, **231**, 274.
69. P. Dumas, D. Moras, C. Florentz, R. Giegé, P. Verlaan, A. van Belkum, and C. W. A. Pleij, *J. Biomol. Struct. Dyn.*, 1987, **4**, 707.
70. B. Felden, C. Florentz, R. Giegé, and E. Westhof, *J. Mol. Biol.*, 1994, **235**, 508.
71. B. Felden, J. F. Atkins, and R. F. Gesteland, *Nature Struct. Biol.*, 1996, **3**, 494.
72. B. Felden, H. Himeno, A. Muto, J. P. McCutcheon, J. F. Atkins, and R. F. Gesteland, *RNA*, 1997, **3**, 89.
73. T. Hermann, B. Schmid, H. Heumann, and H. U. Göringer, *Nucleic Acids Res.*, 1997, **25**, 2311.
74. R. Walczak, E. Westhof, P. Carbon, and A. Krol, *RNA*, 1996, **2**, 367.
75. J.-C. Paillart, E. Westhof, C. Ehresmann, B. Ehresmann, and R. Marquet. *J. Mol. Biol.*, 1997, **270**, 36.
76. M. E. Harris, J. M. Nolan, A. Malhotra, J. W. Brown, S. C. Harvey, and N. R. Pace, *EMBO J.*, 1994, **13**, 3953.
77. E. Westhof, D. Wesolowski, and S. Altman, *J. Mol. Biol.*, 1996, **258**, 600.
78. V. Lehnert, L. Jaeger, F. Michel, and E. Westhof, *Chem. Biol.*, 1996, **3**, 993.
79. L. Jaeger, F. Michel, and E. Westhof, in "Nucleic Acids and Molecular Biology," ed. F. Eckstein, Springer, Berlin, 1996.
80. S. Furois-Corbin and A. Pullman, *Biophys. Chem.*, 1985, **22**, 1.
81. R. Lavery and A. Pullman, *Biophys. Chem.*, 1984, **19**, 171.
82. R. Lavery, A. Pullman, and B. Pullman, *Nucleic Acids Res.*, 1980, **8**, 1061.
83. J. D. Robertus, J. E. Ladner, J. T. Finch, D. Rhodes, R. S. Brown, B. F. C. Clark, and A. Klug, *Nature* 1974, **250**, 546.
84. S. H. Kim, F. L. Suddath, G. J. Quigley, A. McPherson, J. L. Sussman, A. H. J. Wang, N. C. Seeman, and A. Rich, *Science*, 1974, **185**, 435.
85. V. V. Vlassov, R. Giegé, and J.-P. Ebel, *Eur. J. Biochem.*, 1981, **119**, 51.
86. C. Massire, C. Gaspin, and E. Westhof, *J. Mol. Graphics*, 1994, **12**, 201.

87. A. J. Chandler and A. J. Kirby, *J. Chem. Soc., Chem. Commun.*, 1992, 1769.
88. P. Brion and E. Westhof, *Annu. Rev. Biophys. Biomol. Struct.*, 1997, **26**, 113.
89. L. S. Behlen, J. R. Sampson, A. B. DiRenzo, and O. C. Uhlenbeck, *Biochemistry*, 1990, **29**, 2515.
90. D. Michalowski, J. Wrzesinski, and W. Krzyzosiak, *Biochemistry*, 1996, **35**, 10 727.
91. T. Pan and O. C. Uhlenbeck, *Biochemistry*, 1992, **31**, 3887.
92. W. Wintermeyer and H. G. Zachau, *Biochim. Biophys. Acta*, 1973, **299**, 82.
93. A. E. Oleson and M. Sasakuma, *Arch. Biochem. Biophys.*, 1980, **204**, 361.
94. R. Breslow and M. Labelle, *J. Am. Chem. Soc.*, 1986, **108**, 2655.
95. R. Breslow, *Acc. Chem. Res.*, 1991, **24**, 317.
96. P. W. Huber, *FASEB J.*, 1993, **7**, 1367.
97. D. S. Sigman and C.-h. B. Chen, *Annu. Rev. Biochem.*, 1990, **59**, 207.
98. S. Joseph and H. F. Noller, *EMBO J.*, 1996, **15**, 910.
99. K. S. Wilson and H. F. Noller, *Cell*, 1998, **92**, 131.
100. A. Krol and P. Carbon, *Methods Enzymol.*, 1989, **180**, 212.
101. J. D. Puglisi and I. J. Tinoco, *Methods Enzymol.*, 1989, **180**, 304.
102. R. Giegé, M. Frugier, and J. Rudinger, *Curr. Opin. Struct. Biol.*, 1998, **8**, 286.
103. K. Rietveld, R. Van Poelgeest, C. W. A. Pleij, J. H. Van Boom, and L. Bosch. *Nucleic Acids Res.*, 1982, **10**, 1929.
104. C. W. A. Pleij, *Curr. Opin. Struct. Biol.*, 1994, **4**, 337.
105. J. Rudinger, B. Felden, C. Florentz, and R. Giegé, *Bioorg. Med. Chem.*, 1997, **5**, 1001.
106. M. H. Kolk, M. van der Graaf, S. S. Wijmenga, C. W. A. Pleij, H. A. Heus, and C. W. Hilbers, *Science*, 1998, **280**, 434.
107. M. Helm, H. Brulé, F. Degoul, C. Cepanec, J.-P. Leroux, R. Giegé, and C. Florentz, *Nucleic Acids Res.*, 1998, **26**, 1636.
108. H. Moine, C. Cachia, E. Westhof, B. Ehresmann, and C. Ehresmann, *RNA*, 1997, **3**, 255.
109. K. Nishikawa and S. Takemura, *J. Biochem. (Tokyo)*, 1974, **76**, 935.
110. G. W. Fox and C. R. Woese, *Nature*, 1975, **256**, 505.
111. P. J. Romaniuk, I. Leal de Stevenson, C. Ehresmann, P. Romby, and B. Ehresmann, *Nucleic Acids Res.*, 1988, **16**, 2295.
112. C. Brunel, P. Romby, E. Westhof, C. Ehresmann, and B. Ehresmann, *J. Mol. Biol.*, 1991, **221**, 293.
113. H. Moine, B. Ehresmann, C. Ehresmann, and P. Romby, in "RNA Structure and Function," eds. R. W. Simons and M. Grunberg-Manago, Cold Spring Harbor Laboratory Press, Cold Spring Harbor, NY, 1998.
114. S. Lorenz, C. Betzel, E. Raderschall, Z. Dauter, K. S. Wilson, and V. A. Erdmann, *J. Mol. Biol.*, 1991, **219**, 399.
115. C. Betzel, S. Lorenz, J. P. Furste, R. Bald, M. Zhang, T. R. Schneider, K. S. Wilson, and V. A. Erdmann, *FEBS Lett.*, 1994, **351**, 159.
116. S. A. White, M. Nilges, A. Huang, A. T. Brünger, and P. B. Moore, *Biochemistry*, 1992, **31**, 1610.
117. C. C. Correll, B. Freeborn, P. B. Moore, and T. A. Steitz, *Cell*, 1997, **91**, 705.
118. V. Perret, A. Garcia, J. D. Puglisi, H. Grosjean, J.-P. Ebel, C. Florentz, and R. Giegé, *Biochimie*, 1990, **72**, 735.
119. M. H. L. de Bruijn, P. H. Schreier, I. C. Eperon, and B. G. Barell, *Nucleic Acids Res.*, 1980, **8**, 5213.
120. D. Rhodes, *Eur. J. Biochem.*, 1977, **81**, 91.
121. A. R. Bannerjee, J. A. Jaeger, and D. H. Turner, *Biochemistry*, 1993, **32**, 153.
122. L. Jaeger, E. Westhof, and F. Michel, *J. Mol. Biol.*, 1993, **234**, 331.
123. A. Théobald, M. Springer, M. Grunberg-Manago, J.-P. Ebel, and R. Giegé, *Eur. J. Biochem.*, 1988, **175**, 511.
124. D. Otzen, J. Barciszewski, and B. F. C. Clark, *Biochem Mol. Biol. Int.*, 1993, **31**, 95.
125. T. E. LaGrandeur, A. Huttenhofer, H. F. Noller, and N. R. Pace, *EMBO J.*, 1994, **13**, 3945.
126. J. Gabryszuk and W. M. Holmes, *RNA*, 1997, **3**, 1327.
127. A. Hüttenhofer and H. F. Noller, *Proc. Natl. Acad. Sci. USA*, 1992, **89**, 7851.
128. T. Uchiumi and R. Kominami, *J. Biol. Chem.*, 1997, **272**, 3302.
129. A. Hüttenhofer and H. F. Noller, *EMBO J.*, 1994, **13**, 3892.
130. J. Schlegl, V. Gegout, B. Schlager, M. W. Hentze, E. Westhof, C. Ehresmann, B. Ehresmann, and P. Romby, *RNA*, 1997, **3**, 1159.
131. A. Mereau, R. Fournier, A. Gregoire, A. Mougin, P. Fabrizio, R. Luhrmann, and C. Branlant, *J. Mol. Biol.*, 1997, **273**, 552.
132. L. Pearson, C. B. Chen, R. P. Gaynor, and D. S. Sigman, *Nucleic Acids Res.*, 1994, **22**, 2255.
133. D. Brown, J. Brown, C. H. Kang, L. Gold, and P. Allen, *J. Biol. Chem.*, 1997, **272**, 14 969.
134. J. Cavarelli, B. Rees, M. Ruff, J.-C. Thierry, and D. Moras, *Nature*, 1993, **362**, 181.
135. J. Rudinger, J. D. Puglisi, J. Pütz, D. Schatz, F. Eckstein, C. Florentz, and R. Giegé, *Proc. Natl. Acad. Sci. USA*, 1992, **89**, 5882.
136. M. Ares, Jr. and A. H. Igel, *Genes Dev.*, 1990, **4**, 2132.
137. M. Laughrea and J. Tam, *Biochemistry*, 1992, **31**, 12 035.
138. C. Isel, J.-M. Lanchy, S. F. J. Le Grice, C. Ehresmann, B. Ehresmann, and R. Marquet, *EMBO J.*, 1996, **15**, 917.
139. C. Isel, C. Ehresmann, G. Keith, B. Ehresmann, and R. Marquet, *J. Mol. Biol.*, 1995, **247**, 236.
140. M. Schmidt, P. Zheng, and N. Delihas, *Biochemistry*, 1995, **34**, 3621.
141. B. Schmid, L. K. Read, K. Stuart, and H. U. Goringer, *Eur. J. Biochem.*, 1996, **240**, 721.
142. D. Moazed and H. F. Noller, *Nature*, 1987, **327**, 389.
143. S. A. Strobel and K. Shetty, *Proc. Natl. Acad. Sci. USA*, 1997, **94**, 2903.
144. P. Spacciapoli, L. Doviken, J. J. Mulero, and D. L. Thurlow, *J. Biol. Chem.*, 1989, **264**, 3799.
145. E. Skripkin, J.-C. Paillart, R. Marquet, B. Ehresmann, and C. Ehresmann, *Proc. Natl. Acad. Sci. USA*, 1994, **91**, 4945.
146. R. A. Zimmermann, M. J. Gait, and M. J. Moore, In "Modification and Editing of RNA," eds. H. Grosjean and R. Benne, American Socely of Microbiology Press, Washington, DC, 1998.
147. R. Aphasizhev, A. Théobald-Dietrich, D. Kostyuk, S. N. Kochetkov, L. Kisselev, R. Giegé, and F. Fasiolo, *RNA*, 1997, **3**, 893.
148. J. Bullard, M. van Waes, D. Bucklin, M. Rice, and W. Hill, *Biochemistry*, 1998, **37**, 1350.
149. N. Usman and R. Cedergren, *Trends Biochem.*, 1992, **17**, 334.

150. P. F. Agris, A. Malkiewicz, A. Kraszewski, K. Everett, B. Nawrot, E. Sochacka, J. Jankowska, and R. Guenther, *Biochimie*, 1995, **77**, 125.
151. B. E. Eaton and W. A. Pieken, *Annu. Rev. Biochem.*, 1995, **64**, 837.
152. K. Musier-Forsyth and P. Schimmel, *Nature*, 1992, **357**, 513.
153. K. Musier-Forsyth, N. Usman, S. Scaringe, J. Doudna, R. Green, and P. Schimmel, *Science*, 1991, **253**, 784.
154. L. Beigelman, A. Karpeisky, J. Matulic-Adamic, P. Haeberli, D. Sweedler, and N. Usman. *Nucleic Acids Res.*, 1995, **23**, 4434.
155. F. Seela, H. Debelak, N. Usman, A. Burgin, and L. Beigelman, *Nucleic Acids Res.*, 1998, **26**, 1010.
156. H. Aurup, A. Siebert, F. Benseler, D. Williams, and F. Eckstein, *Nucleic Acids Res.*, 1994, **22**, 4963.
157. K. Shah, H. Neenhold, Z. Y. Wang, and T. M. Rana, *Bioconjugate Chem.*, 1996, **7**, 283.
158. R. H. Griffey, B. P. Monia, L. L. Cummings, S. Freier, M. J. Greig, C. J. Guinosso, E. Lesnik, S. M. Manalili, V. Mohan, S. Owens *et al.*, *J. Med. Chem.*, 1996, **39**, 5100.
159. S. T. Sigurdsson, T. Tuschl and F. Eckstein, *RNA*, 1995, **1**, 575.
160. D. J. Earnshaw, B. Masquida, S. Muller, S. Sigurdsson, F. Eckstein, E. Westhof, and M. J. Gait, *J. Mol. Biol.*, 1997, **274**, 197.
161. Y. Kinoshita, K. Nishigaki, and Y. Husimi, *Nucleic Acids Res.*, 1997, **25**, 3747.
162. T. Tuschl, C. Gohlke, T. M. Jovin, E. Westhof, and F. Eckstein, *Science*, 1994, **266**, 785.
163. G. S. Bassi, A. I. H. Murchie, F. Walter, R. M. Clegg, and D. M. J. Lilley, *EMBO J.*, 1997, **16**, 7481.
164. J. A. McCloskey and P. F. Crain, *Nucleic Acids Res.*, 1998, **26**, 196.
165. K. Harbers, R. Thiebe, and H. G. Zachau, *Eur. J. Biochem.*, 1972, **26**, 132.
166. V. S. Zueva, A. S. Mankin, A. A. Bogdanov, and L. A. Baratova, *Eur. J. Biochem.*, 1985, **146**, 679.
167. H. Kasai, N. Shindo-Okada, S. Noguchi, and S. Nishimura, *Nucleic Acids Res.*, 1979, **7**, 231.
168. D. Valenzuela, O. Leon, and L. H. Schulman, *Biochem. Biophys. Res. Commun.*, 1984, **119**, 677.
169. J. Podkowinski, T. Dymarek-Babs, and P. Gornicki, *Acta Biochm. Pol.*, 1989, **36**, 235.

6.06
Chemical RNA Synthesis (Including RNA with Unusual Constituents)

YASUO KOMATSU and EIKO OHTSUKA
Hokkaido University, Sapporo, Japan

6.06.1 INTRODUCTION

The chemical synthesis of RNA has now been developed to the stage of automated large scale production providing the substantial amounts needed for physicochemical and structural studies of RNA. As these synthetic methods are often taken for granted by today's researchers the authors give a brief overview of the methods currently used and discuss their scope and limitations. No effort is made to provide a complete coverage of all the synthetic achievements; rather the authors would like to present an instructive chapter to scientists in related areas.

Although the principle of the coupling reaction is the same in both DNA and RNA syntheses, obtaining a quantity of RNA is more difficult compared to the DNA synthesis because of lower yields probably due to less accessibility of the 3'- 5'-functional group of 2'-O-protected ribo-derivatives during condensation. The yields of oligoribonucleotides depend also on the efficiency in the deprotection and separation from the side products. A variety of combinations of protection methods for the 2'-hydroxyl and 5'-hydroxyl groups have been employed for RNA syntheses. Larger molecules such as tRNAs consisting of approximately 80 nucleotides have been obtained by solid-phase synthesis using the phosphoramidite approach, although the overall yield of the product is much lower than that of DNA of similar size. Practical sizes of RNA for physicochemical studies are still limited to between 10 and 20 nucleotides. Enzymatic syntheses of RNA by transcription or enzymatic ligation of oligoribonucleotides have been used for larger oligoribonucleotides to compensate for the limitations of RNA synthesis.

The two approaches used at present are the solid-phase synthesis of RNA and synthesis in solution. Solid-phase synthesis has become the method of choice as will be described later. Solution-phase synthesis of RNA has some advantages for developing new protecting groups or catalysts for the reaction. The new strategies are important for the synthesis of unusual RNA molecules. There are several approaches to incorporating modified nucleotides into RNA. Suitably protected synthons for the phosphoramidite method are the most straightforward approach, if the modified nucleosides are stable under the conditions used in the synthesis. Although postsynthetic modifications of deoxyribooligonucleotides have been studied, and a few experiments have been reported for the ribo-series, this method needs to be developed further. The replacement of modified RNA fragments is another method of incorporating unusual nucleosides in defined positions, provided that methods for specific cleavage and joining of the RNA fragments are available.

This review will discuss some protecting groups and coupling reagents used in RNA synthesis, and then describe methods for the synthesis of RNA derivatives that are useful for studies of the biological and structural functions of RNA. A few examples of the applications of these RNA derivatives are also included. For the enzymatic incorporation of modified nucleotides, see Chapter 6.14.

6.06.2 SYNTHETIC METHODS FOR OLIGONUCLEOTIDES

In the late 1950s, Khorana and co-workers developed phosphodiester methods, which involved the 5'-phosphomonoester of one nucleoside and the 3'-OH of another, using dicyclohexylcarbodiimide (DCC).[1] This approach was used in the synthesis of the 64 ribo-triplets for elucidation for the genetic code.[2] In the 1970s, as has been described elsewhere,[3] oligonucleotides were synthesized by the phosphotriester method (Scheme 1). Nucleoside phosphodiesters (1) were coupled with a 5'-free nucleoside (2) in the presence of a condensing agent, such as 1-(mesitylene-2-sulfonyl)-3-nitro-1,2,4-triazole (3) (MSNBT).[4] Letsinger and co-workers introduced the phosphite coupling approach using phosphorodichloridite,[5] and several oligoribonucleotides were also synthesized. However, the instability of the intermediate prohibited further application to larger molecules. The phos-phoramidite method was developed by Caruthers and co-workers to solve this problem, by replacing one of the chlorines on the phosphorodichloridite with a secondary amine[6] (Scheme 2). The amidite unit (5) is condensed with (2) in the presence of tetrazole (6). This method has been widely used to synthesize oligodeoxyribonucleotides by solid-phase synthesis, and more recently for the ribo series (see Section 6.06.3.2 for 2'-O-protecting groups). The H-phosphonate method has been improved and found to be suitable for modifying the internucleotidic phosphate.[7,8] As shown in Scheme 3, the 3'-O-(H-phosphonate) (8) is coupled with the 5'-hydroxyl group using a hindered acyl chloride (9). Since the diester (10) is stable under the conditions of the further condensation, the oxidation can be performed at the end of the synthesis. This approach has been applied to the synthesis of biologically active RNA fragments.[9]

The chemical synthesis of oligoribonucleotides requires extra steps, due to the presence of the 2'-hydroxyl group. The 2'-OH of the ribonucleoside has to be continuously protected until the end of the condensation, because a free 2'-OH easily breaks the RNA strand under alkaline conditions, and the 3'-5'-phosphodiester linkage migrates to a 2'-5' linkage under acidic conditions. The 2'-protecting group must be stable during the condensing reactions of the nucleoside units, as well as during the deprotection of the 5'-OH. In the following section, the protecting groups that are frequently used in amidite chemistry are described.

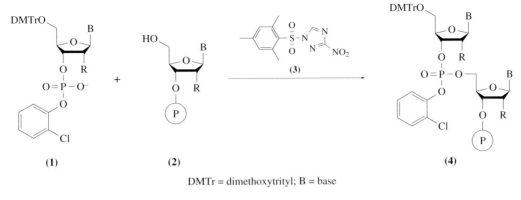

DMTr = dimethoxytrityl; B = base

Scheme 1

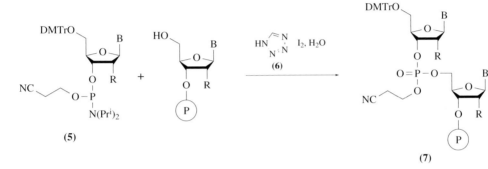

Scheme 2

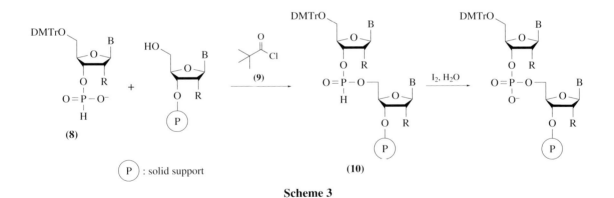

(P) : solid support

Scheme 3

6.06.3 PROTECTING GROUPS

6.06.3.1 5′-OH Protection

Figure 1 shows some protecting groups for the hydroxyl groups. The trityl derivatives (dimethoxytrityl, DMTr (**11**);[10] 9-phenylxanthen-9-yl, (**12**); and 9-*p*-methoxyphenylxanthen-9-yl, Mox (**13**)[11,12,13]) are widely used to protect the 5′-hydroxyl group. Trityl chlorides alkylate the primary 5′-OH preferentially, compared to the secondary hydroxyl groups, because of steric hindrance. These protecting groups are removed by acids. Groups (**12**) and (**13**) are more acid-labile than (**11**). The lipophilicity of the 5′-*O*-terminal dimethoxytrityl derivatives allows rapid purification of the crude tritylated oligonucleotides by reversed-phase chromatography on C-18 silica gel.

When ketal/acetal groups are used to protect the 2′-OH, they are partially cleaved during the detritylation steps, and the yield of the oligoribonucleotides is decreased. To overcome this problem,

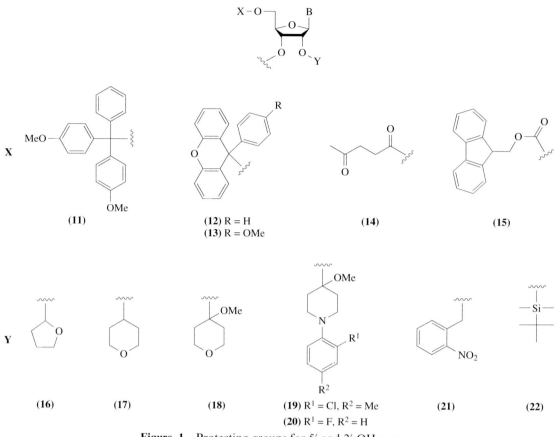

Figure 1   Protecting groups for 5′ and 2′-OH groups.

base-labile groups (levulinyl (**14**); 9-fluorenylmethyl-oxycarbonyl, Fmoc (**15**) are used for the 5′-OH protection. van Boom and co-workers developed the use of 5′-*O*-levulinyl protection for RNA synthesis.[14] 5′-*O*-Levulinyl and 2′-*O*-terahydrofuranyl groups have been combined (X = (**14**), Y = (**16**)).[15] The levulinyl group can be removed after each coupling step by treatment with 0.5 M hydrazine monohydrate in pyridine-acetic acid (3:2 v/v). Gait and co-workers utilized (**15**) for the 5′-*O*-protection and 4-methoxytetrahydropyran-4-yl (**18**) for the 2′-OH (X = (**15**), Y = (**18**)).[16] The Fmoc (**15**) was removed after each coupling step by treatment with 0.1 M DBU in acetonitrile. Since exocyclic amino groups and phosphate are usually protected by base-labile protecting groups, the basic conditions for the deprotection of these groups should be carefully optimized.

6.06.3.2   2′-OH Protection

6.06.3.2.1   Acetal/ketal groups

Acetal and ketal groups have been used as protecting groups for the 2′-OH. These functions are removed under acidic conditions. The ketal groups are deprotected more rapidly than the acetal type and 4-methoxytetrahydropyran-4-yl (**18**) which is removed in milder conditions than tetra-hydropyranyl (**17**) has the advantage of no chirality.[17]

Although the 5′-*O*-dimethoxytrityl group can be deprotected under mild acidic conditions, the acetal and ketal groups are partially removed during the solid-phase synthesis of RNA, which leads to the breakage of the internucleotidic bonds by the following treatment with concentrated ammonia. It is important to prevent the loss of the 2′-OH protecting groups during the deprotection of the 5′-OH. New derivatives 1-(2-chloro-4-methylphenyl)-4-methoxypiperidin-4-yl (**19**) and 1-[(2-fluoro) phenyl]-4-methoxypiperidin-4-yl (Fpmp, (**20**)) were proposed as 2′-OH protecting groups by Reese and co-workers.[18,19] Under strong acidic conditions, the nitrogen atom of the piperidine ring of these protecting groups is protonated, which inhibits cleavage of the 2′-*O*-ketal function. At a lower proton activity (pH2), the nitrogen atom is largely unprotonated, and then the 2′-*O*-ketal function is cleaved easily. Since the dihydro derivative of (**19**) is difficult to prepare, (**20**) is frequently used to protect the 2′-OH. Scheme 4 shows the synthesis of 5′-*O*-Px-2′-*O*-Fpmp amidite units (**26**). The exocyclic amino-protected nucleoside is protected with a bifunctional reagent, 1,3-dichloro-1,1,3,3-tetraisopropyldisiloxane (**27**) (Markiewicz reagent), to give the 3′,5′-TBDMS-nucleoside (**23**).[20,21] The 2′-OH of (**23**) is protected to yield (**20**), and the 3′- and 5′-protecting groups are removed by fluoride anions to yield a 2-*O*-Fpmp nucleoside (**24**). After the 5′-OH of (**24**) is protected by Px-chloride, the 3′-OH of (**25**) is subjected to phosphitylation by 2-cyanoethyl-*N*,*N*-diisopropylchloro-phosphoramidite (**28**).

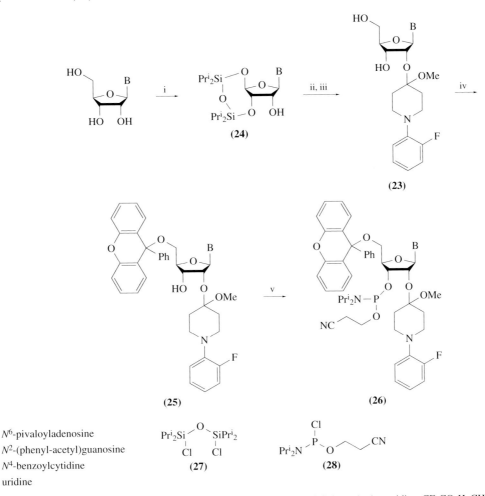

N^6-pivaloyladenosine
N^2-(phenyl-acetyl)guanosine
N^4-benzoylcytidine
uridine

i, (**27**), imidazole, MeCN; ii, 1-(2-fluorophenyl)-4-methoxy-1,2,3,6-tetrahydropyridine, CF$_3$CO$_2$H, CH$_2$Cl$_2$;
iii, Et$_4$NF, MeCN; iv, 9-chloro-9-phenyl-xanthene, pyridine; v, (**28**), THF, EtN(Pr^i)$_2$

Scheme 4

6.06.3.2.2 o-*Nitrobenzyl group*

A physical analysis suggested that the *o*-nitrobenzyl ethers of nucleosides, ONB (**21**), are stable under acidic and alkaline conditions, and can be removed by UV light (280 nm) under slightly acidic conditions. Ikehara and co-workers[22] developed synthetic procedures for 2′-*O*-(ONB) nucleosides, using *o*-nitrobenzyl chloride or the diazomethane. Scheme 5 shows the reactions to prepare 5′-*O*-monomethoxytrityl-2′-*O*-(ONB)-amidite units (**32**). *N*-Benzoylcytidine, *N*-benzoyladenosine, and *N*-isobutyrylguanosine were each treated with *o*-nitrophenyldiazomethane (**33**), obtained from *o*-nitrobenzaldehyde and tosyl hydrazide, to give a mixture of 2′- and 3′-*O*-(*o*-nitrobenzyl) nucleosides

(29) and (30).[22] The isomers can be separated by either precipitation or silica gel chromatography. A physical analysis suggested that the base in the 2′-isomer (29) was stacked with *o*-nitrobenzyl more than the other isomer (30). After the 5′-hydroxyl group being protected using monomethoxytrityl chloride (MMTr-Cl) the 3′-hydroxyl group was phosphitylated for the oligonucleotide synthesis.[23] The *o*-nitrobenzyl groups on the oligonucleotides are deprotected after the removal of the other protecting groups. Although ONB is chemically stable, UV light has to be avoided during the synthesis, and the removal by UV light requires stringent conditions to reduce side reactions.[23–25]

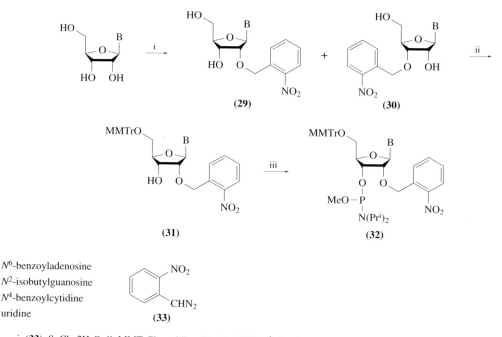

i, (33), $SnCl_2 \cdot 2H_2O$; ii, MMTrCl, pyridine; iii, $MeOP[N(Pr^i)_2]_2$, diisopropylamine-hydroxytetrazole, CH_2Cl_2

Scheme 5

6.06.3.2.3 O-t-*Butyldimethylsilyl group*

Ogilvie and co-workers applied a *t*-butyldimethylsilyl group, TBDMS (22), to the 2′-OH protection in oligoribonucleotide syntheses. They succeeded in synthesizing a 77 ribonucleotide-long RNA with a tRNA sequence by using the solid-phase phosphoramidite method.[26] This protecting group is widely used in RNA synthesis, and protected monomer units are commercially available. Scheme 6 shows the preparation of the 5′-*O*-DMTr-2′-*O*-TBDMS amidite units (37). The *N*-protected nucleosides are converted to the 5′-*O*-DMTr-nucleosides (34), which can be silylated with *t*-butyldimethylsilyl chloride (38) in the presence of imidazole/DMF. A mixture of 2′- and 3′-*O*-TBDMS derivatives (35), (36) is obtained, and (35) is separated by either silica gel chromatography or selective precipitation.[27] The ratio of (35) to (36) depends on the reaction conditions. Silver nitrate enhances the reaction rate and increases the yield of the 2′-isomer.[28] Since (35) migrates to (36) under basic conditions, phosphitylation of the 3′-OH should be performed carefully using tertiary bases, 2,4,6-collidine and *N*-methylimidazole, as catalysts.[29] The purity of the phosphoramidite should be monitored by ^{31}P NMR to detect contamination by side products. After the coupling reactions, the fully protected oligoribonucleotides are cleaved from the solid support, and deblocking of the phosphate and the exocyclic amino groups is carried out with ethanolic ammonia or concentrated ammonia in ethanol. Concentrated ammonia alone results in the loss of the TBDMS and degradation of the RNA. Ethanol increases the solubility of the oligoribonucleotides and suppresses the hydrolysis of the silyl ethers. The TBDMS groups are removed from the 2′-OH by treatment with tetra-*n*-butylammonium fluoride (TBAF) in tetrahydrofuran. However, the deprotection of the TBDMS groups by TBAF is time consuming (16 h) and an improved method using triethylamine/hydrogen fluoride ($Et_3N \cdot 3HF$) has been introduced for the deprotection of silyl groups.[30,31] A mixture of $ET_3N \cdot 3HF$ and *N*-methylpyrrolidinone, which can remove the 2'-*O*-TBDMS at 65 °C in 1.5 h, was also shown to be a convenient reagent.[32]

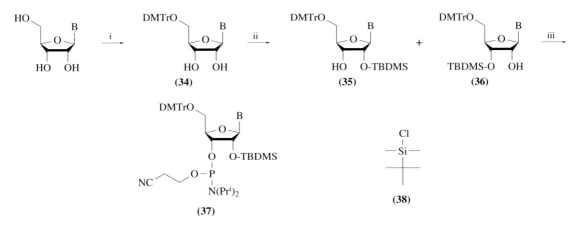

i, DMTr-Cl, THF, pyridine; ii, (38), AgNO$_3$; iii, (28), 2,4,6-collidine, *N*-methylimidazole, THF

Scheme 6

Although the deblocking conditions for the *N*-protecting groups are different, depending on the stability (see the next section), partial desilylation occurs during the deblocking. The resulting shorter side products can be separated by reversed-phase chromatography from the dimethoxy-tritylated product. Alternatively, the completely deblocked oligonucleotide can be purified by either anion-exchange chromatography or gel electrophoresis. Since deblocked RNA fragments are susceptible to ribonucleases and to alkaline hydrolysis, water and glassware have to be sterilized and cleaned completely.

6.06.3.3 Protecting Groups for Exocyclic Aminos

The exocyclic amino groups of nucleosides are usually protected by acyl groups during the condensing reactions to avoid the formation of P-N linkages, as well as to augment the solubility. The *N*-acyl groups are removed by treatment with bases. Since strong basic conditions cause partial removal of the 2'-*O*-silyl ether and result in chain cleavage, more easily removed *N*-protecting groups, such as phenoxy acetyl derivatives (**41–43**), have been introduced. These protecting groups are especially useful for the incorporation of base-labile nucleoside analogues into oligonucleotides by the phosphoramidite method. While groups such as benzoyl (**39**) on N^6-adenosine and N^4-cytidine and isobutyryl (**40**) on N^2-guanosine require 16 h in concentrated ammonium hydroxide : ethanol (3:1) for removal, the phenoxy derivatives require only 4 h.[32,33] Dimethyl-formamidine (**45**) is also used as a labile *N*-protecting group for N^6-adenosine, as well as N^2-guanosine, in the phosphoramidite approach, and can be removed with concentrated ammonium hydroxide : ethanol (3:1) at room temperature for 8 h.[34] Dimethylaminomethylene-protected purine nucleosides have been used in the synthesis by the H-phosphonate method.[9,35]

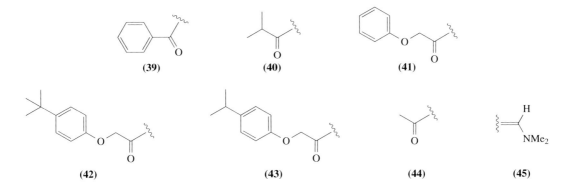

6.06.4 SYNTHESIS OF OLIGONUCLEOTIDE ANALOGUES

6.06.4.1 2′-O-Modified Oligonucleotides as Antisense Nucleic Acids

The 2′-OH of ribonucleotides has various functions in RNA, such as hydrogen bonding, metal ion binding, and protein interactions. The presence of the 2′-OH affects the sugar puckering, and substitutions at the 2′-position yield a wide variety of RNA analogues, which are useful to investigate RNA functions.

6.06.4.1.1 2′-O-Alkylnucleotides

Among 2′-substituted RNA's, 2′-O-alkylderivatives have been synthesized and utilized most frequently as alkaline stable analogues. These analogues maintain the 3′-*endo* puckering and are resistant to ribonucleases. 2′-O-methyl (**46**) and allyl (**47**) RNA derivatives, which have been used to hybridize to DNA and RNA, are illustrated. The methylated nucleosides are also found in tRNAs, rRNA and snRNA. Chemical syntheses of 2′-O-methyloligoribonucleotides were reported using the phosphotriester[36] and phosphoramidite methods.[37] Inoue *et al.* found 2′-O-methyl oligo-nucleotides hybrids with RNA showed stability similar to RNA · RNA and greater stability com-pared with RNA · DNA.[36,38] This stability assists in hybrid formation with chimeric DNAs 3 or 5 mer deoxyoligonucleotides with 2′-O-methyl oligomers on the 3′- and 5′-sides, d(UmGmAATGGAmCm)). In an RNase H reaction, 5′r(ACUUA↓CCUG)3′ was found to be cleaved at the allowed site (**50**) in Figure 2(a).[39] These chimeric antisense oligonucleotides have been applied to the site-specific cleavage of RNA.[40] Figure 2(b) shows the specific cleavage of *E. coli* tRNA^{Gln} for replacement of the 5′-fragment with an inosine-containing oligonucleotide. This experi-ment proved the participation of the 2-amino group of guanosine at position 3 in the reaction with the glutaminyl tRNA synthetase, because the rate of the amino acylation of the modified tRNA was substantially reduced.[41] The 2′-O-methylnucleosides were also incorporated into antisense DNA, and these chimeric antisense oligonucleotides were shown to be efficient, nuclease resistant reagents from the induction of RNase H reactivity against the target RNA.[42]

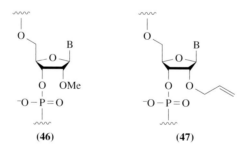

(46)   (47)

2′-O-Alkyloligoribonucleotides serve as good biochemical probes, due to their ability to form highly stable duplexes. 2′-O-Methyl and 2′-O-allyloligoribonucleotides, complementary to the RNA components of snRNPs, inhibited pre-mRNA splicing.[43,44] Biotinylated 2′-O-alkyloligo-ribonucleotides were used to isolate a snRNP from crude extracts by antisense chromatography.[38,45] 2′-O-Allyl probes were shown to have higher specificity to the target RNA than the 2′-O-methyl and 3,3-dimethylallyl probes.[46]

Homopyrimidine oligonucleotides recognize homopurine-homopyrimidine double stranded DNA by triple helix formation. It was also reported that 2′-O-methyl oligoribonucleotides could form triple helices through binding to double stranded DNA.[47,48] Shimizu *et al* reported that 2′-O-methyl pyrimidine oligonucleotides show the highest thermal stability among the various 2′-substituents (2′-deoxy, 2′-fluoro, and RNA). 1,10-Phenanthroline-linked 2′-O-methyl oligo-nucleotides were able to cleave double-stranded DNA at the expected sites.[49]

The biological functions of RNA were examined by partial incorporation of 2′-O-alkylnucleotides into the RNA. The role of the 2′-OH function in hammerhead ribozymes has been studied by the substitution of ribonucleotides with 2′-O-alkylnucleotides (see Chapter 6.14).

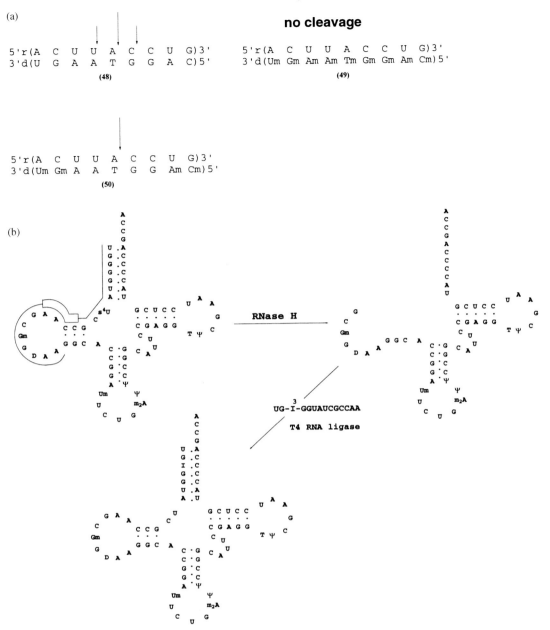

(a)

no cleavage

5'r(A   C   U   U   A   C   C   U   G)3'   5'r(A   C   U   U   A   C   C   U   G)3'
3'd(U   G   A   A   T   G   G   A   C)5'   3'd(Um Gm Am Am Tm Gm Gm Am Cm)5'

(48)   (49)

5'r(A   C   U   U   A   C   C   U   G)3'
3'd(Um Gm A   A   T   G   G   Am Cm)5'

(50)

(b)

RNase H

T4 RNA ligase

3
UG-I-GGUAUCGCCAA

Figure 2 Site specific cleavage of RNA by RNase H using chimeric oligodeoxynucleotides. (a) RNase H cleavage sites in hybrid duplexes. (b) Substitution of guanosine with inosine in the acceptor stem of tRNA[Gln].

6.06.4.1.2 *Other 2'-O-substituted analogues*

The 2'-fluoronucleosides (**51**) have the 3'-endo puckering like ribonucleosides,[50] and have been used in functional RNA, such as hammerhead ribozymes, to prove the importance of the conformation in the catalytic site.[51,52] The 2'-aminonucleosides (**52**) have been incorporated in RNA and further functionalized for cross-linking experiments to investigate higher structures of RNA. Since the 2'-position is located on the surface of the minor groove, intra- and interstrand cross-linking by disulfide bond formation have been carried out in hammerhead (**53**)[53] and *Tetrahymena* ribozymes (**54**).[54]

6.06.4.2 **Phosphate Modification**

Phosphodiester linkages give acidic characteristics to nucleic acids and play important roles in the interactions with proteins and metal ions. The oxygen atoms of the phosphate can be substituted with sulfur atoms to alter the chemical or steric character. The nonbridging oxygen can be derivatized at the oxidation step during the condensing reaction, using phosphite chemistry. In addition to sulfurization of the phosphodiester, phosphoramidates and methylphosphonates are obtained. These derivatives have been extensively studied in the deoxy series, including the stereospecific synthesis. However, in the ribo chemistry, nonbonding phosphorothioates have been synthesized by chemical and enzymatic methods, and are extensively used in mechanistic studies of enzymes as well as ribozymes (see Chapter 6.14).

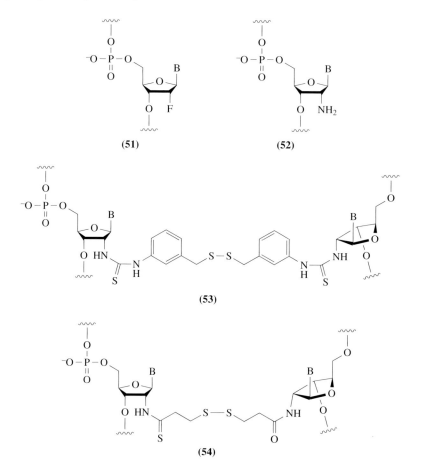

6.06.4.2.1 Substitution of nonbridging oxygen

RNA with a phosphorothioate linkage can be prepared by chemical and enzymatic syntheses. While the enzymatic synthesis gives the phosphorothioate with the Rp-configuration, chemical syntheses yield both Sp (**55**) and RP (**56**) isomers. The nonbridging oxygen of the internucleotidic linkage can be substituted during the oxidation step in the phosphoramidite or H-phosphonate solid-phase synthesis by using sulfurizing agents, such as elemental sulfur in carbon disulfide/2,6-lutidine[55] or tetraethylthiuram disulfide (TETD) in acetonitrile.[56] Since elemental sulfur tends to precipitate in the delivery line, other soluble reagents are preferred for mechanical synthesis. The Rp and Sp isomers can be separated by reverse phase chromatography depending on the sequences and the lengths.

6.06.4.2.2 Substitution of 5'- or 3'-bridging oxygen

The bridging phosphorothioate analogues of DNA have been used to elucidate the cleavage mechanism of nucleases.[57] However, it was difficult to synthesize RNA containing bridging phosphorothioate linkages, due to the reactions for the removal of the protecting groups of the ribose 2'-OH. The deprotecting agents for TBDMS, such as triethylamine/trihydrofluoride or tetrabutyl ammonium fluoride, cause substantial chain cleavage of the thio-RNA. MacLaughlin and co-workers synthesized a chimeric oligonucleotide containing a single thioribonucleotide in an otherwise all-deoxyribonucleotide, using phosphoramidite chemistry.[58,59] They prepared a 2′-*O*-chloroethoxyethyl ribonucleotide amidite (**57**) and a 5′-*S*-triphenylmethyl ribonucleotide amidite (**58**), and then synthesized an oligonucleotide (**59**) containing the 5′-thioribonucleotide. The 5′-triphenylmethyl on the thio-compound was removed by treatment with silver nitrate, and the other trityl derivatives were removed with dichloroacetic acid. It was shown that (**59**), with a 5′-bridging phosphorothioate, was labile under alkaline conditions and in the presence of soft metal ions.[59] These chimeric nucleotides have been used to investigate the cleavage mechanisms of hammerhead ribozymes.[60] Taira and co-workers also proposed a mechanism for the hammerhead ribozyme. They prepared an all ribonucleotide sequence containing one 5′-bridging thiophosphate linkage.[61] Reese *et al.* synthesized uridyl-(3′-5′)-(5′-thiouridine) (**60**) by the H-phosphonate method,[62] using tetrahydropyranyl as a 2′-OH protecting group. They also synthesized a diribonucleotide containing a 3′-bridging phosphorothioate (**61**) by a similar procedure.[63]

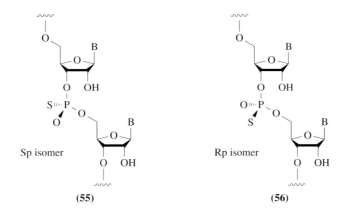

(**55**)   (**56**)

6.06.4.3 Base-modified Nucleotides

The nucleobases are directly involved in the hydrogen bonds and play the key role in the functions of nucleic acids. The exocyclic functional groups on nucleobases can be replaced by the incorporation of properly protected monomer units during solid-phase synthesis. Alternatively, those functions can be modified by postmodification methods, provided that suitable leaving groups are incorporated in the oligonucleotides. For applications of base-modified RNA in studies of ribozyme functions, see Chapters 6.14 and 6.15.

6.06.4.3.1 *Incorporation of modified pyrimidine*

(*i*) *5-Substituted pyrimidines*

RNA containing 5-bromouridine (**62**) was shown to form a stable complex with MS2 coat protein through a transient covalent bond between a cysteine residue via Michael addition.[64] Iwai *et al.* synthesized a Rev-responsive element (RRE) containing (**62**) at the bubble structure to investigate the interaction between RRE and Rev protein.[65] The 5-bromouridine was incorporated by the phosphoramidite method, using TBDMS as the protecting group for the 2′-OH. It has been shown that the recognition of the RRE by Rev requires hydrogen bonds between the protein and the functional groups in the major groove of the distorted RRE.

Bradly and Hanna synthesized 5-thiocyanato-uridine (**63**) phosphoramidites and incorporated them into oligoribonucleotides.[66] After standard deprotection of the acyl and 2′-*O*-TBDMS groups, the thiocyanate group was reduced to a 5-mercapto function, to react with either *p*-azidophenacyl bromide or 5-iodoacetamidofluorescein.

Eaton and co-workers have developed a convenient route that yields 5-carbonyluridine derivatives.[67] 2′,3′-Isopropylidene-5-iodouridine was converted to the 5-carbonyl uridine derivative by using palladium. The 5-carbonyl uridine derivative was converted to the corresponding 5′-triphosphate, and the analogue was found to be a good substrate for T7 RNA polymerase in the transcription of a dsDNA template. The transcripts containing uridine derivatives, (64) and (65), served as templates for avian myeloblastosis virus (AMV) reverse transcriptase to obtain the cDNA. They also applied this method to incorporate 5-pyridylmethylaminocarbonyl-uridine (65) into transcripts for *in vitro* selection of catalytic RNA, and succeeded in the generation of a catalyst for a Diels–Alder cycloaddition.[68] The RNA molecules accelerated the reaction rate by a factor of up to 800 relative to the uncatalyzed reaction. They prepared various 5-position derivatives to select catalytic RNA or the aptamer.[69]

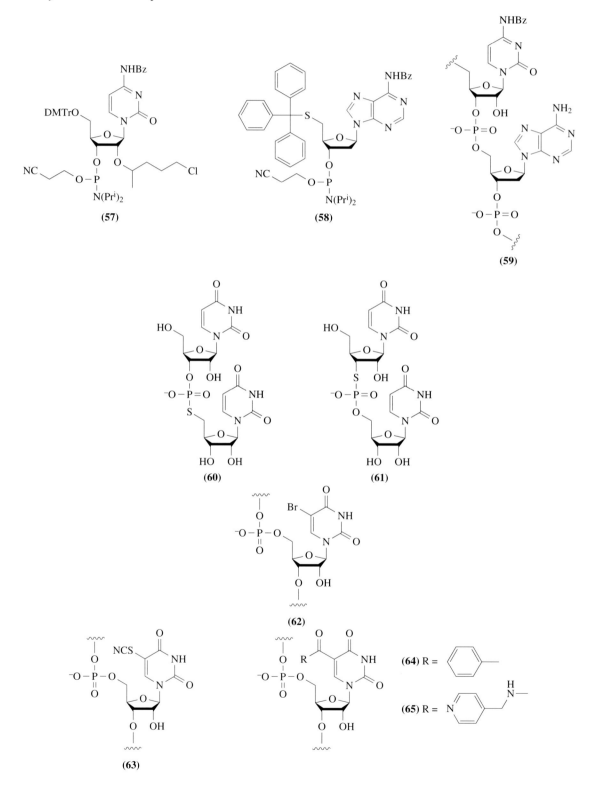

(ii) 4-Substituted pyrimidines

4-Thiouridine (**67**) is known to be a photoreactive compound, and has been used in cross-linking reactions as a probe to investigate the interactions of nucleic acids with proteins and nucleic acids with nucleic acids. Stockley and co-workers prepared a 4-*S*-cyanoethyl-4-thiouridine phosphoramidite unit (**66**) and incorporated it into oligoribonucleotides (Figure 3).[70] Oligoribonucleotides with the *S*-cyanoethyl-4-thiouridine were treated with DBU to remove the *S*- and *O*-cyanoethyl groups, and deprotection of the exocyclic amines was achieved with methanolic ammonia. The 2'-*O*-TBDMS was deprotected using either Et$_3$N · 3HF or TBAF in THF. For the synthesis of the RRE containing *S*-cyanoethyl-4-thiouridine, Fpmp (**20**) was used as the 2'-OH protecting group.[71] Using this probe, specific cross-linking was demonstrated between the Rev protein of HIV-1 and the RRE by irradiation with 350 nm light. 4-Thiouridine could also be incorporated into oligoribonucleotides by postsynthetic modification (see postmodification section). Since 4-thiouridine absorbs 340 nm light, which is a characteristic of the thione chromophore, the presence of 4-thiouridine was detected from the UV-spectrum.

(iii) 2-Substituted pyrimidines

2-Thiouridine is found predominantly in the wobble position of tRNAs. It has been shown that 2-thiouridine favors the 3'-*endo* sugar conformation at the mono- and dinucleotide levels. Kumar and Davis have synthesized a 2-thiouridine phosphoramidite unit (**68**) (Figure 3) and have incorporated it into a pentamer, an anticodon sequence mimic of tRNA^{Lys}.[72] They modified the oxidation step in the solid-phase synthesis such that *t*-butyl hydroperoxide was used instead of I$_2$/H$_2$O. The duplex containing 2-thiouridine (**69**) showed higher stability than RNA containing either normal uridine or 4-thiouridine.

6.06.4.3.2 *Incorporation of modified purines*

Stockley and co-workers synthesized an oligoribonucleotide containing 6-thioguanosine (**71**), using an amidite unit (**70**) (Figure 3).[73] The thio group on the *N*-benzoyl-6-thioguanosine was protected with 3-bromopropionitrile, and the 2'-OH was protected with TBDMS. To deprotect the 2'-*O*-TBDMS, Et$_3$N · 3HF/DMSO (1:1) was used, because TBAF in THF resulted in the degradation of the thiocarbonyl group. They incorporated the 6-thioguanosine into a hammerhead ribozyme strand, and compared the effect of the substitution with that of inosine and 2-aminopurine. Rana and co-workers synthesized a Rev-responsive element containing 6-thiouridine in a specific site, and carried out photo cross-linking experiments in the presence or absence of Rev.[74] Both RNA-RNA and RNA-protein cross-linking reactions were observed.

Other purine analogues, such as isoguanosine and deazapurine nucleosides, have been derivatized and incorporated into ribozymes to investigate the roles of the heterocyclic rings in the catalytic reactions.

6.06.4.4 Postsynthetic Modifications of RNA

Postsynthetic modification methods consist of the conversion of nucleosides with leaving groups to various derivatives on oligonucleotides. Although postsynthetic modifications have been used to obtain DNA derivatives,[75–77] the approach seemed to be difficult to apply to RNA, because of the presence of the 2'-hydroxyl groups. For the postsynthetic modification of the C-4 of pyrimidine, triazol was used as the leaving group.[78] They synthesized a dimer and a heptamer containing 4-triazolylpyrimidone by the triester method. The 4-triazoylylpyridine derivative was converted to 4-dimethylaminopyrimidine and cytidine by treatment with dimethylamine and ammonia, respectively. The phosphoramidite unit of the 4-triazoylylpyridine was prepared from the amidite of uridine by treatment with triazole in the presence of POCl$_3$,[79] and was used for the synthesis of oligoribonucleotides. The 4-triazoylpyridine derivative could be converted to 4-thiouridine by treatment with thiolacetic acid. It has been shown that the 4-*O*-aryloxy group ribonucleoside can act as a leaving group.[80]

Verdine and co-workers have reported the postsynthetic modification of C, A, and G in RNA.

Chemical RNA Synthesis

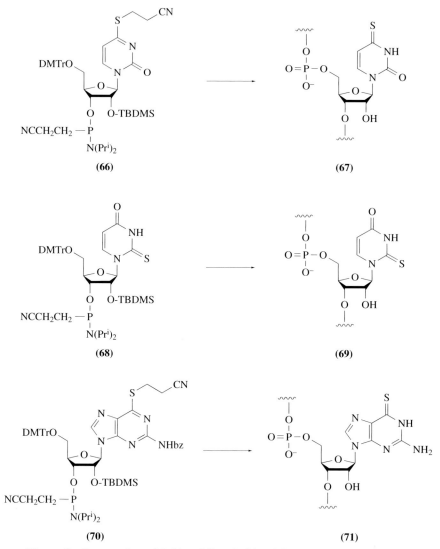

Figure 3 Preparation of 4-thiouridine, 2-thiouridine, and 6-thioguanosine.

4-*O*-(4-Chlorophenyl)uridine and 6-*O*-(4-chlorophenyl)inosine were synthesized as the reactive derivatives.[81] For the functionalization of G, 2-fluoro-6-*O*-(4-nitrophenethyl)inosine was prepared. The protected nucleosides with DMTr and TBDMS, were incorporated into oligoribonucleotides by the phosphoramidite chemistry. The resin-bound oligoribonucleotides, which contained one convertible nucleoside, were treated with alkylamines to give the N^2-alkyl-G, N^4-alkyl-C, and N^6-alkyl-A derivatives.

A 5′-*O*-phosphoramidite can be incorporated into the 5′-ends of synthetic oligo-deoxyribonucleotides, and these were converted into oligonucleotides with a (5′-5′) internucleotidic linkage. Oxidation of the 5′-end nucleoside with periodate yielded dialdehydes, which can be condensed with biotin hydrazide to prepare 5′-biotinylated oligonucleotides by reduction with sodium borohydride.[82] As the length of an oligoribonucleotide increases, the yield decreases in chemical syntheses, so it is difficult to prepare long RNA fragments in a high yield. Wincott and co-workers synthesized two strands of a hammerhead ribozyme separately, and then conjugated the two strands covalently at loop II.[83] The 3′-end of the 5′-half ribozyme strand was oxidized by sodium periodate, and was allowed to react with the 5′-aminohexyl-3′-half ribozyme in the presence of NaBH₃CN. Interestingly, a cyanoborane adduct of the ribozyme was also obtained from the reductive ligation. The postsynthetic ligated ribozymes exhibited activities similar to that of the control ribozyme.

6.06.4.5 Other Analogues

6.06.4.5.1 Transition state analogues

The generation of catalytic antibodies that possess novel phosphodiesterase activities has been a focus of much effort. Janda and co-workers succeeded in eliciting an antibody catalyst that cleaves phosphodiester linkages by using an oxorhenium(V) complex of uridine as a transition state analogue.[84] The screened monoclonal antibodies catalyzed the transesterification of uridine 3'-(p-nitrophenyl phosphate) in the presence of EDTA. Catalytic antibodies that cleave RNA strands at specific sites may be elicited by using transition state analogues with a defined higher structure.

6.06.4.5.2 Peptide nucleic acid

A nucleic acid analogue that contains a peptide backbone, called a "peptide nucleic acid" (PNA), was synthesized by Neilsen *et al.* and was shown to bind to either single or double stranded DNA. The PNA·DNA duplex had a much higher melting temperature than the DNA·DNA duplex.[85] One reason for the extraordinarily high stability of PNA·DNA could be the neutral nature of the PNA backbone. The loss of the negatively charged phosphate backbone decreases the repulsion existing between the polyanionic phosphodiesters of the nucleic acids. The high stability of PNA·DNA was also revealed in triplex (PNA·dsDNA) formation by strand displacement.[86] PNAs have been used as probes for nucleic acid functions and may become potential antisense or antigene drugs.

6.06.5 REFERENCES

1. H. G. Khorana, *Pure Appl. Chem.*, 1968, **17**, 349.
2. R. Lohrmann, D. Söll, H. Hayatsu, E. Ohtsuka, and H. G. Khorana, *J. Am. Chem. Soc.*, 1966, **88**, 819.
3. G. M. Blackburn and M. J. Gait "Nucleic Acids in Chemistry and Biology," 2nd edn., Oxford University Press, Oxford, 1996.
4. S. S. Jones, B. Rayner, C. B. Reese, A. Ubasawa, and M. Ubasawa, *Tetrahedron*, 1980, **36**, 3075.
5. R. L. Letsinger and W. B. Lunsford, *J. Am. Chem. Soc.*, 1976, **98**, 3655.
6. S. L. Beaucage and H. Caruthers, *Tetrahedron Lett.*, 1981, **22**, 1859.
7. B. C. Froehler, P. G. Ng, and M. D. Matteucci, *Nucleic Acids Res.*, 1986, **14**, 5399.
8. P. J. Garegg, I. Lindh, T. Regberg, J. Stawinski, R. Strömberg, and C. Henrichson, *Tetrahedron Lett.*, 1986, **27**, 4051.
9. L. Arnold, J. Smrt, J. Zajicek, G. Ott, M. Schiesswohl, and M. Sprinzl, *Collect. Czech. Chem. Commun.*, 1991, **56**, 1948.
10. M. Smith, D. H. Rammler, I. H. Goldberg, and H. G. Khorana, *J. Am. Chem. Soc.*, 1962, **84**, 430.
11. J. B. Chattopadhyaya and C. B. Reese, *J. Chem. Soc. Chem. Commun.*, 1978, **1978**, 639.
12. M. Kwiatkowski and J. Chattopadhyaya, *Acta Chem. Scand.*, 1984, **B38**, 657.
13. E. Sonveaux, in "Protocols for oligonucleotide conjugates," ed. S. Agrawal, Humana Press, Totowa, NJ, 1994, p. 1.
14. J. H. van Boom and P. M. Burgers, *Tetrahedron Lett.*, 1976, **17**, 4875.
15. S. Iwai and E. Ohtsuka, *Nucleic Acids Res.*, 1988, **16**, 9443.
16. C. Lehmann, Y.-Z. Xu, C. Christodoulou, Z.-K. Tan, and M. J. Gait, *Nucleic Acids Res.*, 1989, **17**, 2379.
17. C. B. Reese, R. Saffhill, and J. E. Sulston, *J., Am. Chem. Soc.*, 1967, **89**, 3366.
18. C. B. Reese, H. T. Serafinowska, and G. Zappia, *Tetrahedron Lett.*, 1986, **27**, 2291.
19. M. V. Rao, C. B. Reese, V. Schelhmann, and P. S. Yu, *J. Chem. Soc., Perkin Trans*, 1993, **1**, 43.
20. E. Ohtsuka, M. Ohkubo, A. Yamane, and M. Ikehara, *Chem. Pharm. Bull.*, 1983, **31**, 1910.
21. W. T. Markiewicz, E. Biala, and R. Kierzek, *Bull. Pol. Acad. Sci. Chem.*, 1984, **32**, 433.
22. E. Ohtsuka, T. Wakabayashi, S. Tanaka, T. Tanaka, K. Oshie, A. Hasegawa, and M. Ikehara, *Chem. Pharm. Bull.*, 1981, **29**, 318.
23. T. Tanaka, S. Tamatsukuri, and M. Ikehara, *Nucleic Acids Res.*, 1986, **14**, 6265.
24. E. Ohtsuka, T. Tanaka, S. Tanaka, and M. Ikehara, *J. Am. Chem. Soc.*, 1978, **100**, 4580.
25. J. A. Hayes, M. J. Brunden, P. T. Gilham, and G. R. Gough, *Tetrahedron Lett.*, 1985, **26**, 2407.
26. K. K. Ogilvie, N. Usman, K. Nicoghosian, and R. J. Cedergren, *Proc. Natl. Acad. Sci. USA*, 1988, **85**, 5764.
27. N. Usman, K. K. Ogilvie, M. Y. Jiang, and R. J. Cedergren, *J. Am. Chem. Soc.*, 1987, **109**, 7845.
28. G. H. Hakimelahi, Z. A. Proba, and K. K. Ogilvie, *Can. J. Chem.*, 1982, **60**, 1106.
29. S. A. Scaringe, C. Francklyn, and N. Usman, *Nucleic Acids Res.*, 1990, **18**, 5433.
30. D. Gasparutto, T. Livache, H. Bazin, A.-M. Duplaa, A. Guy, A. Khorlin, D. Molko, A. Roget, and R. Teoule, *Nucleic Acids Res.*, 1992, **20**, 5159.
31. E. Westman and R. Stromberg, *Nucleic Acids Res.*, 1994, **22**, 2430.
32. F. Wincott, A. DiRenzo, C. Shaffer, S. Grimm, D. Tracz, C. Workman, D Sweedler, C. Gonzalez, S. Scaringe, and N. Usman, *Nucleic Acids Res.*, 1995, **23**, 2677.
33. T. F. Wu, K. K. Ogilvie, and R. T. Pon, *Nucleic Acids Res.*, 1989, **17**, 3501.
34. R. Vinayak, P. Anderson, C. McCollum, and A. Hampel, *Nucleic Acids Res.*, 1992, **20**, 1265.

35. G. Ott, L. Arnold, J. Smrt, M. Sobkowski, S. Limmer, and H.-P. Hofmann, M. Sprinzl, *Nucleosides Nucleotides*, 1994, **13**, 1069.
36. H. Inoue, Y. Hayase, S. Iwai, K. Miura, and E. Ohtsuka, *Nucleic Acids Res.*, 1987, **15**, 6131.
37. B. S. Sproat and A. I. Lamond, in "Oligonucleotides and Analogues," Ed. F. Eckstein, Oxford University Press, New York, 1991, p49.
38. A. I. Lamond and B. S. Sproat, *FEBS Lett.*, 1993, **325**, 123.
39. H. Inoue, Y. Hayase, S. Iwai, and E. Ohtsuka, *FEBS Lett.*, 1987, **215**, 327.
40. Y. Hayase, H. Inoue, and E. Ohtsuka, *Biochemistry*, 1990, **29**, 8793.
41. Y. Hayase, M. Jahn, M. J. Rogers, L. A. Silvers, M. Koizumi, H. Inoue, E. Ohtsuka, and D. Soll, *EMBO J.*, 1992, **11**, 4159.
42. B. P. Monia, E. A. Lesnik, C. Gonzalez, W. F. Lima, D. McGee, C. J. Guinosso, A. M. Kawasaki, P. D. Cook, and S. M. Freier, *J. Biol. Chem.*, 1993, **268**, 14514.
43. A. I. Lamond, B. Sproat, U. Ryder, and J. Hamm, *Cell*, 1989, **58**, 383.
44. A. Mayeda, Y. Hayase, H. Inoue, E. Ohtsuka, and Y. Ohshima, *J. Biochem. (Tokyo)*, 1990, **108**, 399.
45. B. J. Blencowe, B. S. Sprat, U. Ryder, S. Barabino, and A. I. Lamond, *Cell*, 1989, **59**, 531.
46. A. M. Iribarren, B. S. Sproat, P. Neuner, I. Sulston, U. Ryder, and A. I. Lamond, *Proc. Natl. Acad. Sci. USA*, 1990, **87**, 7747.
47. M. Shimizu, A. Konishi, Y. Shimada, H. Inoue, and E. Ohtsuka, *FEBS Lett.*, 1992, **302**, 155.
48. S. H. Wang and E. T. Kool, *Nucleic Acids Res.*, 1995, **23**, 1157.
49. M. Shimizu, H. Morioka, H. Inoue, and E. Ohtsuka, *FEBS Lett.*, 1996, **384**, 207.
50. D. M. Williams, F. Benseler, and F. Eckstein, *Biochemistry*, 1991, **30**, 4001.
51. W. A. Pieken, D. B. Olsen, F. Benseler, H. Aurup, and F. Eckstein, *Science*, 1991, **253**, 314.
52. D. B. Olsen, F. Benseler, H. Aurup, W. A. Pieken, and F. Eckstein, *Biochemistry*, 1991, **30**, 9735.
53. S. T. Sigurdsson, T. Tuschl, and F. Eckstein, *RNA*, 1995, **1**, 575.
54. S. B. Cohen and T. R. Cech, *J. Am. Chem. Soc.*, 1997, **119**, 6259.
55. J. Ott and F. Eckstein, *Biochemistry*, 1987, **26**, 8237.
56. H. Vu and B. L. Hirschbein, *Tetrahedron Letters*, 1991, **32**, 3005.
57. M. Mag, S. Lüking, and J. W. Engels, *Nucleic Acids Res.*, 1991, **19**, 1437.
58. R. G. Kuimelis and L. W. McLaughlin, *Nucleic Acids Res.*, 1995, **23**, 4753.
59. R. G. Kuimelis and L. W. McLaughlin, *Bioorg. Med. Chem.*, 1997, **5**, 1051.
60. R. G. Kuimelis and L. W. Mclaughlin, *J. Am. Chem. Soc.*, 1995, **117**, 11 019–11 020.
61. D.-M. Zhou, N. Usman, F. E. Wincott, J. Matulic-Adamic, M. Orita, L.-H. Zhang, M. Komiyama, P. K. R. Kumar, and K. Taira, *J. Am. Chem. Soc.*, 1996, **118**, 5862.
62. X. H. Lui and C. B. Reese, *Tetrahedron Lett.*, 1995, **36**, 3413.
63. X. H. Liu and C. B. Reese, *Tetrahedron Lett.*, 1996, **37**, 925.
64. S. J. Talbot, S. Goodman, S. R. E. Bates, C. W. G. Fishwick, and P. G. Stockley, *Nucleic Acids Res.*, 1990, **18**, 3521.
65. S. Iwai, C. Pritchard, D. A. Mann, J. Karn, and M. J. Gait, *Nucleic Acids Res.*, 1992, **20**, 6465.
66. D. H. Bradley and M. M. Hanna, *Tetrahedron Lett.*, 1992, **33**, 6223.
67. T. M. Dewey, A. A. Mundt, G. J. Crouch, M. C. Zyzniewski, and B. E. Eaton, *J. Am. Chem. Soc.*, 1995, **117**, 8474.
68. T. M. Tarasow, S. L. Tarasow, and B. E. Eaton, *Nature*, 1997, **389**, 54.
69. B. E. Eaton, L. Gold, B. J. Hicke, N. Janjic, F. M. Jucker, D. P. Sebesta, T. M. Tarasow, M. C. Willis, and D. A. Zichi, *Bioorg. Med. Chem.*, 1997, **5**, 1087.
70. C. J. Adams, J. B. Murray, J. R. P. Arnold, and P. G. Stockley, *Tetrahedron Lett.*, 1994, **35**, 765.
71. A. McGregor, M. V. Rao, G. Duckworth, P. G. Stockley, and B. A. Connolly, *Nucleic Acids Res.*, 1996, **24**, 3173.
72. R. K. Kumar and D. R. Davis, *Nucleic Acids Res.*, 1997, **25**, 1272.
73. C. J. Adams, J. B. Murray, M. A. Farrow, J. R. P. Arnold, and P. G. Stockley, *Tetrahedron Lett.*, 1995, **36**, 5421.
74. Y.-H. Ping, Y. P. Liu, X. L. Wang, H. R. Neenhold, and T. M. Rana, *RNA*, 1997, **3**, 850.
75. A. M. MacMillan and G. L. Verdine, *J. Org. Chem.*, 1990, **55**, 5931.
76. D. A. Erlanson, L. Chen, and G. L. Verdine, *J. Am. Chem. Soc.*, 1993, **115**, 12 583.
77. Y.-Z. Xu, Q. G. Zheng, and P. F. Swann, *J. Org. Chem.*, 1992, **57**, 3839.
78. W. L. Sung, *J. Org. Chem.*, 1982, **47**, 3623.
79. K. Shah, H. Y. Wu, and T. M. Rana, *Bioconjugate Chem.*, 1994, **5**, 508.
80. A. Miah, C. B. Reese, and Q. L. Song, *Nucleosides Nucleotides*, 1997, **16**, 53.
81. C. R. Allerson, S. L. Chen, and G. L. Verdine, *J. Am. Chem. Soc.*, 1997, **119**, 7423.
82. S. Agrawal, C. Christodoulou, and M. J. Gait, *Nucleic Acids Res.*, 1986, **14**, 6227.
83. L. Bellon, C. Workman, J. Scherrer, N. Usman, and F. Wincott, *J. Am. Chem. Soc.*, 1996, **118**, 3771.
84. D. P. Weiner, T. Weimann, M. M. Wolfe, P. Wentworth, and K. D. Janda, *J. Am. Chem. Soc.*, 1997, **119**, 4088.
85. M. Egholm, O. Buchardt, P. E. Nielsen, and R. H. Berg, *J. Am. Chem. Soc.*, 1992, **114**, 1895.
86. P. E. Nielsen, M. Egholm, R. H. Berg, and O. Buchardt, *Science*, 1991, **254**, 1497.

6.07
RNA Editing

MARIE ÖHMAN and BRENDA L. BASS
University of Utah, Salt Lake City, UT, USA

6.07.1 INTRODUCTION TO RNA EDITING

RNA editing, a type of RNA processing, was first discovered by Benne and co-workers in a mitochondrion-encoded mRNA of a kinetoplastid trypanosome.[1] The term RNA editing initially referred only to the process as it occurs in trypanosomes, which involves the post-transcriptional insertion and deletion of uridylate (UMP) within nascent transcripts. The discovery of additional examples of such post-transcriptional sequence alterations led to a broader use of the term. RNA editing is now used to describe the insertion and deletion of nucleotides other than UMP, base deamination, and the cotranscriptional insertion of nonencoded nucleotides. RNA editing has been observed in mRNAs, tRNAs, and rRNAs, in mitochondrial and chloroplast encoded RNAs, as well as in nuclear encoded RNAs. Examples of RNA editing have been found in many metazoa, unicellar eukaryotes such as trypanosomes, and in plants. To date, RNA editing has not been observed in a prokaryote.

Typically, RNA editing reactions are put into two broad catagories based on their reaction mechanisms. One type, insertion/deletion RNA editing, involves the insertion or deletion of nucleotides and actually changes the length of the target RNA. The second type, RNA editing by base modification, changes an encoded nucleotide into a different nucleotide, without changing the overall length of the RNA. In this chapter the authors focus on the most well-characterized examples of these two types of editing, and particularly on those examples where a model for the catalytic mechanism has been described. Details on other examples of RNA editing can be found in several reviews.[2,3]

6.07.2   EDITING BY BASE MODIFICATION

RNA modifications have long been known to occur in tRNAs, rRNAs, and mRNAs.[4,5] Some of these modifications are simple, involving merely the methylation of an existing nucleotide, while others are quite complex, involving multiple enzymes for their synthesis. As additional types of RNA editing are discovered, it becomes harder to distinguish the difference between what has traditionally been called an RNA modification, from what is now categorized as RNA editing. At one point in time the term RNA editing was reserved for processes that involved canonical bases, that is, the insertion or deletion of Watson–Crick nucleotides, or the conversion of one Watson–Crick nucleotide into another. Except for one adenosine modification (see below), for the most part this nomenclature rule still applies. However, some scientists now prefer to use the term RNA editing for any post-transcriptional modification of an RNA.

The two types of base modification that are always categorized as RNA editing both involve deamination reactions. In one case, deamination of cytidine produces uridine[6] (see Figure 1), while in the other case, deamination of adenosine produces inosine[7] (see Figure 3). RNA editing by base modification can change one sense codon to that of a different sense, as well as create (C to U editing) or remove (A to I editing) stop codons. In most cases, a protein with new unique features is produced, and a single genomically-encoded RNA gives rise to a variety of transcripts. Despite their mechanistic similarities, as discussed below, RNA editing by cytidine and adenosine deamination involve very different mechanisms for recognizing their RNA substrates.

6.07.2.1   Editing by Cytidine Deamination

Soon after the first example of editing in kinetoplastid protozoa was discovered, editing was observed within the mammalian apolipoprotein B (apoB) mRNA.[8,9] ApoB mRNA editing was the first example of RNA editing by base modification, as well as the first example of editing of a nuclear-encoded RNA (for review see references 10 and 11). The conversion of a C to a U within apoB mRNA is thought to involve hydrolytic deamination at the C-4 position of cytidine[6] (Figure 1). The cytidine deaminase activity involved in apoB mRNA editing derives from the enzyme APOBEC-1 (apolipoprotein B mRNA editing enzyme, catalytic polypeptide #1),[12,13] but additional factors are required to edit apoB mRNA. However, interestingly, APOBEC-1 can deaminate cytidine mononucleoside in the absence of other factors,[14,15] consistent with the idea that this polypeptide contains the catalytic active site.

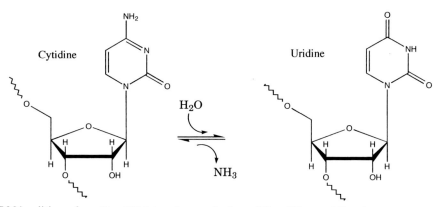

Figure 1   RNA editing of apoB mRNA involves a single cytidine (C) to uridine (U) conversion. As shown, the reaction is thought to proceed by the hydrolytic deamination of the N-4 amino group of cytidine. The deamination is catalyzed by the enzyme APOBEC-1, although additional factors are required for the RNA editing reaction.

Other natural substrates for APOBEC-1 have been looked for, but so far, they have not been found, and the enzyme's only known biological target is apoB mRNA. However, when APOBEC-1 is overexpressed, aberrant C to U editing is observed in additional RNAs.[16] Although not covered here, C to U editing has also been observed in mitochondria and chloroplasts of plants, as well as in the mitochondria of *Physarum polycephalum* (for reviews see references 2 and 3). Interestingly, plant organelles also exhibit U to C editing, the reverse of the deamination reaction. The endogenous factors that catalyze RNA editing in plants and *Physarum* have not yet been identified.

6.07.2.1.1 *Apolipoprotein B mRNA*

Apolipoproteins are required in mammals for the assembly, secretion, and transport of complex lipids; among the various apolipoproteins, apoB is particularly important (reviewed in reference 10). There are two apoB proteins, B-100 and B-48, and each has a different biological function in the metabolism of lipoproteins. Prior to the discovery that the apo-B mRNA was subject to RNA editing, it was suggested that the two proteins were expressed from the same gene because apoB-48 was found to be identical to the amino-terminal half of apoB-100.[17] However, the mechanism of generating the two distinct proteins from a single gene was an enigma.

The mystery was finally solved when cDNA and genomic sequences were compared. In humans, apoB-100 is synthesized mainly in the liver, while apoB-48 is synthesized in the intestine. Comparison of the human genomic sequence of apoB, with human liver and intestinal cDNAs, led to the conclusion that RNA editing was the basis for the production of apoB-48 mRNA. While both the genomic and liver cDNA sequences encoded glutamine (CAA) at codon 2153, the intestinal cDNA had a TAA stop at this codon.[8,9] Thus, a single C to U change at nucleotide 6666 (numbered according to the human sequence) converts the glutamine-2153 codon (CAA) in apoB-100 mRNA to an in-frame stop codon (UAA) in apoB-48 mRNA (Figure 2). The apoB-48 protein is therefore a truncation of apoB-100.

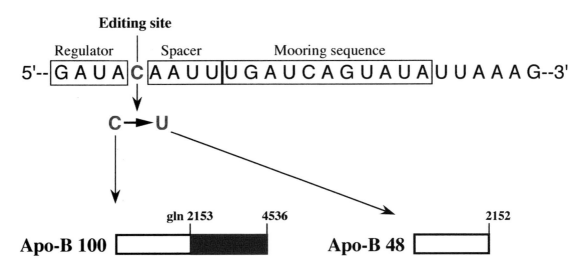

Figure 2 The apoB mRNA editing reaction. The upper part of the figure shows the editing site cytidine (red) in the context of the sequences immediately surrounding the editing site that are important for the reaction. The three different regions that have been determined to be important for editing are boxed (regulator, spacer, and mooring sequence, see text for details). The lower section of the figure illustrates the consequences of editing apoB mRNA. A glutamine codon is changed to a stop codon so that a long (ApoB 100) and a short (ApoB 48) protein can be made from a single open reading frame. The blue rectangle shows sequences in ApoB 100 that are absent in ApoB 48.

So far, editing of apoB mRNA has been found in human, rat, mouse and rabbit, although there is a difference in the extent of editing in liver and intestine between the various organisms. In the mammalian small intestine, the majority (70–95%) of the steady-state mRNA population corresponds to the edited apoB-48 mRNA (reviewed in reference 10). Apo-B mRNA editing occurs in the nucleus and is a post-transcriptional process. This conclusion derives from the observation that a fully processed apoB mRNA isolated from the nucleus is edited to about the same extent as the cytoplasmic apoB mRNA.[18]

RNA editing of apoB mRNA occurs at a single site in the middle of the 14-kilobase mature message. Mutagenesis studies show that a sequence of 22–26 nucleotides surrounding the editing site is important for apoB mRNA editing.[19,20] Within these 22–26 nucleotides, three subdomains have been defined: a *spacer* sequence of 4–6 nucleotides immediately downstream of the editing site; an 11 nucleotide "mooring sequence" (UGAUCAGUAUA) following the spacer, and a "regulator" region immediately upstream of C-6666 (Figure 2; reviewed in references 10 and 11). The mooring sequence is both sufficient and required for editing of the upstream C, as shown by ligating the mooring sequence to an unrelated RNA.[21] Editing is site specific, the hypothesis being that once the editing enzyme (or enzyme complex) binds the mooring sequence, a cytidine at a fixed distance upstream is precisely targeted. The AU-rich regulator region immediately 5′ of the target site appears to stimulate editing, although editing will occur in its absence. Maximal editing appears to require even other, more distal, sequences. For example, a study showed that 139 nucleotides surrounding the rabbit apoB mRNA were required for optimal editing *in vitro*.[22]

6.07.2.1.2 *APOBEC-1 and its catalytic mechanism*

RNA editing of apolipoprotein-B is thought to require multiple proteins, and so far, only one of these, APOBEC-1, has been conclusively identified.[12] APOBEC-1 is the catalytic subunit of the holoenzyme, while one or more, as yet uncharacterized, subunits are proposed to provide the RNA binding functions. When the APOBEC-1 gene is inactivated in mice, editing of apoB mRNA does not occur, and the apoB-48 protein cannot be detected;[23,24] this indicates that there is no genetic redundancy for APOBEC-1 activity.

The APOBEC-1 protein is a 27 kDa protein that contains an N-terminal bipartite nuclear localization signal, a centrally located deamination domain, a possible RNA binding domain that overlaps this catalytic domain, and finally, a C-terminal leucine-rich domain that is presumed to be involved in protein–protein interactions (reviewed in references 10 and 11). The extensive sequence similarities between APOBEC-1 and the *Escherichia coli* cytidine deaminase (ECCDA) suggest that the two enzymes use similar catalytic mechanisms.[25] Thus, since ECCDA requires zinc for catalysis, it has been proposed that APOBEC-1 also requires a catalytic zinc, and *in vitro* experiments support this idea.[15] In addition, certain regions of the APOBEC-1 sequence can be aligned with the amino acids of ECCDA known to be involved in coordinating the catalytic zinc.

Like ECCDA,[26] APOBEC-1 forms a homodimer.[27] Although the function of dimerization is not known, it has been suggested that one of the active sites binds a U within the AU-rich mooring sequence, while the other active site binds and deaminates the target C.[25] This hypothesis is based on the sequence similarity of APOBEC-1 to ECCDA. However, the substrate specificity of the two enzymes differs considerably. ECCDA is specific for nucleoside substrates, while APOBEC-1 recognizes a single C in a specific RNA context.

The missing factor(s) that acts in concert with APOBEC-1 to edit apoB mRNA has long been searched for. A 65 kDa complementing protein has been shown to interact with APOBEC-1 *in vitro* to form an enzymatically active complex.[28] This complementing protein may represent the RNA-binding subunit that docks APOBEC-1, the catalytically active subunit, onto apoB mRNA and in proximity to the target C.

6.07.2.2 Editing by Adenosine Deamination

The second example of RNA editing by base modification involves adenosine deamination. This type of RNA editing requires a specific structure in the RNA substrate, rather than a specific sequence, and has been observed in a number of different RNAs. Adenosine to inosine conversion occurs by a hydrolytic deamination that removes the C-6 amino group from adenosine[7] (Figure 3). A single enzyme called an ADAR (adenosine deaminase that acts on RNA; Figure 3) can catalyze the reaction, without the addition of cofactors.

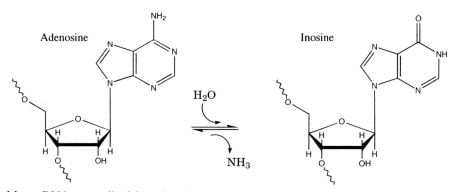

Figure 3 Many RNAs are edited by adenosine (A) to inosine (I) conversion. Enzymes that catalyze the reaction are called ADARs and studies of ADAR1 demonstrate that the reaction occurs by a simple hydrolytic deamination. As shown, the hydrolytic deamination of the amino group at C-6 yields inosine.

The natural RNA substrates for ADARs have been discovered by noticing sequence discrepancies between genes and their cDNAs. Since inosine, like guanosine, prefers to pair with cytidine, adenosine deamination events are indicated when a genomically encoded adenosine appears as a guanosine in a cDNA. Based on such A to G changes, two types of ADAR substrates have been recognized: those that are hypermutated, containing up to 50% A to G changes in a single cDNA, and those that are selectively deaminated, containing less than 10% A to G changes in their cDNAs (reviewed in reference 29). Most of the hypermutated substrates are viral RNAs, and here the function of the adenosine modification is unclear. However, in the selectively deaminated substrates it is clear that adenosine deamination functions to change one codon into another. In this way, ADARs act to regulate gene expression by allowing multiple proteins to be synthesized from a single encoded sequence.

6.07.2.2.1   ADAR proteins and their catalytic mechanism

ADAR activity was first detected when double-stranded RNA (dsRNA) was microinjected into *Xenopus laevis* embryos.[30,31] Since then, the activity has been found in every metazoan assayed (reviewed in reference 29). Two different enzymes with ADAR activity, ADAR1 and ADAR2, have been discovered so far (for review see references 29 and 32). Further, cDNAs have been discovered that, based on sequence similarities, are thought to encode additional ADARs.

Formerly, ADAR1 was called dsRAD or DRADA, while ADAR2 was called RED1.[33] In contrast to the C to U editing that occurs within apo-B mRNA, both ADAR1 and ADAR2 can catalyze deamination within their substrates without additional factors. The ADAR1 protein contains three dsRNA binding motifs (dsRBMs) upstream of a C-terminal catalytic domain that is highly conserved among ADARs (Figure 4); ADAR2 lacks the amino terminal half of ADAR1, and contains only two dsRBMs.[34] ADAR1 has been cloned from several mammals, as well as frogs, while, so far, ADAR2 has only been identified and characterized in mammals. The transcripts for both ADAR1 and ADAR2 are alternatively spliced to generate multiple isoforms.[35-37]

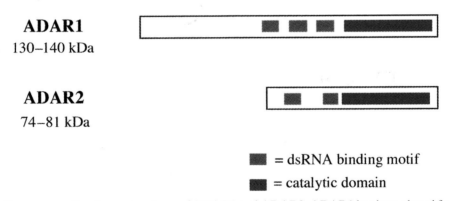

Figure 4   The open reading frame structures of ADAR1 and ADAR2. ADAR1 has been cloned from human, rat, and frog, and ADAR2 from human and rat; the proteins vary slightly from organism to organism so the diagrams are not exact. Red boxes indicate dsRNA binding motifs, and the highly conserved catalytic domain is illustrated with a blue rectangle. Numbers below ADAR1 and ADAR2 correspond to the range of molecular weights observed for the proteins of different organisms.

ADARs catalyze a nucleophilic attack at the C-6 atom of adenine. Since the C-6 atom is buried deep within the narrow major groove of dsRNA, it has been proposed that ADAR gains access to its target by flipping the adenine base out of the RNA helix and into the active site.[38,39] Such base-flipping mechanisms have been observed in DNA methyltransferases (reviewed in references 40 and 41).

6.07.2.2.2   ADAR specificity

The deamination specificity of the founding member of the ADAR family, ADAR1, has been characterized *in vitro*.[38] The enzyme does not have strict sequence requirements, and multiple adenosines can be deaminated in a single RNA. However, deaminations do not occur randomly. ADAR1 has a 5′ nearest-neighbor preference such that adenosines with a 5′ neighbor of A, U, or C are more likely to be deaminated than those with a 5′ G. ADAR1 also disfavors adenosines near 3′ termini (within ~8 nucleotides).

The structure of the RNA substrate is also important for ADAR deamination specificity and dictates whether the RNA is promiscuously deaminated or selectively deaminated. In promiscuously deaminated substrates, ~50% of the adenosines are deaminated at complete reaction, while in selectively deaminated substrates, <10% of the adenosines are deaminated at complete reaction. *In vitro* studies, as well as analyses of endogenous substrates, indicate that the promiscuous type of deamination occurs in the context of long, completely base-paired dsRNA, while the more selective deamination occurs in molecules where base-paired regions are periodically interrupted by mismatches, bulges, and loops (e.g., see Figure 5).

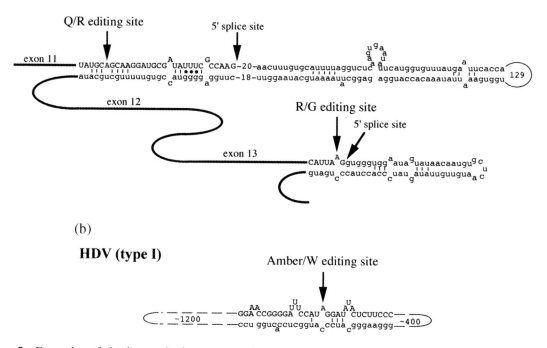

Figure 5 Examples of the base-paired structures that surround ADAR editing sites. (a) The structures required for editing at the Q/R site in exon 11 of gluR-B, and the R/G site in exon 13 of gluR-B, are formed by base pairing between the exon (upper case sequences) and downstream introns (lower case sequences). Editing sites with biological consequences are shown as red capital letters. Only base pairs supported by compensatory mutations are shown (as cited in reference 29). Numbers indicate the length of sequences excluded from the diagram. (b) The sequences immediately surrounding the editing site within the antigenomic RNA of hepatitis delta virus are shown, with symbols and labels as in (a) (reproduced by permission of Elsevier Science Ltd from *Trends Biochem. Sci.*, 1997, **22**, 157).

Although the deamination specificity of ADAR2 has not been extensively studied, data gathered so far clearly show that it is similar to, but distinct from, that of ADAR1. Like ADAR1, ADAR2 can act with high or low selectivity, depending on the structure of the RNA substrate. For example, ~50% of the adenosines in completely base-paired dsRNA are deaminated,[34,35,37] while base-paired regions of biological substrates, such as glutamate receptor (gluR) B mRNA (see below), are deaminated more selectively. Various laboratories have analyzed the ability of ADAR2 to deaminate editing sites within gluR-B pre-mRNA.[34,35,37,42,43] ADAR2 is clearly more efficient at editing the Q/R site, while both enzymes edit the R/G site (see Figure 5(a) and text below). Thus, the two enzymes have distinct but overlapping specificities. Because of this, at present, it is hard to say which enzyme is responsible for which editing event *in vivo*.

6.07.2.2.3 Mammalian glutamate receptor mRNA

Based on pharmacological studies, glutamate receptors in the mammalian central nervous system are classified into three distinct classes: *N*-methyl-D-aspartate (NMDA) receptors, α-amino-3 hydroxyl-5-methyl-isoxazole-4-propionate (AMPA) receptors, and kainate (KA) receptors. Editing by adenosine deamination has been demonstrated in the pre-mRNA of the AMPA receptors, gluR-B, -C, and -D, as well as in the KA receptor subunits gluR-5 and gluR-6. As might be expected, the editing sites are found within sequences that can form base-paired structures. So far, all of the structures required for gluR RNA editing involve base-pairing between exons and introns (see Figure 5(a)).

Editing within gluR-B pre-mRNA has been extensively studied, and two biologically important editing events have been defined within its coding sequence. Since inosine is translated as guanosine,[44] the A to I conversions are synonymous with A to G changes. Editing within exon 11 of gluR-B pre-mRNA converts a glutamine codon (CAG) to an arginine codon (CIG). ADAR editing sites are named according to the amino acid change they produce, and thus, this site is called the Q/R editing site. The Q/R site is edited in 99% of gluR-B pre-mRNAs, at all stages of development. In contrast, gluR-5 and gluR-6 are edited at the Q/R site to 50% and 70%, respectively, in adult rat brain, but to a much lower extent early in embryogenesis (for review see reference 45).

Another editing site has been found in exon 13 of gluR-B, -C, and -D. Here editing converts an arginine codon (AGA) to a glycine codon (IGA) and accordingly, the site is called the R/G editing site. Editing at the R/G site also varies during early development, reaching a maximum in the adult brain of 80–90% for gluR-B and -C.[46] Additional editing sites have also been detected within the coding region of gluR-6, in particular, the I/V and the Y/C site.

Additional studies are required to fully understand the functional consequences of each of the editing events within the various glutamate receptors. However, editing at the Q/R site of gluR-B is particularly well characterized and clearly has a striking effect on function. Specifically, ion channels that contain gluR-B with arginine instead of glutamine at the Q/R site are much less permeable to calcium.[47] Mice that have lower than normal levels of editing at the Q/R site of gluR-B mRNA suffer epilepsy and die about 12 days after birth.[48]

6.07.2.2.4 Hepatitis delta virus antigenomic RNA

Hepatitis delta virus (HDV) is a closed, circular RNA of about 1700 nucleotides. The genome is replicated via an RNA intermediate, the antigenome (for review see reference 49). Both the genome and the antigenome can be folded into rod-shaped, highly base-paired structures, similar to those of viroids. Although the virus expresses only a single open reading frame (ORF), RNA editing by adenosine deamination allows the synthesis of two essential proteins from this single ORF (Figure 6). The shorter protein, HDAg-p24, is synthesized from the encoded ORF, which terminates at an amber stop codon (UAG). The longer protein, HDAg-p27, is synthesized after RNA editing converts the amber stop codon (UAG) to a tryptophan codon (UGG), extending the open reading frame by 19 amino acids. The shorter form of the protein, HDAg-p24, is necessary for replication, while the longer version, HDAg-p27, is required for virus packaging. Editing occurs on the antigenomic RNA at an adenosine referred to as the amber/W editing site[50] and the sequence change is passed to the genome during replication. HDV antigenomic RNA is specifically and efficiently edited *in vitro* at the amber/W site by *X. laevis* ADAR1, consistent with the idea that the ADAR enzyme is responsible for the editing event.[51]

6.07.2.2.5 Other transcripts edited by adenosine deamination

Although only a few examples of ADAR substrates have been described here, other hypermutated as well as selectively deaminated RNAs have been described (as cited in reference 29). So far, the examples of hypermutated RNAs are predominantly viral RNAs, while the selectively deaminated RNAs mainly encode proteins involved in neurotransmission. In addition to the gluR mRNAs, these include mRNAs encoding serotonin receptors[52] and potassium channels.[53] As with other ADAR substrates, in these cases, selective deamination functions to allow multiple proteins to be derived from a single mRNA, while as yet, the function of the hypermutations is unclear.

In contrast to other types of mRNA editing described in this chapter, which involve the conversion of one Watson–Crick nucleotide into another, or the insertion or deletion of Watson–Crick nucleo-

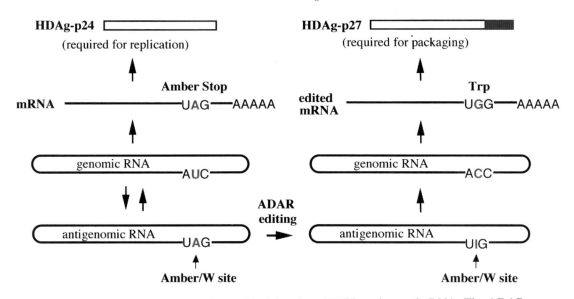

Figure 6 A model for RNA editing of hepatitis delta virus (HDV) antigenomic RNA. The ADAR enzyme acts on the antigenomic RNA at the amber/W site to convert an adenosine (A) to an inosine (I). The sequence change is passed to the genomic RNA during replication. Transcription of the "unedited" genome yields an mRNA that encodes HDAg-p24, a protein required for HDV replication. Transcription of the "edited" genome yields an mRNA that encodes HDAg-p27, a protein required for packaging of the HDV genome and which contains 19 additional amino acids (blue) (reproduced by permission of Macmillan Magazines Ltd from *Nature*, 1996, **380**, 454).

tides, adenosine deamination within an mRNA creates a nucleotide that is not normally found in nascent transcripts. This has allowed a direct measurement of the amount of inosine contained within mRNA of various mammalian tissues.[54] The amounts are quite astonishing and suggest that there are many more inosine-containing RNAs yet to be discovered.

6.07.3 INSERTION/DELETION EDITING

The editing events discussed so far involve the covalent modification of an encoded nucleotide. In this section the second major category of editing is presented, which involves the insertion or deletion of nucleotides that are not genomically encoded. The most well-characterized type of insertion/deletion editing occurs in mitochondrial pre-mRNAs of trypanosomes, a protozoan, and involves the insertion and deletion of UMP. In the current model for this type of editing, several protein enzymes catalyze the requisite cleavage and ligation events, while small RNAs, called guide RNAs (gRNAs), interact with the pre-mRNA to specify where the UMPs are to be added or removed. Other insertion/deletion types of editing, in some cases involving nucleotides other than uridine, have been observed in mitochondria of other organisms as well (reviewed in references 2 and 3). At present it is not clear if these other examples will have mechanistic similarities to the insertion/deletion editing that occurs in trypanosomes.

6.07.3.1 Editing in Trypanosomatid Protozoa

Many mitochondrial-encoded mRNAs of kinetoplastid trypanosomes are edited by insertions or deletions of UMPs (reviewed in references 55–57). The number of editing sites depends on the particular RNA, with some RNAs requiring very few insertions or deletions to create the correct ORF, and others requiring editing at hundreds of sites to create the functional ORF. At the extreme are the pan-edited RNAs, where more than 50% of the nucleotides derive from post-transcriptional U insertions. Overall, deletions are found less frequently, about one-tenth as often as insertions. Editing leads to the creation of new initiation and stop codons, as well as the conversion of one codon into another. Despite the complexity of RNA editing in trypanosomes, functional RNAs are produced. The parasitic trypanosome depends on RNA editing, at least in the stages of the life cycle requiring aerobic respiration.

6.07.3.1.1 *The catalytic mechanism of U insertion/deletion RNA editing*

Overall, editing by U insertion and deletion occurs in a 3′ to 5′ direction along the pre-mRNA. The gRNAs base pair with the unedited pre-mRNA, using GU as well as canonical base pairs, and designate where the uridines should be inserted or deleted. The gRNAs are about 70 nucleotides long and consist of three domains: the *anchor*, a 5′ sequence that is perfectly complementary to the pre-mRNA downstream of the region to be edited; a central domain that is partially complementary to the region to be edited and a 3′ U-tail of 5–24 residues that may stabilize the initial gRNA–mRNA interaction[58-60] (Figure 7). It is the central domain, which is only partially complementary to the unedited pre-mRNA, that actually designates where UMPs are to be inserted or deleted. As shown in Figure 7, positions of mismatched bases mark sites where uridines are to be inserted or deleted.

Over the years, two models have been considered for how insertion/deletion editing is catalyzed. One model, the transesterification model, proposes that insertion/deletion is an RNA-catalyzed reaction, while the other model invokes a series of protein enzymes to catalyze the requisite cleavage and ligation events. Recent evidence all but proves that protein enzymes are involved, and thus, here the authors will focus on the cleavage–ligation model which assumes catalysis by proteins.

An important step in understanding the mechanism of U-insertion/deletion RNA editing was the development of *in vitro* editing systems. Although inefficient, *in vitro* systems for both U-insertion and U-deletion have now been established.[59,61-63] The mechanism for U-insertion, based on the cleavage–ligation model, involves four distinct enzymatic activities: an RNA endonuclease, a uridylyl transferase, an exonuclease, and an RNA ligase[59,61,62] (Figure 7, left panel). All of these enzyme activities have been detected in trypanosomes.[64-67] Initially, the mismatches formed between the central domain of the gRNA and the unedited pre-mRNA direct cleavage by an endonuclease. The anchor sequence of the gRNA binds to the mRNA and cleavage occurs immediately 5′ of the gRNA–mRNA duplex (with respect to the mRNA). The nuclease generates a 3′ hydroxyl on the 5′ cleavage product and a 5′ phosphate on the 3′ cleavage product of the mRNA. Subsequently, terminal uridylyl transferase (TUTase) adds U residues to the 3′-end of the mRNA, using free UTP to incorporate 5′-UMP. Although in the original model,[58] UMP residues were added one at a time, recent evidence suggests that multiple UMPs are added during each round of editing;[68] those that do not base pair with the gRNA are subsequently trimmed by an 3′–5′ exonuclease. The last step is a rejoining of the mRNA by an RNA ligase.

The mechanism for U-deletion resembles U-insertion but probably differs at the initial cleavage step.[63] U-deletion is also initiated by an endonucleolytic cleavage event (Figure 7, right panel), but in contrast to U insertion, requires a U residue immediately upstream of the duplex formed between the pre-mRNA and gRNA anchor sequence. U-deletion requires high concentrations of adenosine nucleotides and therefore differs from U-insertional editing which is inhibited by high ATP or ADP concentrations.[63] The second step of U-deletion editing involves a 3′–5′ exonuclease that removes one or several U residue(s) from the 3′ end of the 5′ half of the pre-mRNA. The U residues are thought to be removed from the RNA individually as UMP. Finally, the RNA is ligated in a manner similar to the last step in U-insertional editing.[60,63,66] The models presented here, for both U-deletion and U-insertion, are based on *in vitro* studies of sequences from natural substrates, such as the pre-mRNA of ATPase subunit 6 in *Trypanosoma brucei* and that of NADH dehydrogenase 7 in *Leishmania tarentolae*.

6.07.4 SUMMARY AND PERSPECTIVES

The discovery that nuclear-encoded mRNAs of metazoa were subject to RNA editing altered the prevailing view that editing was relevant only to those studying mitochondrial gene expression in single-celled eukaryotes. Two types of RNA editing have been found in nuclear-encoded mRNAs, and one type has been observed ubiquitously in every metazoan assayed. The RNA editing observed in nuclear-encoded RNAs involves deamination of encoded nucleotides, rather than the insertion and deletion of nucleotides common in mitochondrial RNA editing; in one case deamination of C

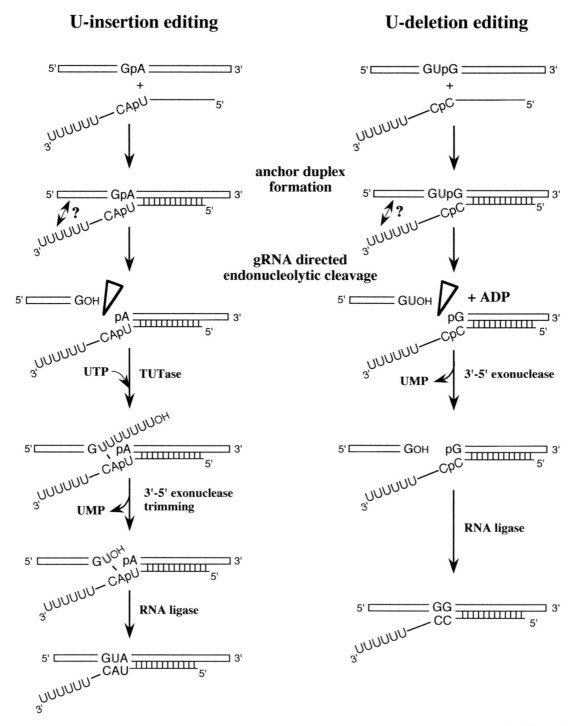

Figure 7 Current models for the mechanism of RNA editing in kinetoplastid mitochondria. The left panel shows a model for U-insertion, and the right panel shows a model for U-deletion. Pre-mRNAs are shown as open boxes, guide RNAs are indicated as lines, and vertical lines indicate base pairs. Open arrowheads indicate cleavage sites, and only sequences immediately surrounding the editing sites are shown. The double arrow indicates a possible interaction between the 3′ U-tail of the gRNA and the pre-edited sequence. Uridines to be inserted or deleted are shown in red. See text for further details.

creates U, and in the other case, deamination of A creates I. While C to U editing of a nuclear-encoded mRNA has only been observed in a single mammalian mRNA, the apoB mRNA, A to I editing has been observed in many RNAs, in many different metazoa. While a single enzyme called an ADAR is sufficient for catalysis of A to I editing, the C to U editing of apoB mRNA requires the enzyme APOBEC-1, as well as additional, as yet unconfirmed, factors.

In the examples of RNA editing within nuclear-encoded mRNAs, editing changes a functional transcript so as to allow a different and additional function. Thus, these deamination reactions are similar to alternative RNA splicing and frameshifting, that is, RNA processing reactions which allow multiple proteins to be generated from a single-encoded sequence. In contrast, the various types of RNA editing that occur in mitochondria can largely be considered as types of repair, that is, processes which create functional RNAs from garbled nascent transcripts. While most of the RNA editing that occurs in mitochondria involves insertion/deletion mechanisms, C to U as well as U to C editing has been observed.

In most cases, the mechanism and macromolecular machinery required for RNA editing in mitochondrial-encoded transcripts has not been defined. However, a great deal of progress has been made in characterizing the editing reaction that occurs in the mitochondria of trypanosomes, the organism where insertion/deletion editing was discovered. Editing in trypanosomal mitochondria requires both protein and RNAs, but recent results indicate the catalytic active sites are contained in the proteins, thus settling a long-standing debate. At least four different enzymatic activities are involved in catalyzing the requisite cleavage and ligation reactions, while the small RNAs, the guide RNAs, designate where the mRNA should be edited by base pairing with the unedited pre-mRNA. To date, the proteins responsible for the enzymatic activities have not been purified to homogeneity.

The focus of future studies on RNA editing will undoubtedly differ for the various types of RNA editing. For most of the mitochondrial systems, studies will be directed towards the definitive identification and characterization of the macromolecules that catalyze the reactions. In these cases, many different RNA substrates have been identified, and once the reactions can be reconstituted *in vitro*, the details of substrate recognition can be worked out. For the RNA editing systems in the nucleus, some of the proteins that catalyze the reactions have been characterized, and thus, here studies will instead focus on the identification of additional substrates, since few are known. Of course, even though some of the proteins that catalyze these reactions have been identified, little is known about how the reactions are regulated *in vivo*, and future studies will also explore the possibility that regulatory factors exist.

6.07.5 REFERENCES

1. R. Benne, J. Van den Burg, J. P. J. Brakenhoff, P. Sloof, J. H. van Boom, and M. C. Tromp, *Cell*, 1986, **46**, 819.
2. H. Grosjean and R. Benne (eds.), "Modification and Editing of RNA," American Society for Microbiology, Washington, DC, 1998.
3. H. C. Smith, J. M. Gott, and M. R. Hanson *RNA*, 1997, **3**, 1105.
4. J. A. McCloskey and P. F. Crain, *Nucleic Acids Res.*, 1998, **26**, 196.
5. P. A. Limbach, P. F. Crain, and J. A. McCloskey, *Nucleic Acids Res.*, 1994, **22**, 2183.
6. D. F. Johnson, K. S. Poksay, and T. L. Innerarity, *Biochem. Biophys. Res. Commun.*, 1993, **195**, 1204.
7. A. G. Polson, P. F. Crain, S. C. Pomerantz, J. A. McCloskey, and B. L. Bass, *Biochemistry*, 1991, **30**, 11 507.
8. L. M. Powell, S. C. Wallis, R. J. Pease, Y. H. Edwards, T. J. Knott, and J. Scott, *Cell*, 1987, **50**, 831.
9. S.-H. Chen, G. Habib, C.-Y. Yang, Z.-W. Gu, B. R. Lo, S.-A. Weng, S. R. Silberman, S.-J. Cai, J. P. Deslypere, M. Rosseneu *et al.*, *Science*, 1987, **328**, 363.
10. L. Chan, B. H. Chang, M. Nakamuta, W. H. Li, and L. C. Smith, *Biochim. Biophys. Acta*, 1997, **1345**, 11.
11. H. C. Smith and M. P. Sowden, *Trends Genet.*, 1996, **12**, 418.
12. B. Teng, C. F. Burant, and N. O. Davidson, *Science*, 1993, **260**, 1816.
13. N. O. Davidson, T. L. Innerarity, J. Scott, H. Smith, D. M. Driscoll, B. Teng, and L. Chan, *RNA*, 1995, **1**, 3.
14. A. J. MacGinnitie, S. Anant, and N. O. Davidson, *J. Biol. Chem.*, 1995, **270**, 14 768.
15. N. Navaratnam, J. R. Morrison, S. Bhattacharya, D. Patel, T. Funihashi, F. Giannoni, B. B. Teng, N. O. Davidson, and J. Scott, *J. Biol. Chem.*, 1993, **268**, 20 709.
16. S. Yamanaka, K. S. Poksay, K. S. Arnold, and T. L. Innerarity, *Genes Dev.*, 1997, **11**, 321.
17. S. G. Young, S. J. Bertics, T. M. Scott, B. W. Dubois, L. K. Curtiss, and J. L. Witztum, *J. Biol. Chem.*, 1986, **261**, 2995.
18. P. P. Lau, W. J. Xiong, H. J. Zhu, S.-H. Chen, and L. Chan, *J. Biol. Chem.*, 1991, **266**, 20 550.
19. J. W. Backus and H. C. Smith, *Nucleic Acids Res.*, 1992, **20**, 6007.
20. R. R. Shah, T. J. Knott, J. E. Legros, N. Navaratnum, J. C. Greeve, and J. Scott, *J. Biol. Chem.*, 1991, **266**, 16 301.
21. D. M. Driscoll, S. Lakhe-Reddy, L. M. Oleksa, and D. Martinez, *Mol. Cell. Biol.*, 1993, **13**, 7288.
22. M. Hersberger and T. L. Innerarity, *J. Biol. Chem.*, 1998, **273**, 9435.
23. K. Hirano, S. G. Young, R. V. Farese, Jr., J. Ng, E. Sande, C. Warburton, L. M. Powell-Braxton, and N. O. Davidson, *J. Biol. Chem.*, 1996, **271**, 9887.
24. J. R. Morrison, C. Paszty, M. E. Stevens, S. D. Hughes, T. Forte, J. Scott, and E. M. Rubin, *Proc. Natl. Acad. Sci. USA*, 1996, **93**, 7154.

25. N. Navaratnam, T. Fujino, J. Bayliss, A. Jarmuz, A. How, N. Richardson, A. Samasekaram, S. Bhattacharya, C. Carter, and J. Scott, *J. Mol. Biol.*, 1998, **275**, 695.
26. L. Betts, S. Xiang, S. A. Short, R. Wolfenden, and C. W. Carter, Jr., *J. Mol. Biol.*, 1994, **235**, 635.
27. P. P. Lau, H.-J. Zhu, A. Baldini, C. Charnsangavej, and L. Chan, *Proc. Natl. Acad. Sci USA*, 1994, **91**, 8522.
28. A. Mehta, S. Banerjee, and D. M. Driscoll, *J. Biol. Chem.*, 1996, **271**, 28 294.
29. B. L. Bass, *Trends Biochem. Sci.*, 1997, **22**, 157.
30. B. L. Bass and H. Weintraub, *Cell*, 1987, **48**, 607.
31. M. R. Rebagliati and D. A. Melton, *Cell*, 1987, **48**, 599.
32. S. Maas, T. Melcher, and P. H. Seeburg, *Curr. Opin. Cell Biol.*, 1997, **9**, 343.
33. B. L. Bass, K. Nishikura, W. Keller, P. H. Seeburg, R. B. Emeson, M. A. O'Connell, C. E. Samuel, and A. Herbert, *RNA*, 1997, **3**, 947.
34. T. Melcher, S. Maas, A. Herb, R. Sprengel, P. H. Seeburg, and M. Higuchi, *Nature*, 1996, **379**, 460.
35. F. Lai, C. X. Chen, K. C. Carter, and K. Nishikura, *Mol. Cell. Biol.*, 1997, **17**, 2413.
36. Y. Liu, C. X. George, J. B. Patterson, and C. E. Samuel, *J. Biol. Chem.*, 1997, **272**, 4419.
37. A. Gerber, M. A. O'Connell, and W. Keller, *RNA*, 1997, **3**, 453.
38. A. G. Polson and B. L. Bass, *EMBO J.*, 1994, **13**, 5701.
39. R. F. Hough and B. L. Bass, *RNA*, 1997, **3**, 356.
40. R. J. Roberts, *Cell*, 1995, **82**, 9.
41. X. D. Cheng and R. M. Blumenthal, *Structure*, 1996, **4**, 639.
42. S. Maas, T. Melcher, A. Herb, P. H. Seeburg, W. Keller, S. Krause, M. Higuchi, and M. A. O'Connell, *J. Biol. Chem.*, 1996, **271**, 12 221.
43. T. Melcher, S. Maas, A. Herb, R. Sprengel, M. Higuchi, and P. H. Seeburg, *J. Biol. Chem.*, 1996, **271**, 31 795.
44. C. Basilio, A. J. Wahba, P. Lengyel, J. F. Speyer, and S. Ochoa, *Proc. Natl. Acad. Sci. USA*, 1962, **48**, 613.
45. P. H. Seeburg, *J. Neurochem.*, 1996, **66**, 1.
46. H. Lomeli, J. Mosbacher, T. Melcher, T. Höger, J. R. P. Geiger, T. Kuner, H. Monyer, M. Higuchi, A. Bach, and P. H. Seeburg, *Science*, 1994, **266**, 1709.
47. M. Hollman, M. Hartley, and S. Heinemann, *Science*, 1991, **252**, 851.
48. R. Brusa, F. Zimmermann, D.-S. Koh, D. Feldmeyer, P. Gass, P. H. Seeburg, and R. Sprengel, *Science*, 1995, **270**, 1677.
49. M. M. Lai, *Annu. Rev. Biochem.*, 1995, **64**, 259.
50. J. L. Casey and J. L. Gerin, *J. Virol.*, 1995, **69**, 7593.
51. A. G. Polson, L. B. Bass, and J. L. Casey, *Nature*, 1996, **380**, 454.
52. C. M. Burns, H. Chu, S. M. Reuter, L. K. Hutchinson, H. Canton, E. Sanders-Bush, and R. B. Emeson, *Nature*, 1997, **387**, 303.
53. D. E. Patton, T. Silva, and F. Bezanilla, *Neuron*, 1997, **19**, 711.
54. M. S. Paul and B. L. Bass, *EMBO J.*, 1998, **17**, 1120.
55. B. Sollner-Webb, *Science*, 1996, **273**, 1182.
56. R. Benne, *Curr. Opin. Genet. Dev.*, 1996, **6**, 221.
57. M. L. Kable, S. Heidmann, and K. D. Stuart, *Trends Biochem. Sci.*, 1997, **22**, 162.
58. B. Blum, N. Bakalara, and L. Simpson, *Cell*, 1990, **60**, 189.
59. M. L. Kable, S. D. Seiwert, S. Heidmann, and K. Stuart, *Science*, 1996, **273**, 1189.
60. S. D. Seiwert, S. Heidmann, and K. Stuart, *Cell*, 1996, **84**, 831.
61. E. M. Byrne, G. J. Connell, and L. Simpson, *EMBO J.*, 1996, **15**, 6758.
62. G. J. Connell, E. M. Byrne, and L. Simpson, *J. Biol. Chem.*, 1997, **272**, 4212.
63. J. Cruz-Reyes, L. N. Rusché, K. J. Piller, and B. Sollner-Webb, *Mol. Cell*, 1998, **1**, 401.
64. K. J. Piller, L. N. Rusche, J. Cruz-Reyes, and B. Sollner-Webb, *RNA*, 1997, **3**, 279.
65. N. Bakalara, A. M. Simpson, and L. Simpson, *J. Biol. Chem.*, 1989, **264**, 18679.
66. J. Cruz-Reyes and B. Sollner-Webb, *Proc. Natl. Acad. Sci. USA*, 1996, **93**, 8901.
67. R. Sabatini and S. L. Hajduk, *J. Biol. Chem.*, 1995, **270**, 7233.
68. J. D. Alfonzo, O. Thiemann, and L. Simpson, *Nucleic Acids Res.*, 1997, **25**, 3751.

6.08
RNA Enzymes: Overview

SIDNEY ALTMAN
Yale University, New Haven, CT, USA

6.08.1 INTRODUCTION

Biochemical catalysis is governed by enzymes. Enzymes are macromolecules, or complexes thereof, that are composed *in vivo* of protein or RNA or both. The vast majority of enzymes characterized to date from biological sources are proteins: a few are RNAs and none, as yet, are DNA or carbohydrates.

Unlike protein enzymes, which have been under investigation for over 150 years, RNA as a biochemical catalyst was first described only in 1982–1983.[1,2] The relatively recent identification of RNA enzymes is undoubtedly partly due to the fact that the biochemistry of individual macromolecular RNAs did not begin to flourish until the 1940s with description of ribosomes (and rRNA) and the 1950s, with the characterization of tRNA. These RNAs are the most abundant RNA species in cells and thus the easiest with which to work. As the technology for the synthesis and identification of individual RNA species progressed, so did the description of RNases, the enzymes that cleave phosphodiester bonds in RNA.

The convergence of both kinds of studies, that is, of RNA substrates and RNases,[3–6] paved the way for the first descriptions of RNases that contained RNA as catalytic entities. These catalysts have a rather narrow functional capability. To date naturally occurring RNA enzymes have only been shown to catalyze the cleavage and ligation of phosphoester bonds.[7,8] RNA species that function other than in the cleavage or biosynthesis of phosphoester bonds have been evolved *in vitro* from random sequences and/or have been shown to function *in vitro* in nonphysiological reactions.[9–14] While these latter studies provide a glimpse of the full spectrum of the catalytic potential of RNA, in this article only RNA enzymes that are known to function in living cells are discussed.

6.08.2 CLASSIFICATION

Enzymes that have a catalytic RNA subunit have been dubbed "ribozymes".[1] While this is a misnomer (unless a totally novel set of terms is created in which we resort to additional, incorrect labels, such as pepzymes, carbozymes, deoxyribozymes), its use has become widespread and thus it is also found in this volume. Irrespective of labels, the known RNA enzymes carry out a limited number of biological functions and have an equally limited number of mechanisms of action. What then is the appropriate way in which to classify and describe these enzymes? Previously, this has been done according to both similarities in RNA structure and the chemistry of the enzymatic mechanisms,[8,15,16] both of which are discussed extensively in the following chapters. However, given the increasing number of ribonucleoprotein complexes implicated in important cellular functions, it may be of interest to list RNA enzymes according to biological function as well (Table 1), in the hope of providing a guide to paths of research that will lead to the discovery of more such enzymes.

Various enzymes listed in Tables 1 and 2 can be grouped together either by a biological or chemical (mode of reaction) classification. The RNA enzymes (self-splicing introns) found in bacteriophage are more readily grouped with similar agents found in lower eukaryotes rather than with the catalytic activities found in single-stranded RNA pathogens. The mechanism of group I introns[8,15] differs radically from that of the so-called "hammerheads" and "hairpins",[18,22,23] which are involved in the replication of plant and human pathogens. RNase P stands alone as the only enzyme that cleaves *in vivo* in *trans*.[2,19] Can the patterns portrayed in the tables be generalized in anticipation of the discovery of more ribozymes?

To date, an RNase A-like mechanism of reaction of catalytic RNAs is restricted to single-stranded RNAs.[17,18,22,23] These are mostly (satellite) pathogens that, for their delivery, depend on some other viral RNA to provide an encapsulating vector. It can reasonably be expected that more such pathogens will be found. Why the catalytic cleavage mechanism is restricted to one that produces 5′ OH and 3′P groups may be related to the requirement for the process of replication that results in circular RNAs that are resistant to RNA exonucleases. Newly discovered single-stranded satellite RNA pathogens may be expected to have similar kinds of catalytic activity. There are no counterpart satellite DNAs. It does well to note that the recently solved crystal structure of the hammerhead ribozyme[22] derived from a satellite RNA and the crystal structure of part of a group I intron RNA[24] lend new insight into our understanding of the mechanisms of action of these enzymes although the reasons why their catalytic capability *in vivo* is limited as outlined above are still a mystery.

Can group II intron activity be expected only in mitochondria of fungi, in "homing" elements, and in plant organelles?[15,16] The RNA subunit of RNase P from several eubacteria has been shown to be catalytic alone *in vitro*[2] but no eukaryotic or archaebacterial counterpart has been shown to be catalytic alone. These observations appear to reflect an origin and distribution of various catalytic RNAs that follows the same lines of the ancient separation between eubacteria and their organelles.

There seems, after 15 years of research on ribozymes, no compelling reason to expect discovery of RNA enzymes with other specificities *in vivo* than those noted above. On the other hand, prior to 1982, there was no compelling reason to expect the discovery of an RNA enzyme. Eubacterial 23S rRNA (see below) may prove to be the first "exception" to this rule.

6.08.3 THE ROLE OF PROTEINS

The turnover number of catalytic RNAs *in vitro* is comparatively low ($\sim$0.1 mol product mol enzyme^{-1} min^{-1}).[7,8,20,25] With the addition of certain proteins, however, the turnover can be increased about 100-fold, in some cases to that approaching what is expected *in vivo*. With RNase P, fungal mitochondrial and phage group I introns, genetic data has been utilized to show that particular proteins must be cofactors of RNA enzyme activity *in vivo*.[15] In fact, the general assumption is made that all catalytic RNAs work in concert with specified (or specific) protein "cofactors" *in vivo*. However, it is also known that other proteins, seemingly unrelated to RNA processing activity, can enhance the turnover numbers of RNA enzymes *in vitro*. Such action can be explained, in part, by charge shielding between enzyme and substrate (*cis* or *trans*), or between domains of the enzyme, itself or by the chaperone-like activity of these proteins. Some group I and group II introns encode proteins that are essential for self-splicing or homing reactions.[8,15,16,26,27,28] In these latter cases the introns serve as vectors for agents essential for the function and propagation of other introns, an example of specialized evolution in the miniworld of mobile genetic elements.

The role of proteins in the action of plant satellite RNA enzymes is not yet clear. We assume that such proteins should exist in their hosts and promote both cleavage and ligation reactions. In the case of δ RNA, proteins are essential for replication and transmission of the RNA[21] but no protein has been identified that directly impinges on the rate of the RNA-catalyzed self-cleavage reaction.

While it is easy to imagine that specific protein factors enhance self-splicing of particular low

Table 1 RNA enzymes.

Enzyme	Source	Location	Function	Protein accessories[a]	Ref.
RNase P	Escherichia coli	—	stable RNA biosynthesis (tRNA, 4.5S RNA, tmRNA)	one protein subunit	7
	Saccharomyces cerevisiae	nucleus	tRNA, rRNA biosynthesis	nine subunits	19, 20
		mitochondria	tRNA biosynthesis	one subunit	
	HeLa cells	nucleus	tRNA, rRNA(?)biosynthesis		18
		mitochondria	tRNA biosynthesis		
Hammerhead and hairpin	plants		replication of pathogens (self-cleavage of multimeric genomic RNA, ligation of ends of monomers)		
δ (ax-head)	human	liver (hepatitis)	replication of pathogens (self-cleavage of multimeric genomic RNA, ligation of ends of monomers)	two	21
VS RNA	Neurospora	mitochondria	?		17
Group I introns	eukaryotes	nucleus	self-splicing	unknown	8, 15
	fungal	mitochondria		cyt-18 (chromosomal): intron-encoded maturase	
	phage	—	several	S12 (host chaperones)	
Group II introns	fungal mitochondria		self-splicing	intron-encoded maturase	15, 16
	bacteria		self-splicing	intron-encoded maturase	

[a] Defined as a protein that enhances the catalytic activity of the RNA or that is essential for some other function of that particular, or another, catalytic RNA.

Table 2   Some properties of catalytic RNAs.

RNA	End groups[a]	Cofactor[b]	Mechanism	Ref.
Group I introns	5′-P, 3′-OH	Yes	transesterification	8
Group II introns	5′-P, 3′-OH	No	transesterification	15, 16
RNase P M1 RNA	5′-P, 3′-OH	No	hydrolysis	7
Viroid/satellite/VS/δ RNAs	5′-OH, 2′,3′-cyclic phosphate	No	transesterification	18

[a] The end groups are those produced during the initial cleavage step of self-splicing reactions or during the usual cleavage reactions of other RNA species.   [b] This column refers to the use of a nucleotide cofactor.

copy number mRNAs, it is even easier to imagine a protein factor that is active in enhancing the rRNA self-splicing reaction of *Tetrahymena* rRNA.[1] Such a protein should be as abundant as the number of pre-rRNA molecules and present a relatively easy task for biochemical purification.

A particularly intriguing case of a catalytic reaction in which RNA must be involved is that of peptidyl transference on ribosomes. In this case, a chemical analogue of the peptidyl transferase reaction has been carried out with purified 23S and 5S rRNA and a small (1% by weight) additional complement of a homogeneous protein fraction from Thermophilus.[29] This case of a reaction that does not involve phosphoester bonds but in which RNA must play a major role has not been carried out in the total absence of protein. Thus, it cannot yet be categorized as either an example of RNA catalysis with a protein cofactor or as a reaction that requires both RNA and protein at the active center. Nevertheless, to reduce the complexity of the function of the large subunit of a ribosome to an ensemble of two RNAs and one protein is still a remarkable feat and, perhaps, creates an opening into an understanding of the transition between an all-RNA and an RNA–protein world.

Many more examples of RNA–protein complexes (ribonucleoproteins) that carry out enzymatic functions (including functions that expend or consume ATP or GTP) exist, particularly in eukaryotes. In all these cases, the RNA has either been shown to perform a template function (telomerase; editing with guide sequences; methylation of snoRNAs) or to require the presence of protein for function of the RNA (eukaryotic RNase P). In no case has the RNA been shown to be catalytic on its own.

6.08.4   PRACTICAL APPLICATIONS

There is a worldwide industry concerned with the application of RNA enzyme technology to practical problems of gene expression. The ability of RNA enzymes to recognize specific RNA structures in solution, determined in part by specific sequences, has led to the design and testing of RNA enzymes that can be used as tools to cleave specific viral or host cell RNAs that are determinants of pathological states. In one case, that of δ RNA, the use of the RNA as a vector for specific "therapeutic" sequences has been contemplated for use in liver disease because of the tropism of δ RNA for hepatocytes.[30]

The basic idea in all the practical methods that use RNA enzymes is the creation of a substrate for cleavage by the enzyme that mimics the structure of a natural substrate for that enzyme.[31] This is accomplished by redesigning, for example, a group I intron, hammerhead, hairpin or δ RNA sequence such that (i) it is divided into two parts with an "enzyme" surface altered in base sequence, (ii) it will then base pair only with the target substrate sequence. In the case of RNase P, which cleaves in *trans* naturally, an external guide sequence (EGS) is designed to hydrogen bond with the target molecule only to create a substrate through mimicry of the structure of natural substrates for the enzyme.

The methods that have been developed to date, impressively successful *in vitro* or in mammalian cells in tissue culture, must still be proven effective in animals: the major obstacle is efficient delivery to the desired tissue or cells of the large (in comparison with currently used antibiotics) RNAs, be they enzymes or guide sequences.

6.08.5   MISCELLANEOUS

Chapters 6.10 and 6.11 in this volume deal with details of the enzymology of catalytic RNAs and some aspects of their biological function. Chapter 6.15 discusses the many approaches to the use of RNA enzymes in biotechnology. It is, perhaps, valuable to make a few additional explanatory comments about what evolution *in vitro* experiments (SELEX methods)[9–12] tell us about RNA catalysis today.

As mentioned above, SELEX methods and other chemical experiments have identified RNA molecules that can carry out functions other than phosphodiester bond cleavage or ligation or have revealed functions in addition to these two that can be carried out *in vitro*. Nevertheless, we should not be overly optimistic about finding these functions in present day RNAs *in vivo*. It is not apparent that Nature has sampled RNA sequence space adequately to select for "all" functions in the sense that sampling (and Darwinian survival) had to depend on the happy coincidence in time of the presence of a particular sequence, its replication, and selection for survival of a particular function. As yet there is no evidence that this improbable convergence occurred for any functions except for those of phosphodiester bond cleavage and ligation. If group I introns or other catalytic RNAs can be made to carry out additional reactions *in vitro*,[8–14] this may be a tribute to the conformational flexibility and reactivity of large RNA sequences rather than to some important facet of biological function.

In writing this chapter, a speculative assumption was made that there might be a way of classifying RNA enzymes that would lend further insight into their occurrence in the contemporary world. On reconsideration, the (perhaps limited) chemical catalytic capabilities of these enzymes, in combination with the diverse biological roles they play in conjunction with auxiliary, or self-encoded proteins, we are struck by a dazzling spectrum of biological variety, one that will undoubtedly be found to be even more remarkable as this field of research is explored further.

ACKNOWLEDGMENTS

I thank Dr Marlene Belfort for comments on the manuscript. Work in the laboratory of the author was supported by grant USPHS GM-19422.

6.08.6 REFERENCES

1. K. Kruger, P. J. Grabowski, A. J. Zaug, J. Sands, D. E. Gottschling, and T. R. Cech, *Cell*, 1982, **31**, 147.
2. C. Guerrier-Takada, K. Gardiner, T. Marsh, N. R. Pace, and S. Altman, *Cell*, 1983, **35**, 849.
3. D. Apirion and A. Miczak, *Bioessays*, 1993, **15**, 113.
4. M. P. Deutscher, in "tRNA: Structure, Biosynthesis and Function," eds. D. Soll and U. L. RajBhandary, American Society for Microbiology, Washington, D.C., 1985, p. 51.
5. C. A. Ingle and S. R. Kushner, *Proc. Natl. Acad. Sci. USA*, 1996, **93**, 12 926.
6. A. Miczak, V. R. Kaberdin, C. L. Wei, and S. Lin-Chao, *Proc. Natl. Acad. Sci. USA*, 1996, **93**, 3865.
7. S. Altman, *Adv. Enzymol*, 1989, **62**, 1.
8. T. R. Cech, in "The RNA World," eds. R. F. Gesteland and J. F. Atkins, CSH Press, 1993, p. 239.
9. L. Gold, C. Tuerk, P. Allen, J. Brinkley, D. Brown, L. Green, S. MacDougal, D. Schneider, D. Tasset, and S. R. Eddy, in "The RNA World," eds. R. F. Gesteland and J. F. Atkins, CSH Press, 1993, pp. 497.
10. R. R. Breaker and G. F. Joyce, *Trends Biotechnol.*, 1994, **12**, 268.
11. J. R. Lorsch and J. W. Szostak, *Acc. Chem. Res.*, 1996, **29**, 103.
12. X. C. Dai, A. de Mesmaeker, and G. F. Joyce, *Science*, 1995, **267**, 237.
13. F. G. Huang and M. Yarus, *Proc. Natl. Acad. Sci. USA*, 1997, **94**, 8965.
14. B. L. Zhang and T. Cech, *Nature*, 1997, **390**, 96.
15. A. M. Lambowitz and M. Belfort, *Annu. Rev. Biochem.*, 1993, **62**, 587.
16. F. Michel and J.-L. Ferat, *Annu. Rev. Biochem.*, 1995, **64**, 435.
17. B. J. Saville and R. A. Collins, *Proc. Natl. Acad. Sci. USA*, 1991, **88**, 8826.
18. R. H. Symons, *Curr. Opin. Str. Biol.*, 1994, **4**, 322.
19. V. Stolc and S. Altman, *Genes Develop.*, 1997, **11**, 2414.
20. J. R. Chamberlain, A. J. Tranguch, E. Pagan-Ramos, and D. R. Engelke, *Progr. Nucleic Acids Res. Mol. Biol.*, 1996, **4**, 87.
21. H.-W. Wang, P.-J. Chen, C-Z Lee, H.-L. Wu, and D.-S. Chen, *J. Virol.*, 1994, **68**, 6363.
22. W. G. Scott and A. Klug, *Trends. Biochem. Sci.*, 1996, **21**, 220.
23. K. R. Birikh, P. A. Heaton, and F. Eckstein, *Eur. J. Biochem.*, 1997, **245**, 1.
24. J. H. Cate, A. R. Gooding, E. Podell, K. Zhou, B. L. Golden, C. E. Kundrot, T. R. Cech, and J. A. Doudna, *Science*, 1996, **273**, 1678.
25. Z. Tsuchihashi, M. Khosla, and D. Herschlag, *Science*, 1993, **262**, 99.
26. T. Coetzee, D. Herschlag, and M. Belfort, *Genes. Develop.*, 1994, **8**, 1575.
27. S. Zimmerly, H. T. Guo, P. S. Perlman, and A. M. Lambowitz, *Cell*, 1995, **82**, 545.
28. M. J. Curcio and M. Belfort, *Cell*, 1996, **84**, 9.
29. H. F. Noller, V. Hoffarth, and L. Zimniak, *Science*, 1992, **256**, 1416.
30. A. Goldberg and H. D. Robertson, personal communication.
31. S. Altman, *Proc. Natl. Acad. Sci. USA*, 1993, **90**, 10 898.

6.09
Ribozyme Selection

ANDREW D. ELLINGTON and MICHAEL P. ROBERTSON
University of Texas at Austin, TX, USA

6.09.1 THE HISTORY OF RIBOZYMES AND RIBOZYMES IN HISTORY

While the discoveries that RNase P and the Group I self-splicing intron could catalyze phosphodiester bond rearrangements popularized the notion that nucleic acids could perform functions beyond their traditional roles as informational macromolecules,[1,2] these experiments had been anticipated almost two decades before by speculations that cellular cofactors and RNAs, such as tRNA, were the lineal descendents of ribozymes. In particular, Crick[3] and Orgel[4] were among the first to suggest that an RNA world may have once existed where ribozymes served in place of protein enzymes. The experimental validation of these early musings further emboldened researchers to suggest that the putative RNA world may have been metabolically complex.[5,6]

Further support for the RNA world hypothesis came from studies of both natural and unnatural RNA catalysts. Soon after the discovery of RNA catalysis, several additional natural ribozymes were identified and characterized. Small hairpin and hammerhead ribozymes were found to catalyze the cleavage of specific phosphodiester bonds to yield 5'-hydroxyl and 2',3'-cyclic phosphate termini (reviewed in Ref. 7). The sequences and structures of these small ribozymes bore no resemblance to their larger counterparts. Moreover, their mechanisms were vastly dissimilar: RNase P used water as a nucleophile rather than a 2' hydroxyl to cleave pre-tRNAs, yielding 5' phosphates and free 3' hydroxyls. In contrast, the Group I ribozymes used the 3′ hydroxyl of an exogenous guanosine cofactor to initiate a splicing cascade in which the 3′ hydroxyl of one exon was eventually joined to the 5′ phosphate of a second exon and the intervening intron was concomitantly released as a linear RNA. The observed diversities of ribozyme sequences and mechanisms begged the conclusion that either these few remaining ribozymes had descended from a complex bestiary of catalysts, or that ribozyme catalysis was relatively easy to invent. In either instance, the more ribozymes that were found, the more likely it was that an RNA world may have once existed. However, all of the natural ribozymes catalyzed the scission, formation, or rearrangement of phosphodiester bonds. If ribozymes were the progenitors of protein enzymes, then it was likely that they had also once catalyzed a wider variety of reactions, such as those involved in energy-yielding metabolism (e.g., redox reactions) or in the biochemical transformations of simple organic skeletons (e.g., carbon–carbon bond cleavages and formations). These hypothesized ribozymes might now be extinct, but there was potentially a way to generate their doppelgangers: *in vitro* selection.

6.09.2 *IN VITRO* SELECTION OF RIBOZYMES

In vitro selection has been applied to the study of both natural and unnatural ribozymes.[90] *In vitro* selection is governed by the same principles that are operative during natural selection: first, a pool of heritable diversity is generated. In the case of natural selection, these are the mutations that arise during replication. In the case of *in vitro* selection, the pool can be chemically synthesized with the desired degree of randomness. Second, variants within the pool are selected for function. In the case of natural selections, organismal survival is frequently linked to molecular function; some bacteria can survive in the presence of an antibiotic because they have an enzyme that can hydrolyze that antibiotic. Similarly, in the case of *in vitro* selection, molecular survival is made dependent on the ability of ribozymes to catalyze reactions. Finally, the functional variants are preferentially amplified. In the case of natural selection, organismal variants that are more functional are frequently better able to reproduce. It is also possible, of course, that organismal variants can survive based on their reproductive capacities alone, without attendant functionality. In the case of *in vitro* selection, functional ribozymes gain preferential access to reproductive resources (enzymes and nucleotides) and are amplified. In both natural and *in vitro* selection, multiple cycles of selection and amplification can optimize the functions of molecules in successive generations. However, in natural selection the variation that exists is episodic, with relatively few new mutations being introduced into each generation, while during *in vitro* selection a much larger range of sequence variation is introduced at the beginning of an experiment and winnowed to a few functional species over multiple cycles. For this reason, it is safe to conclude that *in vitro* selection moves at a much more rapid clip than natural selection and can functionally optimize molecules with much greater efficiency.

In greater detail, *in vitro* selection generally starts with a random sequence nucleic acid population (Figure 1). The nucleic acid population is chemically synthesized as a single-stranded DNA oligonucleotide, initially amplified by the polymerase chain reaction (PCR) into double-stranded DNA templates, and *in vitro* transcribed into RNA via promoters appended to the PCR primers. There are several types of randomization that can be utilized to examine or generate ribozymes. Partial randomization is used to generate RNA pools that are centered on the wild-type molecule. Depending on the degree of mutagenesis, ribozyme populations that contain anywhere from all possible single to all possible hextuple substitutions can be generated. The level of partial randomization (or "doping") can be varied by varying the composition of the phosphoramidite mix used for chemical synthesis. For example, ribozymes can be synthesized that contain 70% of a wild-type residue (e.g., G) and 10% of each nonwild-type residue (e.g., A,U,C). PCR mutagenesis or other mutational techniques can also be used to generate doped pools, albeit at much lower substitution rates (typically 96–99% wild-type, with an uneven skew of mutations). Such doped pools are useful for delimiting and defining functional sequences and structures. For example, if a given residue at a given position is essential for ribozyme function, then prior to selection the wild-type residue may be present only 70% of the time, while following selection the wild-type residue may be present up to 100% of the time. Conversely, if a given residue is neutral for ribozyme function, then there will be little difference between the proportion of the wild-type residue in the pre- and postselection RNA populations. Similarly, residues that form base pairs may coordinately change during selection. For example, if a ribozyme contains a G:C base pair that forms a functional secondary structure motif, but neither the G nor the C is specifically required for function, then in a selected population the G:C base pair may be replaced with A:U, U:A, or C:G pairings. These alternative pairings will appear as covariations in the primary sequence data, and can be used to map or confirm secondary structural features and motifs of ribozymes. Covariations are not limited to Watson–Crick pairings, but can also identify or confirm other isosteres such as wobble pairings or purine:purine interactions.

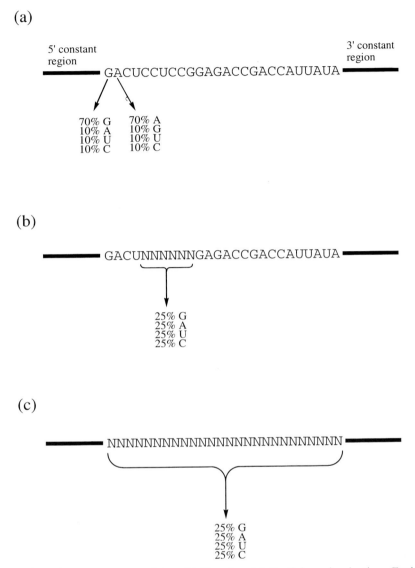

Figure 1 Types of random sequence nucleic acid libraries. (a) Partial randomization. Each position in a constant sequence region is "doped" with mutations. The level of doping can be preset by the experimenter. (b) Segmental or "shallow" randomization. A portion of a constant sequence region is completely randomized. The random sequence positions can be grouped (as shown here) or dispersed. Most shallow random pools contain all possible sequences. (c) Complete or "deep" randomization. Most or all of the residues that lie between the constant regions are completely randomized. Most deep random pools do not contain all possible sequences.

Functional ribozymes can also be selected from completely randomized populations. While the degree of randomization will be absolute (25% of each of the four canonical nucleotides), the length of the random sequence tract can vary. Functional variants of known ribozymes can be selected from RNA pools in which short segments are completely randomized. Results from selection experiments with such "shallow" or segmentally random pools will be similar to results from selection experiments with partially randomized pools, but will be restricted to the region that was initially randomized. The relative advantages of partial vs. shallow random pools depend on the length of the ribozyme and on whether functional sequence or functional structure is being probed. Partial randomization allows residues important for function to be quickly sieved from those neutral to function over the entire length of a molecule. However, sequence covariations necessary to establish or confirm structural features are relatively rare. For example, in a pool that contains 70% wild-type residues, a wild-type base pair will be found in roughly 50% of the population (0.7×0.7), while nonwild-type base pairs will be found in only 3% of the population ($3 \times 0.1 \times 0.1$). In contrast, shallow but complete randomization will equally represent all base pairs at a given position.

Finally, novel ribozymes with no natural or known counterparts can be selected from "deep random" pools that contain from 30 to over 200 randomized positions. Deep random pools typically span from 10^{13} to 10^{16} different variants, and hence completely cover a sequence space of from 22 ($4^{22} \times 10^{13}$) to 27 ($4^{27} \times 10^{16}$) residues in length (the actual coverage will also be dependent on the length of the random sequence region, a factor not taken into account in this simple calculation (see also Ref. 8). Many ribozymes will have more than 30 functional residues, and thus deep random pools may contain only a vanishingly small fraction of all the possible ribozymes that can perform a given function. For example, selections for ribozyme ligases have recovered a variety of novel functional motifs, but at least one of these motifs was predicted to be present at a level of only one in every 2.5×10^{18} different sequences examined.[9]

While there are numerous, different ribozyme activities that can be selected for, the selection experiments themselves fall into one of three classes: cleavage selections, ligation selections (Figure 2), and selections for binding. Cleavage selections remove information (sequence, structural, or chemical) from a ribozyme, while ligation selections add information to a ribozyme. As an example of a cleavage selection, hammerhead ribozyme variants have been tethered to a solid support, and those variants that could chew themselves away from the support were eluted and subsequently amplified.[10] In a variation on this theme, the hammerhead ribozyme population could be conjugated to biotin, and those variants that removed the biotin label would not be captured by a streptavidin affinity matrix. As an example of a ligation selection, Group I self-splicing intron variants have been engineered to carry out a reversal of the first step in the splicing cascade and append an oligonucleotide tag to themselves.[11] The ligation tag was designed to correspond to a primer sequence for the polymerase chain reaction, and the successful variants were directly amplified from the starting population (see also Figure 9). Again, the oligonucleotide tag could be biotinylated, and the successful Group I variants could be removed from solution and subsequently amplified. Ligation selections can also encompass the addition of more than just oligonucleotide tags: for example, alkyl transferases that successfully transferred an activated biotin moiety to themselves have been selected from random sequence populations.[12] In some instances, cleavage and ligation selections can be combined. For example, Burke and co-workers[13] have developed an ingenious method for selecting hairpin ribozyme variants (Figure 2(c)). In this system the 5′ end of a substrate RNA that contained the cleavage site was covalently attached to the 3′ end of the hairpin ribozyme through a polycytidylate linker, generating a ribozyme that was covalently linked to its substrate. Molecules that were active for both cleavage and ligation were selected by first generating cleaved ribozyme–substrate chimeras that contained a 2′,3′ cyclic phosphate, and then incubating these cleavage products with a new substrate that could participate in the reverse reaction to generate a new 3′ priming site for reverse transcription.

RNA molecules cannot only be directly selected for catalysis, but indirectly selected as well. RNA binding species (aptamers) have been selected from random sequence populations. One of the old saws of bioorganic chemistry is that enzymes catalyze reactions by binding to the transition state of the reaction. Therefore, RNA molecules selected to bind transition state analogues should catalyze the corresponding reaction, just as antibody molecules screened for binding to transition state analogues have proven to be catalytic. Aptamers that could bind to a transition state analogue of an isomerization reaction were selected from an RNA pool that contained 128 randomized positions.[14] One of the aptamers isolated from this selection bound the analogue with a K_i of 7 μM and could catalyze the isomerization of the corresponding biphenyl substrate at a rate 88-fold greater than background. However, such strategies are not uniformly successful: aptamers selected to bind to a Diels–Alder transition state analogue could not catalyze the corresponding Diels–Alder cycloaddition reaction.[15] There are only a few examples of indirect selection for ribozyme catalysis, and in general the catalysts that have been derived seem to be much slower than those derived by direct selection. However, the strength of indirect selection methods is that they are more likely to yield "true" catalysts that can turn over substrates, as opposed to ribozymes that perform a single reaction cycle. While directly selected catalysts have been converted to multiple turnover ribozymes, the separated substrates have tended to be oligonucleotides rather than small molecules such as the biphenyl cited above.

As has been alluded to above, selected ribozyme variants are typically amplified by a combination of reverse transcription, the polymerase chain reaction, and *in vitro* transcription (usually with

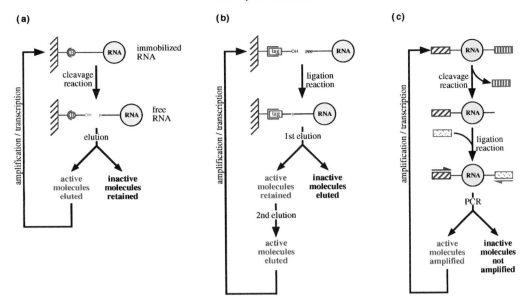

Figure 2 Types of ribozyme selection experiments. These variations on a theme were chosen to demonstrate the versatility of selection protocols. (a) Cleavage selection. An RNA pool is immobilized via a biotin conjugated to its 3′ end. Those RNA molecules that can cleave themselves away from the column are eluted, amplified, and rederivatized for further cycles of selection. (b) Ligation selection. A "tag" sequence within an oligonucleotide is immobilized via base pairing to an oligonucleotide affinity column. A random sequence RNA pool is mixed with the immobilized oligonucleotide, and those RNA molecules that can covalently append themselves to the primer are separated from the remainder of the population. The ligated products are eluted from the affinity column, amplified, and further selected for ligation function. (c) Cleavage/ligation selection. Variants in an RNA pool that can cleave away a constant region are isolated and mixed with a different constant region. Those variants that can also perform the reverse reaction, ligation, or otherwise capture the oligonucleotide are then preferentially amplified using primers specific for the second constant region. These reactions all take place in solution, away from a solid support (as could the selections cited in (a) and (b), depending on what isolation and amplification steps are employed).

T7 RNA polymerase). Other amplification procedures have also been used, such as isothermal amplification or 3SR,[16] which mimics the retroviral life cycle. During amplification, mutations arise and can provide fodder for subsequent rounds of selection. Since relatively few of the total number of possible variants are examined in most selection experiments, mutagenic amplification procedures, such as mutagenic PCR, are often incorporated during the amplification step.

6.09.3 IDENTIFYING FUNCTIONAL SEQUENCES AND STRUCTURES

A variety of natural ribozymes, ranging from the small hairpin and hammerhead ribozymes to the large Group I self-splicing intron and ribonuclease P, have been segmentally randomized and selected for catalytic function. Data from these selection experiments, such as the locations of functionally important residues and sequence covariations, has been used for the development of ribozyme structural models. The authors consider several particularly illuminating examples.

The relatively small size (30–50 residues) of the hairpin and hammerhead ribozymes precludes the formation of extensive tertiary structural contacts, and thus their secondary structures are the key to their overall folds. A secondary structure had been proposed for the hairpin ribozyme (Figure 3). However, initial attempts at confirming this structure were limited by the dearth of natural sequences available for comparative analysis (only three hairpin ribozymes were known, and these contained little sequence variation). An *in vitro* cleavage–ligation selection was therefore used to generate an artificial phylogeny of functional ribozymes.[17] Initially, secondary structural features of the hairpin ribozyme were established by selecting functional ribozymes from populations in which each of the helices was randomized in turn.[18,19] Helices 1 and 2 flank the cleavage–ligation junction and may be involved in positioning catalytically important residues. These helices were separately but completely randomized, and active variants were selected. The artificial phylogeny of selected sequences confirmed almost all of the previously proposed pairing interactions for helices 1 and 2 and, in addition, suggested a catalytic role for particular residues. Similarly, selected sequences supported the pairing scheme for helix 3. The existence of helix 4 had not previously been conclusively established, and secondary structural models ranged from those that contained no helix to those that contained helices of variable length. Analysis of selected ribozymes revealed that three predicted base pairs contributed to catalytic activity.

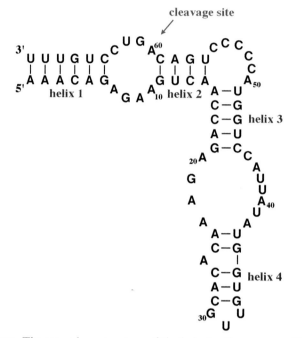

Figure 3   Hairpin ribozyme. The secondary structure of the hairpin ribozyme used by Burke and co-workers for cleavage/ligation selection experiments is shown. The helices are labeled; single-stranded regions are named according to which helices they fall between (e.g., J1/2 falls between helices 1 and 2, reading in the 5′ direction. J2/1 also falls between helices 1 and 2, but in the opposite orientation). The cleavage junction is indicated by an arrow.

The participation of formally single-stranded joining regions in secondary structural interactions has also been probed.[17] A four-nucleotide region that spanned the cleavage–ligation junction was completely randomized, and functional variants were again selected. A guanosine at a position immediately 3′ of the cleavage–ligation junction was identified as essential for activity. Similar experiments yielded four invariant nucleotides within the regions joining helices 3 and 4. These residues were thought to play critical roles in either structure or catalysis, but the lack of covariation prevented further analysis.

Further analyses of the hairpin ribozyme focused on locating tertiary structural interactions.[18] It was thought that tertiary structural interactions would most likely reside in the single-stranded joining regions of the molecule. A series of pools was constructed in which the internal loop sequences J1/2, J2/1, J3/4, and J4/3 were segmentally randomized, either alone or pairwise. For example, in one pool J1/2 was randomized, in another pool J1/2 and J3/4 were simultaneously randomized, then J1/2 with J2/1 or J4/3, and so forth. Sequence analysis of selected variants revealed that the same results were obtained regardless of whether single-stranded regions were randomized independently or pairwise. The lack of observed covariations between positions in internal loop regions was interpreted as implying a lack of interactions between bases within these regions. However, these results could also be interpreted to imply either that wild-type tertiary interactions are structurally and functionally invariant, or that tertiary structural interactions involve ribose or the phosphate backbone.

Finally, in order to identify functional residues and interactions that may have been dispersed throughout the hairpin ribozyme, a doped pool was constructed that contained, on average, five substitutions per molecule. After selection for active variants, a double substitution (G21U, U39C) was identified that had higher activity than one of the single substitutions (G21U) in isolation. This result could be interpreted to mean that positions 21 and 39 were involved in a tertiary structural contact. To better intepret this result, a series of site-specific mutations were constructed at positions 21 and 39. The activities of the substituted ribozymes were inconsistent with a tertiary structural interaction, but were consistent with the hypothesis that U39C was a nonspecific suppressor of multiple, different substitutions of G21.

The hammerhead ribozyme (Figure 4) has also been examined using an *in vitro* cleavage selection. In an attempt to optimize utilization of an RNA substrate containing AUA at the cleavage site, Nakamaye and Eckstein[20] completely randomized three residues in the core of a hammerhead ribozyme found in barley yellow dwarf virus. After three rounds of selection and amplification, two isolates showed slight improvements in the efficiency of cleavage of the AUA-containing substrate. However, it proved difficult to identify variants with much higher cleavage efficiencies when the entire core of the hammerhead ribozyme was randomized. Similarly, when 10 residues within the core of a UCA-cleaving hammerhead ribozyme were randomized and active variants were selected, only two sequence changes in the consensus core were noted: A9U and A7U.[21] Interestingly, both of these changes actually decreased the speed and efficiency of the hammerhead ribozyme but were selected because they could be more readily amplified than the wild-type hammerhead consensus. Additional experiments that randomized only a portion of stem II of the hammerhead ribozyme,[22] that replaced stem II with a random hexanucleotide string,[23] or that randomized all 14 nucleotides in the core[10] also failed to find variants with dramatically improved reaction kinetics or efficiencies. Tang and Breaker[24] also randomized 14 residues in the core of the hammerhead ribozyme, including all residues known to be critical for function, and selected variants that could cleave themselves. However, in contrast to previous selection experiments, Tang and Breaker inhibited cleavage reactions that occurred during transcription by designing an "allosteric trigger" into the ribozyme. An anti-ATP aptamer was appended to stem II of the hammerhead ribozyme in such a way that cleavage was inhibited in the presence of ATP and encouraged in its absence. After six rounds of selection and amplification, the 10^{13} molecules in the initial pool had been winnowed to one major variant, the consensus hammerhead sequence (a U7C substitution was also observed, but had previously been shown to occur in natural ribozyme sequences as well). In addition to the hammerhead ribozyme itself, two other minor sequence classes were observed that did not appear to be mere variants of the hammerhead ribozyme, but were in fact completely different sequence classes. Both of these classes of ribozymes showed much less dependence on the engineered allosteric trigger, and could readily cleave themselves even in the presence of ATP. Both clases of ribozymes were roughly as active as the hammerhead ribozyme in 20 mM $MgCl_2$.

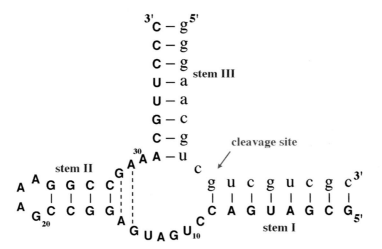

Figure 4 Hammerhead ribozyme. The secondary structure of a hammerhead ribozyme similar to those used in the selection experiments that have been cited is shown. The helices are labeled. Sequential, non-Watson–Crick G:A pairings are represented by dashed lines. The cleavage junction is indicated by an arrow.

However, under lower, more physiologically relevant magnesium concentrations, both classes showed enhanced cleavage activity relative to the selected hammerhead ribozymes. Consistent with these findings, additional rounds of selection in either the presence of ATP or in the presence of low concentrations of magnesium (2 mM) allowed the novel Class I ribozymes (Figure 5) to predominate over the consensus hammerhead in the selected population. The Class I ribozymes cleaved their RNA substrates at the phosphodiester linkage immediately 5′ to the phosphodiester linkage cleaved by the hammerhead ribozyme. Finally, Vaish *et al.*[25] randomized not only the core residues of the hammerhead but the residues corresponding to stem II as well (22 residues in total). In addition, these authors also altered the cleavage site from GUC to AUG, a sequence not normally recognized by the natural hammerhead. After 13 cycles of cleavage selection and amplification, several different sequence classes were observed. One of the most populous and active classes was predicted to form a three-helix junction separated by short single-stranded joining regions. This ribozyme was not only hammerhead like in terms of its secondary structure, but the core sequences were similar to those found in the hammerhead. The major difference between the selected ribozyme and the natural hammerhead appeared to be the numbers of residues in the joining regions. The selected ribozyme could efficiently ($k_{cat} \times 0.5$ min^{-1}) cleave the phosphodiester bond following the unnatural AUG and other noncanonical triplets.

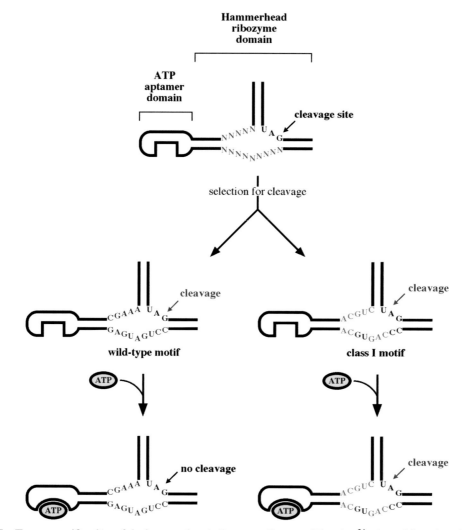

Figure 5 Transmogrification of the hammerhead ribozyme. Tang and Breaker[24] selected functional ribozymes from a randomized population based on the hammerhead ribozyme. Some of the selected variants (left, blue) were similar in sequence to the wild-type hammerhead and were responsive to the "allosteric trigger" built into the selection. Other variants (right) showed limited similarity to the hammerhead ribozyme (blue) but appeared to represent novel sequence motifs (green). Note that while the hammerhead cleaved following A that the novel motif cleaved following the previous residue, U. The novel ribozymes also were not responsive to the allosteric trigger. Orientation of the hammerhead and hammerhead-like ribozymes is as in Figure 4.

In vitro ligation selections with the Group I ribozyme (Figure 6) have also proven useful in establishing or confirming secondary structural features. In initial experiments, residues that abutted a pseudoknot structure in the catalytic core were probed.[11] Nine nucleotides within the J3/4 and J7/3 joining regions and the P3 stem were completely randomized, and active ribozyme variants were selected in three cycles. Following selection, only wild-type residues were found at positions

J3/4-1 and J3/4-2, suggesting a critical structural or catalytic role for these residues. Other residues were found to co-vary in a manner that was consistent with the existence of direct interactions. At position P3-7, a potential C:G pairing predominated in natural Group I intron phylogeny, but either a G:C base pair or an A:U base pair was predicted to form in selected variants. At P3-8, the wild-type A:U base pair could be functionally replaced with A:A, A:G, or U:A alternatives. These results were interpreted as evidence for a reverse Hoogsteen geometry or a requirement for conformational flexibility.

Long-range tertiary structural interactions, such as contacts between the coaxial stacks of the

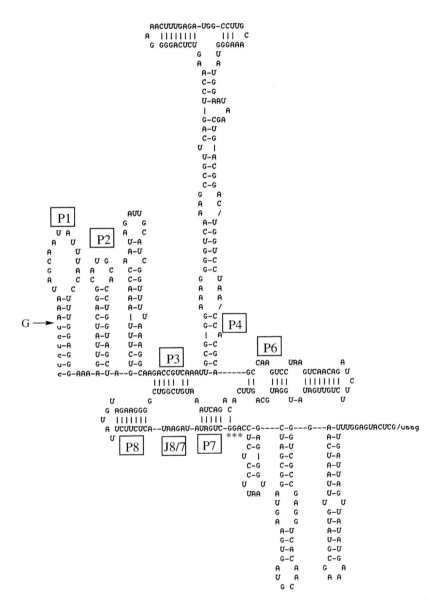

Figure 6 Tetrahymena Group I self-splicing intron. The sequence and secondary structure of the ribozyme are shown. Intron sequences are in upper-case letters, while exon sequences are in lower case. Paired stems or helices are denoted in order as P1, P2, and so forth. Single-stranded joining regions lie between some of the paired stems and are denoted as Jn/n, where n indicates the stems 5′ and 3′ of the joining region, respectively (e.g., J8/7 is labeled). Individual residues can be labeled according to their positions within paired stems (e.g., P3-8, the eighth residue within P3) or within joining regions (e.g., J6/7-3, the third residue within J6/7). The site of attack by an exogenous guanosine on P1 that initiates the splicing cascade is indicated by an arrow. The approximate positions of residues 312, 313, and 314, residues that mutated in response to a selection for utilization of DNA as a substrate, are shown by ***.

Group I ribozyme, have also been verified by *in vitro* selection experiments.[26] Sequence covariations observed in the natural Group I intron phylogeny had previously prompted Michel *et al.*[27] to propose that triple base pairs formed between P4 and J6/7, and between P6 and J3/4. The existence of some of these triple base pairs was confirmed by site-directed mutational analysis, but proof for others remained elusive. In addition, the results of the site-directed mutational analysis implied that the base triples might interact with one another in a complex manner. To determine the rules for base-triple formation in more detail and to potentially gain insights into the multiple interactions within this region, an *in vitro* selection experiment was designed based on the Group I self-splicing intron sunY. The initial pool contained 14 randomized residues scattered through P4, P6, J3/4, and J6/7 and spanned 2.7×10^8 different variants. After three rounds of selection, the ligation ability of the selected population was actually 1.5-fold better than that of the wild-type ribozyme. Sequences from the most active clones in the final population identified four invariant residues: J3/4-1 and J3/4-2 (as in the previous selection), and the base pair P6-2. Sequence covariations were also recovered that could be modeled as identical or isosteric to base triples that had previously been predicted to form between J6/7-1 and P4-1, and J6/7-2 and P4-2.

The aforementioned selection experiments were carried out with the hope of defining the Group I structure. Additional experiments have sought to use the Group I ribozyme as a model for better understanding general principles of RNA folding and structure. Michel and co-workers had previously used phylogenetic and biochemical analyses to identify a common interaction between GNRA tetraloops and common sequence motifs in Group I and Group II introns.[28-30] Tetraloops with the sequence GAAA interacted preferentially with an 11 nucleotide motif CCUAAG... UAUGG that formed an internal loop secondary structure, while tetraloops with the sequences GURA interacted with two contiguous base pairs in a continuous helix: GUAA interacted with C:G, C:G pairs, while GUGA interacted with C:G, U:A pairs (Figure 7). The GAAA docking site has been resolved in atomic detail,[31] while the GURA docking sites have been most easily rationalized by a model in which the third and fourth residues of the tetraloop formed minor groove base-triples with the corresponding base pairs, with the third residue of the tetraloop discriminating between the second base pair.

To better determine whether the rules for tetraloop recognition that have been gleaned from phylogenetic analyses of natural ribozyme variants were complete or merely represented a subset of possible rules, Costa and Michel[32] generated an artificial phylogeny of tetraloop recognition sequences. As a model for the tetraloop–receptor interface, Costa and Michel chose an interaction between L2 and P8 that had been proposed to occur in a variety of Group I ribozymes. As a starting point for selection experiments, the bacteriophage T4 td self-splicing intron was divided into two pieces: a P1–P2 substrate that contained the tetraloop, and a catalytic core into which the substrate could dock. The P8 stem of the catalytic core was replaced with 21 random sequence residues, and variants that could catalyze the attack of the 3′ terminal guanosine on the 5′ intron–exon junction of the substrate (the reverse of the second step in splicing) were selected. Since the L2–P8 interaction had been shown to be essential for efficient catalysis, all and only those variants that could repair this interaction should have been selected. Two different constant sequence substrates were used: one containing a GAAA tetraloop, and one containing a GUGA tetraloop. Following selection, the predominant class of ribozymes that could interact with the GAAA tetraloop contained the 11 nucleotide internal loop motif previously identified by phylogenetic analysis. Similarly, the predominant class of ribozymes that could interact with the GUGA tetraloop contained consecutive C:G and U:A base pairs in a continuous stem. These results supported the notion that natural selection is an efficient search algorithm for optimal interactions. However, additional, unnatural binding motifs were also identified. For example, a variant of the 11 nucleotide internal loop motif efficiently bound GAAA tetraloops. Similarly, ribozymes containing helices with G:C, C:G stacks followed by short internal loops were found to bind GUGA tetraloops and, unlike their natural counterparts, could discriminate strongly against GCGA tetraloops. While some of these novel tetraloop receptors were found to have natural counterparts, many of the selected sequences had no correspondents even though they were as kinetically competent as the natural receptors.

6.09.4 IMPROVING AND CHANGING CATALYSIS

In vitro selection can be used to alter the chemical and structural properties of natural ribozymes from their norms. In the most complete set of experiments to date, Joyce and co-workers have

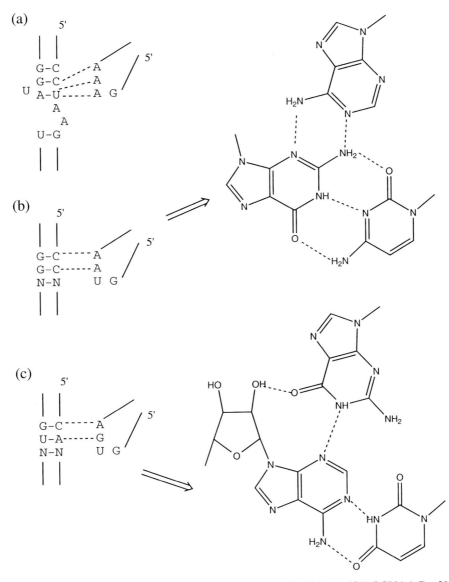

Figure 7 Tetraloop receptors. (a) Interactions between the 11 nucleotide motif 5′ CCUAAG... UAUGG and a GAAA tetraloop (after Cate *et al.*[31]) The A–U base pair in the helix is a reverse Hoogsteen pairing. Hydrogen bonds are indicated by dashed lines. (b) Proposed interactions between a helical tetraloop receptor and a GUAA tetraloop (after Jaeger *et al.*[29] and Costa and Michel.[30]). (c) Proposed interactions between a helical tetraloop receptor and a GUGA tetraloop.

subjected several aspects of the Group I ribozyme mechanism to artificial selection pressure. To evolutionarily modify the self-splicing intron, the Scripps researchers developed a ligation selection scheme in which a primer binds to the 5′ end of the ribozyme (mimicking the 5′ intron–exon junction) and is attacked by the 3′ terminal guanosine. This engineered reaction is essentially the reverse of the second step in the splicing cascade.

Initially the chemical specificity of the Group I ribozyme was changed. The wild-type self-splicing intron cleaves specific phosphodiester bonds within RNA substrates by using at least one, and probably two, magnesium ions to polarize the 3′ oxygens of endogenous and exogenous guanosine substrates.[33] Starting from a small pool of deletion variants, Robertson and Joyce[34] were able to identify a ribozyme that could better utilize a DNA primer as a substrate. However, the rate enhancement that was achieved was quite small, and appeared to be due to a general improvement in catalysis rather than to a specific improvement in DNA cleavage ability. To explore whether additional sequence changes might further improve catalysis with DNA substrates, Beaudry and

Joyce[35] partially randomized 140 positions in the catalytic core of the Tetrahymena ribozyme at a rate of 5% and prepared a starting population that completely spanned all four-error substitutions. After 10 cycles of selection and amplification the selected molecules had an aggregate activity that was 30-fold greater than the wild-type parent toward DNA; the most active individual clones could cleave DNA substrates up to 100-fold more efficiently.

While there were some common functional themes among the evolved, DNA-cleaving Group I variants, different ribozymes chose different evolutionary strategies. Catalytic improvements could be attributed to changes in both K_m for the DNA substrate and k_{cat} for the reaction. Further, most of the catalytic improvement was due to changes in only four residues, although substitutions at a number of other positions also contributed to the altered activities. The most active clones have a double mutation, GA to UG at positions 313 and 314, that disrupts a pairing in the wild-type ribozyme that would normally position the terminal guanosine adjacent to the 3′ intron–exon junction. These ribozymes not only cleave DNA more readily, but also decrease the rate of hydrolysis at the 3′-intron–exon junction (by threefold in the aggregate, and up to 10-fold in individuals). In contrast, other ribozyme variants had sacrificed fidelity for activity and could splice themselves into different positions on the DNA primer.

Because the DNA cleavage activity of the ribozyme population looked as though it was still improving by generation 10, Tsang and Joyce[36] continued the selection for an additional 17 generations. To increase the stringency of the selection, the concentration of the DNA substrate was decreased by 50-fold; this modification of the protocol should have favored those catalysts that could bind substrate most tightly and/or those catalysts that could more efficiently react following substrate binding. In fact, an examination of the kinetic and substrate binding properties of individual variants and the evolving pool, respectively, showed that by generation 18 K_m had dropped about 50-fold relative to generation 9, while K_d had decreased about 10-fold. At this stage the variants were saturated with substrate, and further improvements in catalysis were encouraged by reducing the time allowed for the splicing reaction to occur by 12-fold. Again, the population responded, this time by increasing k_{cat} about 50-fold between generations 18 and 27. Overall, the catalytic efficiency (k_{cat}/K_m) of individual generation 27 ribozymes had improved by 1000-fold relative to generation 9 and by 100 000-fold relative to the wild-type parent.

However, while the evolved ribozymes had drastically improved DNA-cleaving abilities, they also retained the ability to cleave RNA. In fact, the generation 27 ribozymes still cleaved RNA approximately 10-fold better than DNA. To further improve the DNA-cleaving abilities of these ribozymes and to reverse their substrate preference, additional rounds of selection were performed in which the DNA substrate was mixed with an RNA oligonucleotide corresponding to the product of the DNA cleavage reaction.[37] Those ribozymes that could preferentially bind the DNA substrate relative to the RNA product analogue could also potentially cleave the DNA substrate and capture the sequence tag necessary for amplification. In contrast, those ribozymes that could not discriminate between DNA and RNA oligonucleotides would likely be inhibited by binding of the RNA product analogue. After 36 additional generations of *in vitro* selection, the ribozymes accumulated additional mutations that allowed them to cleave the DNA oligonucleotide 50-fold better than the generation 27 ribozymes (k_{cat}/K_m), and to utilize the DNA substrate roughly 100-fold better than the corresponding RNA substrate. However, comparisons between individual ribozymes randomly chosen from generations 27 and 63 were less impressive: the DNA-cleavage ability of a single generation 63 ribozyme had improved only fivefold, and its relative preference for DNA by only 20-fold. While the generation 63 ribozymes had on average 28 mutations relative to the wild-type intron, much of the improvement could be traced to two covarying mutations: U271C and G312AA. Although the mutations at 313 and 314 dominated the population at generations 27 and 36, these mutations had been completely displaced by the new mutations at 271 and 312 by generation 45, and these new mutations remained fixed in the population through the remainder of the selection.

Unexpectedly, the generation 27 ribozymes could also carry out the cleavage of linkages other than phosphodiester bonds.[38] As with the original, engineered ribozyme, substrates that could pair with the internal guide sequence were positioned adjacent to the activated terminal guanosine of the ribozyme. However, the evolved ribozyme had apparently either increased the nucleophilicity of the hydroxyl moieties on the guanosine and/or reduced the specificity of the attack, since a phosphodiester bond between deoxyribose and arabinose could be efficiently cleaved. Moreover, when an amide bond was introduced between a 3′ terminal amine and a 5′ terminal carboxylate it could also be cleaved by the ribozyme, which formed a transient ester linkage to the cleavage product. The amide linkage could be introduced between two modified DNA molecules, or between DNA and an amino acid or peptide. While the rate of amide hydrolysis was only ca. 10^{-5}–10^{-6} min^{-1} this is still roughly 1000–10 000-fold faster than the background reaction.

6.09.5 INVENTING CATALYSIS: THE ROAD TO THE RIBOSOME

It has been hypothesized that ribozymes preceded protein enzymes during evolution. If so, then it is likely that ribozymes invented translation, and that ribozymes should be able to carry out many of the reactions that occur during translation. This hypothesis has received support from *in vitro* selection experiments that have successfully recapitulated the activities of some of the individual steps in translation (Figure 8; for a review see Ref. 39). First, while contemporary ribozymes have proven adept at phosphodiester bond transfers, ancient ribozymes in any nascent metabolism would have also had to build and manipulate high-energy phosphoanhydride bonds. For example, the synthesis and utilization of ATP would likely have been critical to almost any aspect of metabolism. Lorsch and Szostak[40] have selected ribozymes that can carry out reactions similar to cellular kinases, which typically transfer the gamma phosphate of ATP to a substrate. An ingenious selection scheme was devised in which an RNA pool was first mixed with an ATP analogue (ATP-γS) that contained a sulfur in place of one of the terminal, nonbridging oxygen atoms. Those RNA molecules that could self-phosphorothiolate themselves could also form disulfide bonds and were captured on a thiol affinity column. Several major classes of kinases were identified that utilized either the 5′ hydroxyl or internal 2′ hydroxyl residues as nucleophiles. The rate acceleration of one of the fastest ribozymes was estimated to be an astounding 10^9-fold. While several of the kinases could also utilize ATP as a substrate, they preferred the thiol analogue by roughly two orders of magnitude. Further kinetic analysis of one of the classes indicates that the ribozyme mechanism is relatively simple: the ATP cofactor binds adjacent to a 5′ hydroxyl and the enzyme promotes the formation of a metaphosphate-like transition state.[41] Similarly, Huang and Yarus[42] have selected RNA molecules that catalyze the conjugation of phosphate-containing ligands to a 5′ alpha phosphate, as opposed to a 5′ hydroxyl. These ribozymes were isolated by first selecting pool RNAs that could hydrolyze their initiating 5′ triphosphates to monophosphates. The monophosphorylated species were subsequently circularized, separated from linear species by gel electrophoresis, and preferentially amplified. The nascent pyrophosphatase activity derived from this selection proved to be relatively nonspecific, and nucleophiles other than water, such as pyrophosphate, could also attack the 5′ triphosphate and displace the 5′ beta and gamma phosphates. The pyrophosphate exchange reaction was further enhanced by selecting for RNA molecules that could immobilize themselves on a UTP column (forming a tetraphosphate linkage). Immobilized species were amplified *in situ*, purified, and reselected. After a total of 20 cycles of the first selection method and four cycles of the second selection method, a predominant family of catalysts was identified; this family enhanced the rate of pyrophosphate exchange by at least 500 000-fold compared to unselected RNAs. Moreover, when assayed with a variety of substrates that contained terminal phosphates, the pyrophosphate-exchanging ribozyme could catalyze its conjugation to virtually all of them, from nucleoside monophosphates (forming a 5′-5′ pyrophosphate bond) to nucleoside triphosphates (forming a tetraphosphate linkage) to Nϵ-phosphate arginine (forming a pyrophosphoramidate bond).[43] The remarkable versatility of this ribozyme indicates that it is possible that primordial ribo-organisms could have generated and manipulated a wide variety of metabolic intermediates containing high-energy phosphoanhydride bonds.

Second, the energy held in phosphoanhydride or phosphoester bonds could have been used to generate aminoacyl tRNAs or other activated amino acid intermediates. In this respect, Yarus and co-workers have demonstrated that self-aminoacylating RNAs can be selected from random sequence pools.[44] An oligonucleotide pool that contained 50 random sequence positions was incubated with phenylalanyl-AMP, the same mixed anhydride intermediate that occurs during the biosynthesis of the ester bond of phenylalanyl tRNA. RNA molecules that attached a phenylalanyl residue to themselves were further derivatized with a naphthoxyacetyl moiety, isolated on a reverse-phase column based on their increased hydrophobicity, and subsequently amplified and reselected. After 11 cycles of selection and amplification, individual clones were analyzed and assayed for their ability to acylate themselves. One of the clones was found to be highly reactive, and was estimated to accelerate the rate of ester formation by roughly 250 000-fold (however, see Ref. 44 for corrections). Interestingly, the clone is aminoacylated on its 2′ (3′) terminus, just as tRNA is. Further examination of the self-aminoacylating species revealed that multiple different species were catalytic, and that the catalytic core of the most active species can be pared down to a mere 43 residue RNA of which only 17 residues were in the initial random sequence population.[45] While the ribozyme was specific for the AMP leaving group, it differed from cellular tRNA synthetases in that it showed little preference for its cognate amino acid phenylalanine and could also conjugate itself to serine or alanine.[46] Nonetheless, the plethora and simplicity of the selected catalysts bolster the contention that aminoacylation activity should have been accessible even to primordial RNA molecules.

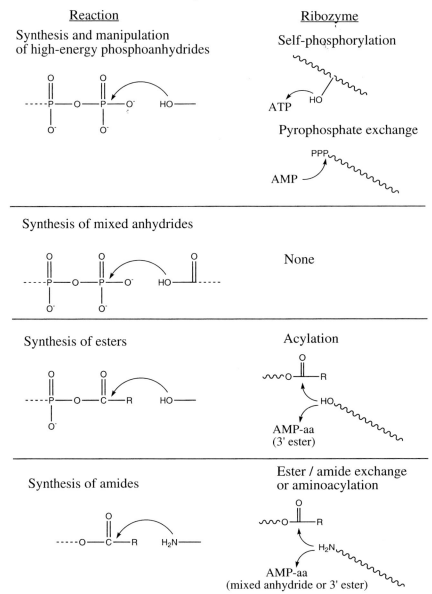

Figure 8 Ribozymes mimic the chemistry of translation. Discrete chemical steps would have been necessary for the metabolic invention of translation. Selected ribozymes (described in greater detail in the text) can mimic many of these. The abbreviations in this diagram include: R, a generic symbol for a sidechain; aa = amino acid, linked to AMP via either a (2′) 3′ ester linkage or a mixed anhydride bond, and PPP = the 5′ triphosphate of a RNA.

Third, activated amino acids could have undergone additional chemical transformations, such as the conversion of ester linkages of aminoacyl tRNAs to the amide linkages of peptide backbones. A number of unnatural ribozymes have been isolated that can perform such reactions. For example, Lohse and Szostak[47] have selected a ribozyme with acyl transferase activity. An oligonucleotide that was esterified at its 2′ (3′) hydroxyl with a methionine residue that was in turn biotinylated via its alpha amino group was annealed to a random sequence population. Those RNA species that catalyzed the transfer of the methionine to themselves in turn captured the biotin moiety and were immobilized on a streptavidin column. The selected RNA species were then amplified and re-selected for acyl transferase activity. After 11 cycles of selection and amplification, the activity of the pool had increased 10 000-fold, and the predominant class of selected ribozymes catalyzed the self-esterification of their 5′ hydroxyls. The acyl transferase ribozyme is reminiscent of a DNA ligase, in that the substrate for the reaction is an esterified oligonucleotide aligned with a hydroxyl nucleophile via base pairing to a complementary template. Just as isolated ribosomes have been shown to transfer activated amino acids to either hydroxyl or amine groups, the selected ribozyme can also use a 5′ amine as a nucleophile to generate an amide bond. Zhang and Cech[48] have selected a ribozyme that can also synthesize an amide (peptide) bond. While Lohse and Szostak relied on an endogenous hydroxyl to act as a nucleophile, Zhang and Cech appended an amine moiety to their pool by tethering the amino acid phenylalanine to a 5′ guanosine monophosphorothioate (see Figure 14). The derivatized pool was again mixed with a suitably activated amino acid, a biotinylated methionine that formed an ester with the (2′) 3′ hydroxyl of adenosine monophosphate. Ribozymes that could form a peptide bond and displace the AMP leaving group were immobilized, amplified, rederivatized, and again reacted with the activated methionine. After 19 cycles of selection and amplification, the RNA pool exhibited substantial activity, and individual ribozymes were cloned and sequenced. In contrast to the results of Lohse and Szostak, Zhang and Cech observed several different classes of catalysts. One of the most populous and active catalysts had a k_{cat} of 0.05 min^{-1}, and could accelerate peptide bond formation by a million-fold relative to the uncatalyzed reaction. The selected catalyst could use several different activated amino acids as substrates, but showed an absolute preference for the AMP leaving group. The relative preference of the ribozyme for different amino acid side chains intriguingly mimicked that of the ribosome for short, aminoacylated RNAs. Although the ribozymes selected in both these experiments can so far only catalyze amino acid transfers in a *cis* configuration, the fact that amide bond formation can be carried out by a relatively short, unmodified RNA provides experimental support for the hypothesis that ribosomal RNA may have arisen in an RNA world.

6.09.6 ASSESSING THE BREADTH AND DEPTH OF SELECTED CATALYSTS: THE LIGASES

While the examples cited above are remarkable both in terms of the range of reactions that are catalyzed and in terms of catalytic and kinetic properties of the ribozymes themselves, it is difficult to draw generalized conclusions from the individual experiments. The failure to develop a more general understanding of "ribozymology" is in part a result of the fact that until recently true comparative enzymology has been all but impossible. There were only a few natural ribozymes, and the unnatural ribozymes were selected from different pools, under different buffer conditions and stringencies, to catalyze quite different reactions. However, as more results have accumulated it has become possible to compare the catalytic and kinetic properties of at least one class of selected nucleic acid enzymes, the ribozyme and deoxyribozyme ligases. Bartel and Szostak[49] originally selected ribozyme ligases from a long and complex random sequence population. The initial pool spanned 220 random sequence positions and contained over 10^{15} different species. The selection procedure was conceptually simple but technically challenging: RNA sequences that could append a constant sequence "tag" to their 5′ ends were preferentially removed from the population by affinity purification and preferentially amplified by reverse transcription and PCR (Figure 9). In order to avoid aggregation, individual RNA species were immobilized on an affinity column specific for their 3′ constant regions. Following selection, the putative catalysts could be regenerated by nested PCR, and the cycle could be carried out iteratively. After four cycles of selection and amplification, ligase activity could be detected in the selected population. An additional three cycles of selection were carried out in which additional mutations were incorporated into the population via error-prone PCR, and a final three rounds of selection were carried out under increasingly stringent, low magnesium conditions. The population continued to improve in each round, so that by round 10 the population could catalyze the ligation of RNA strands over 7 million-fold faster than the uncatalyzed reaction. The number of selected ribozymes and corresponding complexity of the selected population was roughly assessed by looking for restriction polymorphisms. Interestingly, variants that predominated at early rounds of the selection were often displaced in later rounds, suggesting either that the programmed or spontaneous introduction of mutations led to the production of faster ribozyme variants during the course of the selection or that alteration of the selection conditions or some other restriction of the population caused a change in the fundamental criterion for selection.

Individual ribozymes were cloned from the round 10 population, and their sequences and kinetic properties were determined.[9,50] Of 66 clones that were sequenced, 45 were from a single predominant, homologous family, while the remaining 21 could be divided into six additional, evolutionarily-related families. Several families showed primary and secondary structural similarities, and could in turn be grouped into three diverse classes of ligases. There was a single representative of Class I,

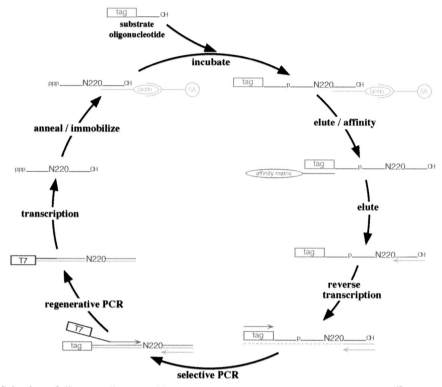

Figure 9 Selection of ribozyme ligases. This scheme was employed by Bartel and Szostak[49] to isolate ribozyme ligases from an RNA pool that spanned 220 random sequence positions. A biotinylated oligonucleotide complementary to the 3′ end of the pool was immobilized on a streptavidin column and was used in turn to immobilize pool RNAs. Immobilization prevented aggregation in the presence of high concentrations of magnesium. Those species that could ligate a substrate ribo-oligonucleotide to themselves were captured by affinity chromatography and preferentially amplified using primers specific for a "tag" sequence embedded in the substrate. A second round of PCR amplification was required to append the T7 RNA polymerase promoter to the double-stranded DNA templates. Transcription of the PCR products gave rise to RNA molecules that could be introduced into subsequent rounds of selection.

three other families formed Class II, and the two remaining families were independent examples of Class III (Figure 10). Despite the fact that analysis of the round 6 population indicated that the predominant ligase activity in the population formed 3′-5′ phosphodiester bonds, both Class II and Class III ligases formed 2′-5′ phosphodiester bonds. This may indicate that the more stringent, lower magnesium selection conditions that were used during the later rounds (8–10) of selection favored 2′-5′ phosphodiester bond formation and led to the displacement of most of the nascent 3′-5′ ligases. The Class I and Class II ligases could be engineered to act on substrates provided in the *trans* configuration, but their predicted secondary structures suggested that this was only because they could bind an alternative conformation of the hairpin stem originally provided in the *cis* configuration. This hypothesis was further bolstered by the fact that the Class II ligases were rate limited by product release.

The selected ligases were surprisingly complex, both in terms of their functional primary sequences and their predicted secondary structures. The Class I ligase was partially randomized and reselected for ligation function.[9,50] As discussed in Section 6.09.2, some residues were partially or completely conserved following selection and were presumed to be important for function; at least 93 residues are present in the core catalytic domain. Other residues covaried with one another following selection, verifying a predicted secondary structure that included several apparently stacked helical junctions and two pseudoknots (Figure 10). Several independent Class II ligases were selected and in general could be folded to form a long helical stem interrupted by bulge loops. A Class III ligase was also randomized and reselected for function. Analysis of selected sequences revealed that upwards of 19 residues were partially or completely conserved, and that the Class III ligases folded into a branched pseudoknot structure.

Because of its speed and complexity, the Class I ligase has been used as a starting point for additional experimentation. For example, Wright and Joyce[51] have developed a scheme for the

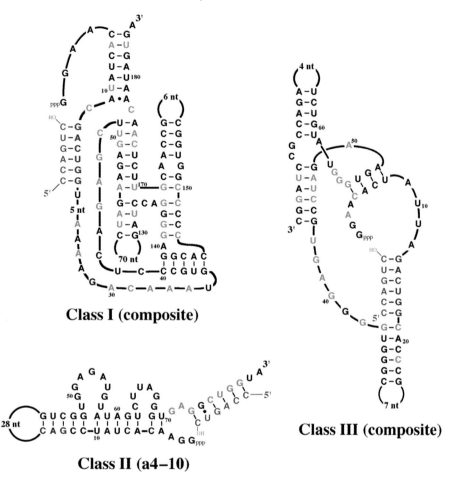

Figure 10 Selected ligases. Bartel and Szostak[49] discovered three major classes of ribozyme ligases by *in vitro* selection. The predicted secondary structures of these classes are shown. The Class I and Class III secondary structures are composites based on deletion and mutational analyses, while a single deletion variant from the Class II ligases, a4–10, is shown. Residues in red are the substrates for ligation, while residues in blue are the substrate binding sites originally programmed into the pools. Residues in green are highly conserved between functional mutational variants or independent isolates.

continuous evolution of the Class I ribozyme (Figure 11). In this scheme, the ligation junction of the Class I ligase has been altered so that it spans a T7 RNA polymerase promoter site. In the absence of ligation, the RNA cannot be amplified by isothermal amplification, because no promoter exists to generate new RNA. In contrast, variants that ligate effectively supply their own promoter and are preferentially amplified. Because isothermal amplification is a continuous process, any new RNA molecules that are generated by transcription of a ligated template, including any mutational variants, are simultaneously and iteratively subjected to this stringent requirement for survival. The continuously evolving ribozymes could be propagated by simple serial dilution of the population into new "food" (substrates, enzymes, and nucleotides), just as with a growing population of bacteria, rather than by the sequential isolation of ligated RNA, amplified DNA, and amplified RNA intermediates. Thus, over time it is expected that the fastest and/or most fecund species should predominate. After 100 cycles of continuous evolution, the catalytic efficiency of at least one member of the population had improved by over 10 000-fold, and its exponential growth rate had improved by over nine orders of magnitude. The continuously evolved ribozyme ligase was found to be much faster than Bartel and Szostak's[49] original ribozyme, and even faster than Ekland *et al.*'s[9] optimized ligase (albeit under slightly different reaction conditions).

It should be noted that use of the Class I ligase was essential to the success of these experiments. Previous selections for ligation activity had been carried out using a group II ribozyme and isothermal amplification, yet these did not yield a continuously evolving quasispecies.[52] The speed of the natural ribozyme was slow relative to the speed of the amplification scheme; so slow, in fact,

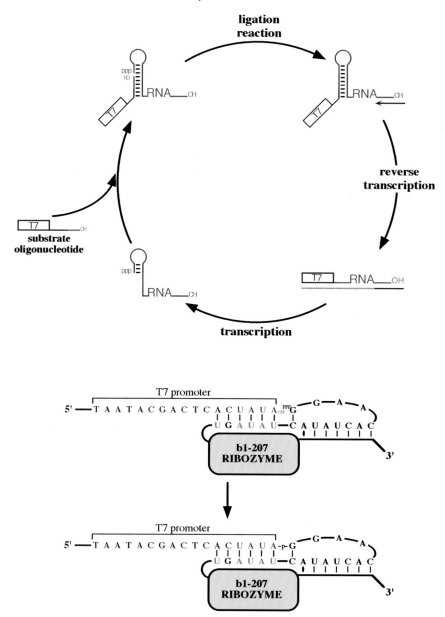

Figure 11 Continuous evolution of ligase activity. The scheme for continuous evolution is similar to the original scheme for the selection of ligase activity (Figure 9). However, the key difference between these schemes is that the ligation junction has been altered to correspond to the T7 RNA polymerase promoter. Residues in blue are DNA in the ligation substrate, while residues in red are RNA. Residues in green have been mutated in the "core" ribozyme to accommodate the new ligation substrate. Upon ligation and reverse transcription, a hybrid molecule is created in which a double-stranded DNA that includes the T7 RNA polymerase promoter abuts an RNA–DNA duplex. This hybrid can be used as a template for transcription, and thus the entire cycle of selection and amplification that is normally carried out in discrete steps can be carried out continuously.

that most ribozymes may have been converted by reverse transcription into double-stranded (and hence inactive) molecules before they could catalyze primer addition. The relatively slow and inefficient amplification of the natural ribozyme abetted the evolution of an interesting parasite, "RNA Z," which could more efficiently feed off of the isothermal amplification scheme.[52] This fate could potentially also have befallen the Class I ligase: the variant that Wright and Joyce started with was a factor of 10 000 slower than its parent, and was also originally too slow to participate in the continuous evolution scheme. Therefore, a population of sequence variants centered on the Class I ligase was generated, and more efficient ligators were identified by 15 rounds of conventional selection. The resultant population was roughly 100-fold more active than the parent ribozyme, but variants in this population were still not active enough to undergo continuous evolution. Ligation activity was further increased by carrying out 100 cycles of a simplified selection protocol, in which ribozymes were allowed to react for five minutes and were then transferred to the amplification reaction mixture. The quasispecies generated by this "manual" evolution scheme was finally adequate to kick-start continuous evolution. What is particularly remarkable is that the "capacity" of the Bartel ligase for functional evolution was great enough to withstand adaptation to a completely new ligation junction and a 10 000-fold decline in activity.[92]

Finally, Ekland and Bartel[53] were able to convert the Class I ligase into a template-directed polymerase by separating the enzyme into three domains. The template domain could pair with the catalytic domain, and the substrate domain could in turn pair with the template (Figure 12). Following this division, it was found that a single nucleotide could be added to the substrate in a template-directed manner at a position corresponding to the original ligation junction. Amazingly, when additional, unpaired residues were inserted between the substrate–template and template–catalyst pairings the substrate could nonetheless be serially and faithfully extended in a template-directed manner.

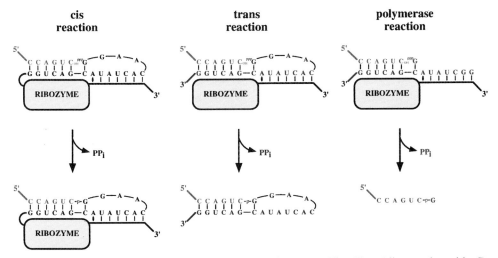

Figure 12 Converting a ribozyme ligase to a ribozyme polymerase. The Class I ligase selected by Bartel and Szostak[49] can be engineered to carry out a limited, template-directed polymerization reaction.[53] The original ribozyme (left, *cis* reaction) ligates an oligonucleotide substrate to itself with concomitant displacement of pyrophosphate. The ribozyme's template can be split away from the remainder of the ribozyme (middle, *trans* reaction). The core ribozyme can bind the template in *trans* and can catalyze ligation of the substrate and template to form a hairpin stem. Finally, the ligation junction can be sheared away and the core ribozyme will still catalyze the template-directed addition of nucleoside triphosphates. If additional residues are included in the template between the substrate:template and template:ribozyme pairings then the sequential, template-directed addition of mononucleotides can be observed.

While the various Bartel ligases are perhaps the best studied of the selected ribozymes, other selection experiments have yielded additional ligases with novel sequences and mechanisms (Figure 13). Chapman and Szostak[54] selected a ribozyme ligase that formed a $5',5'P^1,P^4$-tetraphosphate linkage. The original pool spanned 90 random sequence positions, and the ligation substrate was a short, biotinylated oligophosphoramidate with an imidazole leaving group. Members of the original pool that could append themselves to the ligation substrate were sieved from the remainder of the population using a streptavidin column, and were subsequently amplified and reselected. After eight cycles of selection and amplification a major class of ligases could be identified, but numerous unique clones remained in the population. The major class of selected ligases had a catalytic core of 54 residues, formed a pseudoknot structure with an internal binding site for the ligation substrate, and could utilize the 5′ triphosphate of the RNA transcript as a nucleophile. The internal substrate binding site could be altered to accommodate alternative ligation substrates. Similarly, Hager and Szostak[55] selected a ribozyme ligase from a pool that spanned 210 random sequence positions and that initiated with a 5′-5′ pyrophosphate linkage (AppG). The ligation substrate could pair with the 5′ constant region of the pool to form a hairpin stem that juxtaposed the substrate's 3′ hydroxyl with the pool's 5′ pyrophosphate. This ligation junction was designed to mimic the reaction intermediate normally formed by T4 DNA ligase, and the oligonucleotide pair was in fact an efficient substrate for the protein ligase. The similarity to the T4 ligase reaction mechanism was further emphasized by the inclusion of an ATP binding site (anti-adenosine aptamer) in the core of the pool. After only four cycles of selection and amplification the population was dominated by a single clone. Six additional rounds of selection were carried out using a mutagenic amplification regime. In parallel, a minor variant from round 4 was separated from the majority population and further selected for ligase activity. These secondary selections yielded ribozymes related to those originally observed in the round 4 populations, but with roughly 10–100-fold increases in ligase activity over the majority clone from round 4. All of the final variants that were tested formed 3′-5′ phosphodiester bonds and were specific for the 5′-5′ pyrophosphate linkage but not for the AMP leaving group. Finally, DNA as well as RNA can act as a catalyst (as we will see in greater detail in the next section), and a deoxyribozyme has been selected that can ligate DNA to itself.[56] A constant sequence substrate was activated with a phophorimidazolide and mixed with a random sequence DNA pool, and those species that acquired the constant sequence tag were preferentially amplified. After nine cycles of selection and amplification the ligation activity of the pool had been substantially enhanced. While there were several different deoxyribozymes still present in the pool, most of these catalysts could be folded to form a secondary structure that brought the 5′ hydroxyl of the deoxyribozyme into close promixity with the 3′ phosphorimidazolide of the substrate. The deoxyribozyme was relatively slow (0.07 min^{-1}, k_{cat}) but was still 3400-fold faster than the templated background reaction.

6.09.7 METALLORIBOZYMES

Although it is apparent that the catalytic parameters of known ribozymes can be substantially altered by mutation and that novel ribozymes can be selected from random sequence populations, there are limits to what the chemistries inherent to the four canonical nucleotides can accomplish. Ribozymes composed only of G, A, U, and C lack acids or bases with pK_a values near 7 and have few strong nucleophiles. In consequence, many of the natural RNA catalysts have been shown to be metalloenzymes, possibly because metals are much better nucleophiles than most of the chemical moieties found on RNA. Since the choice and positioning of metals for catalysis is strongly dependent on RNA sequence and structure, these properties have also proven to be fodder for selection experiments. The ability of the hammerhead ribozyme to utilize magnesium in catalysis has been optimized.[57] A small random sequence library that contained from one to four random sequence positions in stem II was selected for its ability to cleave a target RNA in the presence of sequentially smaller concentrations of magnesium. Active complexes were separated from inactive by gel mobility shift, and at the conclusion of the selection variants were identified that could cleave the target 20-fold faster than the initial pool. The Group I self-splicing intron has been used again as an excellent model system for evolutionary manipulations, this time to alter the metal specificity of ribozyme catalysis. Although the Tetrahymena ribozyme cannot normally utilize calcium for RNA cleavage, Lehman and Joyce[58] started from the same doped sequence population used to isolate DNA-cleaving ribozymes and selected variants that could function in the presence of calcium and only trace amounts of magnesium. After eight cycles of selection and amplification, the selected population was 170-fold more active in the presence of calcium than the wild-type; individual clones were up to 300-fold more active. However, the selected ribozymes were still 1000–10 000-fold less active in calcium than the wild-type ribozyme was in magnesium. As was the case with ribozymes selected for DNA cleavage, the activity of the selected variants was due largely to specific changes at relatively few (six) positions. When the selection for calcium utilization was carried out on an additional four generations,[59] the population improved an additional threefold in a standard assay. The calcium-dependent ribozymes could now cleave RNA to the same extent as the wild-type enzyme with magnesium, although they were still kinetically slower. As a result of additional selection, one of the generation 8 mutations was displaced from the selected population, and another three rose to prominence. The preponderant changes from the DNA cleavage and the calcium dependence selections do not overlap except for one to two residues, indicating that in both instances specific functional mutations are being fixed, rather than just general "up" mutations. Similarly, Frank and Pace[60] selected RNase P variants from a partially randomized population that could cleave an appended tRNA substrate from themselves in the presence of calcium rather than the normative magnesium. Amazingly, a single point mutation in the highly conserved P4 pseudoknot structure confered a 90-fold alteration in the metal specificity of the ribozyme. These results provide strong supporting evidence for the existence of specific binding sites for catalytic metals in ribozymes.

Not only can the metal binding properties of natural ribozymes be modified, but artificial metalloribozymes can also be generated. Pan and Uhlenbeck[61] set out to map the residues and structures involved in a lead-catalyzed tRNA self-cleavage event that had previously been observed.

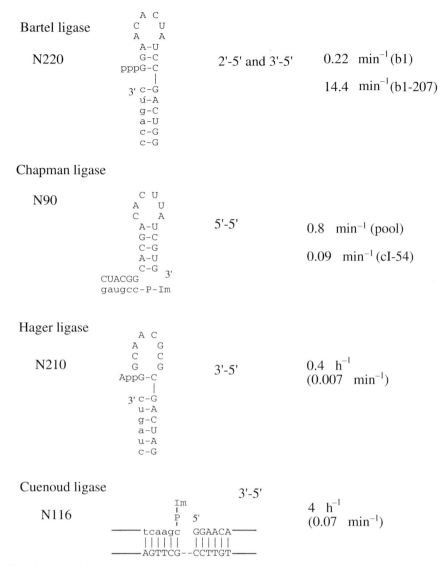

Figure 13 Chemistries of selected ligases. The programmed ligation junctions for the four different ligases described in this review are shown. None of these ligation junctions were actually used by the selected nucleic acid enzymes. In the first three instances, a 3′ hydroxyl present at the termini of an RNA pool was meant to displace a leaving group from the 5′ end of a phosphorylated substrate; the Chapman ligase departed from this scheme and catalyzed the formation of a 5′–5′ tetraphosphate linkage. For the Cuenoud ligase, a 5′ hydroxyl present at the termini of a DNA pool was meant to displace a leaving group from the 3′ end of a phosphorylated substrate. Other characteristics of the selected ligases, including the lengths of the starting random sequence pools, the types of linkages generated by ligation, and rates, are shown. The values reported for the rates of catalysis may only apply to a few members of a selected population.

The tRNA molecule was segmentally randomized, circularized, and those variants that could linearize themselves were selected and amplified. After six cycles of selection and amplification, catalysts were identified that performed lead-dependent cleavage reactions at a variety of sites. Surprisingly, most of the selected RNAs had evolved to form structures that were unlike that of the parental tRNA. One variant could be paired to an active stem–internal loop–stem structure, and the lead ion bound and cleaved the RNA within the internal loop. The cleavage reaction is highly specific for lead, and appears to use lead hydroxide as a base.[62] As expected, the mechanism of cleavage involves the formation of a 2′,3′ cyclic phosphodiester bond.[63] Subsequently, however, the lead:RNA complex also catalyzes resolution of the cyclic phosphate into a 3′ monophosphate ester, just as ribonuclease A does. Similarly, Williams *et al.*[64] have isolated novel ribozymes that can catalyze self-cleavage in the presence of magnesium. A random sequence 100-mer was embedded within a variant of yeast tRNA[Leu3] and self-cleaving species were isolated using a classic cleavage selection that separates substrates from products by gel electrophoresis. Numerous different species that could cleave their constant sequences at different locales were identified. As expected, the reacted ribozymes contained 2′,3′ cyclic phosphates at their termini. The new ribozymes could frequently be reduced to internal loop motifs. One of these motifs was engineered to act in a bimolecular format and, consistent with its structural simplicity, was a rather slow catalyst with a cleavage rate of $0.2 \; h^{-1}$. This rate was roughly 200-fold faster than the background reaction but around 100-fold slower than a comparable hammerhead ribozyme.

The importance of metals for nucleic acid catalysis is even more apparent when we examine the selection of DNA enzymes (deoxyribozymes or "dinozymes"). Although DNA is best known as an informational macromolecule, rather than as a progenitor of shape and catalytic functionality, the secondary and tertiary structural interactions that DNA is capable of should be almost as complex as those already observed in structured RNAs. To this end, several groups have selected "deoxy-ribozymes" or "dinozymes" from random sequence populations (reviewed Refs. 65–67). Initially, Breaker and Joyce[68] selected a DNA enzyme that could cleave an RNA substrate. A DNA pool was synthesized that contained a single ribotide at the junction between the constant and random sequence regions. The pool was immobilized via a terminal biotin moiety, the cleavage reaction was initiated by the addition of lead cation (Pb^{2+}), and those variants that could cleave themselves from the column were eluted, amplified, and reselected. After only five cycles of selection and amplification, the population was significantly enriched in self-cleaving variants. A major variant was identified, and as expected performed self-scission at the labile ribose linkage. The major variant could be engineered to act in *trans* as a mere 38-nucleotide deoxyribozyme. Despite the fact that the catalyst was both short and composed entirely of DNA, it exhibited a rate enhancement of roughly 100 000-fold compared to the uncatalyzed reaction (in the presence of lead). Mechanistic studies of the lead-dependent deoxyribozyme in which lanthanides were substituted for lead indicated that a single metal ion was involved in catalysis and that binding to the metal occurred almost exclusively via the 2′ hydroxyl of the unique ribotide.[69] Using the same selection scheme, Breaker and Joyce[70] selected deoxyribozymes that could cleave the phosphodiester linkage adjacent to the ribose in the presence of either magnesium, manganese, or zinc. The deoxyribozymes selected in the presence of manganese and zinc exhibited at least some activity in the presence of noncognate metals, with the exception of magnesium. Clones from these selected populations frequently showed sequence similarity to the deoxyribozyme originally selected in the presence of lead. In contrast, the deoxy-ribozyme selected in the presence of magnesium was uniquely active with magnesium and was dominated by a single species whose sequence and structure were unlike those of the lead cleavage motif. The magnesium cleavage motif contained 15 conserved nucleotides flanked by three stems, two of which hold the cleavage substrate in place, and was more complex than the lead cleavage motif, which contained only six conserved nucleotides. Although the nascent rate of magnesium-dependent cleavage of the phosphodiester linkage adjacent to ribose was three orders of magnitude slower than the rate of lead-dependent cleavage, the magnesium cleavage motif gave a catalytic improvement (k_{cat}) of 100 000-fold, similar to the improvement seen with the lead cleavage motif. Faulhammer and Famulok[71,72] also selected deoxyribozymes that could cleave a phosphodiester bond adjacent to a single ribose. The selections were similar to those carried out by Breaker and Joyce, but the magnesium concentration was slightly lower (0.5 mM vs. 1.0 mM) and histidine was also present in the elution buffer. After 10 cycles of selection and amplification, a variety of magnesium-dependent deoxyribozymes were isolated. The rates of the selected metalloribozymes were similar to those observed by Breaker and Joyce, ca. $0.01–0.1 \; min^{-1}$ (k_{cat}). As was the case with other selected deoxyribozymes, no marked preference for a single or cognate metal was noted. However, one of the deoxyribozyme classes showed an unexpected catalytic improvement in the presence of calcium, despite the fact that hydrated calcium has a pK_a even higher than that of magnesium. The observed preference for calcium is unique amongst natural and selected ribozymes and deoxyribozymes (for example, recall the Group I ribozyme selected to utilize calcium). It was hypothesized that the deoxyribozyme fortuitously formed a tight ($K_{d,app} = 70 \; \mu M$) binding pocket for calcium, even though the DNA species were not exposed to calcium at any point during the selection procedure.

Building on these results, Carmi et al.[73] selected deoxyribozymes that could cleave an all-DNA substrate. While the selection scheme was similar to that previously employed by Breaker and Joyce,[68] the reaction was initiated with copper (Cu^{2+}) rather than lead cations. After seven cycles of selection and amplification, two major sequence classes of catalysts were identified. Both classes attacked sequences in their constant regions, but appeared to utilize different cleavage chemistries. Both ribozymes appeared to rely on radicals to initiate cleavage, but the first class (CA1) required ascorbate (included during the original selection) for radical generation, while the second class

(CA3) could cleave in the presence of copper alone. Both classes of deoxyribozymes generated a range of cleavage products, consistent with localized radical generation and diffusion from an enzyme active site. Although the maximal rates of the selected catalysts were roughly 10–100-fold slower than the rates observed with the deoxyribozymes that cleaved an RNA–DNA linkage (above), the rate enhancement was greater (10^6-fold above background) because of the greater stability of the all-DNA substrate. The second class of ribozymes was further optimized by additional rounds of selection and amplification, and was engineered to act as a restriction endonuclease for single-stranded DNA.[74] The hybridizing arms of the optimized catalyst could be altered to recognize and form duplexes with different DNA substrates. Surprisingly, some cross-reaction between different substrates was observed, due in part to the fact that the deoxyribozyme recognizes its substrates not only via duplex formation but also via the presentation of a triplex strand.

Finally, Santoro and Joyce[75] selected a deoxyribozyme that can quickly and efficiently cleave RNA of almost any sequence. Instead of including a single ribonucleotide in the constant region, an RNA dodecamer was incorporated into the 5' end of a DNA pool as a potential cleavage target, and the reaction was initiated by the addition of magnesium. After 10 cycles of selection and amplification, a wide variety of sequences were found that could cleave the RNA oligonucleotide. Two deoxyribozymes were identified that could form obvious base pairs with their RNA substrates; outside of the hybridizing arms, these deoxyribozymes contained only 13 and 15 residues. As expected, these deoxyribozymes were dependent on magnesium to facilitate RNA cleavage. The deoxyribozyme that contained a catalytic core of 15 deoxyribonucleotides (10–23) could cleave its RNA substrate in *trans*, could be rationally altered to target other RNA sequences, and had a second-order rate constant (k_{cat}/K_m) in excess of 10^8 under multiple turnover conditions, the same order of magnitude as a protein enzyme such as ribonuclease A.

The metalloribozymes that have so far been considered in this section have catalyzed phosphodiester bond rearrangements, primarily cleavage reactions. However, metals can also aid in other types of catalysis and can even themselves be substrates for catalysis. Conn *et al.*[76] selected an RNA molecule from a random sequence population that could noncovalently bind to a porphyrin, *N*-methylmesoporphyrin IX (NMM). This compound is a distorted porphyrin analogue and is thought to conformationally resemble the transition state for porphyrin metallation. Antibodies that bind NMM can catalyze porphyrin metallation, and it was hypothesized that anti-NMM aptamers might also catalyze porphyrin metallation. In fact, a small (35-nucleotide) aptamer could catalyze the copper metallation of porphyrins. The k_{cat}/K_m values for the artificial catalyst rivaled those of protein ferrochelatases. Li and Sen[77] selected anti-NMM aptamers from DNA libraries and these also proved to be deoxyribozyme chelatases. The rate enhancement was approximately 1700-fold over the background reaction (k_{cat}) and was only 40-fold less efficient than the corresponding catalytic antibody (k_{cat}/K_m). The deoxyribozyme could catalyze the metallation of meso-, deutero, and protoporphyrin IX, while the catalytic antibody was specific for mesoporphyrin alone. Further optimization of the deoxyribozyme[78] reduced its size to only 24 nucleotides, while optimization of buffer conditions resulted in an improvement in catalysis to 3700-fold above background (k_{cat}/K_m), a value comparable to that of catalytic antibodies and natural ferrochelatases. Based on its high guanosine content and dependence on potassium, the deoxyribozyme was thought to form some manner of G-quartet structure. Further analysis of the deoxyribozyme by UV and fluorescence spectroscopy has revealed that the mesoporphyrin IX (MPIX) substrate has an increased basicity of 3–4 pH units in the deoxyribozyme active site, in part because of an induced distortion of the porphyrin conformation.[79] It is also possible that the polyanionic nucleic acid stabilizes the developing positive charge in the porphyrin complex, although this mechanism may be less intrinsically likely given that no metal binding site was apparent on the deoxyribozyme.

6.09.8 MODIFIED CATALYSTS

While the folded structure of a ribozyme may create an environment that augments the chemistries of the individual bases, sugars, or phosphate backbone, a quick comparison between the building blocks of ribozymes and the building blocks of protein enzymes illustrates why most of the RNA world was supplanted roughly 3.5 billion years ago. To the extent that metal ions enhance ribozyme catalysis, it might be expected that other chemical moieties would also yield improvements in either the versatility or speed of ribozymes. Chemical modifications can be introduced into random sequence pools in two ways: incorporation at the termini of RNA or DNA, or incorporation internally via modified nucleotides. Both methods have yielded functional catalysts.

It has been previously seen that an amino group incorporated at the termini of an RNA led to the selection of an amide synthase (Figure 14).[48] The monophosphorothioate used in these experiments as a handle for amino acid conjugation also has unique chemical properties of its own. Wecker *et al.*[80] synthesized an RNA pool that was initiated with a guanosine monophosphorothioate. When this pool was incubated with an activated variant of the peptide bradykinin (*N*-bromoacetyl-bradykinin), individual RNA species were alkylated at their 5′ ends and could be separated from the remainder of the population on [β-acryloylamino)phenyl]mercuric chloride polyacrylamide gels. In this gel system, RNA molecules with unmodified thiolates are retarded relative to RNA species with blocked thiolates, such as those that have appended bradykinin to their 5′ ends. After 12 cycles of selection via gels and affinity chromatography on thiol columns, the RNA population showed roughly 200-fold improvement in alkylation activity. The sequences of individual variants were generally dissimilar from one another except for the presence of a short sequence motif that could pair with and position the 5′ phosphorothiolated termini of the RNA molecules. More detailed analysis of one of the variants supported the hypothesis that the enhanced alkylation reaction was due primarily to the positioning of the reactive bradykinin adjacent to the terminal thiolate rather than to activation of the thiolate by the catalyst. The fact that free, unmodified bradykinin specifically inhibited the sulfur alkylation reaction was further proof for the hypothesized entropic trapping mechanism.

A similar strategy was employed by Wiegand *et al.*[81] for the selection of amide synthases. These

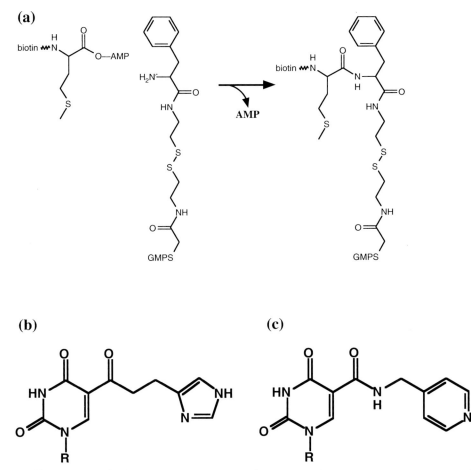

Figure 14   Ribozyme modifications. (a) Derivatization of the 5′ end of a ribozyme. Zhang and Cech[48] initiated transcripts with guanosine monophosphorothioate (GMPS), and alkylated the single sulfur moiety on these transcripts with *N*-bromoacetyl-*N*′-phenylalanyl cystamine. The final products thus contained an amino acid (phenylalanine) joined to RNA via a reversible disulfide bond. The derivatized pool was mixed with biotinylated methionine esterified to the (2′)3′hydroxyl of AMP and ribozymes that could exchange the external ester for an internal amide were selected. (b) The modified base 5-imidazole-UTP used by Wiegand *et al.*[81] for the selection of ribozyme amide synthases. (c) The modified base 5-pyridylmethylcarboxamide-UTP used by Tarasow *et al.*[82] for the selection of ribozyme Diels–Alder synthases. R = ribose 5′ triphosphate.

authors conjugated an amine to the beginning of RNA molecules by ligating a modified DNA oligonucleotide to the RNA pool. In addition, though, the RNA pool contained a modified uridine residue, 5-imidazole uridine, in place of uridine (Figure 14). The modified pool was mixed with AMP-biotin, and biotinylated RNAs were separated on streptavidin, amplified, and reselected. After 16 cycles of selection and amplification, numerous catalysts remained in the population but, as was the case for the sulfur alkylases described above, many of these catalysts contained a common, short sequence motif. In this case, however, the motif did not constrain the 5′ end of the RNA but rather formed a stem-bulge structure that may have assisted in catalysis. The rate enhancements of the selected catalysts were of the order of 10^4–10^5-fold, similar to the 10^6-fold rate enhancement claimed by Zhang and Cech[48] for their amide ligases. The ribozyme recognized the biotin moiety that it attached to itself, but did not appear to recognize the leaving group, as UMP, AMP, and ribose monophosphate derivatives were all effective donors. These experiments not only reinforced the notion that amide synthases could have existed in a putative RNA world, but also showed that chemical augmentation could significantly influence RNA catalysis. At least one of the selected catalysts was absolutely dependent on the modified nucleotide for activity, although the precise role of the modification in catalysis was unknown. The characterized ribozyme could function in the absence of divalent metals, but its reaction rate was enhanced by divalent ions such as magnesium and calcium and was most significantly increased by the addition of Cu^{2+}. Interestingly, the copper ions appeared to contribute to substrate binding rather than chemical catalysis.

Eaton and co-workers have also used modified nucleotides to move ribozymes into new catalytic realms. Tarasow *et al.*[82] report the selection of ribozymes that can catalyze carbon–carbon bond formation, enhancing the rate of a Diels–Alder reaction. One-half of the product, the diene, was conjugated to the end of an RNA pool via ligation. The pool again contained a modified uridine, 5-pyridylmethylcarboxamide-uridine (Figure 14), in place of the natural nucleotide. A dienophile conjugated to biotin was mixed with the pool, and biotinylated species were selected over 12 cycles. The selected sequences fell into several families, but the families were again related by a core motif. One of the selected species enhanced the rate of the Diels–Alder cycloaddition by 800-fold. As was observed with the modified amide synthases, catalytic activity was dependent on the presence of the modified nucleotide in transcripts. As was observed with the modified amide synthases, catalytic activity was enhanced by divalent metals, but was most significantly enhanced by (indeed, was absolutely dependent on) copper ions.

6.09.9 CONCLUSIONS

6.09.9.1 Natural Ribozymes and Selection: Structure Determination

The benefits of artificial phylogenetic analyses for nucleic acid structure prediction become apparent when RNA and protein catalysts are compared. The primary sequence of both types of biopolymers determines their global architecture, which in turn determines their catalytic functions. However, protein folds are determined partially by the formation of secondary structural elements, and largely by the establishment of tertiary packing interactions in a hydrophobic core. The secondary structures of proteins can sometimes be predicted from primary sequences, but the further identification of tertiary structural interactions is difficult. In consequence, predicting the three-dimensional structure of a protein from the primary sequence remains a daunting problem.

In contrast, some of the features that make nucleic acid catalysts less structurally diverse than proteins make their structures far easier to predict. Whereas protein catalysts are derived from a repertoire of 20 amino acids with a wide array of chemical properties, ribozymes are built from a cast of four chemically similar nucleotides: adenosine, guanosine, cytidine, and uridine. Whereas the interactions between amino acids in protein cores are idiosyncratic and highly dependent on context, the interactions between nucleotides in functional RNAs tend to occur along a single plane and are much more independent of context. In short, a G:C base pair in one helix is structurally similar to a G:C base pair in another. Moreover, the global folds of ribozymes are determined largely by secondary structural elements, and only partially by tertiary contacts. Therefore, establishing a solid secondary structure is a major step towards determining the overall three-dimensional structure of a ribozyme.

We have seen that selection was useful for confirming the proposed secondary structure of the hairpin ribozyme. It might also have proven useful for determining or confirming the secondary structure of the Group I ribozyme, except for the fact that abundant natural phylogenetic data already existed. In consequence, artificial phylogenetic analyses have been most useful in determining the structure of new ribozymes. The Class I Bartel ligase is as large and easily as complex as large, natural ribozymes such as the Group I ribozyme. The secondary structure of the Class I ligase was established almost solely based on sequence covariations that were observed following partial randomization and reselection for function.

6.09.9.2 Natural Ribozymes and Selection: Nature's Search Engine

An interesting question that can be uniquely explored by *in vitro* selection experiments is to what extent are natural ribozymes functionally optimal. Or, in other words, are there other ribozymes with different sequences that are functionally superior to natural ribozymes? Admittedly, a precise answer to this question is unobtainable, since both function and survival are highly dependent on the environment in which a catalyst is observed. It would be impossible in a test tube setting to recapitulate the millions to billions of years of selection pressures that have yielded Group I ribozymes. Conversely, few ribozymes selected in test tubes have been optimized for function in organisms, and it is unlikely that unnatural ribozymes will be released into the wild solely to observe how they fare. Nonetheless, as long as we are cognizant of the caveat that the natural and unnatural worlds have different criteria for survival, then we can compare the results of natural and unnatural selection experiments.

From the vantage of the experimentalist, Nature's search engine appears to be an astoundingly good one. Even when all 14 residues in the core of the hammerhead ribozyme are randomized and the population is selected for cleavage function, the selected ribozymes "relax" back to the wild-type sequence. The 11 nucleotide tetraloop receptor motif in the Group I ribozyme can be recovered by *in vitro* selection. Similarly, Frank *et al.*[83] randomized 22 residues in and around the P4 (pseudo-knotted) stem of ribonuclease P. After 10 cycles of selection for cleavage away from a tethered tRNA molecule, every position within the randomized region had reverted to wild-type!

Of course, there is also ample evidence that there are many apparently unnatural catalysts that can function as well or better than their natural counterparts. For example, Costa and Michel[32] recovered both wild-type and nonwild-type tetraloop receptors in the same experiment. Tang and Joyce (1997) found nonhammerhead motifs in the same cleavage selection that yielded hammerhead ribozymes. Cleavage selections in which metals other than magnesium were included in the reaction have yielded unnatural motifs.

How can Nature's apparent optimality be reconciled with the apparently limited scope of Nature's catalysts? By assuming that Nature can identify local rather than global optima. While the hammerhead may in fact be a globally optimum solution to the problem of how to cleave double-stranded RNA that contains an UH (H = not G) target, it is interesting to note that the selection experiments that did not return the hammerhead ribozyme either had an altered stem II (Tang and Breaker[24]) or a randomized stem II[25] in addition to a randomized hammerhead core. Similarly, while limited, segmental randomization of the Group I ribozyme or RNase P followed by selection can yield wild-type catalysts, selections for similar or equivalent function from deep random pools generally yield quite different catalysts. In the most dramatic example of how natural ribozymes are not necessarily globally optimal catalysts, Tuschl *et al.*[84] synthesized random sequence libraries that contained U2 and U6 spliceosomal RNAs at their core. The libraries were selected for their ability to catalyze the ligation of an oligonucleotide substrate to themselves. It was hoped that this selection would yield spliceosomal-like catalysts that were independent of proteins, and that analysis of these catalysts might yield clues to the origin of the spliceosome. However, the catalysts that were eventually selected bore little or no resemblance to the spliceosome either in terms of sequence or reaction mechanism. Four activities were identified: the first was self-cleavage to form a 2',3' cyclic phosphate followed by ligation of the constant sequence oligonucleotide substrate. The second was the attack of a 2' hydroxyl within the substrate on the 5' termini of the ribozyme, releasing pyrophosphate. The third class of ribozymes also catalyzed self-cleavage, but additionally cleaved the substrate oligonucleotide and finally ligated the substrate cleavage product to the 2',3' cyclic phosphate, a three-step reaction. The fourth class of ribozymes could catalyze the formation of an unusual 2',3' branch between a 2' hydroxyl of the substrate and an internal phosphate. This bestiary of catalysts was completely unanticipated, given that largely wild-type RNAs were the starting point for the selection.

These diverse results can all be reconciled by assuming that ribozymes are sparsely represented in any given sequence space. During the course of evolution Nature may have chanced upon a given functional peak, and climbed to the top of that peak by point mutation and selection. However, precisely because Nature optimizes catalysts largely by point mutation it may have proven difficult to visit other, widely separated peaks that had equivalent function. At a guess, Nature may have visited at least all possible 15-mers in sequence space (e.g., the hammerhead) during the course of evolution but not all possible 30-mers or greater. Interestingly, at least one back of the envelope calculation suggests that the current sequence content of the biosphere should be of the order of all possible 28-mers: if one assumes that genes are the same within a species, but different between species, then there are roughly 10^9 different species, each of which has roughly 10^5 genes, each of which has roughly 10^3 bases. These gross approximations yield an estimate of 10^{17} different sequences in the biosphere ($\times 4^{28}$).

6.09.9.3 Unnatural Ribozymes and Selection: The Role of Metals

Many, but not all, of the ribozymes and deoxyribozymes we have considered are metalloenzymes. Since selection experiments to some extent examine the universe of possible catalysts (however, see "You get what you select," below), then by reviewing whether and how metals are adopted by selected catalysts the general importance of metals in nucleic acid catalysis can be assessed. Two questions are fundamental: first, are metals important for catalysis, and second, which metals are important for catalysis.

It is frequently assumed that divalent cations are critical for ribozyme catalysis.[85] Such an assumption at first glance seems reasonable, given the limited chemistries available to nucleic acids via the functional groups of the nucleotides. However, ribozymes may have other catalytic mechanisms available to them that obviate the role of metals in catalysis, such as the conformational stabilization of transition states or activated intermediates. Thus, while it is surprising, it is nonetheless reasonable that experiments have revealed that natural ribozymes are not necessarily metalloenzymes. Murray *et al.*[86] have assayed the hammerhead, hairpin, and VS ribozymes in buffers devoid of divalent cations, and found that these ribozymes can catalyze self-cleavage in the presence of 4 M LiCl at rates that are identical to those in the presence of magnesium.

Unnatural ribozymes also do not seem to absolutely require divalent metal ions for catalysis. In the same experiments in which Faulhammer and Famulok[72] selected a calcium-preferring deoxyribozyme, they also identified deoxyribozymes that could cleave a phosphodiester bond adjacent to a ribose that required no divalent metal ion for catalysis. The rates of cleavage for the metal-independent deoxyribozymes were similar to those of the metal-dependent deoxyribozymes (ca. 0.005 min^{-1}). The deoxyribozymes selected by Geyer and Sen[69] were also metal-independent and also cleaved the targeted bond at a similar rate (ca. 0.01 min^{-1}). These values are 1–3 orders of magnitude slower than those observed for natural ribozymes such as the hammerhead, and for other selected deoxyribozymes.[69] In contrast, Jayasena and Gold[91] have selected self-cleaving ribozymes from a random sequence population using a selection method similar to that used to select lead-dependent, tRNA-like ribozymes.[61] Thus, with the proviso that monovalent salts or general acids and bases can to some extent "repair" deficits in catalysis resulting from a lack of divalents, divalent metal ions may be worth factors of 100–1000 in the overall rates of nucleic acid catalysis.

Selection experiments can also reveal whether some metals are more amenable to certain types of catalysis than others. For example, several selections for catalysts have been carried out using a mixture of metals, yet the selected catalysts have a decided preference for one or more of the individual metals. The deoxyribozyme ligase[56] was isolated in the presence of both magnesium and zinc, yet required only zinc for activity. The capping ribozyme[42,43] was isolated in the presence of magnesium, calcium, zinc, and manganese, yet required only calcium for activity. The aminoacylase[44] was isolated from a buffer that contained calcium, copper, iron(III), magnesium, manganese, and zinc, yet required only calcium and magnesium for activity. The novel reactions catalyzed by the Diels–Alder synthase[82] and the amide synthase[81] required copper for activity, although a number of other transition metals were originally present in the selection buffer. It was hypothesized that copper may provide Lewis acid sites for catalysis. While it is interesting that selected catalysts show preferences for individual metals, the preferences are not always absolute. For example, the "leadzyme" motif selected by Breaker and Joyce[52,68,70] could utilize a variety of other metals for catalysis, while the deoxyribozyme ligase could utilize copper in place of zinc.

6.09.9.4   Unnatural Ribozymes and Selection: The Role of Modifications

One of the newest innovations in ribozyme chemistry is the inclusion of modified bases in selection reactions. While the modified nucleotides that were incorporated into the selections for the Diels–Alder synthase and amide synthases were clearly essential for the catalytic activity of the ribozymes, it is unclear whether similar ribozymes could have been selected in the absence of the modified nucleotides and what the kinetic parameters of such unmodified ribozymes might be. It has been claimed that since previous selections for a Diels–Alder synthase with all-RNA pools had failed,[15] that it was the presence of a modified nucleotide that led to the later, successful selection of catalysts.[82] However, another more critical distinction between these experiments was that the all-RNA selection was an indirect selection for transition-state affinity, while the modified RNA selection was a direct selection for catalysis. A different, more tenable comparison between all RNA and modified RNA catalysts might be the comparison between an all-RNA amide synthase[79] and a modified RNA amide synthase.[81] Although the amine conjugates, selection procedures, and metal specificities of the selected ribozymes also differed, the substrates used for selection were similar and the two catalysts have surprisingly similar k_{cat}/K_m values. Further doubt regarding the superiority of chemically modified nucleic acid catalysts comes from a selection that attempted to produce an amide synthase but instead yielded an ester synthase. Jenne and Famulok[89] synthesized an RNA pool that was similar to those previously used by Zhang and Cech[48] and Wiegand *et al.*[81] for the selection of amide synthases: the pool containing a dipeptide, citrulline-cysteine, conjugated to its 5′ end. The delta amino group of citrulline would have been expected to have a nucleophilicity similar to the alpha amino group of the phenylalanine at the termini of the pool generated by Zhang and Cech[48] and the alkyl amino group of the pool generated by Wiegand *et al.*[81] However, after 13 cycles of selection using a (2′) 3′ AMP ester of biotinylated phenylalanine as a substrate, at least one of the selected ribozymes catalyzed the transfer of the amino acid ester to an internal 2′ hydroxyl rather than to the presumably much more nucleophilic terminal amine. Despite the ambiguity of these results, it is intuitively resonant, if not proven, that augmenting an evolving biopolymer with more or better chemical groups should yield faster, more specific catalysts.

The chemical augmentation of ribozymes need not occur via covalent modification or the incorporation of modified bases. Like protein enzymes, ribozymes could presumably noncovalently bind cofactors that would assist in catalysis. However, it has proven difficult to select for such catalysts. Faulhammer and Famulok[71,72] attempted to select for histidine-dependent deoxyribozyme cleavage of a phosphodiester bond adjacent to ribose, but identified only metal-dependent or metal-independent catalysts. In contrast, Roth and Breaker[87] have recently reported the selection of histidine-dependent deoxyribozymes using almost exactly the same selection procedure. The reaction is very specific for histidine and accelerates the rate of phosphodiester bond cleavage by roughly 10-million-fold.

6.09.9.5   Unnatural Ribozymes and Selection: You Get What You Select

One caveat to the interpretation of the results of selection experiments is that while these experiments are carefully designed to produce catalysts with a desired activity, the catalysts that are eventually generated must in fact adhere to only one imperative: survival. In general, attempts to direct the reactivities of selected ribozymes are successful. For example, experiments and pools designed to select for ligase activity typically yielded ligases, not phosphatases. However, it is interesting to note that design features that are not implicitly required for survival are often blatantly ignored by selected ribozymes. For example, the Bartel ligases were selected from a pool that could form a hairpin stem at the ligation junction, just as the Group I ribozyme does (Figures 10 and 13). However, none of the selected catalysts used the designed stem in the original pool, and instead chose to generate new ligation junctions. The Class I ligase breaks up the designed stem by forming a pseudoknot, and only one residue holds the 3′ portion of the ligation junction in place; the Class II ligase does not use Watson–Crick pairing at the junction, instead relying on a series of purine residues that may form noncanonical pairings; and the Class III ligase uses Watson–Crick pairing for the 5′ portion of the ligation junction but not the 3′ portion. The Class I and III ligases use their original constant regions to bind the substrate for ligation, but the Class II ligase generates an internal template for the substrate as well. Moreover, two of these three classes of ribozymes unexpectedly formed 2′–5′ phosphodiester bonds rather than 3′–5′ bonds. It was not until after the 2′–5′ junctions were experimentally identified that Lorsch *et al.*[88] proved that reverse trancriptase might read through this junction. It was also in part the unexpected promiscuity of reverse transcriptase that led Tuschl *et al.*[84] to select multiple, diverse, multistep catalysts from a supposedly constrained pool. As another example, although the 5′–5′ Chapman ligases were selected from a pool that could poise the 3′ terminal hydroxyl adjacent to the phosphorimidazolide of the ligation substrate, the selected catalysts eschewed the designed substrate-binding site and instead generated their own internal templates for poising the ligation substrate. Finally, the ligases selected by Hager and Szostak[55] and Cuenoud and Szostak[56] also generated their own internal substrate-binding sites from random sequences despite the fact that constant sequence ligation junctions were in each instance designed into the selections.

The tendency for selected ribozymes to develop their own unique catalytic mechanisms rather than acceding to the wishes of their designers is also apparent when nonoligonucleotide substrate binding sites are considered. Although the Lorsch kinases utilized ATP as a substrate, they did not coherently utilize the antiadenosine aptamer included within the original random sequence pool. Multiple mutations accumulated in the antiadenosine aptamers programmed into the original pool, and the K_m of the ribozymes for ATP was much higher than the original K_d of the aptamer. One of the best characterized aptamers had a K_m of 3 mM for ATP, while the original antiadenosine aptamer had a K_d of 1 μM for ATP.[41] Similarly, although the Hager ligases utilized AMP as a leaving group, they appear to have ignored the antiadenosine aptamer originally provided them. GMP is also efficiently used as a leaving group, and the final, selected ligases contained multiple mutations in their internal antiadenosine aptamers.

Other steps in the selection procedure besides catalysis can also influence the nature of the survivors. Selections that involve immobilization run the risk of selection for binding species rather than catalysts. While this has not been reported, Jenne and Famulok[89] had hoped to select for amide synthases by first capturing an activated biotin then eluting catalysts from a streptavidin column by reversing the disulfide conjugation of a 5′ terminal amine. However, they noted that as the selection progressed, increasingly higher concentrations of disulfide (up to 2 M) was required to elute captured species. Upon closer examination, it became apparent that the selected ribozymes were transacylases and that their elution was likely due to disruption of biotin–streptavidin interactions rather than reduction of a (nonessential) disulfide bond.

Finally, there are multiple ways in which the selection process can be skewed, diverted, or thwarted during the amplification procedure. For example, replication "parasites" can arise during selections for catalysis, such as the RNA Z species that aborted the original attempts to develop a continuous selection scheme. While such parasites are an extreme example of how amplification can skew the course of a selection, the amplification protocol may influence the nature of selected catalysts as much as the selection protocol. In the selections for the Bartel and Hager ligases, the ligation junction was also a template for reverse transcription. In contrast, the selection that produced the Chapman ligases did not require that the ligation junction serve as a template for reverse transcription. Thus, in the selections for Bartel and Hager ligases only replicable (3′–5′ and the surprising 2′–5′) ligation junctions were identified, while in the selection for the Chapman ligases nonreplicable (5′–5′) ligation junctions were identified. Similarly, in the selections for deoxyribozyme nucleases,[72] cleavage to release active species could have occurred anywhere along the strand, yet only those species that were both cleaved and retained enough of their constant region to bind a PCR primer would be subsequently amplified and reselected. In this respect, it is interesting to note that one family of nucleases, CA3, cleaved itself at two sites, within the primer binding region and internally.

6.09.9.6 The RNA World: Ribozymes vs. Protein Enzymes

Table 1 shows the rates and relative rate enhancements for many of the ribozymes and deoxyribozymes discussed in this chapter. What is surprising is the relative uniformity of nucleic acid catalysis over a variety of catalysts, selection procedures, pools, metal ions, and chemistries. When all is said and done, ribozymes and deoxyribozymes are not very fast compared to their protein counterparts. Overall, most nucleic acid catalysts have k_{cat} values of 0.01–1.0 min^{-1} and have k_{cat}/K_m values of roughly 100–1000 compared with values that are typically three orders of magnitude or more higher for proteins. The apparent exceptions to this rule, such as the Lohse transacylase, have "artificially" high k_{cat}/K_m values that are driven by the extremely low K_m values for oligonucleotide substrates, or else catalyze reactions that were extremely facile to begin with, such as metallation of porphyrins.

What are the reasons behind this unflattering comparison between ribozymes and proteins? One might suggest that it is just that proteins have had so much longer to evolve, that billions of years

Table 1   Characteristics of selected nucleic acid enzymes.

Ribozyme	Rate[a]	Acceleration[b]	Metal	Ref.
Natural ribozymes				
Group I DNA cleavage	3×10^7 (m.M)$^{-1}$ (k/K)	100 000 rt wt	Mg	36
Group I Ca utilization		300 rt wt	Ca or Mg	58, 59
Unnatural ribozymes				
Self-kinase	6000 (m.M)$^{-1}$ (k/K)	10^9 rt hydrolysis	Mg	40
Pyrophosphate exchange	0.03 min^{-1} (assay)	500 000 rt Rd0	Ca	42, 43
Aminocylation	70 (m.M)$^{-1}$ (2nd order)	250 000 rt Rd0	Mg/Ca	44, 46
Transacylation	4×10^6 (m.M)$^{-1}$ (k/K) 0.19 min^{-1} (k)	10 000 rt Rd0	Mg	47
	350 (m.M)$^{-1}$ (k/K)		Mg	89
Amide bond formation	260 (m.M)$^{-1}$ (k/K)	10^6 rt uncatalyzed	Mg	48
	290 (m.M)$^{-1}$ (k/K)	100 000 rt uncatalyzed	Mg/Ca Cu	81
Ligases				
Bartel Class I	100 min^{-1} (mt) 1×10^7 (k/K)	10^9 rt templated, uncatalyzed	Mg	9
Chapman	0.1 min^{-1} (k)	1000 rt uncatalyzed	Mg	54
Hager	0.007 min^{-1} (st)	500 000 rt templated, uncatalyzed	Mg	54
Cuenoud	0.07 min^{-1} (mt)	3400 rt templated 100 000 rt untemplated	Zn or Cu	55
Nucleases				
RNA cleavage by DNA	0.2 min^{-1} (st)	100 000 rt uncatalyzed	Pb	52, 68
	0.02 min^{-1} (st)		Mg	70
	0.006 min^{-1} (st)		Mg or Ca	72
	3.4 min^{-1} (mt)	10^9 rt uncatalyzed	Mg	75
	0.01 min^{-1} (st)		—	69
DNA cleavage by DNA	0.04 min^{-1} (st)	10^6 rt uncatalyzed	Cu	73
Chelatases				
Ribozyme	125 000 (m.M)$^{-1}$ (k/K)	460 rt uncatalyzed	Cu	76
Deoxyribozyme	40 000 (m.M)$^{-1}$ (k/K)	3700 rt uncatalyzed	Cu or Zn	78
Modified ribozymes				
Diels–Alder synthase	240 (m.M)$^{-1}$ (k/K)	800 rt uncatalyzed	Mg/Ca or Cu	82

[a] $k = k_{cat}$; $k/K = k_{cat}/K_m$; m.M = minutes.molar; mt = multiple turnover; Rd0 = Round 0; rt = relative to; st = single-turnover. [b] The rates and rate enhancements that are reported are generally taken directly from the primary literature. Usually only one of several rates or an average rate is reported. No attempt is made to reconcile apparently contradictory values (e.g., the differences between estimates for rate acceleration by Conn *et al.*[76] and Li and Sen[77] for similar reactions) or to judge the relevance of the uncatalyzed reaction (round 0 pools vs. templated reactions vs. untemplated reactions).

in the wild will of necessity produce catalysts that are far more optimized and elegant than those that are produced within weeks or months in a laboratory. However, it should be recognized that the selective constraints present in any *in vitro* selection experiment are likely to be much more stringent than the constraints on molecular evolution that are present in the wild. There can be little or no neutral evolution during the initial rounds of a selection experiment: there is purely a highly streamlined selection for function. Moreover, as we have previously seen, the number of molecules that are queried for function during an *in vitro* selection experiment dwarfs the number of variants that may be present in the wild. Thus, even though it is likely that Nature has carried out more generations of selection than experimentalists, it is unlikely that the catalysts that have been generated by *in vitro* selection are any less optimal.

A more optimistic caveat may be that the ribozymes of the ancient or modern RNA world will receive enormous assistance from compounds that contain functional groups with better pK_a's and nucleophilicity, such as Eaton's modified bases or Breaker's histidine cofactor. It will be interesting during the next several years to observe whether or not the speed and specificity of selected catalysts improve along with improvements in nucleotide chemistry.

A final constraint on the speed of selected catalysts may be the way in which the selection experiments are carried out. While it is true that the selections are stringent, they are, with few exceptions, direct selections, selections for single-turnover reactions. Since the nucleic acid catalysts are not selected for more than a single turnover, it is possible that this stricture in turn imposes an unseen limitation on the rate of catalysis that can be obtained.

6.09.9.7   The RNA World: Is Bigger Better?

The natural ribozymes range in size from the tiny hammerhead (ca. 20 nucleotides in length) to the giant Group I and Group II self-splicing introns (> 200 nucleotides in length). Selected ribozymes span a similar size range: the Yarus aminoacylase had only 17 selected nucleotides in a core of 43 nucleotides, while the Bartel Class I ligase could not be reduced to less than 112 nucleotides in length. It might be *a priori* expected that the functional complexity of a catalyst should be related to its informational or structural complexity; that is, an enzyme that performs a difficult chemical reaction might be expected to require more residues arrayed in a more complex architecture than an enzyme which performs a simple chemical reaction. This supposition may be roughly true for the natural ribozymes: the small hammerhead ribozyme relies on an internal nucleophile (2′ hydroxyl) to cleave RNA, while the significantly larger RNase P positions an external nucleophile (water). If this supposition should prove more general, it would have implications for evolution of the RNA world. As self-replicators grew in length, they would have been able to acquire more or better catalytic functionality.

Unfortunately, it is difficult to correlate the sizes and functional complexities of selected nucleic acid catalysts, since the reactions that are selected for are frequently quite different and often contrived. The hypothetical metric that relates size and complexity might be expected to be most easily observed within a given class of ribozymes: for example, a small ribozyme performing a chemical reaction might be expected to be less efficient than a much larger ribozyme performing the same reaction. This hypothesis has been bolstered by the careful analysis of the complexities of selected ribzoyme ligases carried out by Bartel and co-workers. Upon initial selection, the Class III ribozyme ligase contained at least 19 functionally important residues and could catalyze self-ligation at a rate of ca. 0.005 min^{-1}. The Class I ribozyme ligase contained upwards of 93 structurally or functionally important residues and could catalyze self-ligation at a rate of ca. 0.22 min^{-1}. More importantly, partial randomization and reselection of the Class III ligase yielded a 38-fold functional improvement, while optimization of the Class I ligase yielded a 700-fold functional improvement (Bartel, personal communication). Based on these results, it can be tentatively concluded that the size of the Class I ligase relative to the Class III ligase led not only to an initial selective advantage but also to a much more robust evolutionary potential.

Interestingly, the Class I ribozyme ligase should not even exist. Based on the number of conserved and semiconserved positions that were found following partial randomization and reselection, the probability of the Class I ribozyme ligase existing in a pool of 220 random sequence nucleotides would have been ca. 4×10^{-19}. Since the initial pool contained only ca. 1.4×10^{15} sequences, the Class I ligase should only have been found once in every 2000 experiments. While it is possible that Bartel and Szostak[49] were merely extremely lucky, the more likely explanation is also more intriguing: catalysts as complex as, but different in sequence than, the Class I ligase must be relatively abundant, at least in the universe of 220-mers.

To numerically address the relationship between pool size and catalytic activity, Sabeti *et al.*[8] appended long (148 residue) random sequence pools to the Class II and Class III Bartel ligases and assayed the catalytic activities of the resultant constructs. Depending on the placement of the pool relative to the ribozyme, the effects on the activities of the ribozymes ranged from zero- to 18-fold inhibition. When individual clones were analyzed, the median effect was fivefold inhibition. The primary conclusion of this study is that longer sequences should not suppress the selection of shorter catalysts that may be included within. Given the previous observation that the selection of the Class I ribozyme ligase was an improbable event, it is safe to conclude that long pools are inherently better for the selection of complex catalysts.

However, this conclusion is not the same as proving that higher order sequence spaces of necessity contain numerous catalysts whose information content and complexity would rival that of the Class I ligase. This hypothesis strongly predicts that if the selection that produced the Class I ligase were to be carried out again using a different pool of the same length and complexity, then an equally functional, equally complex ribozyme with a completely different sequence and structure should be selected. While to our knowledge this precise experiment has not been carried out, Robertson and Ellington[93] have selected ribozyme ligases from a random sequence population that spanned 90 positions and that also contained ca. 10^{15} different sequence variants. Like the Bartel ligases, the Robertson ligases displace pyrophosphate from the 5′ end of transcripts. The optimized, selected Robertson ligase had a self-ligation rate of only 0.2 min^{-1} and had roughly 25 functionally important residues within its catalytic core. Similarly, the Chapman ligases were also selected from a random sequence population that spanned 90 positions and that contained ca. 10^{15} different sequence variants. While the Chapman ligases displace an imidazole rather than a pyrophosphate from their 5′ termini, the ligation reactions are chemically similar. The catalytic core of the best 5′–5′ Chapman ligase had a self-ligation rate of approximately 0.09 min^{-1} and comprised only 54 positions. Making the tenuous assumption that these experiments were carried out in a similar fashion, either Bartel and Szostak were lucky, other researchers were unlucky, or the length of the pool does in fact have a profound effect on the complexity of selected catalysts. Arguing against this analysis, the Hager ligases were selected from a random sequence population that spanned 210 positions and that contained ca. 10^{15} different sequence variants. The Hager ligases displace a nucleoside monophosphate rather than a pyrophosphate from their 5′ termini but again the reactions are chemically similar. The selected Hager ligases could not be significantly foreshortened by deletion analysis, suggesting that they might be internally complex. However, in contrast to the Bartel ligases the Hager ligases were extremely slow, with the most active carrying out self-ligation at a rate of only ca. $0.4\ h^{-1}$ ($0.0067\ min^{-1}$).

6.09.9.8 The RNA World: What Was It Like?

As was pointed out in Section 6.09.1, comparative biochemistry can be used to make a strong case for the existence of a metabolically complex RNA world. This case has now been amply supported by the discovery of ribozymes that can catalyze many different reactions that may have occurred in the RNA world. However, at least two intellectual hurdles remain in trying to fully conceptualize what occurred during the early evolution of life. First, it is unclear how the RNA world came to be. Reactions that mimic prebiotic chemistry at best yield short oligonucleotides, not long random sequence populations. Thus, it is unclear where the fodder for the earliest *in vivo* selection experiments may have come from. One scenario is that short, self-replicating oligonucleotides (or oligonucleotide-like molecules, such as peptide nucleic acids, PNAs) grew and diversified until they could acquire catalytic function. The nascent evidence that suggests a correlation between ribozyme and ribozyme complexity bolsters this hypothesis. It is possible that beyond some "critical length" many different catalytic functions became accessible to otherwise simplistic replicators. Second, it is unclear how long the RNA world may have existed. While it can be argued that the ribosome and tRNA are remnants of the reign of RNA, microfossils that are strikingly similar to modern organisms existed more than three billion years ago. Assuming that organismal morphology is a reliable marker of identity (a tenuous assumption), then cells substantially similar to those that exist today, and that included a full complement of proteinaceous catalysts, may have existed as much as 3.5 billion years ago. Since the Earth's environment would not have been clement for abiogenesis until roughly four billion years ago, this leaves a narrow window (relatively speaking) for the evolution of replication, cellularization, and complex metabolism. The results summarized in this chapter may provide a basis for reconciling our view of the RNA world with these caveats. Despite the best efforts of experimentalists, RNA catalysts are slow and relatively nonspecific, and seem to operate best on their own kind. If the metabolism of ancient organisms were populated with the best and brightest ribozymes from the present, then these organisms would have replicated slowly, trapped chemical energy grudgingly, and incorporated carbon inefficiently. Overall, the RNA world would have been short and metabolically brutish. However, the very features of RNA catalysts that make them unacceptable as organismal mainstays may have fortuitously rendered them into optimal interlopers between origins and translation. We have seen that even short RNA molecules can catalyze reactions, albeit slowly, and that such short RNA molecules can work with a variety of substrates. As a particularly good example, the pyrophosphate exchange enzyme selected by Huang and Yarus[43] can in effect explore the combinatorial chemistry of 5′ end modification. The amide synthase of Wiegand *et al.*[81] can carry out a similar range of derivatization reactions given the presence of a terminal amine. Thus, the one advantage of an early RNA world would have been that many, many different molecular species would have been invented, most inadvertently. Irrespective of the error rate of replication, the error rate of metabolism, the synthesis of unintended or unwanted compounds, would likely have been huge. Thus, the RNA world would have inexorably slouched towards the invention of molecules with greater catalytic abilities, proteins.

6.09.10 REFERENCES

1. K. Kruger, P. J. Grabowski, A. J. Zaug, J. Sands, D. E. Gottschling, and T. R. Cech, *Cell*, 1982, **31**, 147.
2. C. Guerrier-Takada, K. Gardiner, T. Marsh, N. Pace, and S. Altman, *Cell*, 1983, **35**, 849.
3. F. H. Crick, *J. Mol. Biol.*, 1968, **38**, 367.
4. L. E. Orgel, *J. Mol. Biol.*, 1968, **38**, 381.
5. S. A. Benner, A. D. Ellington, and A. Tauer, *Proc. Natl. Acad. Sci. USA*, 1989, **86**, 7054.

6. S. A. Benner, M. A. Cohen, G. H. Gonnet, D. B. Berkowitz, and K. P. Johnsson, in "The RNA World," eds. R. F. Gesteland and J. F. Atkins, Cold Spring Harbor Laboratory Press, Cold Spring Harbor, NY, 1993, p. 27.
7. T. Tuschl, J. B. Thomson, and F. Eckstein, *Curr. Opin. Struct. Biol.*, 1995, **5**, 296.
8. P. C. Sabetti, P. J. Unrau, and D. P. Bartel, *Chem. Biol.*, 1997, **4**, 767.
9. E. H. Ekland, J. W. Szostak, and D. P. Bartel, *Science*, 1995, **269**, 364.
10. M. Ishizaka, Y. Ohshima, and T. Tani, *Biochem. Biophys. Res. Commun.*, 1995, **214**, 403.
11. R. Green, A. D. Ellington, and J. W. Szostak, *Nature*, 1990, **347**, 406.
12. C. Wilson and J. W. Szostak, *Nature*, 1995, **374**, 777.
13. B. Sargueil and J. M. Burke, *Methods Mol. Biol.*, 1997, **74**, 289.
14. J. R. Prudent, T. Uno, and P. G. Schultz, *Science*, 1994, **264**, 1924.
15. K. N. Morris, T. M. Tarasow, C. M. Julin, S. L. Simons, D. Hilvert, and L. Gold, *Proc. Natl. Acad. Sci. USA*, 1994, **91**, 13 028.
16. J. C. Guatelli, K. M. Whitfield, D. Y. Kwoh, K. J. Barringer, D. D. Richman, and T. R. Gingeras, *Proc. Natl. Acad. Sci. USA*, 1990, **87**, 1874.
17. A. Berzal-Herranz, S. Joseph, and J. M. Burke, *Genes and Dev.*, 1992, **6**, 129.
18. A. Berzal-Herranz, S. Joseph, B. M. Chowrira, S. E. Butcher, and J. M. Burke, *EMBO J.*, 1993, **12**, 2567.
19. S. Joseph, A. Berzal-Herranz, B. M. Chowrira, S. E. Butcher, and J. M. Burke, *Genes and Dev.*, 1993, **7**, 130.
20. K. L. Nakamaye and F. Eckstein, *Biochemistry*, 1994, **33**, 1271.
21. N. K. Vaish, P. A. Heaton, and F. Eckstein, *Biochemistry*, 1997, **36**, 6495.
22. J. B. Thomson, S. T. Sigurdsson, A. Zeuch, and F. Eckstein, *Nucleic Acids Res*, 1996, **24**, 4401.
23. D. M. Long and O. C. Uhlenbeck, *Proc. Natl. Acad. Sci. USA*, 1994, **91**, 6977.
24. J. Tang and R. R. Breaker, *RNA*, 1997, **3**, 914.
25. N. K. Vaish, P. A. Heaton, O. Federova, and F. Eckstein, *Proc. Natl. Acad. Sci. USA*, 1998, **95**, 2158.
26. R. Green and J. W. Szostak, *J. Mol. Biol.*, 1994, **235**, 140.
27. F. Michel, A. D. Ellington, S. Couture, and J. W. Szostak, *Nature*, 1990, **347**, 578.
28. F. Michel and E. Westhof, *J. Mol. Biol.*, 1990, **216**, 585.
29. L. Jaeger, F. Michel, and E. Westhof, *J. Mol. Biol.*, 1994, **236**, 1271.
30. M. Costa and F. Michel, *EMBO J.*, 1995, **14**, 1276.
31. J. H. Cate, A. R. Gooding, E. Podell, K. Zhou, B. L. Golden, A. A. Szewczak, C. E. Kundrot, T. R. Cech, and J. A. Doudna, *Science*, 1996, **273**, 1696.
32. M. Costa and F. Michel, *EMBO J.*, 1997, **16**, 3289.
33. L. B. Weinstein, B. C. Jones, R. Cosstick, and T. R. Cech, *Nature*, 1997, **388**, 805.
34. D. L. Robertson and G. F. Joyce, *Nature*, 1990, **344**, 467.
35. A. A. Beaudry and G. F. Joyce, *Science*, 1992, **257**, 635.
36. J. Tsang and G. F. Joyce, *Biochemistry*, 1994, **33** 5966.
37. J. Tsang and G. F. Joyce, *J. Mol. Biol.*, 1996, **262**, 31.
38. X. Dai, A. De Mesmaeker, and G. F. Joyce, *Science*, 1995, **267**, 237.
39. A. J. Hager, J. D. Pollard, and J. W. Szostak, *Chem. Biol.*, 1996, **3**, 717.
40. J. R. Lorsch and J. W. Szostak, *Nature*, 1994, **371**, 31.
41. J. R. Lorsch and J. W. Szostak, *Biochemistry*, 1995, **34**, 15 315.
42. F. Q. Huang and M. Yarus, *Biochemistry*, 1997, **36**, 6557.
43. F. Q. Huang and M. Yarus, *Proc. Natl. Acad. Sci. USA*, 1997, **94**, 8965.
44. M. Illangasekare, G. Sanchez, T. Nickles, and M. Yarus, *Science*, 1995, **267**, 643.
45. M. Illangasekare, O. Kovalchuke, and M. Yarus, *J. Mol. Biol.*, 1997, **274**, 519.
46. M. Illangasekare and M. Yarus, *J. Mol. Biol.*, 1997, **268** 631.
47. P. A. Lohse and J. W. Szostak, *Nature*, 1996, **381**, 442.
48. B. Zhang and T. R. Cech, *Nature*, 1997, **390**, 96.
49. D. P. Bartel and J. W. Szostak, *Science* 1993, **261** 1411.
50. E. H. Ekland and D. P. Bartel, *Nucleic Acids Res.*, 1995, **23**, 3231.
51. M. C. Wright and G. F. Joyce, *Science*, 1997, **276**, 614.
52. R. R. Breaker and G. F. Joyce, *Proc. Natl. Acad. Sci. USA*, 1994, **91**, 6093.
53. E. H. Ekland and D. P. Bartel, *Nature*, 1996, **382**, 373.
54. K. B. Chapman and J. W. Szostak, *Chem. Biol.*, 1995, **2**, 325.
55. A. J. Hager and J. W. Szostak, *Chem. Biol.*, 1997, **4**, 607.
56. B. Cuenoud and J. W. Szostak, *Nature*, 1995, **375**, 611.
57. M. Zillmann, S. E. Limauro, and J. Goodchild, *RNA*, 1997, **3**, 734.
58. N. Lehman and G. F. Joyce, *Nature*, 1993, **361**, 182.
59. N. Lehman and G. F. Joyce, *Curr. Biol.*, 1993, **3**, 723.
60. D. N. Frank and N. R. Pace, *Proc. Natl. Acad. Sci. USA*, 1997, **94**, 14 355.
61. T. Pan and O. C. Uhlenbeck, *Biochemistry*, 1992, **31**, 3887.
62. T. Pan, B. Dichtl, and O. C. Uhlenbeck, *Biochemistry*, 1994, **33**, 9561.
63. T. Pan and O. C. Uhlenbeck, *Nature*, 1992, **358**, 560.
64. K. P. Williams, S. Ciafre, and G. P. Tocchini-Valentini, *EMBO J.*, 1995, **14**, 4551.
65. P. Burgstaller and M. Famulok, *Angew. Chem., Int. Ed. Engl.*, 1995, **34**, 1189.
66. J. K. Bashkin, *Curr. Biol.*, 1997, **7**, R286.
67. R. R. Breaker, *Nature Biotech.*, 1997, **15**, 427.
68. R. R. Breaker and G. F. Joyce, *Chem. Biol.*, 1994, **1**, 223.
69. C. R. Geyer and D. Sen, *J. Mol. Biol.*, 1998, **275**, 483.
70. R. R. Breaker and G. F. Joyce, *Chem. Biol.*, 1995, **2**, 655.
71. D. Faulhammer and M. Famulok, *Angew. Chem., Int. Ed. Engl.*, 1996, **35**, 2837.
72. D. Faulhammer and M. Famulok, *J. Mol. Biol.*, 1997, **269**, 188.
73. N. Carmi, L. A. Shultz, and R. R. Breaker, *Chem. Biol.*, 1996, **3**, 1039.
74. N. Carmi, S. R. Balkhi, and R. R. Breaker, *Proc. Natl. Acad. Sci. USA*, 1998, **95**, 2233.

75. S. W. Santoro and G. F. Joyce, *Proc. Natl. Acad. Sci. USA*, 1997, **94**, 4262.
76. M. M. Conn, J. R. Prudent, and P. G. Schultz, *J. Am. Chem. Soc.*, 1996, **118**, 7012.
77. Y. Li and D. Sen, *Nature Struct. Biol.*, 1996, **3**, 743.
78. Y. Li and D. Sen, *Biochemistry*, 1997, **36**, 5589.
79. Y. Li and D. Sen, *Chem. Biol.*, 1998, **5**, 1.
80. M. Wecker, D. Smith, and L. Gold, *RNA*, 1996, **2**, 982.
81. T. W. Wiegand, R. C. Janssen, and B. E. Eaton, *Chem. Biol.*, 1997, **4**, 675.
82. T. M. Tarasow, S. L. Tarasow, and B. E. Eaton, *Nature*, 1997, **389**, 54.
83. D. N. Frank, A. D. Ellington, and N. R. Pace, *RNA*, 1996, **2**, 1179.
84. T. Tuschl, P. A. Sharp, and D. B. Bartel, *EMBO J.*, 1998, **17**, 2637.
85. M. Yarus, *FASEB J.*, 1993, **7**, 31.
86. J. B. Murray, A. A. Seyhan, N. G. Walter, J. M. Burke, and W. G. Scott, *Chem. Biol.*, 1998, 587.
87. A. Roth and R. R. Breaker, *Proc. Natl. Acad. Sci. USA*, 1998, **95**, 6027.
88. J. R. Lorsch, D. P. Bartel, and J. W. Szostak, *Nucleic Acids Res*, 1995, **23**, 2811.
89. A. Jenne and A. Famulok, *Chem. Biol.*, 1998, **5**, 23.
90. S. Baskerville, D. Frank, and A. D. Ellington, in "RNA Structure and Function", Cold Spring Harbor Laboratory Press, Cold Spring Harbor, NY, 1998, p. 203.
91. V. K. Jayasena and L. Gold, *Proc. Natl. Acad. Sci., USA*, 1997, **94**, 10 612.
92. M. P. Robertson and A. D. Ellington, *Curr. Biol.*, 1997, **7**, R376.
93. M. P. Robertson and A. D. Ellington, *Nature Biotechnol.*, 1999, **17**, 62.

6.10
Ribozyme Enzymology

JULIANE K. STRAUSS-SOUKUP and SCOTT A. STROBEL
Yale University, New Haven, CT, USA

6.10.1 INTRODUCTION

The identification of RNA molecules that serve as catalytic machines was a monumental discovery. In the early 1980s, it was demonstrated that an intron within the pre-rRNA of *Tetrahymena thermophila* had the ability to splice itself out of the transcript, independent of proteins.[1] Soon afterward, the RNA subunit of ribonuclease P (RNase P) was identified as the catalytic portion of this ribonucleoprotein complex.[2] These discoveries led to the theory that biology may have evolved from an "RNA world" where RNA could take on the dual roles of information storage and biocatalyst.

This chapter presents an overview of the reaction mechanisms used by RNA enzymes. Most biocatalysts are proteins and they perform this role efficiently due to the chemistry available from their 20 amino acid side chains. RNA lacks this functional group diversity, and therefore it was originally not thought to be a likely candidate for a catalytic molecule. The authors discuss the ways in which RNA may achieve efficient catalysis despite the lack of diversity in its nucleoside residues. The next sections of the chapter describe the secondary and tertiary structural elements of RNA enzymes, as well as the role metal ions play in RNA catalysis. Finally, a thermodynamic and kinetic profile of two well-characterized RNA enzymes, the *Tetrahymena* group I intron and the hammerhead ribozyme, are reviewed.

6.10.2   TYPES OF RNA CATALYSIS AND REACTION MECHANISMS

6.10.2.1   Group I/Group II Nucleophilic Attack

RNA enzymes use different reaction mechanisms for catalysis (Figure 1). Group I and group II splicing involves a two-step transesterification reaction, where an exchange of phosphate diesters occurs with no net change in the number of ester linkages.[3] Group I introns use an external nucleophile and group II introns use an internal nucleophile in the first step of splicing (Figure 1(a), (b)).[3] In the first step of group I splicing, the 3'-OH of an exogenous nucleophile attacks the 5' splice site. The comparable step in group II splicing is carried out by the 2'-OH of an internal bulged A nucleotide. Both reactions generate a 3'-OH group at the end of the 5' exon, which acts as a nucleophile for the second transesterification reaction, resulting in exon ligation. The group I and group II introns are single-turnover enzymes, and therefore they do not act as true catalysts *in vivo*.[3] The *Tetrahymena* group I intron has a rate constant for the chemical step of ~ 350 min^{-1}, which corresponds to a rate enhancement of $\sim 10^{11}$-fold relative to the solution reaction.[4] The self-splicing ai5γ group II intron has a slower chemical step (~ 0.03 min^{-1}), and it provides $\sim 10^{7}$-fold rate enhancement over solution hydrolysis.[5]

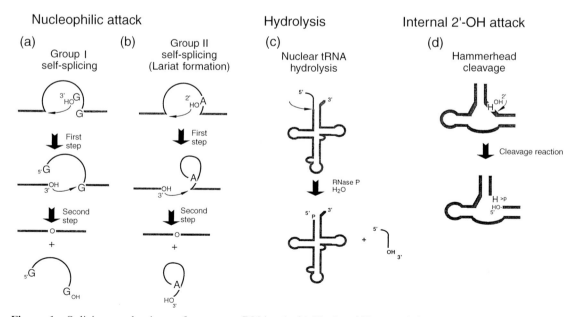

Figure 1   Splicing mechanisms of precursor RNAs. (a–b) Nucleophilic attack by an external (a) or internal (b) nucleotide. Thin lines indicate introns. Thick lines indicate flanking exons. (c) Hydrolysis. Thick line indicates mature tRNA. Thin line denotes pre-tRNA that is removed. (d) Internal 2'-OH attack. Thick line indicates ribozyme (after Cech[3]).

6.10.2.2 Hydrolysis by External H₂O or OH⁻

The second major ribozyme reaction mechanism involves hydrolysis by an external H_2O or OH^- (Figure 1(c)). RNase P and in some cases group I and group II introns use H_2O or OH- as a nucleophile. RNase P catalyzes the removal of a block of nucleotides at the 5′ end of a pre-tRNA molecule to produce the mature 5′ terminus (reviewed in reference 3). RNase P cleaves to generate a 5′-phosphate and 3′-OH termini.[3] RNase P is the only known catalytic RNA that can act as a true enzyme, with multiple turnover kinetics *in vivo*.[2,6-8] RNase P is made up of both protein and RNA components.[9] The RNA component has been identified as the catalytic subunit, but the protein is needed for activity *in vivo*.[3] RNase P processes the precursors to all the cellular tRNAs (~20 molecules), which indicates that it can recognize related structural elements. The RNase P RNA has a rate constant of ~ 180 min^{-1}, which corresponds to a rate enhancement of $\sim 10^{11}$-fold relative to the solution reaction.[10]

6.10.2.3 Internal 2′-OH Attack

The third class of reaction mechanisms involves attack by an internal 2′-OH (Figure 1(d)). The small ribozymes which include the hammerhead, hairpin, hepatitis delta virus, and *Neurospora* Varkud satellite (VS) RNAs, all use an internal 2′-OH adjacent to the cleavage site as the nucleophile (reviewed in reference 11). The RNA enzymes have different secondary and tertiary structures, but they react by the same mechanism. They cleave to generate 2′,3′-cyclic phosphates and 5′-OH termini. Small ribozymes are responsible for processing plant pathogenic RNA or human virus RNA during rolling circle replication in order to generate monomer length RNAs.[12] The hammerhead ribozyme has a rate constant for the chemical step of ~ 1 min^{-1}, which corresponds to a rate enhancement of $\sim 10^8$-fold relative to the solution reaction.[13]

6.10.3 SECONDARY AND TERTIARY STRUCTURES OF RNA ENZYMES

The global structure of a ribozyme plays a major role in its ability to perform catalysis. RNA enzymes fold into compact structures that involve secondary and tertiary structural contacts. Two-dimensional models of the group I, group II, RNase P, hammerhead, and hairpin ribozymes are shown in Figure 2. These models have been determined by comparative sequence analysis. The formation of an active site for the chemical reaction is dependent on tertiary interactions that can form the catalytic core. Helix packing allows regions of the RNA enzymes to fit closely together in order to correctly position those groups involved in catalysis.

6.10.3.1 Secondary Structure

6.10.3.1.1 Group I introns

In the case of group I introns, the tertiary structure of the RNA is important for binding of the substrate and nucleotide cofactor, and for the cleavage reaction. A secondary structure of the *Tetrahymena* intron was proposed by Michel and Davies (Figure 2(a)).[14-16] The intron forms nine paired regions, P1–P9, with joining regions between the helices.[14,16] The ribozyme has two main components, the P4–P6 domain and the P3–P7–P8 domain, which requires P4–P6 for folding.[17,18] These two domains form the active site of the enzyme, into which the P1 helix docks. Paired region P1 includes the 5′ splice site and is formed by base pairing of the 3′ end of the 5′ exon with the internal guide sequence (IGS), contained within the intron.[19] The interactions between domains P4–P6 and P3–P7–P8 and the P1 helix create the correct catalytic core for RNA splicing.[20-22] Group I introns also contain a guanosine binding site in the major groove of the P7 helix,[23] a terminal guanosine that defines the 3′-splice site,[23-25] and joiner regions J4/5 and J8/7 that are important for alignment of the P1 helix into the active site for nucleophilic attack by guanosine.[26-30] Tertiary hydrogen bonding, stacking interactions, and metal ions allow for close packing of the RNA into the active site of this highly compact ribozyme. The P2, P5, and P9 helices are considered peripheral domains located outside of the ribozyme core.

The global structure of the *Tetrahymena* intron was probed using Fe^{II}-EDTA to distinguish the

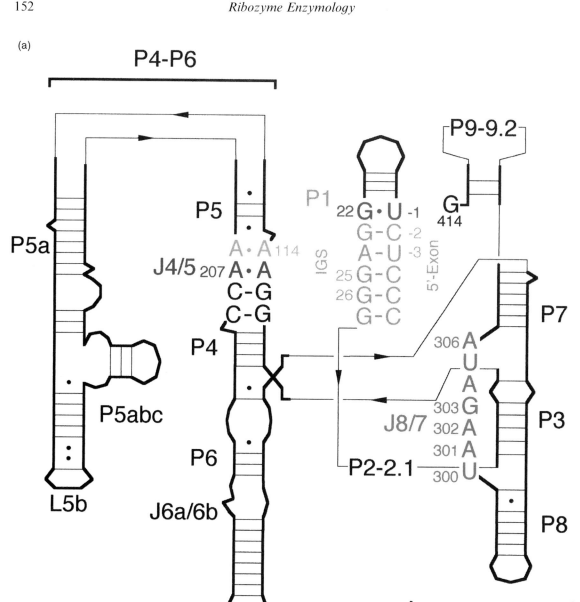

Figure 2 Secondary structure of (a) *Tetrahymena* group I intron, P1 helix (gray), G22·(−1) base pair (green), J8/7 strand (red), sheared A·A pairs in J4/5 (blue and yellow); (b) group II intron; (c) RNase P RNA; (d) hammerhead ribozyme, H = A, U or C nucleotides; and (e) hairpin ribozyme.

inside from the outside of the RNA.[20,31] In the absence of Mg^{2+}, cleavage occurred throughout the intron, indicating that the RNA backbone was unfolded and accessible.[31] After adding Mg^{2+}, areas of the RNA backbone were protected, which indicated formation of tertiary structure. These areas included regions within P4, P6, P3, P7, and P8 that are phylogenetically conserved.

6.10.3.1.2 *Group II introns*

Group II introns are organized into a set of six helical domains that emanate from a central wheel[14] (Figure 2(b)). The secondary structure was formed using phylogenetic data and mutational and chemical modification studies.[32] Domains I and V are essential for catalytic function and the

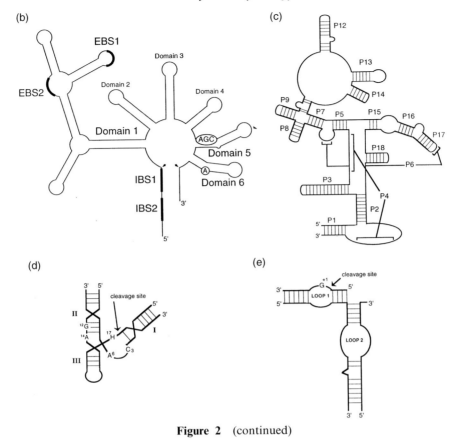

Figure 2 (continued)

branched adenosine within domain VI contains the nucleophile for the transesterification reaction.[33] Domain I contains two sequence segments, exon-binding sites 1 and 2 (EBS1 and EBS2), that form tertiary contacts to the intron-binding sites 1 and 2 (IBS1 and IBS2), which are found in the 5′ exon.[34,35] The two EBS–IBS interactions are essential for 5′ splice site selection, lariat formation, and stability. Domain V is the "catalytic core" of the intron[36,37] and includes the chemically essential and phylogenetically conserved AGC triad.[38] Domains II, III, and IV are peripheral domains that enhance catalysis and contribute to the overall structure. High salt concentrations are needed *in vitro* for catalytic activity, therefore it is thought that a protein cofactor is utilized *in vivo*.[33]

6.10.3.1.3 *RNase P*

The holoenzyme of RNase P from *E. coli* is made up of a 377 nucleotide RNA (M1 RNA) and a 13.8 kDa protein (the *C*5 protein).[9] C5 is important *in vivo* for maximal efficiency of cleavage at physiological salt concentrations.[9] The secondary structure of the RNA was determined by comparative sequence analysis with other eubacterial M1-like RNAs (Figure 2(c)).[39,40] Some of the helices have been identified by phylogenetic comparisons. The RNA has 18 paired regions. Two of these helices, P4 and P6, are formed between nonadjacent single-stranded regions.[41] Helix P4 and the immediately adjacent nucleotides are the most highly conserved structural elements in the RNA.[42] This region is thought to be the catalytic center of the RNA enzyme. Tetraloop–receptor interactions and pseudoknots are two of the tertiary interactions identified in the global structure of RNase P RNA.[43] A global structure model was proposed that included three multihelix domains surrounding the universally conserved P4 helix.[41]

6.10.3.1.4 *Hammerhead*

The hammerhead ribozyme is one of the smallest catalytic RNAs, comprising just 30 nucleotides. The secondary structure (Figure 2(d)) of the hammerhead involves three Watson–Crick base-paired helices (numbered I–III).[13,44,45] The helical regions are joined by conserved single-stranded segments. Mutagenesis experiments have demonstrated that two of these nonhelical segments are required for catalysis (the single-stranded regions, C_3–A_9 and G_{12}–A_{14}).[44,46] In addition, the nucleotide at the cleavage site must be unpaired.

6.10.3.1.5 *Hairpin*

The hairpin ribozyme is composed of two pairs of helix–loop–helix segments (Figure 2(e)).[47] The smaller symmetrical loop contains the reactive phosphodiester.[48] The secondary structure has been determined by data from mutagenesis, phylogenetic comparisons, *in vitro* selections, cross-linking, and chemical protection experiments.[47,49–51] The ribozyme interacts with its substrate through two short helices on either side of loop 1.[51] Active molecules all contain G at the +1 position.[51] The working model for catalysis by this ribozyme is that the two looped regions assemble to form an active site. Although no tertiary interactions have been identified between the two loops, several have been proposed.[52,53]

6.10.3.1.6 *Hepatitis delta virus*

The smallest hepatitis delta virus (HDV) ribozyme found to have catalytic activity *in vitro* is ∼85 nucleotides long.[54] The secondary structure involves four paired regions, P1–P4. Tertiary interactions between P1 and P2 form a type of pseudoknot.[25] Support for this model comes from mutagenesis studies and the secondary structure generated from active *trans*-acting ribozymes with novel sequences.[55] The cleavage site resides in P1, and therefore, this helix is involved in cleavage site selection.[55,56] P1 and P3 have length requirements, whereas the lengths of P2 and P4 can be varied since they only play structural roles.[55] The crystal structure of the HDV ribozyme has verified many of the proposed secondary interactions and identified an additional short helix within the structure. This results in a double-nested pseudoknot in the L3 loop with the joining region J4/2.[57]

6.10.3.1.7 Neurospora *VS RNA*

Neurospora contain VS RNA, which is an 881 nucleotide single-stranded circular molecule.[58] A secondary structure was proposed for the self-cleaving region of the VS RNA, nucleotides 617–881, based on site-directed mutagenesis and chemical modification structure probing.[59] This catalytic RNA segment contains six paired regions with intervening single-stranded loops and bulges. In helices II–VI, site-directed base substitutions decreased or abolished activity, while compensatory substitutions restored activity.[59] The site of cleavage is in the single-stranded region of helix I. Mutations at certain positions in helix I inactivated the ribozyme and compensatory mutations did not restore activity. This suggested a conformational change during formation of the active structure that involved these nucleotides.[59] Helix VI contains a GNRA tetraloop, which could be involved in a long-range tertiary interaction. Damage selection and modification experiments have also identified nucleotides within the junction regions that are important for structure and function.[59,60]

6.10.3.2 Tertiary Structure

6.10.3.2.1 **Tetrahymena** *group I intron*

The three-dimensional structure of a ribozyme is essential for its ability to act catalytically. Requirements such as substrate and cofactor binding, positioning of the scissile phosphate, and the actual catalytic reaction, are not possible without an intricate tertiary structure. While three-dimensional models have been proposed for many of the RNA enzymes, crystal structures of only two ribozymes and one ribozyme domain are available. The three-dimensional structures of both the hammerhead and HDV ribozymes and the P4–P6 domain of the *Tetrahymena* group I intron have been determined by X-ray crystallography.

A crystal structure of the P4–P6 domain of the *Tetrahymena* intron has provided detailed information about the tertiary structure of a tightly packed RNA fold.[61] The P4–P6 domain is made up of two arrays of stacked helices that are packed against one another at two distant regions of the structure.[61] These helices are held together by a GNRA tetraloop in one helix that interacts through hydrogen bonds and base stacking with an 11-nt tetraloop receptor in another helix (Figure 2(a)).[61] At three points within the P4–P6 domain, adenosine platforms were seen. In the platform, the adenosines are side by side and form a pseudo base pair.[61,62] Platforms allow the minor groove to widen and become more accessible for base pairing and stacking.

An A-rich bulge within P5a bridges two parallel helical stacks (Figure 2(a)).[61] The backbone of the A-rich bulge makes a corkscrew turn. Its bases are flipped out and they interact with residues of the P4 helix on one side and the three-helix junction within P5abc on the other face.[61] Formation of ribose zippers and metal ion binding allows close interactions between RNA helices.[61] Pairs of riboses interact through hydrogen bonds to form ribose zippers in the A-rich bulge and in the GAAA tetraloop. The close packing of phosphates in nearby helices are also mediated by hydrated magnesium ions.[63,64]

The J4/5 region was proposed to interact with the P1 substrate helix within the catalytic core, and the G·U wobble pair (positions 22 and −1, respectively, within P1) has been shown to cross-link to adenosines within this region (Figures 2(a) and 3).[22,28–30] In the crystal, adjacent adenosines in J4/5 forms sheared homopurine base pairs.[15,21,22,61] This configuration forces the functional groups of the adenosines into the minor groove, where they are accessible to interact with the P1 helix.

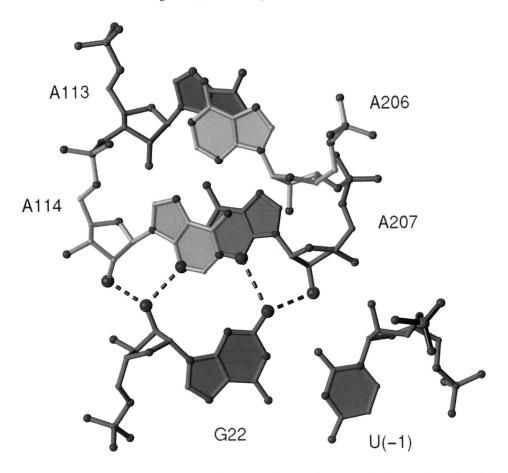

Figure 3 P1 helix docking into the active site of the *Tetrahymena* ribozyme. Tertiary hydrogen bonds between G22 and the sheared A·A pairs are shown; G22·(−1) base pair (green), sheared A·A pairs in J4/5 (blue and yellow).

The structure of the P4–P6 domain served as a starting point for the modeling of other elements within the catalytic core based upon biochemical tertiary constraints. Nucleotide analogue interference mapping (NAIM) has been used to identify important functional groups within the *Tetrahymena* intron.[65] These studies identified an RNA helix packing motif between the G·U wobble pair at the cleavage site in P1 and the two sheared A–A pairs in J4/5, termed the wobble pair receptor (Figure 3).[28] This interaction is important for 5′ splice site selection.

Using the NAIM results, nucleotide analogue interference suppression (NAIS) experiments were designed. NAIS aids in the identification of specific hydrogen bonding partners within an RNA tertiary structure.[28,65] Using this approach, if a tertiary interaction is disrupted by deletion or alteration of one functional group, then no energetic penalty will result from deletion or alteration of the second functional group.[28] This approach confirmed the hydrogen bonding partners between the G·U wobble pair and the sheared A·A pairs in J4/5.[28] Additional experiments using NAIS demonstrated that the P1 helix and the J8/7 single-stranded region form an extended minor groove triple helix within the catalytic core of the ribozyme.[66] These data revealed the importance of the J8/7 strand in organizing several helices within the active site (Figure 4).

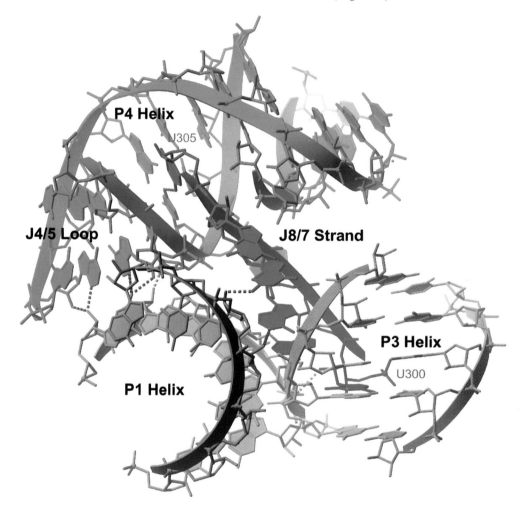

Figure 4 Model of the *Tetrahymena* intron catalytic core, composed of helices P1 (gray), P3 (pink), P4 (blue), and joining regions J4/5 (blue) and J8/7 (green). Tertiary hydrogen bonds are indicated by dotted lines.

6.10.3.2.2   *Hammerhead*

Three crystal structures have been solved for the hammerhead ribozyme.[67–70] In one structure, a DNA strand containing the unpaired nucleotide at the cleavage site was base-paired to an RNA strand to form helices I and III.[67] The DNA strand acted as an inhibitor since it lacked the 2′ OH at the cleavage site. The first all-RNA crystal structure contained an RNA enzyme bound to a substrate that had a 2′-methoxy group at the cleavage site.[69,70] The second all-RNA structure used a substrate with a *tallo*-5′-C-methyl ribose modification at the cleavage site that created a kinetic bottleneck at the final step of the cleavage reaction.[68] This system resulted in a crystal structure of the conformation during self-cleavage. In the crystals, the four conserved nucleotides, 5′-$C_3U_4G_5A_6$-3′, form a sharp turn that is almost identical to the conformation of the uridine turn in tRNA

(Figure 2(d)).[71,72] The rest of the catalytic core forms a non-Watson–Crick duplex with a divalent ion binding site. In the first two structures solved, the cleavage site phosphate was in a standard helical conformation. This does not allow for in-line attack of the 2′-OH, which is the mechanism used by the ribozyme.[67,69,70] The *tallo*-5′-C-methyl ribose substrate resulted in a conformation in the crystal structure where the intermediate is correctly oriented for in-line attack.[68]

6.10.4 THE ROLE OF METALS IN RNA CATALYSIS

Most ribozymes are considered metalloenzymes because divalent metal ions are important for activity. Cationic metals shield the negative charge of the RNA phosphate backbone, which allows them to fold into complex structures.[73,74] In addition, metal ions can bind specific sites within an RNA and play a direct role in catalysis. Therefore, metal ions play a key role in forming the intricate shape of an active ribozyme.

Many ribozymes use divalent ions, usually Mg^{2+}, for RNA folding and catalysis. Mg^{2+} binds six ligands, possesses octahedral geometry, and has a high preference for binding to oxygen relative to Mn^{2+}, another divalent ion.[75] RNA molecules have a number of sites that can interact with metal ions in solution.[76] These sites include oxygen atoms on the ribose sugar, including the 2′-OH and 5′- and 3′-oxygens. Metal sites also include phosphate oxygens and base carbonyls.

6.10.4.1 Metal Ions and RNA Folding

6.10.4.1.1 *Site-specific metal ion binding*

The identification of metal binding sites within large RNAs is a challenging biochemical problem, though the development of two techniques, metal rescue and metal ion cleavage, has assisted in this endeavor. Interference assays are performed in the presence of different metal ions to determine the sites of ion coordination. RNA molecules are transcribed with a small percentage of ribonucleotide analogues that contain phosphorothioate diester linkages and the active RNAs are selected from the overall population by their ability to splice in the presence of Mg^{2+}.[77] The phosphorothioate linkages are cleaved with iodine and the cleavage products are resolved by electrophoresis.[78] The active fractions will exhibit radioactive bands at all the positions where a phosphorothioate linkage was unimportant for activity. Phosphorothioate interference corresponds to sites of RNA–metal or RNA–RNA interactions.[64,77,79,80] The role of phosphates can be distinguished by repeating the experiment in the presence of Mn^{2+}, which binds sulfur with an affinity similar to oxygen.[81] Phosphate oxygens important for catalysis that coordinate metals are tolerant of the phosphorothioate linkage when Mn^{2+} is present. Those positions that can be "rescued" by Mn^{2+} suggest that phosphates are directly coordinated to a metal ion. This type of experiment was performed on the *Tetrahymena* group I intron and several sites within the catalytic core were identified as potential metal binding sites.[77] A similar approach has been used to map metal sites within other RNAs.[42,77]

Metal ion cleavage experiment can also identify sites where metal ions may be present. At high pH, Mg^{2+} can cleave RNA because the main species of Mg^{2+} is $Mg(OH)_2$.[82] At neutral pH, $Pb(OH)_2$ can perform a similar reaction.[83,84] The metal hydroxides act as bases to remove protons from the 2′-OHs, which causes the remaining oxygen to nucleophylically attack the phosphodiester linkage and produce a 2′,3′-cyclic phosphate and 5′-OH termini. Although this technique cannot identify specific sites of metal ion binding, it can define a vicinity within which metal ions are likely to be found. For example, the site of Pb^{2+} cleavage in tRNA was similar to the site of metal binding in the crystal.[83,85,86] Metal ion cleavage performed on the *Tetrahymena* intron identified a group of potential metal sites within the catalytic core.[87] In addition, the potential metal binding sites of RNase P were determined by this method.[82,88]

Metal binding sites within RNA have also been mapped using Fenton chemistry since Fe^{2+} can occupy Mg^{2+} ion binding sites.[89] Incubation of RNAs with Fe^{2+}, ascorbate, and hydrogen peroxide yields hydroxyl radicals that cleave the RNA near sites of metal binding. In contrast to Fe^{2+}/EDTA footprinting, which maps the solvent-accessible surface of an RNA, this technique identifies metal sites on the inside of an RNA.[89]

In addition to nonspecific binding, metal ions can also bind to specific sites within an RNA molecule and aid in the formation of tertiary contacts. Metal ions can inner-sphere coordinate to specific functional groups or are fully hydrated and bind the RNA by outer-sphere coordination. It was first shown that metals may bind specific sites when crystals of the tRNA[Phe] revealed sites of Mg^{2+} binding.[90–92] The crystal structure of the P4–P6 domain from *Tetrahymena* includes three types of metal binding sites.[61,63,64,93] First, hydrated magnesium ions bind to the major groove of $G \cdot U$ pairs by outer-sphere coordination. Second, Mg^{2+} ions bind to RNA through direct inner-sphere coordination. In these examples, Mg^{2+} coordinates to specific functional groups of the bases, such as phosphate oxygens and the O-6 carbonyls. These interactions help form a metal ion core that may drive folding.[63,64] Finally, three sites of monovalent ion binding have been identified by X-ray crystallography and NAIM.[93] These ions make inner-sphere contacts to functional groups below the A–A platforms.

In the hammerhead structure a metal ion is bound to a *pro*-R_p phosphoryl oxygen and to a ring nitrogen.[67–70] When a phosphorothioate was incorporated at the *pro*-R_p oxygen there was a decrease in catalytic activity in the presence of Mg^{2+}.[44,94] This result can be explained if the Mg^{2+} plays a role in structure or catalysis.

6.10.4.2 Role of Metal Ions in Catalysis by *Tetrahymena* Group I Intron

Although metal ions are required for RNA folding, they also play an essential role in the actual catalytic reactions of most ribozymes. A two-metal ion mechanism for catalytic RNAs was modeled after the catalytic mechanism of protein enzymes, such as *E. coli* DNA polymerase, that employ metals for activity.[95] Steitz and Steitz hypothesized that there are specific binding sites for two metals at ribozyme active sites, and that these metals may play a major role in stabilizing the transition state of the reaction. Large ribozymes catalyze attack on the scissile phosphate by either an exogenous guanosine (group I), water (RNase P and sometimes group II), or an internal adenosine (group II). In the first step of splicing of group I and II introns, the R-OH group is activated for nucleophilic attack by a Mg^{2+} ion (metal A). A second Mg^{2+} ion (metal B) is proposed to act as a Lewis acid to stabilize the leaving oxyanion and the transition state intermediate. During the reaction, the phosphorus at the active site goes through a trigonal bipyrimidal coordination state. The reactions result in a substantial negative charge on the leaving group, which is neutralized by direct metal ion coordination. In the second step of group I splicing, the Mg^{2+} ions reverse roles. Mg^{2+} ion (A) facilitates the 3'-oxyanion leaving group of the terminal G, and Mg^{2+} ion (B) activates the attacking hydroxyl, the 3'-OH of the 5'-exon. Unlike the small ribozymes, large ribozymes do not use an adjacent 2'-OH for catalysis. Therefore, the size of the large ribozymes may be required for the formation of a complex structure that can bind nucleophiles and the essential metal ions.

The *Tetrahymena* ribozyme undergoes splicing in the presence of Mg^{2+}.[96–98] The L-21 ScaI ribozyme, was constructed that can perform a reaction analogous to the first step of splicing and act as a multiple turnover enzyme (see below).[4,99,100] A 3'-phosphorothiolate linkage was placed at the scissile phosphate in order to determine the role of metal coordination to the 3' oxygen leaving group.[101] Cleavage of this substrate in the presence of Mg^{2+} was undetectable. Catalytic activity was partially restored when the reaction was carried out in Mn^{2+}, which is a thiophilic metal. This result identified a metal ion binding site in the catalytic core, and suggested that this metal ion directly coordinates to the 3' leaving oxyanion in the transition state (Figure 5(a), metal #1).

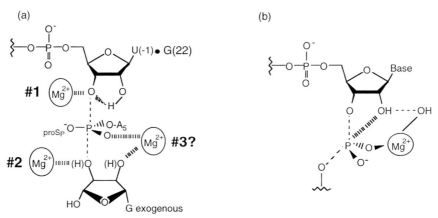

Figure 5 Proposed catalytic metal ions in the *Tetrahymena* group I intron (a) and the hammerhead ribozyme (b).

A second metal ion was shown to be involved in catalysis by the *Tetrahymena* intron using a ribozyme that could perform the reverse of the first step of splicing with the substrate IspU, 3'-(thioinosylyl)-(3'-5')-uridine.[102] This substrate, with a 3'-bridging sulfur substitution, was active only in the presence of thiophilic metals (Cd^{2+} or Mn^{2+}) due to the special linkage. This result indicated that again stabilization of the leaving group was important. The principle of microscopic reversibility argued that in the forward reaction this metal site must be activating the 3'-OH of the nucleophile (Figure 5(a), metal #2). This supported the idea that metal ions at two different sites were involved in catalysis.

A third possible metal ion was postulated to interact with the 2'-OH of the exogenous guanosine (Figure 5(a), metal #3).[103] This metal ion may play a role in catalysis, since it has an effect on the rate of transesterification. However, there may be some overlap in that one metal ion may be able to coordinate to both the 2'- and 3'-OHs of the exogenous guanosine.

6.10.4.3 Role of Metal Ions in Catalysis by Hammerhead Ribozyme

The small ribozymes catalyze intramolecular attack of an adjacent 2'-OH group on the phosphodiester to produce a 2',3'-cyclic phosphates and 5'-OH termini. Metal ions have been found to be important for the catalytic reactions of most members of this class. A metal-coordinated hydroxyl group may act as a general base to deprotonate the 2'-OH (Figure 5(b)).[83] Metal ions may also be directly coordinated to the phosphoryl oxygen (Figure 5(b)).[104] This could stabilize the increased negative charge of the oxyanion in the transition state or it may cause the phosphorus to be more susceptible to attack. The metal ion may also help stabilize the developing negative charge on the leaving group.

Metal ions are important for both structure and catalysis in the hammerhead ribozyme. NMR and nuclease protection experiments have determined that divalent ions are not required for folding, and that monovalent or higher valent polycations can compensate for the absence of Mg^{2+}.[105,106] Divalent ions are, however, absolutely required for catalytic activity. Experiments were performed to measure the pH-dependent catalytic cleavage rate as a function of metal ion identity.[107] For all the metals tested, the rate increased with pH, which indicated that general base catalysis was occurring. Therefore, it was proposed that a metal hydroxide in the catalytic core deprotonates the 2'-OH at the cleavage site, promoting nucleophilic attack on the scissile phosphate (Figure 5(b)). Metal rescue experiments indicated that the phosphoryl oxygen at the cleavage site was directly coordinated to a metal ion in the transition state of the reaction.[108] This indicated two distinct roles for metal ions in hammerhead catalysis.[109]

In the crystal structures of the hammerhead, sites of metal binding have been identified.[67–70] All three structures show divalent ion binding to sites important for structure. Two of the structures[68–70] also show divalent metal ions bound to sites important for catalysis. In one structure, all of the Mn^{2+} sites identified reappeared as Mg^{2+} sites.[70] In addition, another Mg^{2+} bound directly to the *pro*-R phosphate oxygen adjacent to the scissile bond. This additional Mg^{2+} sites was only found in the conformational intermediate structure, indicating its relevance to the cleavage mechanism. This structure is adopted before the conformation amenable to in-line attack forms. The mechanism of cleavage was then thought to involve either: (i) two Mg^{2+} ions that initiate cleavage, one bound to the *pro*-R phosphate oxygen and the other involved in a metal hydroxide that attacks the cleavage site 2'-OH; or (ii) a single Mg^{2+} ion that binds to the *pro*-R phosphate oxygen and also provides the hydroxide that initiates cleavage.

Work by Lott *et al.* favors the two-metal ion mechanism.[109] Using a hammerhead ribozyme construct, the rate of cleavage of an RNA substrate was studied in the presence of a constant amount of Mg^{2+}, with La^{3+} as a competing metal ion. k_{obs} was measured as a function of La^{3+} concentration. Bound Mg^{2+} was displaced by La^{3+} as its concentration was increased. Because a kinetic parameter was monitored, the resulting activity titration as a function of La^{3+} ion concentration defined the relative amounts of the metal ions in the transition state of the rate-limiting step. The two metals had very different effects at the two binding sites. It was postulated that the metal ion at the first binding site coordinates the attacking 2'-oxygen and lowers the pK_a of the attached proton. The metal ion at the second binding site helps catalyze the reaction by absorbing the negative charge that accumulates on the leaving group 5'-oxygen in the transition state.

6.10.4.4 Metal Ions Play a Passive Role in the Hairpin Ribozyme

Somewhat surprisingly, data has shown that metal ions play a passive role in the cleavage reaction of the hairpin ribozyme.[110,111] These data are the exception to the rule that metal ions are required for catalysis. In this instance it is thought that metal ions are merely present for structural purposes. It has been shown that cobalt hexammine can substitute for Mg^{2+} in cleavage reactions of the hairpin ribozyme without a change in catalytic activity.[110,111] This result indicated that inner-sphere coordination was not essential to the hairpin ribozyme. Substrates were used that had an equal mixture of R_p- and S_p-phosphorothioate linkages at the scissile phosphate.[110] In the presence of Mg^{2+}, these substrates had activity similar to that of the wild-type substrate, which indicated that Mg^{2+} is not directly bound to a phosphorus center and that it is not absolutely required for catalysis.[110] In another set of experiments, two substrates were synthesized, one had an R_p-phosphorothioate at the scissile phosphate and a second had an S_p-phosphorothioate, and no thio effects were detected.[111] Furthermore, addition of monovalent cations enhanced activity in reactions with cobalt hexammine, but inhibited activity in reactions with magnesium.[110] Since cobalt hexammine is a poor hydrogen bond donor, the addition of monovalent ions may allow specific sites to be occupied by hydrated cations required for catalysis. Activity may be inhibited with magnesium and monovalent ions present because of competition between the two for the Mg^{2+} binding sites.

This section has concentrated on describing ribozymes as metalloenzymes. Several experiments have demonstrated that metal ions, especially divalent ions, are required for RNA molecules to fold into their catalytically active conformations. Metal ions shield the negative charge on RNA and bind to specific sites within the molecule. In addition to their structural role, metal ions also contribute to the catalytic reactions of ribozymes. Because RNA molecules lack functionalities with a pK_a near neutrality, metal ions may help in stabilizing the developing negative charge on the leaving oxyanions and the attacking hydroxyls.

6.10.5   KINETICS OF THE *TETRAHYMENA* AND HAMMERHEAD RIBOZYME REACTIONS

6.10.5.1   *Tetrahymena* Group I Intron

It was first reported in 1981 that RNA molecules were capable of self-splicing; in other words that RNA could act as an enzyme.[1] The first such catalytic RNA was the group I intron of *Tetrahymena* rRNA. Since then a vast amount of data on the kinetics of the self-splicing reactions have been collected on this intron. This includes kinetic analysis of both steps of splicing and the development of intron constructs capable of multiple turnover catalysis.

6.10.5.1.1   *Reaction components*

Since the discovery of catalytic RNA, numerous experiments have been performed to decipher the necessary components and chemical mechanism of catalysis. It was initially determined that divalent cations and guanosine were essential for splicing.[1,112,113] ATP, CTP, and UTP could not substitute for GTP, but guanosine, 5′-GMP, and 5′-GDP could be utilized.[1] This indicated that GTP did not provide an energy source since the number of phosphates was unimportant. The apparent lack of an energy requirement for splicing suggested a phosphoester transfer mechanism in which each cleavage step was coupled to a ligation step.[1] It was also determined that guanosine analogues lacking a 2′- or 3′-OH could not initiate splicing and that guanosine was added to the 5′ end of the intervening sequence (IVS) during its excision.[1]

6.10.5.1.2   *Self-splicing mechanism*

The splicing reaction proceeds by inversion of configuration at the active site phosphorus, using an S_N2 reaction or "in-line attack" mechanism.[114,115] This type of pathway is used by protein phosphotransferases and phosphatases.[116,117] S_N2 reactions proceed through a trigonal bipyramidal configuration in the transition state.[116] The stereochemical course of RNA catalysis was determined using a stereo specific phosphorothioate substitution at the cleavage site and snake venom phosphodiesterase (which has a 1700-fold greater activity on R_p isomers).[114,115] It was determined that the ribozyme reaction proceeded with inversion of configuration at the phosphorus. In reactions with substrates that contained a phosphorothioate at the R_p oxygen, the rate of the reaction was decreased 1000-fold relative to unsubstituted substrates. The presence of sulfur may have interfered with coordination of Mg^{2+} to the phosphate oxygen in the chemical transition state.

The phosphorothioate substrates were also used to identify the rate-limiting step of splicing. Sulfur was substituted for the *pro*-R_p (nonbridging) phosphoryl oxygen at the scissile phosphate.[118] Although there was no effect on the equilibrium binding constant of the substrate, there was a 2.3-fold decrease for the reaction with guanosine and a seven-fold decrease for hydrolysis.[118] These "thio-effects" indicated that chemistry was the rate-limiting step because of the inherent ability of phosphorothioate linkages to be cleaved. No rescue of activity was seen in the presence of Mn^{2+}, indicating that the *pro*-R_p oxygen is not directly coordinated to Mg^{2+}. Substitution of the *pro*-S_p oxygen with sulfur gave a thio effect of $\sim 10^3$-fold, indicating that there probably exists a direct contact between the *pro*-S_p oxygen of the substrate and a metal ion or a functional group in the ribozyme active site.

6.10.5.1.3 Kinetics of **Tetrahymena** *splicing—first step of splicing*

After self-splicing of the *Tetrahymena* rRNA precursor, the intervening sequence undergoes cyclization and site-specific hydrolysis to produce an intron lacking the first 19 nucleotides, termed the L-19 ribozyme.[99,112,119] The cyclization reaction proceeds after the L1 loop of P1 base pairs to the IGS.[120] The terminal G then nucleophilically attacks the 5′ end of the intron to release the first 19 nucleotides.[120] To investigate the first step of splicing, a form of the intron lacking the first 21 nucleotides and the terminal G at the 3′-end was designed, termed the L-21 ScaI ribozyme.[121] This RNA cannot undergo cyclization or the second step of splicing, but it can perform the first step of splicing with multiple turnover kinetics in the presence of a substrate oligonucleotide and exogenous G.[4,122] The oligonucleotide binds to the complementary IGS to form the P1 helix which is docked into the catalytic core of the enzyme.[4,122]

The L-21 ScaI ribozyme was used to determine a number of kinetic parameters for the first step of splicing.[4,122] Pulse-chase experiments were performed using various amounts of enzyme (RNA), substrate, and guanosine cofactor in order to determine individual rates and binding constants. Initially, binding of guanosine cofactor and substrate to the enzyme was shown to be independent and random.[4] More careful studies demonstrated that there is a modest level of cooperativity between guanosine and substrate binding and anticooperativity between product and guanosine binding to the enzyme.[123,124] In reactions where the concentration of guanosine was saturating, and the concentration of substrate was subsaturating, the rate-limiting step was substrate binding (k_{on}^S).[4] Under these $(k_{cat} \cdot K_m^{-1})^S$ conditions, every RNA substrate that bound was cleaved because the energy barrier for substrate release was higher than the energy required to proceed with the chemical step. Under k_{cat} conditions, where the concentrations of both substrate and guanosine were saturating, the release of product from the active site was rate limiting.[4,118] Finally, under single turnover $(k_{cat} \cdot K_m^{-1})^G$ conditions, where all of the substrate was bound to the enzyme and G was also saturating the chemical step (k_c) is rate limiting at lower pH. At higher pH, substrate binding is rate limiting. Under multiple turnover conditions, the rate-limiting step becomes release of the product.[4] Mutations within the ribozyme or substrate weaken E·S complex binding, thereby increasing the multiple turnover rate.[25]

There is a pH dependence for the single turnover reaction, $E \cdot S + G \rightarrow$ products, with a pK_{app} of 6.9.[126] The titratable groups of RNA have pK_a values $< \sim 4$ and $> \sim 9$. When the 2′ position of the uridine at the cleavage site in the substrate was modified to -H or -F, the substrate reacted at very different rates at low pH, but achieved the same limiting rate at high pH.[126] Substitution of the *pro*-R_p oxygen by sulfur resulted in a "thio effect" only at pH below 6.9. This suggested that the chemical cleavage step is rate limiting at low pH and that a conformational step is rate limiting at higher pH. The pH dependence also indicated that a proton may be lost from the E·S·G ternary complex prior to chemical cleavage.

6.10.5.1.4 Kinetics of **Tetrahymena** *splicing—second step of splicing*

The L-21 G414 ribozyme construct was used to study the second step of splicing.[127] This intron is missing the first 21 nucleotides, but it has the 3′ terminal guanosine G414. Reactions with this ribozyme were investigated in the forward and reverse directions, where G414 acts as a leaving group in the forward reaction and as a nucleophile in the reverse reaction.[127] The rate constant of the chemical step for the reaction with the L-21 G414 construct was identical to that obtained for the reaction of the L-21 ScaI ribozyme with exogenous G.[4,127,128] This result supported the idea that there is a single G binding site and that the orientation of the bound G is the same for both steps of splicing. The results also proposed that the L-21 G414 ribozyme exists primarily with the terminal G docked into the G-binding site. This docking was destabilized ~100-fold when an electron-withdrawing group was attached to G414.

The chemical step of the exon ligation reaction reached an internal equilibrium near 1, which suggested that the bound substrate and product are thermodynamically matched and that there is energetic symmetry within the active site.[127] Slow dissociation of the 5' exon analogue relative to a ligated exon analogue suggested a kinetic mechanism to ensure efficient ligation of exons.[127]

6.10.5.1.5   P1 helix docking

As mentioned above, docking of the P1 helix into the catalytic core is essential for catalysis. Binding of substrate RNA to the ribozyme has been shown to occur by a two-step process. The P1 helix docks into tertiary interactions in different registers after binding of the substrate.[129,130] This indicated that the ribozyme could act processively. The ability of the P1 helix to change registers without dissociation of the substrate led to a two-step model for substrate binding. In this model, an open complex is formed first through base-pairing interactions within P1, then a closed complex forms involving tertiary interactions.[129] The same conclusions were drawn from fluorescence-detected stopped-flow experiments using a pyrene-labeled RNA substrate. Pyrene experienced three different microenvironments during binding, indicating conformational changes as the P1 helix was formed.[131,132] These correspond to the open and closed complexes, while the third conformation is that of the cleavage step.

The $G22 \cdot U(-1)$ wobble pair at the cleavage site is one of the only universally conserved nucleotide pairs outside the catalytic core.[26,133] This pair contributes significantly to P1 helix docking and reaction specificity, which is essential for catalysis. The importance of $G \cdot U$ base pair conservation was shown by *in vitro* mutagenesis.[26] Single base changes in either the IGS or the 5'-exon (excluding mutations in the $G \cdot U$ pair) disrupted splicing, but the compensatory mutations restored activity. Only the $A \cdot C$ wobble pair retained a significant fraction of the wild-type activity.[29,133,134] The secondary and tertiary structural elements associated with the $G \cdot U$ wobble pair were found to be essential for splice site selection since mutation to a G–C pair reduced catalytic activity and this substrate was bound ~10–20-fold weaker than one with a $G \cdot U$ wobble pair.[133,135,136]

Unnatural nucleotide analogues were used to probe the contributions made by the $G \cdot U$ pair. It was demonstrated that the 2'-OH and the N-2 exocyclic amine of G were important for hydrogen bonding to the catalytic core.[29,134] The amine is the only sequence-specific contact within the helix that contributes to 5' splice site selection and transition state stabilization. The amine makes its contribution (2.5 kcal mol^{-1}) in the context of all three wobble pairs tested, but does not contribute (<0.8 kcal mol^{-1}) when presented with a Watson–Crick geometry.[29] The amine also makes a small contribution to chemical transition state stability (1.0 kcal mol^{-1} relative to an I–U pair) that is largely independent of the stability given by the 2'-OH of $U(-1)$.[134] This indicates that tertiary contacts between the exocyclic amine of guanine and its hydrogen bonding partner in the active site are improved during the chemical transition. Results from NAIM indicate that the 2'-OH and N-2 exocyclic amine of the G22 hydrogen bond into the minor groove of two consecutively stacked sheared $A \cdot A$ pairs in J4/5.[28] This wobble–wobble receptor interaction may be the only sequence-specific tertiary contact between P1 and the catalytic core.[26]

Six sequence nonspecific 2'-OH groups participate in P1 helix docking (Figure 2(a)).[30,137–139] C(-2) and U(-3) are found in the substrate strand and G22, G23, G25, and G26 are in the IGS. The P1 helix conforms to A-form geometry, therefore all the 2'-OH groups lie within the minor groove on the same face of the helix.[76] This region of the helix is about a quarter turn away from the exocyclic amine of the $G22 \cdot U(-1)$ pair. The 2'-OHs at positions G22 and G25 played a critical role in docking P1 into the catalytic core, contributing 2.6 and 2.1 kcal mol^{-1}, respectively.[30] Docking was moderately stabilized (~1 kcal mol^{-1}) by G23 and G26 within the IGS (see positions in Figure 2(a)).[30,127]

By contrast, the 2'-OH at the (-1) position had very little effect on ground state binding, but had a large effect on the chemical step.[137] The 2'-OH at the cleavage site of the substrate was replaced with other substituents to investigate their effect on the chemical step.[128] The total transition state stability by the ribozyme was 4.8 kcal mol^{-1} greater when a ribose was present at the cleavage site compared to a deoxyribose.[128] This effect is specific to the chemical transition state because a series of 2′ substituents followed a linear free energy relationship between the rate of the chemical step and the pK_a of the leaving group in this order: 2′-F$_2$ > -F > -H.[137] It was assumed that electron withdrawing groups helped accelerate the rate. Stability from the 2′-OH at U(−1) is proposed to result from donation of an intramolecular hydrogen bond to the incipient 3′-oxyanion in the transition state.[128] This is specific to the transition state because it is not until the bond between the 3′ oxygen atom and the phosphorus atom begins to break that there is an accumulation of negative charge on the 3′-oxygen.[128]

The hydrogen bonding partners for many tertiary interactions within the catalytic core have been identified. A tertiary interaction with the U(-3) position was mapped to a phylogenetically conserved adenine (A302) in J8/7 (Figure 4).[15,140] A hydrogen bond between the 2′-OH of U(-3) and the N-1 of A302 was demonstrated by DMS footprinting and mutagenesis.[27] NAIS demonstrated that the 2′-OH of G25 and the 2′-OH of A301 were energetically coupled, probably through a direct hydrogen bond (Figure 4).[66] The C(-2) and G26 2′-OHs also contribute moderately to P1 docking (~ 1 kcal mol^{-1}),[30,139,141] and based upon NAIS experiments they form hydrogen bonds with the N-2 amine of G303 and the 2′-OH of U300, respectively (Figure 4).[66]

These biochemical data, in conjunction with the crystal structure of the P4–P6 domain, have allowed modeling of most of the catalytic core of the *Tetrahymena* intron.[66] In the model, the J8/7 single strand forms a 4 bp triple helix in the minor groove of P1 and it interacts with both strands of the P1 helix (Figure 4). Each base triple involves at least one 2′-OH, so DNA molecules cannot form this complex. In addition, mutational analysis has identified a base triple between U305 and a major groove face of P4.[142,143] Inclusion of this interaction into the J8/7 triplex placed U305 in a non-A form geometry. A304 within J8/7 bridged the interactions between P1 and P4.[66] The J8/7 region has also been shown by NAIM and mutational analysis to interact with P3.[15,80] Residues U300 and A301 may make hydrogen bonding contacts to P1 and P3, placing J8/7 in the middle of a group of tightly packed helices (Figure 4). J8/7 is also the joining region between P7 and P8, so it is actually in close contact with at least five helices within the catalytic core of the group 1 intron.

6.10.5.1.6 *Guanosine binding*

A guanosine cofactor is required for 5′ splice site cleavage and exon ligation, and it binds to a specific site within the ribozyme with a K_m of $\sim 60–90$ μM.[113,124] Other guanosine analogues exhibited higher K_m values. Based on the free energy of binding values determined, it seemed as though the intron could discriminate between different cofactors.[113] Indeed, deoxyguanosine and dideoxyguanosine are competitive inhibitors of the splicing reaction, indicating the importance of the 2′-OH.[98]

The guanosine binding site was identified using guanosine analogues and mutant ribozymes.[23] The highly conserved G264·C311 base pair within P7 was specifically mutated to A–U.[23] The A–U mutation greatly reduced the rate of 5′ and 3′ cleavage, but there was only a slight further decrease when the 3′ terminal G was changed to an A.[23] The A–U mutant preferred 2-amino purine (2-AP) as a cofactor over G. This showed that G was bound to the G264·C311 base pair, since 2-AP bound to an A–U pair is isosteric to G bound to a G–C pair. These results indicated there is a single G binding site since the same mutation (A–U) affected the G residue requirement for both 5′ and 3′ splicing.[23]

The guanosine cofactor binds in a nonplanar (or axial) configuration with respect to the G264·C311 base pair.[144] The activity of mutant ribozymes with substrates which had hydrogen bond donors and acceptors switched provided proof of an interaction between guanosine and A265 (right below the binding site G264·C311 base pair). This additional interaction was only possible with axial binding of G.

Identification of the guanosine binding site made it possible to demonstrate that both splicing reactions were catalyzed within the same active site.[25] Mutation of the G264·C311 base pair to C264·G311 resulted in preference for adenosine as the substrate for cleavage at the 5′ splice site.[25] The exon ligation step could not be performed with these mutations. However, the second step of splicing could occur if the G at the 3′ splice site was changed to an A.[25] This result indicated that a single determinant specified nucleoside binding for both steps of splicing. In addition, the rate of the chemical step for reactions of the L-21 ScaI intron with bound exogenous guanosine and the L-21 G414 intron with its 3′ terminal guanosine were approximately the same. This supported a single G-binding site model and led to the proposal that the overall active site architecture and orientation remained the same in both steps. Subsequent experiments identified a positive entropy change for guanosine binding and for the chemical step of the ribozyme reaction.[145] The increase in disorder in the ground state could be due to a conformational change or to the release of bound water. The positive entropy for reaching the transition state of the chemical step was probably due to the release of bound water from a catalytic Mg^{2+} ion(s).

6.10.5.2 Hammerhead Ribozyme

6.10.5.2.1 *Hammerhead constructs*

The *in cis* hammerhead ribozyme was dissected into a catalytic domain and its complementary substrate to generate an *in trans* cleaving ribozyme that was used to determine rate constants for the reaction.[146-148] The hammerhead constructs studied contained a large ribozyme RNA, with most of the conserved nucleotides and a small substrate RNA with the cleavage site.[148] The constructs had the same essential unpaired nucleotides at the junctions of the helices and the same helix II (see Figure 2(d)). The constructs differed in the lengths and sequences of helices I and III.[148]

6.10.5.2.2 *Kinetics of hammerhead cleavage*

Under multiple turnover conditions, ribozyme constructs with fewer than six base pairs in each of the substrate binding helices I and III exhibited substrate binding and dissociation rates that were fast relative to k_{cat}.[147,148] Product release was also fast and therefore k_{cat} represented the rate of phosphodiester bond cleavage, which was ~ 1–2 min^{-1}. This corresponded to a rate enhancement of $\sim 10^8$ over the solution reaction.[13,147,148] Lengthening of helices I and III increased stability and in turn drastically reduced the dissociation rates of substrate and product. k_{cat} decreased such that a hammerhead ribozyme with five base pairs in both helices I and III had a k_{cat} of ~ 1.4 min^{-1}, but a construct with eight base pairs in each helix had a k_{cat} of only ~ 0.008 min^{-1}.[147,148] Substrate binding strength had a modest effect on the rate constant of the reaction. Increases in helix stability resulted primarily in decreased dissociation rates with little effect on association rates. Mutations to the ribozyme or substrate that eliminated catalysis increased the stability of the hammerhead. Therefore, substrate destabilization may play a role in hammerhead catalysis. The equilibrium and kinetic properties of the ribozyme–product complexes resembled those expected for a simple RNA duplex.[148] Substrate binding rate constants fell within the range expected for the formation of simple RNA duplexes. Product release was faster than the rate-limiting step of chemistry for a hammerhead construct that contained five base pairs between the product and ribozyme. For constructs where products bound by seven base pairs to the ribozyme, the rate of product dissociation was rate limiting.[148] With these small RNAs, dissociation rate constants were directly proportional to the stability of the helix.

One specific hammerhead construct, the HH16 ribozyme, was studied in particular detail.[147] This ribozyme had eight potential base pairs in each of the substrate recognition helices. This construct was designed to be stable enough to determine product binding and to investigate reversal of the cleavage step. HH16 cleaved bound substrate with a rate constant of ~ 1 min^{-1}, similar to what was found in the previous experiments.[147] The rate of ligation of the 5′ product and 3′ product to form substrate was only ~ 0.008 min^{-1}. Therefore, HH16 preferred to maintain products on the ribozyme. Product and substrate association rates were between 10^7 and 10^8 M^{-1} min^{-1}, which are values similar to those for simple helices.[147] The stabilities of ribozyme–product complexes did not indicate any additional tertiary interactions. Finally, the dissociation constant for the binding of substrate to the ribozyme is $\sim 10^{-17}$ M, indicating tight binding between these two molecules.

In order to determine which phosphates are important for catalytic activity, phosphorothioate linkages were incorporated. These substitutions identified four phosphates in the conserved central core that play a role in the self-cleavage reaction.[94] In addition, substitution of sulfur for a non-bridging oxygen at the cleavage site reduced the affinity of an essential Mg^{2+} ion involved in efficient cleavage.[94] The reaction proceeded normally in the presence of Mn^{2+}. These results suggested that a divalent ion is coordinated directly to the phosphate at the cleavage site.

A 1998 crystal structure of the hammerhead captured a conformation that is in agreement with the geometry needed for in-line attack, the mechanism thought to be used by the hammerhead ribozyme for catalysis.[68] Previous crystal structures have identified structures in the ground state which did not contain the proper positioning for an S_N2 reaction.[69,70] This result suggested that a conformational change must take place prior to cleavage, allowing the in-line geometry to form.[68] In addition, the presence of metal ions in the newest structure is also in agreement with the kinetic pathway thought to be utilized. This structure lends support to the biochemical assays that have studied the kinetics of hammerhead self-cleavage.

6.10.6 OTHER CATALYTIC RNAS

Using *in vitro* selection, a number of ribozymes have been identified for their ability to catalyze reactions at phosphodiester bonds. For example, a leadzyme was identified from a yeast tRNA^{Phe} selection.[83,149] The best Pb^{2+}-dependent ribozyme exhibited a cleavage rate 20-fold higher than that of the original yeast tRNA^{Phe} molecule and had a substantially different geometry than the original tRNA. RNA ligases have also been isolated by selection from a random library.[150] These ribozymes catalyze the ligation of a 5′ triphosphate with a 3′-OH (a 5′ → 3′ ligase).[151,152] The rate enhancement ($\sim 10^{10}$-fold over background) exhibited by this ribozyme is the highest for a selected catalytic RNA.[151] Another ribozyme was selected for its catalytic activity as an RNA polynucleotide kinase.[153] The reaction involved transfer of a γ-thiophosphate from ATP-γS to either a 5′-OH or an internal 2′-OH of an RNA molecule. The best ribokinase had a K_m for ATP-γS of ~ 40 nM.[153] This value is close to that determined for the protein enzyme, T4 polynucleotide kinase. In addition, the k_{cat} was only $\sim 10^4$-fold slower than that of the protein enzyme.

In vitro selection has also identified ribozymes that are able to catalyze reactions at carbon centers. The unmodified *Tetrahymena* ribozyme weakly catalyzed the hydrolysis of an aminoacyl ester bond at a rate about 5–15 times greater than the uncatalyzed rate.[154] The carboxylate ester was targeted to the catalytic core of the ribozyme by covalently attaching an amino acid to an oligonucleotide that bound to the internal guide sequence. Another RNA was selected for its ability to aminoacylate its own 2′ (3′) terminus when phenylalanyl-monophosphate was present.[155] This reaction is similar to the aminoacyl group transfer performed by protein aminoacyl-transfer RNA synthetases. Further selections have identified RNAs capable of binding to non-nucleotide substrates and performing self-alkylation reactions.[156] In addition, RNAs were evolved that had acyl transferase activity and could transfer an RNA-linked amino acid to their own 5′-amino-modified terminus, analogous to peptidyl transfer on the ribosome.[157] In 1997, a ribozyme was selected for its ability to perform the same peptidyl transferase reaction as the ribosome.[158] These RNAs could join amino acids by a peptide bond when given two substrates, one that had an amino acid (*N*-blocked methionine) esterified to the 2′ (3′)-oxygen of adenosine and the other which was an acceptor amino acid (phenylalanine) with a free amino group.

6.10.7 CONCLUSIONS

This description of ribozyme enzymology has demonstrated the importance of structure and metal ions for catalytic activity. While the group I and hammerhead ribozymes are the most fully characterized examples of catalytic RNAs, the principles learned from these examples are likely to be applicable to other structured RNAs including the other examples of catalytic RNAs.

ADDENDUM

A 5.0 Å crystal structure of the P3–P9 regions of the *Tetrahymena* group I intron has recently been solved. There appear to be binding pockets for both the 5′-splice site helix and the guanosine cofactor. Therefore, the ribozyme seems to contain a preorganized active site.[159]

6.10.8 REFERENCES

1. T. R. Cech, A. J. Zaug, and P. J. Grabowski, *Cell*, 1981, **27**, 487.
2. C. Guerrier-Takada, K. Gardiner, T. Marsh, N. Pace, and S. Altman, *Cell*, 1983, **35**, 849.
3. T. R. Cech, in "The RNA World," eds. R. F. Gesteland and J. F. Atkins, Cold Spring Harbor Laboratory Press, Cold Spring Harbor, NY, 1993.
4. D. Herschlag and T. R. Cech, *Biochemistry*, 1990, **29**, 10 159.

5. W. J. Michels and A. M. Pyle, *Biochemistry*, 1995, **34**, 2965.
6. C. Guerrier-Takada and S. Altman, *Science*, 1984, **223**, 285.
7. T. Ikemura, Y. Shimura, H. Sakano, and H. Ozeki, *J. Mol. Biol.*, 1975, **96**, 69.
8. P. Schedl and P. Primakoff, *Proc. Natl. Acad. Sci. USA*, 1973, **70**, 2091.
9. S. Altman, *Adv. Enzymol. Rel. Areas Mol. Biol.*, 1989, **62**, 1.
10. J. A. Beebe and C. A. Fierke, *Biochemistry*, 1994, **33**, 10 294.
11. T. Pan, D. M. Long, and O. C. Uhlenbeck, in "The RNA World," eds. R. F. Gesteland and J. F. Atkins, Cold Spring Harbor Laboratory Press, Cold Spring Harbor, NY, 1993.
12. G. A. Prody, J. T. Bakos, J. M. Buzayan, I. R. Schneider, and G. Bruening, *Science*, 1986, **231**, 1577.
13. O. C. Uhlenbeck, *Nature*, 1987, **328**, 596.
14. F. Michel, A. Jacquier, and B. Dujon, *Biochimie*, 1982, **64**, 867.
15. F. Michel and E. Westhof, *J. Mol. Biol.*, 1990, **216**, 585.
16. R. W. Davies, R. B. Waring, J. A. Ray, T. A. Brown, and C. Scazzocchio, *Nature*, 1982, **300**, 719.
17. D. W. Celander and T. R. Cech, *Science*, 1991, **251**, 401.
18. E. A. Doherty and J. A. Doudna, *Biochemistry*, 1997, **36**, 3159.
19. R. B. Waring, P. Towner, S. J. Minter, and R. W. Davies, *Nature*, 1986, **321**, 133.
20. D. W. Celander and T. R. Cech, *Biochemistry*, 1990, **29**, 1355.
21. J.-F. Wang and T. R. Cech, *Science*, 1992, **256**, 526.
22. J.-F. Wang, W. D. Downs, and T. R. Cech, *Science*, 1993, **260**, 504.
23. F. Michel, M. Hanna, R. Green, D. P. Bartel, and J. W. Szostak, *Nature*, 1989, **342**, 391.
24. T. Inoue, F. X. Sullivan, and T. R. Cech, *J. Mol. Biol.*, 1986, **189**, 143.
25. M. D. Been and A. T. Perrotta, *Science*, 1991, **252**, 434.
26. F. L. Murphy and T. R. Cech, *Proc. Natl. Acad. Sci. USA*, 1989, **86**, 9218.
27. A. M. Pyle, F. L. Murphy, and T. R. Cech, *Nature*, 1992, **358**, 123.
28. S. A. Stobel, L. Ortoleva-Donnelly, S. P. Ryder, J. H. Cate, and E. Moncoeur, *Nat. Struc. Biol.*, 1998, **5**, 60.
29. S. A. Strobel and T. R. Cech, *Science*, 1995, **267**, 675.
30. S. A. Strobel and T. R. Cech, *Biochemistry*, 1993, **32**, 13 593.
31. J. A. Latham and T. R. Cech, *Science*, 1989, **245**, 276.
32. G. Chanfreau and A. Jacquier, *Science*, 1994, **266**, 1383.
33. A. M. Pyle, in "Catalytic RNA," eds. F. Eckstein and D. M. J. Lilley, Springer, New York, 1996.
34. A. Jacquier and M. Rosbash, *Science*, 1986, **234**, 1099.
35. A. Jacquier and F. Michel, *Cell*, 1987, **50**, 17.
36. K. A. Jarrell, R. C. Dietrich, and P. S. Perlman, *Mol. Cell. Biol.*, 1988, **8**, 2361.
37. F. Michel, K. Umensono, and H. Ozeki, *Gene*, 1989, **82**, 5.
38. C. L. Peebles, M. Zhang, P. S. Perlman, and J. S. Franzen, *Proc. Natl. Acad. Sci. USA*, 1995, **92**, 4422.
39. B. D. James, G. J. Olsen, J. Liu, and N. Pace, *Cell*, 1988, **52**, 19.
40. E. S. Haas, A. B. Banta, J. K. Harris, N. R. Pace, and J. W. Brown, *Nucleic Acids Res.*, 1996, **24**, 4775.
41. M. E. Harris, A. V. Kazantsev, J.-L. Chen, and N. R. Pace, *RNA*, 1997, **3**, 561.
42. M. E. Harris and N. R. Pace, *RNA*, 1995, **1**, 210.
43. C. Massire, L. Jaeger, and E. Westhof, *RNA*, 1997, **3**, 553.
44. D. E. Ruffner, G. D. Stormo, and O. C. Uhlenbeck, *Biochemistry*, 1990, **29**, 10 695.
45. R. H. Symons, *Annu. Rev. Biochem.*, 1992, **61**, 641.
46. A. C. Forster and R. H. Symons, *Cell*, 1987, **49**, 211.
47. A. Hampel and R. Tritz, *Biochemistry*, 1989, **28**, 4929.
48. L. A. Hegg and M. J. Fedor, *Biochemistry*, 1995, **34**, 15 813.
49. A. Hampel, R. Tritz, M. Hicks, and P. Cruz. *Nucleic Acids Res.*, 1990, **18**, 299.
50. L. Rubino, M. E. Tousignant, G. Steger, and J. M. Kaper, *J. Gen. Virol.*, 1990, **71**, 1897.
51. A. Berzal-Herranz, S. Joseph, and J. M. Burke, *Genes Dev.*, 1992, **6**, 129.
52. P. A. Feldstein and G. Bruening, *Nucleic Acids Res.*, 1993, **21**, 1991.
53. Y. Komatsu, M. Koizumi, N. Sekiguchi, and E. Ohtsuka, *Nucleic Acids Res.*, 1993, **21**, 185.
54. M. D. Been, *Trends Bio. Sci.*, 1994, **19**, 251.
55. M. D. Been, A. T. Perrotta, and S. P. Rosenstein, *Biochemistry*, 1992, **31**, 11 843.
56. A. T. Perrotta and M. D. Been, *Nucleic Acids Res.*, 1990, **18**, 6821.
57. A. Ferré-D'Amaré, K. H. Zhou, and J. A. Doudna, *Nature*, 1998, **395**, 567.
58. B. J. Saville and R. A. Collins, *Cell*, 1990, **61**, 685.
59. T. L. Beattie, J. E. Olive, and R. A. Collins, *Proc. Natl. Acad. Sci. USA*, 1995, **92**, 4686.
60. T. L. Beattie and R. A. Collins, *J. Mol. Biol.*, 1997, **267**, 830.
61. J. H. Cate, A. R. Gooding, E. Podell, K. Zhou, B. L. Golden, C. E. Kundrot, T. R. Cech, and J. A. Doudna, *Science*, 1996, **273**, 1678.
62. J. H. Cate, A. R. Gooding, E. Podell, K. Zhou, B. L. Golden, A. A. Szewczak, C. E. Kundrot, T. R. Cech, and J. A. Doudna, *Science*, 1996, **273**, 1696.
63. J. H. Cate and J. A. Doudna, *Structure*, 1996, **4**, 1221.
64. J. H. Cate, R. L. Hanna, and J. A. Doudna, *Nat. Struc. Biol.*, 1997, **4**, 553.
65. S. A. Strobel and K. Shetty, *Proc. Natl. Acad. Sci. USA*, 1997, **94**, 2903.
66. A. Szewczak, L. Ortoleva-Donnelly, S. Ryder, E. Moncoeur, and S. Strobel, *Nat. Struct. Biol.*, 1998, **5**, 1037.
67. H. W. Pley, K. M. Flaherty, and D. B. McKay, *Nature*, 1994, **372**, 68.
68. J. B. Murray, D. P. Terwey, L. Maloney, A. Karpiesky, N. Usman, L. Beigelman, and W. G. Scott, *Cell*, 1998, **92**, 665.
69. W. G. Scott, J. T. Finch, and A. Klug, *Cell*, 1995, **81**, 991.
70. W. G. Scott, J. B. Murray, J. R. P. Arnold, B. L. Stoddard, and A. Klug, *Science*, 1996, **274**, 2065.
71. G. J. Quigley and A. Rich, *Science*, 1976, **194**, 796.
72. D. B. McKay, *RNA*, 1996, **2**, 395.
73. M. Record and T. Lohman, *Biopolymers*, 1978, **17**, 159.
74. G. Manning, *Q. Rev. Biophys.*, 1978, **11**, 179.

75. J. Huheey, in "Inorganic Chemistry," Harper & Row, New York, 1985.
76. W. Saenger, in "Principles of Nucleic Acid Structure," Springer-Verlag, New York, 1984.
77. E. L. Christian, M. Yarus, *J. Mol. Biol.*, 1992, **228**, 743.
78. G. Gish and F. Eckstein, *Science*, 1988, **240**, 1520.
79. E. L. Christian and M. Yarus, *Biochemistry*, 1993, **32**, 4475.
80. L. Ortoleva-Donnelly, A. A. Szewczak, R. R. Gutell, and S. A. Strobel, *RNA*, 1998, **4**, 498.
81. V. L. Pecoraro, J. D. Hermes, and W. W. Cleland, *Biochemistry*, 1984, **23**, 5262.
82. S. Kazakov and S. Altman, *Proc. Natl. Acad. Sci. USA*, 1991, **88**, 9193.
83. R. S. Brown, J. C. Dewan, and A. Klug, *Biochemistry*, 1985, **24**, 4785.
84. L. S. Behlen, J. R. Sampson, A. B. DiRenzo, and O. C. Uhlenbeck, *Biochemistry*, 1990, **29**, 2515
85. J. R. Rubin and M. Sundaralingam, *J. Biomol. Struct. Dyn.*, 1983, **1**, 639.
86. G. Dirheimer, J. P. Ebel, J. Bonnet, J. Gangloff, G. Keith, B. Krebs, B. Kuntzel, A. Roy, J. Weissenbach, and C. Werner, *Biochimie*, 1972, **54**, 127.
87. B. Streicher, U. von Ahsen, and R. Schroeder, *Nucleic Acids Res.*, 1993, **21**, 311.
88. J. Ciesiolka, W. Hardt, J. Schlegl, V. A. Erdmann, and R. R. Hartmann, *Eur. J. Biochem.*, 1994, **219**, 49.
89. C. Berens, B. Streicher, R. Schroeder, and W. Hillen, *Chem. Biol.*, 1998, **5**, 163.
90. A. Jack, J. E. Ladner, D. Rhodes, R. S. Brown, and A. Klug, *J. Mol. Biol.*, 1977, **111**, 315.
91. S. R. Holbrook, J. L. Sussman, R. W. Warrant, G. M. Church, and S.-H. Kim, *Nucleic Acids Res.*, 1977, **4**, 2811.
92. G. J. Quigley, M. M. Teeter, and A. Rich, *Proc. Natl. Acad. Sci. USA*, 1978, **75**, 64.
93. S. Basu, R. P. Rambo, J. Strauss-Soukop, J. H. Cate, A. R. Ferré-D'Amaré, S. A. Strobel, and J. A. Doudna, *Nat. Struct. Biol.*, 1998, **5**, 986.
94. D. E. Ruffner and O. C. Uhlenbeck, *Nucleic Acids Res.*, 1990, **18**, 6025.
95. T. A. Steitz and J. A. Steitz, *Proc. Natl. Acad. Sci. USA*, 1993, **90**, 6498.
96. K. Kruger, P. J. Grabowski, A. J. Zaug, J. Sands, D. E. Gottschling, and T. R. Cech, *Cell*, 1982, **31**, 147.
97. C. A. Grosshans and T. R. Cech, *Biochemistry*, 1989, **28**, 6888.
98. B. L. Bass and T. R. Cech, *Biochemistry*, 1986, **25**, 4473.
99. A. J. Zaug and T. R. Cech, *Science*, 1986, **231**, 470.
100. A. J. Zaug and T. R. Cech, *Biochemistry*, 1986, **25**, 4478.
101. J. A. Piccirilli, J. S. Vyle, M. H. Caruthers, and T. R. Cech, *Nature*, 1993, **361**, 85.
102. L. B. Weinstein, B. C. N. M. Jones, R. Cosstick, and T. R. Cech, *Nature*, 1997, **388**, 805.
103. A. S. Sjogren, E. Pettersson, B. M. Sjoberg, and R. Stromberg, *Nucleic Acids Res.*, 1997, **25**, 648.
104. A. M. Pyle, *Science*, 1993, **261**, 709.
105. H. A. Heus and A. Pardi, *J. Mol. Biol.*, 1991, **217**, 113.
106. R. A. J. Hodgson, N. J. Shirley, and R. H. Symons, *Nucleic Acids Res.*, 1994, **22**, 1620.
107. S. C. Dahm, W. B. Derrick, and O. C. Uhlenbeck, *Biochemistry*, 1993, **32**, 13 040.
108. S. C. Dahm and O. C. Ohlenbeck, *Biochemistry*, 1991, **30**, 9464.
109. W. Lott, B. Pontius, and P. von Hippel, *Proc. Natl. Acad. Sci. USA*, 1998, **95**, 542.
110. A. Hampel and J. A. Cowan, *Chem. Biol.*, 1997, **4**, 513.
111. K. J. Young, F. Gill, and J. A. Grasby, *Nucleic Acids Res.*, 1997, **25**, 3760.
112. A. J. Zaug, P. J. Grabowski, and T. R. Cech, *Nature*, 1983, **301**, 578.
113. B. L. Bass and T. R. Cech, *Nature*, 1984, **308**, 820.
114. J. A. McSwiggen and T. R. Cech, *Science*, 1989, **244**, 679.
115. J. Rajagopal, J. A. Doudna, and J. W. Szostak, *Science*, 1989, **244**, 692.
116. J. R. Knowles, *Annu Rev. Biochem.*, 1980, **49**, 877.
117. F. Eckstein, *Annu Rev. Biochem.*, 1985, **54**, 367.
118. D. Herschlag, J. A. Piccirilli, and T. R. Cech, *Biochemistry*, 1991, **30**, 4844.
119. A. J. Zaug, J. R. Kent, and T. R. Cech, *Science*, 1984, **224**, 574.
120. M. D. Been and T. R. Cech, *Cell*, 1986, **47**, 207.
121. A. J. Zaug, C. A. Grosshans, and T. R. Cech, *Biochemistry*, 1988, **27**, 8924.
122. D. Herschlag and T. R. Cech, *Biochemistry*, 1990, **29**, 10 172.
123. P. C. Bevilacqua, K. A. Johnson, and D. H. Turner, *Proc. Natl. Acad. Sci. USA*, 1993, **90**, 8357.
124. T. S. McConnell, T. R. Cech, and D. Herschlag, *Proc. Natl. Acad. Sci. USA*, 1993, **90**, 8362.
125. B. Young, D. Hershlag, and T. R. Cech, *Cell*, 1991, **67**, 1007.
126. D. Hershlag and M. Khosla, *Biochemistry*, 1994, **33**, 5291.
127. R. Mei and D. Herschlag, *Biochemistry*, 1996, **35**, 5796.
128. D. Herschlag, F. Eckstein, and T. R. Cech, *Biochemistry*, 1993, **32**, 8312.
129. D. Herschlag, *Biochemistry*, 1992, **31**, 1386.
130. S. A. Strobel and T. R. Cech, *Nat. Struc. Biol.*, 1994, **1**, 13.
131. P. C. Bevilacqua, R. Kierzek, K. A. Johnson, and D. H. Turner, *Science*, 1992, **258**, 1355.
132. P. C. Bevilacqua, Y. Li, and D. H. Turner, *Biochemistry*, 1994, **33**, 11 340.
133. J. A. Douna, B. P. Cormack, and J. W. Szostak, *Proc. Natl. Acad. Sci. USA*, 1989, **86**, 7402.
134. S. A. Strobel and T. R. Cech, *Biochemistry*, 1996, **35**, 1201.
135. A. M. Pyle, S. Moran, S. A. Strobel, T. Chapman, D. H. Turner, and T. R. Cech, *Biochemistry*, 1994, **33**, 13 856.
136. D. Knitt, G. Narlikar, and D. Herschlag, *Biochemistry*, 1994, **33**, 13 864.
137. D. Herschlag, F. Eckstein, and T. R. Cech, *Biochemistry*, 1993, **32**, 8299.
138. P. C. Bevilacqua and D. H. Turner, *Biochemistry*, 1991, **30**, 10 632.
139. A. M. Pyle and T. R. Cech, *Nature*, 1991, **350**, 628.
140. T. R. Cech, S. H. Damberger, and R. R. Gutell, *Nat. Struct. Biol.*, 1994, **1**, 273.
141. G. J. Narlikar, M. Khosla, N. Usman, and D. Herschlag, *Biochemistry*, 1997, **36**, 2465.
142. M. A. Tanner, E. M. Anderson, R. R. Gutell, and T. R. Cech, *RNA*, 1997, **3**, 1037.
143. M. A. Tanner and T. R. Cech, *Science*, 1997, **275**, 847.
144. M. Yarus, M. Illangesekare, and E. Christian, *J. Mol. Biol.*, 1991, **222**, 995.
145. T. S. McConnell and T. R. Cech, *Biochemistry*, 1995, **34**, 4056.

146. J. Haseloff and W. Gerlach, *Nature*, 1988, **334**, 585.
147. K. Hertel, D. Herschlag, and O. Uhlenbeck, *Biochemistry*, 1994, **33**, 3374.
148. M. Fedor and O. C. Uhlenbeck, *Biochemistry*, 1992, **31**, 12 042.
149. T. Pan and O. C. Uhlenbeck, *Biochemistry*, 1992, **31**, 3887.
150. D. Bartel and J. Szostak, *Science*, 1993, **261**, 1411.
151. E. Ekland, J. Szostak, and D. Bartel, *Science*, 1995, **269**, 364.
152. E. Ekland and D. Bartel, *Nucleic Acids Res.*, 1995, **23**, 3231.
153. J. Lorsch and J. Szostak, *Nature*, 1994, **371**, 31.
154. J. A. Piccirilli, T. S. McConnell, A. J. Zaug, H. F. Noller, and T. R. Cech, *Science*, 1992, **256**, 1420.
155. M. Illangasekare, G. Sanchez, T. Nickles, and M. Yarus, *Science*, 1995, **267**, 643.
156. C. Wilson and J. W. Szostak, *Nature*, 1995, **374**, 777.
157. P. Lohse and J. Szostak, *Nature*, 1996, **381**, 442.
158. B. Zhang and T. R. Cech, *Nature*, 1997, **390**, 96.
159. B. L. Golden, A. R. Gooding, E. R. Podell, and T. R. Cech, *Science*, 1998, **282**, 259.

6.11
Viroids

ROBERT H. SYMONS
University of Adelaide, Glen Osmond, SA, Australia

6.11.1 INTRODUCTION

6.11.1.1 Nature of Viroids

Viroids are the smallest pathogenic agents yet described. They are single-stranded circular RNA molecules which vary in length from 246 to 463 nucleotides and are found only in plants. Of the 27 viroids characterized so far (Table 1; Figure 1), 25 infect dicotyledonous plants and the other two infect monocotyledonous plants. The diseases caused by some of these viroids are of considerable agricultural importance. The coconut cadang cadang viroid found in the Philippines, which consists of only 246 nucleotides kills most of the palms it infects. For all other viroids, death of infected plants is unusual, and phenotypic effects vary from essentially no symptoms to various degrees of debilitation and dwarfing.

Table 1 Classification of viroids.

Viroid-group	Viroid-subgroup	Viroid	Abbreviation	Length (nucleotides)
A. ASBV-group	ASBV-subgroup	Avocado sunblotch viroid	ASBV	246–250
		Chrysanthemum chlorotic mottle viroid	CChMV	398–399
		Peach latent mosaic viroid	PLMV	337
B. PSTV-group	B1. PSTV-subgroup	Chrysanthemum stunt viroid	CSV	354–356
		Citrus exocortis viroid	CEV	370–463
		Citrus viroid IV	CV IV	284
		Coconut cadang cadang viroid	CCCV	246–346
		Coconut tinangaja viroid	CTiV	254
		Columnea latent viroid	CLV	370–373
		Hop latent viroid	HLV	256
		Hop stunt viroid	HSV	297–303
		Iresine viroid	IrV	370
		Mexican papita viroid	MPV	360
		Potato spindle tuber viroid	PSTV	359–360
		Tomato apical stunt viroid	TASV	360–363
		Tomato planta macho viroid	TPMV	360
	B2. ASSV-subgroup	Apple scar skin viroid	ASSV	329–330
		Apple dimple fruit viroid	ADFV	306
		Australian grapevine viroid	AGV	369
		Citrus bent leaf viroid	CBLV	318
		Citrus viroid III	CV III	294–297
		Grapevine yellow speckle viroid-1	GYSV-1	366–368
		Grapevine yellow speckle viroid-2	GYSV-2	363
		Pear blister canker viroid	PBCV	315–316
	B3. CbV-subgroup	Coleus blumei viroid 1	CbV 1	248–251
		Coleus blumei viroid 2	CbV 2	301
		Coleus blumei viroid 3	CbV 3	361–364

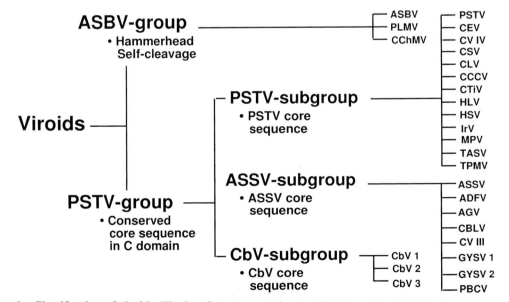

Figure 1 Classification of viroids. The key features used for the division of viroids into two groups and of the PSTV-group into three subgroups are indicated. Viroid abbreviations are defined in Table 1.

All evidence indicates that viroids do not code for any proteins. Consequently they must rely completely on normal plant processes for their replication and spread throughout the plant. The host processes involved in their replication are slowly being characterized but essentially nothing is known about how they cause disease. Because of their small size and intriguing biological properties, they have attracted considerable interest, yet there are only a few groups worldwide who are characterizing them at the molecular level.

The reader will find references 1–12 a rich source of historical and background information on viroids. The book by Diener[1] is the classic text and remains a valuable reference work. References 2–4 are multiauthor volumes, while references 5–12 are review articles that cover the more molecular aspects of viroid structure, function, and evolution.

6.11.1.2 An Historical Perspective

Viroid-caused diseases are a phenomenon of the twentieth century in the sense that these diseases were recognized and described in agricultural crops, and later shown to be of viroid aetiology, only in the twentieth century. In 1967, Diener and Raymer[13] concluded that PSTV is a free, low molecular weight RNA on the basis of its low sedimentation rate and sensitivity to ribonuclease, a result confirmed by Semancik and Weathers[14] for CEV. Accumulating evidence further confirmed the small, protein-free, RNA nature of these infectious agents and the ability of RNA eluted from a specific RNA band on a polyacrylamide gel to be infectious when inoculated on a susceptible host plant.[15,16] The term viroid was proposed as a generic term for PSTV in 1971[15] and is now widely accepted. The first sequence of a viroid was published[17] in 1978 (PSTV) and the sequences of other viroids appeared progressively thereafter.

6.11.2 CLASSIFICATION OF VIROIDS

Table 1 lists the 27 viroids together with an abbreviation used for each one and their size in number of nucleotides. It is becoming increasingly common to add a small "d" at the end of the abbreviation to indicate a viroid; the original abbreviations are used in this review. In many viroids, the length can vary from one to several nucleotides, as indicated in Table 1, while in two viroids, CCCV and CEV, terminal repeats of nucleotides can substantially increase the size (see Section 6.11.5.2).

On the basis of comparative sequence analysis between the different viroids as well as other features, the 27 viroids can be divided into two main groups, the ASBV-group with three members and the PSTV-group for the remainder. There is a further subdivision of the PSTV-group into three subgroups (Figure 1). Many aspects of their properties relevant to this classification are considered throughout this review.

6.11.3 ISOLATION, PURIFICATION, AND SEQUENCING OF VIROIDS

All the viroids listed in Table 1 were originally identified in field samples taken from plants showing symptoms of diseases, especially of agricultural crops. Most, but not all, viroids can be readily maintained in the greenhouse on herbaceous or woody host plants. For example, tomato is a very useful host for PSTV, CEV, and TPMV and chrysanthemum for CSV and CChMV. Some viroids of woody species have a very limited host range: CCCV and CTiV only infect members of the palm family, GYSV-1 and GYSV-2 are only found in grape vines, and ASBV is limited to avocado and related species. In contrast, CEV, which was first isolated from citrus as its name implies, has a host range that includes species in several different families of plants, and PSTV can infect many species in at least 12 different families. Appearance of symptoms on inoculated plants usually occurs within a month for herbaceous hosts. However, in coconut seedlings inoculated with CCCV and avocado seedlings infected with ASBV symptoms may not appear for one to two years. In all cases, it is preferable to maintain day temperatures of inoculated plants at 25–30 °C since higher than usual greenhouse temperatures favor viroid replication.

Infection of plants for experimental purposes is most commonly done by rubbing crude extracts of viroid-infected plants, partly purified extracts, or purified viroid on leaves of uninfected host plants, or, in the case of more woody species, by slashing through a drop of extract placed on a stem of a host plant. Infectious cDNA clones of viroids have proven of great benefit for the characterization of viroid replication and pathogenicity since they can also be used for inoculation; rubbing purified cDNA clones on the leaves of susceptible plants usually suffices. The great advantage here is that a single sequence variant can be used for inoculation, in contrast to RNA extracts of infected plants, which may contain many sequence variants. Obviously, also, cDNA clones can be used for mutagenesis experiments to investigate structure/function relationships in viroids.

Many viroids are present in infected plants at levels which allow their purification by gel electrophoresis of crude or partly purified RNA extracts. In the case of viroids that are present at low concentrations, for example GYSV-1 and GYSV-2, larger amounts of starting material and extra concentration steps are needed. The fractionation of RNAs by chromatography on CF11 cellulose is commonly used here.

Viroids were originally sequenced by partial RNase digestion, and gel electrophoresis purification of individual fragments followed by enzymatic sequencing using several nucleases of known specificity.[18,19] The method is reasonably rapid but problems may arise when different sequence variants exist as a population in an infected plant (see Section 6.11.5). The preparation of cDNA clones of a viroid is now obviously the method of choice but some initial sequence information is necessary to define appropriate oligonucleotide primers for the synthesis of first-strand cDNA.

6.11.4 DOMAIN MODEL FOR PSTV GROUP OF VIROIDS

From an examination of sequence homology among members of the PSTV-subgroup of viroids that had been sequenced prior to 1985, a domain structural model[20] for viroids was developed (Figure 2) which has stood the test of time. The domain model was based on the rod-like secondary structure of viroids as predicted from sequence comparisons and was consistent with the early electron microscope images of PSTV and CCCV.[21–23] The three members of the ASBV-group of viroids do not conform to this model. However, as more members of this group are described, it is possible that an appropriate domain model will be developed.

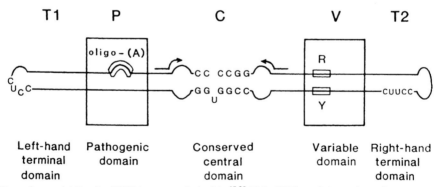

Figure 2 Domain model for the PSTV-group of viroids.[18,20] This 1985 model was based on sequences of seven viroids known at that time in the PSTV-subgroup. Boundaries between the domains were determined by marked changes in sequence homology on pairwise comparisons between viroids. Conserved nucleotides are indicated. R and Y indicate a short oligopurine-oligopyrimidine helix. Arrows depict an inverted repeat sequence with the potential to form a hairpin stem. Later models also contained a similar inverted repeat sequence in the bottom strand of the C domain (Figures 3–5).

The boundaries between neighboring domains of the PSTV-group of viroids were determined by comparative sequence analysis between pairs of viroids. In many, but not all, such comparisons, the boundaries between domains were defined by sharp changes in sequence homology, from high to low or vice versa. Different pairwise viroid comparisons were consistent in defining the exact position of the boundaries. The pathogenic domain P was so called because it is associated with symptom expression, at least for the PSTV-subgroup members at the time when the model was developed.[7,18-20] Likewise, the variable domain V was so named because it showed the greatest sequence variability between closely related viroids. The C or central conserved domain is the most highly conserved domain in pairwise comparisons between viroids. The U-bulged helix and the inverted repeat sequence, indicated by the arrows in Figure 2, are highly conserved in the PSTV-subgroup members.

The publication of sequences for members of the ASSV-subgroup and CbV-subgroup viroids confirmed that the domain model developed for the PSTV-subgroup viroids is directly applicable to them also and this allowed a further elaboration of the viroid classification scheme[24] with the result shown in Figure 1.

The C domains of three PSTV-subgroups are compared in Figure 3. They each show inverted repeat sequences, indicated by the arrows, and a central core sequence specific to each subgroup; this core sequence is often referred to as the central conserved region (CCR) within the C domain. Another conserved feature of the C domain is the potential ability of the conserved sequences in the top strand of the C domain to form palindromic structures[25,26] (Figure 4) in tandem monomeric repeats. Likewise, the presence of the inverted repeat sequences allows the potential formation of two hairpin-loop structures, as indicated schematically in Figure 5. Whether such structures are biologically significant is not known; this is discussed further in Section 6.11.8.

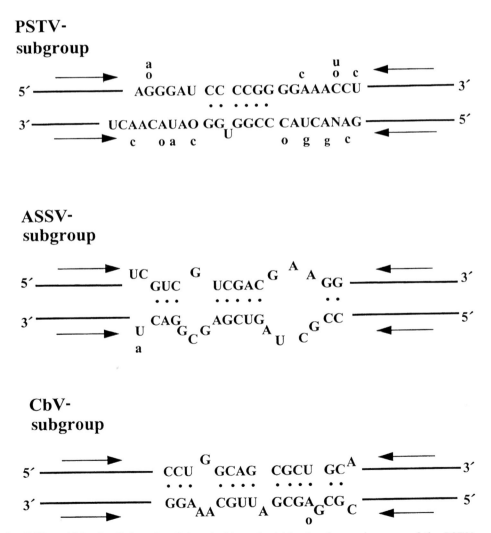

Figure 3 CCRs within the C domain of the viroids and within the three subgroups of the PSTV-group of viroids (see Figures 1 and 2). The sequences given for the PSTV- and ASSV-subgroups are those of PSTV and ASSV, respectively. Variations of sequence in one or more of the subgroup members are given in lower case above or below the relevant residue. O indicates a single base deletion.

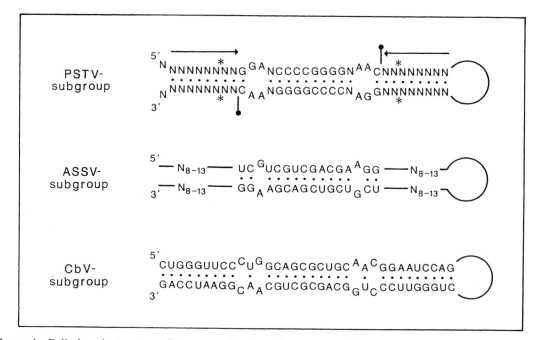

Figure 4 Palindromic structures that are possible in tandem monomeric repeats of members of each subgroup of the PSTV-group of viroids.[25,26] Sequences are from the top strand of the C domain while the specific sequences are from the CCR within the C domain except for CbV where the CCR sequence is yet to be determined since CbV-1, CbV-2, and CbV-3 have high sequence homology in the C domain.[27] Reverse arrows on the top strand of PSTV-subgroup indicate location of inverted repeat sequences for the three subgroups of viroids. N is a nonconserved nucleotide. In HSV, a U is inserted at the positions marked by a vertical bar. Asterisks denote the sites of insertion of two unpaired nucleotides in HLV. (Reproduced by permission of CRC Press from "Viroids: Pathogens at the Frontiers of Life," 1991.)

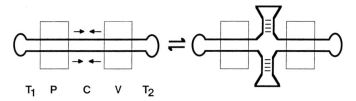

Figure 5 Schematic diagram of potential alternative hairpin structures within the C domain of members in the PSTV-group of viroids. Left-hand side: domain model of Figure 2 with two pairs of inverted repeat sequences indicated by arrows. Right-hand side: potential stem-loop structures formed in the C domain involving the inverted repeat sequences. (Reproduced by permission of Oxford University Press from *Nucleic Acids Res.*, 1997, **25**, 2683.)

6.11.5 SEQUENCE PATTERNS AND VARIATION IN VIROID SEQUENCES

6.11.5.1 Sequence Variants within Viroid Populations

The ability to produce full-length cDNA clones of purified viroids followed by the sequencing of many individual clones from a single viroid isolate has provided us with important information on the dynamics of viroid populations. Viroids, being RNA pathogens, are likely to mutate during replication over a period of time and on transfer from one host to another. Not all mutants so produced will be viable but it would be expected that some will be and that these will replicate to a level detectable in a viroid preparation. Hence, it can be predicted that a purified viroid sample will contain more than one sequence variant where a sequence variant is defined as an individual viroid molecule of defined sequence.

These predictions have been amply verified by comprehensive sequence analysis of isolates of several viroids. Perhaps the most extensive analyses have been done with CEV.[7,29,30] Some isolates seem to contain only one sequence variant, others contain two, whereas another CEV isolate contained nine different sequence variants in a total of 20 cDNA clones sequenced. The latter result indicates a much greater number of sequence variants than the nine sequenced in that CEV population, which was isolated from a single field orange tree. Relevant here is that, if a particular sequence variant is present at a level of 5% of the total variants, then at least 20 cDNA clones would have to be sequenced to provide a reasonable possibility of detecting it. The sequence variations found in 15 Australian sequence variants of CEV as well as two Californian variants were mostly located in the P and V domains and were concentrated into two smaller regions within the P and V domains.

In an attempt to define the region of the viroid molecule which determines the severity of symptom expression, cDNA clones were prepared from two isolates of CEV, one which gave severe symptoms on tomato and the other which gave mild symptoms.[31] The left-hand part of a severe isolate was joined with the right-hand part of a mild isolate through the C domain, and vice versa. Infectivity of such constructs on tomato seedlings indicated that severe symptoms were determined by the P domain.

There is no information about how viroids produce symptoms in infected plants (Section 6.11.9) and hence no molecular explanation for the results of domain-swapping experiments. In the case of CEV and CCCV, viroid replication appears to occur in the nucleus where the viroids accumulate[32] so that symptom expression is probably a consequence of some interference with normal nuclear function.

Sequence variants have been reported in populations of other viroids of the PSTV-group, e.g., PSTV, HSV, CCCV, and CSV, and hence this appears to be a common feature of viroid infection. In the case of the ASBV-group of viroids, 16 sequence variants were found in isolates prepared from three different Australian avocado trees.[33] Sequence length varied from 246 to 251 nt with sequence variation occurring mainly in the region of the left and right terminal loops.

6.11.5.2 Terminal Repeat Sequences in at least Two Viroids

In 1982, the sequencing of CCCV provided the first evidence for sequence duplications in the T2 domain.[34] It was already known that longer forms of CCCV accumulated in infected coconut palms as disease symptoms progressively become worse.[35] The 246/247 nt CCCV present early in infection is gradually replaced by one or more larger forms in which the T2 region and part of the V domain are duplicated (Figure 6).[7,34,35] Four forms have been identified with duplications of 41, 50, 55, and 100 (2×50) nucleotides. The mechanism by which such duplications are produced is not known but it is possible that a jumping RNA polymerase switching from one template strand to another during RNA synthesis may synthesize specific repeat sequences. Obviously, such sequence variants are viable but it is feasible that many other variants are produced which are not viable or are only replicated poorly and thus accumulate at below detectable levels.

An unusual variant of CEV, CEV D-92, was detected when an inoculum maintained on *Gynura aurantiacum* was used to infect a hybrid tomato (*Lycopericon esculentum* × L. *peruvianum*).[30] This new variant, of length 463 nt, contains an extra 92 nucleotides made up essentially of two 46 nt repeats which begin within the V domain at a site comparable to that found for the repeat sequences of CCCV. Hence, very similar duplications can occur in viroids which infect monocot and dicot hosts. For the CEV 463 nt variant, there was a dramatic reduction in symptom expression as compared with the 371 nt CEV parent when it was inoculated back onto *Gynura*.

6.11.5.3 RNA Rearrangements and their Potential Role in Viroid Evolution

The domain model described suggests that viroids evolve by the rearrangement of domains between two or more viroids infecting the same cell, followed by further mutation. For example, the T1 and T2 domains of CLV show high sequence homology to the corresponding domains in PSTV and TASV, respectively, and the boundaries are sharply defined by pairwise comparisons between members of the PSTV-subgroup[36] (Figure 7). Domain exchange at precise boundaries provides a nice model for recombination events but subdomain lengths also appear capable of exchanging as indicated by the presence in CLV of subdomain lengths of TPMV sequences in the P domain and of HSV and PSTV sequences in the C domain (Figure 7).

Examples of even more scrambled mixtures of sequences are found in the ASSV-subgroup of viroids.[37] For example, AGV contains sequences which appear to have been derived from CEV,

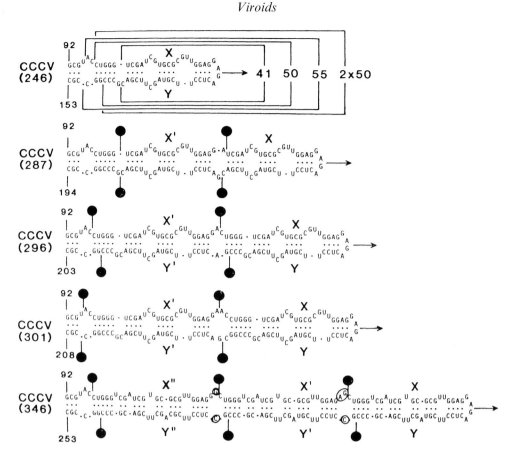

Figure 6 Partial sequence duplications of the T2 domain of CCCV.[7] The two sequences X and Y in CCCV (246) with a total of 41, 50, 55, and 2 × 50 nucleotides are duplicated as indicated. The arrows pointing to the right mark the boundaries of the X and Y sequences, while the filled circles mark the boundaries of the duplicated sequences. Circled nucleotides are sites of mutation in sequence variants. (Reproduced by permission of CRC Press from "Viroids: Pathogens at the Frontiers of Life," 1991.)

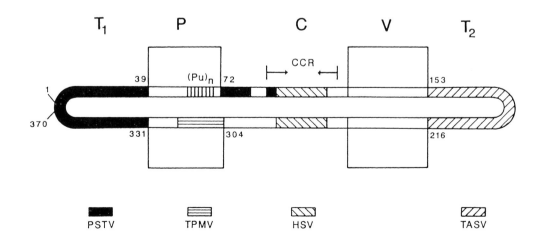

Figure 7 Schematic diagram of CLV showing regions of high sequence homology to other viroids. Residue numbers at boundaries of domains are given for this 370 nt viroid. (Reproduced by permission of CRC Press from "Viroids: Pathogens at the Frontiers of Life," 1991.)

PSTV, ASSV, and GYSV. These segments have sequence homologies which vary from about 50 to 100% of those in the putative parent viroids. Not only does AGV provide another example of RNA rearrangements within domains but it was also the first example in which rearrangements appear to have taken place between viroids belonging to two separate PSTV-subgroups.

6.11.5.4 Poly (pur) and Poly (pyr) Bias in Viroid Sequences

All evidence indicates that viroids can replicate in plant cells without the assistance of any helper virus; this is in contrast to viral single-stranded circular and linear satellite RNAs of roughly the same size range which are dependent on a helper. We have little understanding of why these two groups of RNAs differ this way. However, one aspect which may be relevant is the pattern of purine and pyrimidine tracts, which can make up substantial portions of the viroid genomes and are much less frequent in helper virus-dependent RNAs.[38] Intriguingly, a single-stranded circular human RNA of 1700 nt, hepatitis delta virus RNA with some viroid-like features, also has much greater than average distribution of these purine and pyrimidine tracts.[38]

The biological significance of this difference in the extent of distribution of these poly (pur) and poly (pyr) tracts is not known but it has been suggested that it may be related to differences in the replication strategies of the two groups.[38] For example, all evidence points to a role of the host nuclear DNA dependent RNA polymerase II in the replication of the PSTV-group of viroids and the hepatitis delta virus RNA, whereas the helper dependent satellite RNAs are replicated by RNA polymerases coded for by the helper viral RNAs.[39]

6.11.6 REPLICATION OF VIROIDS

6.11.6.1 Rolling Circle Mechanism of Replication

There seems to be universal agreement that viroids are replicated by a rolling circle mechanism, as first proposed in 1984[40] (Figure 8), and the finer aspects of their replication are slowly being unraveled. In one of the two variations of the rolling circle mechanism (Figure 8(a)), the infectious plus RNA is copied continuously by an RNA polymerase to produce a long minus RNA strand. Specific cleavage of the ($-$) strand, either enzymatically or RNA catalyzed, produces monomeric ($-$) strands which are circularized by a host RNA ligase. Continuous copying of this ($-$) RNA produces a continuous ($+$) strand which is cleaved at monomeric lengths that are then circularized to provide the circular ($+$) RNAs which accumulate in the infected cell.

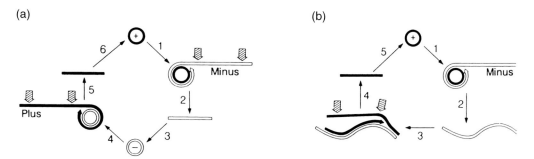

Figure 8 Two rolling circle models for the replication of viroids.[40] (a) Pathway followed by three members of the ASBV-group where both minus and plus RNAs self-cleave by the hammerhead structure. (b) Pathway followed by members of the PSTV-group of viroids where only the plus RNA is processed.

The three members of the ASBV-group are the only viroids which follow this pathway, as they have the ability to self-catalyze specific cleavage of both the ($+$) and ($-$) strands and multimeric ($-$) RNA does not accumulate in infected plants, at least for ASBV.[41-43]

In the other variation of the rolling circle model (Figure 8(b)), which is predicted to be the pathway for all members of the PSTV-group of viroids, the long linear (−) strand is not cleaved and can be detected in infected plants.[44] This (−) strand is then copied to provide the linear (+) strand which is specifically cleaved to monomers that are then circularized to give the progeny (+) viroid. Support for this model is indicated by the presence of high molecular weight (−) RNA in infected plants and the absence of unit length (−) species.

6.11.6.2   Host Enzymes involved in Viroid RNA Synthesis

There is no evidence that any viroids can code for any functional polypeptides. Only short potential open reading frames are found in these RNAs, which indicates that viroid replication must be mediated by host RNA polymerases. Inhibition of viroid replication, at least for PSTV and CEV, by alpha-amanitin points to a role for the nuclear host-encoded DNA polymerase RNA II in viroid replication.[11,39,45] There are no detailed studies as to how an enzyme that usually copies DNA is induced to copy an invading viroid RNA. Eukaryotic RNA polymerases are complex, multicomponent enzymes and no-one has yet accepted the challenge of characterizing their viroid RNA copying properties *in vitro*. One significant aspect of this problem would be trying to replicate *in vitro* the structural forms of the (+) and (−) viroid RNAs *in vivo* where they are likely to be quite different to the strongly folded RNAs isolated from infected plants.

Of the three viroids identified in the ASBV-group, ASBV is found in chloroplasts (see Section 6.11.6.3) so that its replication is predicted to be by the more prokaryote-like chloroplast DNA dependent RNA polymerase. The location of PLMV and CChMV in infected plants has yet to be determined but it seems reasonable to expect that this will also be in the chloroplasts.

6.11.6.3   Where does Viroid Replication occur within the Cell?

Unraveling the mechanism of viroid replication obviously requires knowledge of the site of replication within the cell. *In situ* hybridization approaches with analysis at both the electron microscope and confocal levels can identify sites of accumulation which may or may not also be the sites of synthesis. Sorting out the fine details of synthesis and accumulation possibilities will be a difficult task, and has yet to be done for any viroid.

Earlier studies have shown the inhibition of replication of PSTV and CEV by alpha-amanitin,[45] thereby implicating the host nuclear DNA-dependent RNA polymerase II in viroid synthesis. The first definitive *in situ* hybridization studies were carried out by Harders *et al.*[46] using nuclei purified from PSTV-infected tomato plants and biotinylated RNA probes followed by fluorescence detection by confocal laser scanning microscopy. Viroids were detected in 6–18% of infected nuclei and were homogeneously distributed throughout the nucleoli with minor distribution in the nucleoplasm.

The first study to localize viroids by *in situ* hybridization at the electron microscope level made use of biotinylated RNA probes for CCCV and ASBV, and detection using gold-labeled monoclonal antibiotin antibodies.[47] CCCV was located in the nucleus and nucleoplasm of cells of infected leaves of oil palm (*Elaeis guineensis*). Both PSTV and CCCV are in the same PSTV-subgroup of viroids so that the viroid distribution in nuclei is reasonably consistent between monocot and dicot plants. However, ASBV, the type member of the ASBV-group, was located within the chloroplasts, mostly on the thylakoid membranes of cells from infected leaves of avocado (*Persea americana*) (Figure 9), indicating a fundamental difference between the two group of viroids.[47] An independent study also localized ASBV in chloroplasts.[48]

In a more detailed study, the tissue and intracellular distribution of CCCV and CEV, both members of the PSTV-subgroup, was determined by *in situ* hybridization and both confocal laser scanning and transmission electron microscopy.[49] Both viroids were found in the vascular tissues as well as in the nuclei of mesophyll cells of infected oil palm and tomato, respectively. At the subnuclear level, however, CEV was distributed across the entire nucleus (Figure 10) in contrast to CCCV which was mostly concentrated in the nucleolus with the remainder distributed throughout the nucleoplasm.

The differences between the reports of the two laboratories on the distribution within the nucleus of the closely related viroids CEV and PSTV need to be considered. The whole nuclear distribution of CEV in sections of infected tomato tissue[47] contrasts with the nucleolar localization of PSTV, also in tomato.[46] It is possible that the purification of nuclei prior to *in situ* hybridization may have

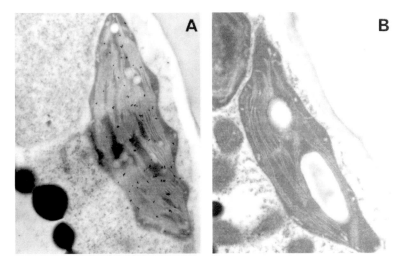

Figure 9 Localization of ASBV on thylakoid membranes of avocado chloroplasts by *in situ* hybridization with biotinylated RNA probes.[47] Hybridized probes detected by direct immunogold localization using 15 nm diameter gold-labeled antibiotin monoclonal antibodies. Section of mature mesophyll leaf cell from ASBV-infected avocado. (Reproduced in part by permission of BIOS Scientific Publishers from *Plant J.*, 1994, **6**, 99.)

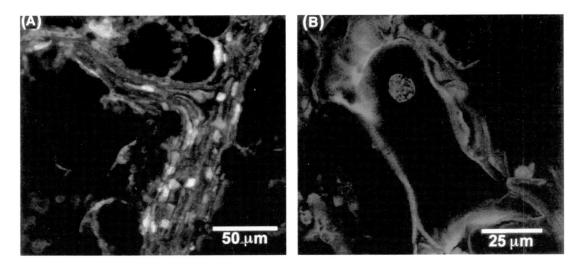

Figure 10 Localization by *in situ* hybridization of CEV in infected tomato by confocal microscopy. (a) Branched vascular bundle (blue) from CEV-infected tomato leaf tissue containing abundant CEV signal (red/orange). (b) Single mesophyll cell from CEV-infected tomato leaf showing cell nucleus with viroid signal (red/orange) and cell structure by autofluorescence (green). (Reproduced in part by permission of BIOS Scientific Publishers from *Plant J.*, 1996, **9**, 457.)

caused a redistribution of the PSTV into the nucleolus, especially since there is no fundamental reason why PSTV and CEV should show different nuclear distributions. A more comprehensive side-by-side repetition of the two experiments is needed to resolve these differences.

Obviously, we have a long way to go to gain a better appreciation of what is happening in the nucleus for the PSTV-group of viroids. Studies of nuclear distribution as a function of time after infection would be a good start as well as the inclusion of another couple of viroids from the PSTV-group which also infect tomato.

6.11.7 PROCESSING REACTION *IN VITRO* IN THREE VIROIDS VIA THE HAMMERHEAD SELF-CLEAVAGE REACTION

6.11.7.1 Hammerhead Self-cleavage Structure

The hammerhead self-cleavage structure was first identified in both the (+) and (−) forms of ASBV[41] and so named and further defined during the characterization of the self-cleavage of the 324 nt virusoid or viroid-like satellite RNA of lucerne transient streak virus (vLTSV)[50-52] and the (+) form of the 359 nt satellite RNA of tobacco ringspot virus.[53] Hammerhead self-cleavage has also been identifed in two other members of the ASBV-group of viroids, PLMV and CChMV, where both the (+) and (−) RNAs self-cleave.[42,43] The history and experimental approaches involved in the discovery and characterization of the hammerhead self-cleavage reaction is covered in several reviews.[4-6,54,55]

Two-dimensional hammerhead structures for the plus strand of ASBV as well as of the RNA transcript of the newt satellite II DNA, one of only two nonpathogenic RNAs so far identified that self-cleaves, are shown in Figures 11(a) and 11(b). The boxed nucleotides are highly conserved in the hammerhead self-cleaving viroid and satellite RNAs identified so far. The two base pair stem III of (+) ASBV is inherently unstable and that of the newt hammerhead structure, which has only a two base hairpin loop, even more so. However, the (−) ASBV hammerhead stem III, which has a three base pair stem III and a three base hairpin loop, is more stable.

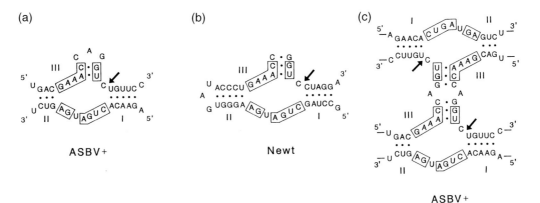

Figure 11 Single (a) and (b) and double (c) hammerhead structures. The newt double-hammerhead structure has the same arrangements as for plus ASBV. (Reproduced by permission of Elsevier Science from *Trends Biochem. Sci.*, 1989, **14**, 445.)

For the two other viroids which contain (+) and (−) hammerhead structures, PLMV and CChMV, the stem III contains from six to eight base pairs and hence should be quite stable.[42,43] The potential instability of stem III in the (+) ASBV and newt hammerhead structures led to the prediction that a double hammerhead structure might be involved in self-cleavage (Figure 11(c)). This was verified for (+) ASBV by mutational studies *in vitro*,[56] described in Section 6.11.7.2. In the case of the newt hammerhead structure, support for a double-hammerhead structure *in vitro* was indicated by a concentration dependence on the rate of self-cleavage.[56]

6.11.7.2 Double-hammerhead Structure in the Processing of the Plus Sequence of ASBV

Experimental proof of the double-hammerhead structure for ASBV was provided[56] using single base mutations of cDNA prepared from a tandem dimer repeat of the 247 nt ASBV in a SP6 RNA polymerase transcription vector (Figure 12). Either or both of the conserved GAAAC sequences (GAAAC-(A) or (B)) were mutated to GAAC in the dimeric cDNA. Mutation of both GAAACs, as predicted, abolished self-cleavage at both sites, SC-1 and SC-2. When only GAAC-(A) was mutated, self-cleavage at site SC-1 was unaffected, whereas self-cleavage at site SC-2 was abolished. Likewise, the mutation of GAAAC-(B) to GAAC abolished self-cleavage at SC-1 and not at SC-2. Both results are completely consistent with the double-hammerhead models shown in Figures 11(c) and 12(c).

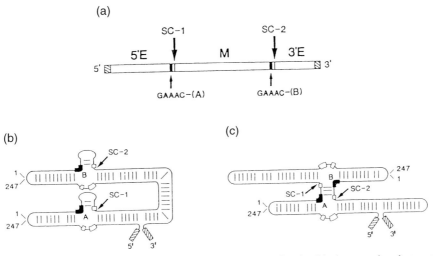

Figure 12 Experimental approach to demonstrate the presence of a double-hammerhead structure in RNA transcripts of a dimeric cDNA clone of ASBV.[56] (a) Schematic diagram of RNA transcript with self-cleavage sites SC-1 and SC-2. The conserved GAAAC in the hammerhead structure was mutated to GAAC in position A and/or B. (b) and (c) Single- and double-hammerhead arrangements, respectively, of dimeric RNA transcripts of plus ASBV. Self-cleavage occurs between residues 55 and 56 in plus ASBV and between residues 69 and 70 in minus ASBV.[44] (Reproduced by permission of CRC Press from *Crit. Rev. Plant Sci.*, 1991, **10**, 189.)

The differences between the two dimeric structures and the interpretation of the mutation data can be best visualized in the diagrams of Figure 12(b), containing two single hammerheads in the dimeric transcript, and the double-hammerhead structure in Figure 12(c). Note, for example, that the GAAAC-(A) sequence in Figure 12(b) (thick black section, labeled A) is juxtapositioned next to SC-1 in the single-hammerhead structure, whereas it is juxtapositioned next to SC-2 in the double-hammerhead structure of Figure 12(c). Hence, the GAAC-(A) mutation in Figure 12(b) directly inhibits self-cleavage at site SC-1 whereas, in the double-hammerhead structure of Figure 12(c), it inhibits self-cleavage at site SC-2.

An interesting variation of the result just described was observed[57] when the same kinds of transcriptions were carried out with dimeric cDNA clones of (−) ASBV, where the three base pair stem III and three base loop (not shown) should provide greater stability to stem III than the two base pair stem III of (+) ASBV. However, transcription of the minus cDNA clones showed that self-cleavage was still by the double-hammerhead structure. During these transcription reactions, the RNA would be expected to fold progressively into tertiary structures as it became separated from the RNA polymerase.

When gel-purified full-length dimeric transcripts of the (+) and (−) dimeric RNA were incubated under self-cleavage conditions, the (+) RNA still cleaved via a double-hammerhead structure. However, the (−) ASBV now self-cleaved via a single-hammerhead structure, indicating that stem III of the (−) ASBV hammerhead is, as predicted, more stable than that of the (+) ASBV RNA.[57]

In the case of the newt RNA hammerhead, with an even less stable stem III, a short (40-mer) RNA transcript containing the newt hammerhead sequence was incubated over a range of concentrations under self-cleavage conditions. The extent of self-cleavage was dependent on RNA concentration and showed roughly second-order kinetics, consistent with a bimolecular, double-hammerhead reaction.[56,58]

6.11.7.3 Crystallization of the Hammerhead

The simplistic two-dimensional structure of the hammerhead[41,50] has served its purpose well for the design of many successful experiments to characterize the self-cleavage reaction. The challenge of determining its three-dimensional structure was accepted by the crystallographers and nine years after the initial publications of the self-cleavage reaction, the first analysis of a crystal structure was published.[59] For this work, a 34-mer ribozyme and a 13-mer deoxynucleotide substrate were hybridized together and crystallized, and the X-ray crystallographic structure determined at 2.6 Å. This was followed soon after by the structure of an all-RNA ribozyme where self-cleavage was blocked by a 2′-methyl group on the 2′-hydroxyl at the cleavage site.[60] Finally, the crystal structure of an unmodified hammerhead RNA in the absence of divalent metal ions was solved and it was shown that this ribozyme can cleave itself in the crystal when divalent ions are added.[61]

The three-dimensional structures developed from the three approaches are essentially the same, and a diagrammatic representation of the structure is given in Figure 13. Stem III forms the stem of a Y-shaped structure where the arms are formed by stems I and II. The reader is referred to the original papers[59-61] for details.

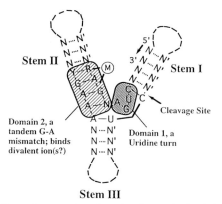

Figure 13 Schematic drawing of one hammerhead molecule as derived from its crystal three-dimensional structure.[58-61] Non-Watson–Crick base pairs are shown as thin lines. M, divalent cation. (Reproduced by permission of Spring-Verlag from *Nucleic Acids Mol. Biol.*, 1996, **10**, 161.)

6.11.7.4 Self-cleavage Hammerhead Reaction *in vivo*

Because ASBV cleaves itself *in vivo*, it seemed likely that this self-cleavage reaction will have a central role in the rolling circle replication of ASBV *in vivo*. Providing proof of this is much more difficult. The host range of ASBV is very narrow and essentially limited to avocados and related woody species which do not lend themselves to simple experiments using infectious cDNA clones. In experiments done many years ago to show the infectivity of purified ASBV on young avocado seedlings, it took a year for symptoms to appear. Likewise, PLMV offers no better prospects because of its very limited host range, again with a difficult woody species. However, in 1997, the characterization of CChMV with its herbaceous chrysanthemum[43] host raised the prospect of repeating the experiment described below where a viroid-like satellite RNA or virusoid of vLTSV was used to provide evidence for the involvement of the hammerhead self-cleavage reaction *in vivo*.

The approach used in this experiment[63] was to mutate an infectious cDNA clone of vLTSV such that the (−) RNA self-cleavage was eliminated and that of the (+) RNA left intact. In the wild-type situation where both the (+) and (−) RNAs self-cleave, only monomeric and short polymeric species of (+) and (−) RNAs are found in infected plants in the presence of the specific helper virus.[44] When the (−) RNA self-cleavage is eliminated by appropriate mutation, it can be predicted that high molecular weight polymeric (−) species will replace the monomeric and short polymeric (−) sequences of the wild type. Such high polymeric (−) species are found in plants infected with three virusoids where the (+) RNAs self-cleave *in vitro* via the hammerhead structure but the (−) RNAs do not self-cleave[64] and have no known self-cleavage activity.[41] Hence, three cDNA clones of vLTSV were mutated at three sites to eliminate self-cleavage *in vitro* in the (−) RNA.[63] When the excised inserts from these clones were co-inoculated with the helper virus LTSV on an appropriate herbaceous host plant, high molecular weight (−) vLTSV RNA was detected as expected while the control nonmutated vLTSV gave only the wild-type lower molecular weight species.

It was a surprise, however, to find that the mutated virusoids also produced some monomeric (−) vLTSV progeny in infected plants.[63] Sequence analysis of these progeny showed that 8–20% contained reversions and pseudoreversions of the introduced mutations which allowed self-cleavage in these progeny. Overall, the results provide convincing evidence for a role of hammerhead self-cleavage in the replication of at least the (−) RNA, but, as in many biological experiments, the experimental data are never as precise as the models on which we design our experiments.

In a less direct approach, the structure of a series of RNAs extracted from ASBV-infected avocado were characterized.[65] Sequence analysis of different (+) and (−) forms of ASBV were consistent with the symmetric rolling circle mechanism (Figure 8) and the predicted role of the hammerhead self-cleavage reaction.

6.11.7.5 Hammerhead Self-cleavage *in trans*—Application to Cleavage of Target RNAs

In the original single-hammerhead structure of (+) and (−) ASBV, the sequences making up the two halves of each structure originate from opposite strands in the central region of the molecule,[41] and self-cleavage *in trans* was demonstrated using two large fragments of the (−) ASBV molecule.[52] It was Uhlenbeck[66] who first recognized the potential for hammerhead self-cleavage *in trans* and who used a variety of short oligonucleotides to study the reaction and to use part of the hammerhead domain as a nuclease to cleave sequences embedded in different RNAs. The simplest construct of 19 nt RNA could cause rapid and specific cleavage of a 24 nt substrate and it had all the properties of an RNA enzyme or ribozyme, especially as it catalyzed multiple turnover of substrate.

Haseloff and Gerlach[67] extended this approach by the design of small ribozymes based on the hammerhead structure for the specific cleavage at three sites of an 835 nt RNA transcript of the bacterial chloramphenicol acetyl transferase gene *in vitro*. The approach was to find GUC sequences in the target RNA to define the cleavage site 3′ to the C residue and then to design the ribozyme around the core sequence of the (+) hammerhead domain of the 359 nt satellite RNA of tobacco ringspot virus which had been characterized by Bruening and his colleagues.[53,54] When the three ribozymes were in molar excess of the substrate, almost complete cleavage of the substrate could be obtained after 60 min at 50 °C. When the RNA substrate was in excess, each of the ribozyme constructs participated in greater than 10 cleavage events at each of the three sites in 75 min.[67]

The publication in 1988 of this work stimulated considerable effort by others to develop similar ribozyme constructs for their own particular system *in vitro* and to extend the approach to *in vivo* situations. There has been an explosion of effort in this area and the reader is referred to the three chapters on ribozymes in this volume for further details (Chapters 6.09, 6.10, and 6.15).

6.11.8 WHAT IS THE PROCESSING REACTION DURING ROLLING CIRCLE REPLICATION IN THE PSTV GROUP OF VIROIDS?

6.11.8.1 The Processing Reaction is likely to be RNA Catalyzed

In contrast to the well-characterized hammerhead self-cleavage reaction in the (+) and (−) RNAs of the three members of the ASBV-group of viroids, as well as evidence indicating a role for such reactions in rolling circle replication, very little is known about the processing event in the PSTV-group of viroids. In the few cases where such viroids have been investigated, the presence of low levels of high molecular (−) RNA species and essentially only monomeric (+) species strongly indicates that a processing reaction is only involved in the (+) species.[44,68,69] It can be reasonably safely predicted that this will be the situation for all members of the PSTV-group of viroids because of the conserved features of this viroid group.

Mainly because of the absence of any type of RNA catalyzed processing reaction in members of the PSTV-group of viroids, there has been a tendency to assume that processing of the multimeric (+) species is enzyme catalyzed although no such potential enzyme has been identified.[70,71] There has been one exception in a preliminary report[72] showing a 1–5% conversion of a dimeric RNA transcript of PSTV to monomeric linear PSTV plus two end fragments when the dimeric RNA was incubated under conditions used for Group I splicing reactions. The cleavage site was placed between residues 250–270 which is in the bottom strand of the C domain of PSTV and in a highly prospective region for a processing site. No further information has been published on this initial observation.

In view of the hammerhead self-cleavage reactions in the three members of the ASBV-group of viroids and in a number of circular single-stranded satellite RNAs associated with plant viruses,[11,43] it seems reasonable to predict that the processing reaction in the rolling circle replication of members of the PSTV-group of viroids will be RNA catalyzed also. The most likely cause of lack of success in providing clear evidence of such a reaction is the ability of single-stranded RNA to fold into multiple conformations *in vitro*, none of which are the active form required for self-cleavage. The author's own experience with the 324 nt viroid-like satellite RNA or virusoid of vLTSV provides a good example.[50-52] Various length RNA transcripts only self-cleaved after heat denaturation and snap cooling followed by assembly of the reaction mixture plus Mg^{2+} on ice. No self-cleavage occurred when the heated and snap cooled RNA was allowed to warm up prior to the addition of Mg^{2+}. In other words, conditions had to be empirically determined to form an active conformation capable of self-cleavage. Hence, it was considered likely that success in demonstrating a specific processing reaction in members of the PSTV-group of viroids would be crucially dependent on developing folding procedures where at least a small fraction of the population of RNA molecules would be in an active conformation in the presence of a divalent or multivalent cation.

6.11.8.2 Initial Indications of a Specific Processing Site

The CCR within the C domain of viroids, is a likely site for any type of specific processing reaction. The author's initial experiments exploring this region employed mutagenesis and infectivity studies on longer than unit length transcripts of cDNA clones of CEV. A potential *in vivo* processing site was identified in the upper strand of the CCR which corresponds to G96–G98 of the PSTV sequence.[25] However, this site was not unique in such experiments as would be necessary for a specific self-cleavage site, since mutation in other parts of the CEV molecules showed that the basic requirements for infectivity appeared to be the ability of RNA transcripts to form a short double-stranded region of viroid and vector sequences at the junction of the two termini.[33] Presumably, some processing reaction *in vivo* allowed the elimination of the vector sequences within the double-stranded region and restoration of the circular CEV.

6.11.8.3 A Common Mechanism in the CCR for all PSTV-group Viroids?

Given that there will be a common processing mechanism for all PSTV-group viroids, then it is highly likely to occur in a region conserved between its three subgroups (Figure 3). The most conspicuous feature they have in common is the potential for their inverted repeat sequences to form the two hairpin structures depicted in Figure 5. The sequences of the CCR in each subgroup vary (Figure 3) but the general two-dimensional structure of their hairpin loops is similar. That there is something unusual about these sequences, and hence about the structure of the CCR, is indicated by the specific cross-linking in PSTV of G98 to U260 that occurs on irradiation of the purified viroid with UV-light (Figure 14).[73] Obviously, there must be a juxtaposition of the two bases, presumably in a coplanar structure, for such cross-linking to occur. The specific UV-cross-linking also occurred[74] when a 55 nt RNA transcript constructed from sequences within the C domain was used; this RNA contained the sequence G88 to U114 coupled to the sequence G245 to A272 (see Figure 14). Since neither hairpin loop of Figure 14 is possible in the construct, a rod-like structure appears to be required for cross-linking to occur.

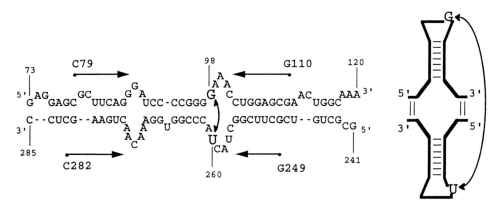

Figure 14 Sequence of the C domain of PSTV in the rod-like structure (left) and the potential hairpin stem-loop structure that can be formed from it (right). The four arrows represent the inverted repeat sequences,[73,74] the base residues of which are numbered. The UV-induced cross-linking between G98 and U260 is shown by a double-headed arrow. (Reproduced by permission of Oxford University Press from *Nucleic Acids Res.*, 1997, **25**, 2683.)

Obviously, we have too little structural information to explore possible conserved three-dimensional structures within the C domains of the three PSTV-subgroups by model building. At some point it may be possible for the RNA crystallographers to turn their attention to the sequences within the C domain of the PSTV-group viroids.

6.11.8.4   Is the Self-cleavage Site in the Bottom Strand of the CCR of the PSTV-subgroup of Viroids?

The author's group have searched for a specific self-cleavage site in the RNA sequences which comprise the C domain and have used CCCV for this work because it is the smallest member of the PSTV-subgroup. The approach was to prepare cDNA clones of the sequences in the C domain; thus, one clone contained the top strand of the C domain connected via a small loop to the bottom strand and the other contained the bottom strand connected to the top strand. A wide range of denaturation and reannealing approaches were explored using RNA transcripts prepared from these cDNA clones as well as various divalent and polyvalent cations, to see if self-cleavage could be detected by polyacrylamide gel electrophoresis.

One set of specific conditions identified a self-cleavage site (about 5% self-cleavage) on the left-hand side of the bottom hairpin loop[75] in Figure 5. The site is certainly in a conserved region within the members of the PSTV-subgroup and must be considered as a potential *in vivo* processing site. Proving that it actually is such a site is much more difficult and an important approach here would be mutagenesis analysis of infectious cDNA viroid clones. A different viroid would have to be used for this work since CCCV is quite impractical as its host range is restricted to the coconut and oil palms and related species in the palm family. Infectious cDNA clones of CEV are available with tomato seedlings as the experimental host plant. Initial experiments would involve single base mutations on either side of, or near to, the putative processing site and determination of their effect on infectivity. Two-dimensional structure modeling around this site-would permit a more rational design of mutations.

6.11.9   HOW DO VIROIDS EXERT THEIR PATHOGENIC EFFECTS?

6.11.9.1   What do we know about Pathogenicity?

Viroids do not code for any proteins or polypepetides which implies that pathogenic effects must be mediated in some way by interaction with one or more cellular constituents. Beyond this there are essentially no data on the molecular basis of viroid-induced pathogenic responses in plants. Speculation has centered around base-pair interactions with small nuclear and cytoplasmic RNAs, activation of protein kinases as a triggering event in viroid pathogenesis, the inhibitory binding of viroids to specific proteins, etc.[12,76] Obtaining concrete experimental evidence is likely to remain very difficult, especially as it will be necessary to distinguish between primary and secondary events.

Symptom expression can vary widely, depending on the particular viroid and the host plant. CCCV is essentially lethal to coconut and oil palms in the Philippines, with death coming several years after the appearance of the first symptoms. The smallest viroid certainly packs the biggest punch. Whether this highly virulent effect is related to the very limited monocot host range of CCCV is not known. ASBV also has a very limited host range within the dicot family Lauraceae where it causes symptoms ranging from severe to so mild that some trees are asymptomatic. Symptoms can also vary widely in the same tree from year to year.

We also have no molecular explanation as to how small sequence variations in a viroid can convert a severe PSTV variant into a mild one, except to suggest that sequence variation may modulate interactions with particular host components. It has, however, been possible to map pathogenic effects to particular domains within members of the PSTV-subgroup of viroids. Sequence analysis of sequence variants of PSTV, CEV, and HSV showed that essentially all symptom-affecting sequence differences for each viroid are located in the P and V domains.[8] In the case of CEV, sequence analysis of 17 sequence variants showed that sequence variation occurred within regions within the P and V domain,[7] further narrowing the part of CEV molecule likely to influence pathogenicity.

Unraveling the molecular basis of pathogenicity remains a most difficult task. Naturally occurring isolates of viroids, defined as viroid isolated from a single plant, usually contain a mixture of more than one sequence variant. It is therefore essential for such work to use infectious cDNA clones of the viroid under study to ensure that plants are inoculated with only one sequence variant. This approach does not remove the problem, it only delays it, since mutants of the inoculated viroid appear in progeny viroids. The site of replication of each viroid in the cell is obviously important in the details of pathogenicity but, as considered earlier (Section 6.11.6.3), the site of accumulation may not be the site of synthesis.

6.11.9.2 What Determines the Host Range of Viroids?

As for pathogenicity, we really have no explanations at the molecular level as to what determines host range, mainly because we know so little about the specificity of the interactions between the invading viroid and host processes. The three viroids in the ASBV-group all have very limited host ranges: ASBV to some members of the family Lauraceae, PLMV to a few species in the family Rosaceae, and CChMV to only the chrysanthemum in the very large Compositae family. A common feature of these three viroids is that both the plus and minus RNAs of each viroid self-cleave via the hammerhead structure. There is no information as to whether there is any functional correlation between limited host range and hammerhead self-cleavage. ASBV accumulates on the thylakoid membranes of avocado chloroplast;[47] it will obviously be of considerable interest to determine if both PLMV and CChMV also accumulate on the thylakoid membranes of their host plants.

Within the PSTV-subgroup of viroids, the host range varies from a very limited range for CCCV, as considered in Section 6.11.9.1, to quite wide ranges for PSTV, CEV, and HSV. Members in the ASSV- and CbV-subgroups seem to have restricted host ranges, but this could change as more members of these groups are discovered and characterized.

6.11.10 REFERENCES

1. T. O. Diener, "Viroids and Viroid Diseases," Wiley, New York, 1979.
2. T. O. Diener (ed.), "The Viroids," Plenum Press, New York, 1987.
3. J. S. Semancik (ed.), "Viroids and Viroid-like Pathogens," Academic Press, New York, 1987.
4. R. H. Symons (ed.), "Seminars in Virology," W. B. Saunders, Philadelphia, PA, 1990, vol. 1, issue 2.
5. R. H. Symons, *Crit. Rev. Plant Sci.*, 1991, **10**, 189.
6. R. H. Symons, *Annu. Rev. Biochem.*, 1992, **61**, 641.
7. P. Keese, J. E. Visvader, and R. H. Symons, in "RNA Genetics," eds. E. Domingo, J. Hollard, and P. Ahlquist, CRC Press, Boca Raton, FL, 1988, vol. 111, p. 71.
8. R. H. Symons, *Mol. Plant Microbe Interact.*, 1991, **4**, 111.
9. D. Riesner, *Mol. Plant Microbe Interact.*, 1991, **4**, 122.
10. T. O. Diener, *Virus Genes*, 1996, **11**, 47.
11. R. H. Symons, *Nucleic Acids Res.*, 1997, **25**, 2683.
12. R. Flores, F. D. Serio, and C. Hernandez, *Semin. Virol.*, 1997, **8**, 65.
13. T. O. Diener and W. B. Raymer, *Science*, 1967, **158**, 378.
14. J. S. Semancik and L. G. Weathers, *Phytopathology*, 1968, **58**, 1067.
15. T. O. Diener, *Virology*, 1971, **45**, 411.
16. J. S. Semancik and L. G. Weathers, *Nature New Biology*, 1972, **237**, 242.
17. H. J. Gross, H. Domdey, C. Lossow, P. Jank, M. Raba, H. Alberty, and H. L. Sänger, *Nature*, 1978, **273**, 203.
18. P. Keese and R. H. Symons, in "Viroids and Viroid-like Pathogens," ed. J. S. Semancik, Academic Press, New York, 1987, p. 1.
19. P. Keese and R. H. Symons, in "The Viroids," ed. T. O. Diener, Plenum Press, New York, 1987, p. 37.
20. P. Keese and R. H. Symons, *Proc. Natl. Acad. Sci. USA*, 1985, **82**, 4582.
21. H. L. Sänger, G. Klotz, D. Riesner, H. J. Gross, and A. K. Kleinschmidt, *Proc. Natl. Acad. Sci. USA*, 1976, **73**, 3852.
22. J. M. Sogo, T. H. Koller, and T. O. Diener, *Virology*, 1973, **55**, 70.
23. J. W. Randles and T. Hatta, *Virology*, 1979, **96**, 47.
24. A. M. Koltunow and M. A. Rezaian, *Intervirology*, 1989, **30**, 194.
25. J. E. Visvader, A. C. Forster, and R. H. Symons, *Nucleic Acids Res.*, 1985, **13**, 5843.
26. T. O. Diener, *Proc. Natl. Acad. Sci. USA*, 1986, **83**, 58.
27. R. L. Spieker, *J. Gen. Virol.*, 1996, **77**, 2839.
28. J. L. McInnes and R. H. Symons, in "Viroids: Pathogens at the Frontiers of Life," ed. K. Maramoroch, CRC Press, Boca Raton, FL, 1991, p. 21.
29. J. E. Visvader and R. H. Symons, *Nucleic Acids Res.*, 1985, **13**, 2907.
30. J. S. Semancik, J. A. Szychowski, A. G. Rakowski, and R. H. Symons, *J. Gen. Virol.*, 1994, **75**, 727.
31. J. E. Visvader and R. H. Symons, *EMBO J.*, 1986, **5**, 2051.
32. R. G. Bonfiglioli, D. R. Webb, and R. H. Symons, *Plant J.*, 1996, **9**, 457.
33. A. G. Rakowski and R. H. Symons, *Virology*, 1989, **173**, 352.
34. J. Haseloff, N. A. Mohamed, and R. H. Symons, *Nature*, 1982, **299**, 316.
35. J. S. Imperial, J. B. Rodriguez, and J. W. Randles, *J. Gen. Virol.*, 1981, **56**, 77.
36. R. Hammond, D. R. Smith, and T. O. Diener, *Nucleic Acids Res.*, 1989, **17**, 10 083.

37. M. A. Rezaian, *Nucleic Acids Res.*, 1990, **18**, 1813.
38. A. D. Branch, S. E. Lee, O. D. Neel, and H. D. Robertson, *Nucleic Acids Res.*, 1993, **21**, 3529.
39. I. M. Schindler and H. P. Mühlbach, *Plant Sci.*, 1992, **84**, 221.
40. A. D. Branch and H. D. Robertson, *Science*, 1984, **223**, 450.
41. C. J. Hutchins, P. D. Rathjen, A. C. Forster, and R. H. Symons, *Nucleic Acids Res.*, 1986, **14**, 3627.
42. C. Hernandez and R. Flores, *Proc. Natl. Acad. Sci. USA*, 1992, **89**, 3711.
43. B. Navarro and R. Flores, *Proc. Natl. Acad. Sci. USA*, 1997, **94**, 11 262.
44. C. J. Hutchins, P. Keese, J. E. Visvader, P. D. Rathjen, J. L. McInnes, and R. H. Symons, *Plant Mol. Biol*, 1985, **4**, 293.
45. R. Flores and J. S. Semancik, *Proc. Natl. Acad. Sci. USA*, 1982, **79**, 6285.
46. J. Harders, N. Lukacs, M. Robert-Nicoud, T. M. Jovin, and D. Riesner, *EMBO J.*, 1989, **8**, 3941.
47. R. G. Bonfiglioli, G. I. McFadden, and R. H. Symons, *Plant J.*, 1994, **6**, 99.
48. M. I. Lima, M. E. N. Fonseca, R. Flores, and E. W. Kitajima, *Arch. Virol.*, 1994, **138**, 385.
49. R. G. Bonfiglioli, D. R. Webb, and R. H. Symons, *Plant J.*, 1996, **9**, 457.
50. A. C. Forster and R. H. Symons, *Cell*, 1987, **49**, 211.
51. A. C. Forster and R. H. Symons, *Cell*, 1987, **50**, 9.
52. A. C. Forster, A. C. Jeffries, C. C. Sheldon, and R. H. Symons, *Cold Spring Harbor Symp. Quant. Biol.*, 1987, **52**, 249.
53. G. A. Prody, J. T. Bakos, J. M. Buzayan, I. R. Schneider, and G. Bruening, *Science*, 1986, **231**, 1577.
54. G. Bruening, *Semin. Virol.*, 1990, **1**, 127.
55. K. R. Birikh, P. A. Heaton, and F. Eckstein, *Eur. J. Biochem.*, 1997, **245**, 1.
56. R. H. Symons, *Trends Biochem. Sci.*, 1989, **14**, 445.
57. A. C. Forster, C. Davies, C. C. Sheldon, A. C. Jeffries, and R. H. Symons, *Nature*, 1988, **334**, 265.
58. C. Davies, C. C. Sheldon, and R. H. Symons, *Nucleic Acids Res.*, 1991, **19**, 1893.
59. H. W. Pley, K. M. Flaherty, and D. B. McKay, *Nature*, 1994, **372**, 68.
60. W. G. Scott, J. T. Finch, and A. Klug, *Cell*, 1995, **81**, 991.
61. W. G. Scott, J. B. Murray, J. R. P. Arnold, B. L. Stoddard, and A. Klug, *Science*, 1996, **274**, 2065.
62. D. G. McKay, *Nucleic Acids Mol. Biol.*, 1996, **10**, 161.
63. C. Davies, C. C. Sheldon, and R. H. Symons, *Nucleic Acids Res.*, 1991, **19**, 1893.
64. C. C. Sheldon, A. C. Jeffries, C. Davies, and R. H. Symons, *Nucleic Acids Mol. Biol.*, 1990, **4**, 227.
65. J. A. Daros, J. F. Marcos, C. Hernandez, and R. Flores, *Proc. Natl. Acad. Sci. USA*, 1994, **91**, 12 813.
66. O. C. Uhlenbeck, *Nature*, 1987, **328**, 596.
67. J. Haseloff and W. L. Gerlach, *Nature*, 1988, **334**, 585.
68. A. D. Branch, H. D. Robertson, and E. Dickson, *Proc. Natl. Acad. Sci. USA*, 1981, **78**, 6381.
69. R. A. Owens and T. O. Diener, *Proc. Natl. Acad. Sci. USA*, 1982, **79**, 113.
70. G. Steger, T. Baumstark, M. Mörchen, M. Tabler, M. Tsagris, H. L. Sänger, and D. Riesner, *J. Mol. Biol.*, 1992, **227**, 719.
71. T. Baumstark and D. Riesner, *Nucleic Acids Res.*, 1995, **23**, 4246.
72. H. D. Robertson, D. L. Rosen, and A. D. Branch, *Virology*, 1985, **142**, 441.
73. A. D. Branch, B. J. Benenfeld, and H. D. Robertson, *Proc. Natl. Acad. Sci. USA*, 1985, **82**, 6590.
74. C. P. Paul, B. J. Levine, H. D. Robertson, and A. D. Branch, *FEBS Lett.*, 1992, **305**, 9.
75. Y. Liu and R. H. Symons, *RNA*, 1998, **4**, 418.
76. T. O. Diener, R. W. Hammond, T. Black, and M. G. Katze, *Biochimie*, 1993, **75**, 533.

6.12
Structural Elements of Ribosomal RNA

STEVEN T. GREGORY, MICHAEL O'CONNOR, and
ALBERT E. DAHLBERG
Brown University, Providence, RI, USA

6.12.1 INTRODUCTION

Ribosomal RNAs (rRNAs) represent an immense challenge to structural biochemists and molecular biologists who attempt to define their role as the structural and functional core of the translational machinery. All ribosomes consist of two subunits of unequal mass, the smaller containing a 16S–18S rRNA molecule, the larger containing a 23S–28S rRNA molecule in addition to, in most organisms, a small 5S rRNA molecule. Complexed with numerous ribosomal proteins, these massive rRNAs must fold into compact ribonucleoprotein particles, interact with factors for initiation, elongation and termination, bind and differentiate among numerous tRNA molecules based on their interaction with mRNA, and move the tRNA–tRNA–mRNA complex through the ribosome in a precise and concerted fashion. Deciphering the three-dimensional structure of the large rRNAs promises to reveal not only many new underlying principles that govern RNA higher order structure and folding, but also how large RNA molecules change conformation to achieve the coordinated movement of multiple substrates.

While the three-dimensional structures of either 16S or 23S rRNA have yet to be determined at high resolution, the substantial effort directed toward this objective has nevertheless proven fruitful. There is great confidence in the highly detailed secondary structure models derived by the method of comparative sequence analysis and a number of tertiary interactions are now being detected by this approach.[1] Comparative sequence analysis has also identified secondary structure motifs, such as tetraloops and E loops, which are likely to play key roles in establishing and maintaining the tertiary structure of the ribosome. The role of tetraloops and their interaction with specific docking sites (receptors) in the formation of tertiary structure of smaller RNAs has been examined in detail,[2] and their participation in rRNA architecture is expected to be equally important. In addition, high-resolution structure determination of other secondary structure building blocks and functionally important structural elements will continue to reveal new modes of tertiary interaction.

The accumulated data from cross-linking studies and footprinting studies places a sufficient number of constraints on the three-dimensional folding of 16S rRNA for the construction of several models.[3-5] Remarkable advances are also being made using X-ray crystallography and cryoelectron microscopy which, it is anticipated, will culminate in no less than an atomic resolution structure of the ribosome.[6] In this chapter, the authors review several aspects of what is currently known about rRNA secondary and tertiary structure, and describe in some detail the structure of several important functional sites that have been characterized at high resolution.

6.12.2 SECONDARY STRUCTURE ELEMENTS OF rRNAs

6.12.2.1 Secondary Structure Models Derived by Comparative Sequence Analysis

Ribosomal RNA secondary structure models have been derived primarily via comparative sequence analysis. The principle of the comparative approach, first applied to prediction of the secondary structures of tRNAs, has more recently been applied to other RNAs including rRNAs.[1] The underlying principle of this method is the presumption that homologous RNA molecules, having experienced similar selective pressures throughout their evolution, will adopt similar three-dimensional structures, despite having very different nucleotide sequences. Thus, while the nucleotide sequence of any individual RNA molecule is compatible with any of a vast number of alternative secondary structures, only one of these structures will be compatible with the nucleotide sequences of all homologous RNAs. In practice, two bases are inferred to engage in hydrogen-bonded base pairing if they covary in a concerted fashion among a large number of RNA molecules. The greater the number of phylogenetically independent exchanges, the greater the certainty of the base pairing interaction. Examples of nonconcerted changes are taken as evidence against base pairing. Formally, mechanisms other than direct base–base interaction can be invoked to explain covariation, although base pairing is the most intuitively satisfying for a series of consecutive covariations that follow Watson–Crick pairing rules. In addition to the standard Watson–Crick exchanges (i.e., A–U to G–C, G–C to C–G, etc.), noncanonical exchanges (such as U–U to C–C, U–G to C–A, A–G to G–A, etc.) are also detectable by covariation analysis. Thus, covariation is not restricted *a priori* by any specific set of pairing rules based on the hydrogen bonding arrangements of the pairing bases. Nevertheless, the vast majority of helices in rRNAs are constructed of antiparallel, canonical Watson–Crick or G–U wobble pairs (roughly 90%[1]) with occasional noncanonical pairing interactions embedded within standard helices. In some cases these noncanonical pairings comprise parts of conserved motifs that may eventually be found to participate in RNA–RNA, RNA–protein, or RNA–ligand interactions (Figure 1).

The secondary structures of all rRNAs have been divided (somewhat arbitrarily) into domains; three for 16S rRNAs and six for 23S rRNAs. It is unknown how structurally or functionally independent these divisions are, although the 3′ minor domain of 16S has been assembled *in vitro* with ribosomal proteins into a discrete particle,[7] while individual domains of 23S rRNA have been found to vary in their ability to stimulate peptide bond formation.[8]

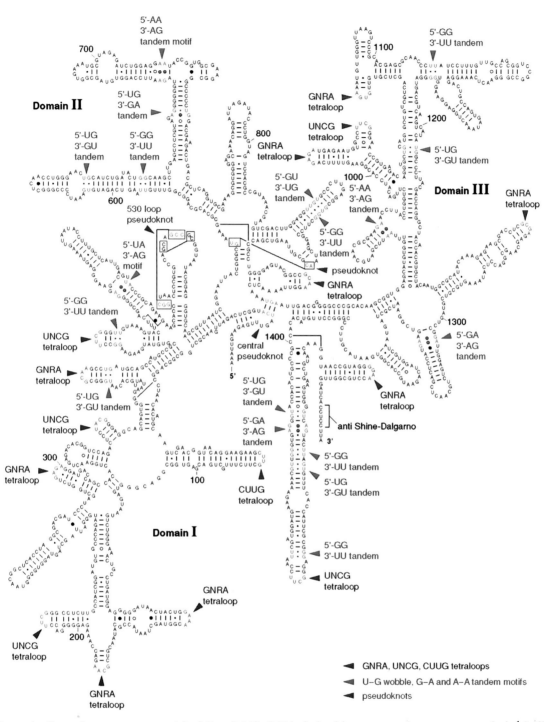

Figure 1 Secondary structure model of *E. coli* 16S rRNA derived by comparative sequence analysis.[1] Indicated are GNRA, UNCG, and CUUG tetraloops (blue), G–U and purine–purine base pairing motifs (green), and pseudoknots (red). The structure has been modified to include G–U and purine–purine base pairing interactions described in Refs 9 and 13.

The universal sequence conservation of certain regions of rRNAs makes them resistant to structural modeling via comparative sequence analysis. Unfortunately, it is the structures of these regions that are of the greatest interest as their conservation implies a critical association with ribosome functions. Establishing the structures of such sites will require high-resolution biophysical techniques.

6.12.2.2 Secondary Structure Elements and Motifs

The secondary structures of rRNAs are composed largely of short Watson–Crick helices that are either connected to one another by internal loops and multihelix junctions, or capped by terminal loops (see Figure 1). The three-dimensional structures of multihelix junctions have not been solved, whereas the structures of a number of internal loops have been examined by NMR spectroscopy or inferred from related structures in other RNA molecules. Internal loops can also be thought of as a series of consecutive mismatches (paired or unpaired), and can be symmetric (an equal number of residues on each strand) or asymmetric (an unequal number on each strand), the latter possibly resulting in bulged residues.

Several internal loop or mismatch motifs have been identified by comparative sequence analysis. A high frequency of purine–purine mismatches exist at certain locations in rRNAs.[9] These mismatches frequently occur in tandems and in nonrandom arrangements. Thus, while 5′-GA/3′-AG, 5′-AA/3′-AG, and 5′-AA/3′-GG occur at high frequency, 5′-AG/3′-GA is extremely rare. These patterns are rationalized as resulting from the formation of a type II, or "sheared," base pairing configuration. The type II sheared GA pair consists of hydrogen bonds from N2G to N7A and N6A to N3G. This configuration positions the G toward the major groove and the A toward the minor groove, with an underwinding of the helix. In contrast with a single sheared G-A pair, two consecutive G-A pairs facing one another in the 5′-GA/3′-AG arrangement can be embedded within a Watson–Crick helix without substantially distorting the neighboring backbone conformation. The structure of this motif has been observed directly in model duplexes,[10] the hammerhead ribozyme,[11] and the *Tetrahymena* ribozyme as the related 5′-AA/3′-AA motif.[2] A highly conserved 5′-GA/3′-AG tandem occurs in 16S rRNA at positions 1305–1306/1332–1331, flanked by two mismatches creating the structure 5′-GGAU/3′-AAGU (Figure 1[9]). A related motif, 5′-UA/3′-AG, consists of a U–A reverse Hoogsteen pair followed by a type II sheared A–G pair.[9,12] This motif is highly conserved at positions 486–487/448–447 of 16S rRNA (Figure 1) and at several locations in 23S rRNA.[9]

Secondary structure motifs involving G-U pairings have also been identified by comparative sequence analysis.[13] These include 5′-UG/3′-GU, 5′-GG/3′-UU, and 5′-UG/3′-GA tandems. It is expected that such motifs most often involve G–U pairs in the wobble configuration. In the case of 5-UG/3′-GU, opposing wobble pairs may "compensate" for deficiencies in stacking caused by the wobble configuration. In the case of 5′-UG/3′-GA, the G moved toward the minor groove by the wobble configuration may enhance stacking of the minor groove adenosine in the sheared configuration. Consecutive G–U pairs in the wobble configuration project successive N-2 amino groups into the minor groove. Such an arrangement might serve as a protein recognition site or an anchoring site for RNA–RNA interaction. 5′-GA/3′-AG and 5′-UG/3′-GU motifs occur in a nonrandom fashion and are likely to play an important, albeit as yet undefined, role in rRNA structure and function.

Several consecutive noncanonical base pairing interactions combine to form an asymmetric internal loop found in 5S and 23S rRNAs. Called the E loop motif after the E loop of eukaryotic 5S rRNA,[12] this structure consists of a type II sheared A–G pair, a U–A reverse Hoogsteen pair (N3U–N7A, O2U–N6A), a bulged G residue, and a symmetric (N-6–N-7) A–A pair. While one backbone strand conforms closely to A form geometry, the other strand displays a striking local reversal of direction to accommodate the bulged G residue and the symmetric A–A pair (both of which are *anti*). It has been proposed that the eukaryotic E loop belongs to a larger family of loops, the S loops, which might serve as docking modules for long-range RNA–RNA interactions.[14] Interestingly, the E loop of *E. coli* 5S rRNA[15] differs substantially from that of eukaryotic 5S rRNA. In addition to a sheared A–G pair and a U–A reverse Hoogsteen pair, the loop also contains a G–G (N-1, N-2-O-6) pair, a head-to-head type I A–G (N-6A–O-6G) pair, and a nonwobble G–U (N-1, N-2G–O-4U) pair. The bacterial loop also lacks the bulged G residue and backbone reversal characteristic of the eukaryotic loop.

The most frequently found single class of terminal loops found in rRNA are the tetranucleotide loops or tetraloops[16] (see Figure 1). While originally identified in rRNAs, they are also found in many other RNAs and are likely to be one of the most common secondary structure motifs in rRNAs. Three main classes of tetraloops, GNRA, UNCG, and CUUG, comprise half the terminal loops in rRNAs. These loops have in common a highly ordered and compact structure with extensive hydrogen bonding. In particular, UNCG tetraloops are highly stable thermodynamically and may function as nucleation sites for RNA folding. GNRA tetraloops, on the other hand, are important tertiary structure elements, docking into the minor groove of RNA helices at highly specific tetraloop receptor sequences.[2,17]

Interestingly, the GNRA fold has also been found in larger loops,[18] suggesting that it may exist hidden in a variety of terminal and internal loops and at multihelix junctions. The GAAA tetraloop–receptor complex can also withstand substantial variation in both the loop and receptor structures, suggesting that a wide range of equivalent or similar docking interactions may exist.[19] This variability in the structure of such complexes will make them more difficult to recognize by sequence analysis, but may prove to be a fundamental building block for RNA architecture.

6.12.3 HIGH-RESOLUTION STRUCTURES

Recent advances in methodologies for the synthesis of RNA molecules and their analysis by NMR spectroscopy have led to the structure solutions for several rRNA analogues. As described in previous chapters, UNCG,[20] GNRA,[21] and CUUG[22] classes of tetraloops comprise half the terminal loops in rRNAs. More recently, the structures of loops and helices unique to specific functional sites of rRNAs have also been solved. These studies reveal some of the remarkable variety of conformations that RNAs can adopt, while at the same time providing insight into some of the common themes of rRNA secondary structure.

The practical limitations of the methodologies used to derive high-resolution structures prevent their accounting for tertiary interactions or interactions with ribosomal proteins. Thus, it is possible, and perhaps likely, that the conformation of particular elements differ in some details in the context of an intact functioning ribosome. In addition, the structure of an oligonucleotide analogue may represent only one of several possible alternate conformations adopted in the ribosome during the translation cycle. Its relevance to ribosome structure can, to some degree, be measured by its compatibility with results from mutagenesis, structure probing, and other biochemical methodologies using intact ribosomes. Disagreements between structural data derived with analogues and data from intact ribosomes may be indicative of the existence of tertiary interactions or conformational dynamics *in situ*.

Each of the structures described below was analyzed with the goal of revealing features important to a particular aspect of ribosome function. While some of these structures, such as the decoding center and P loop analogues, probably do not represent common secondary structure motifs used in rRNA architecture, they do illustrate the potential for RNA to adopt unusual conformations to form sites of interaction with specific ligands. The GUAAUA hexaloop may be found to be representative of a class of six base loops, while the dimethyl A stem loop is a common GNRA tetraloop with a dramatically different conformation due to base methylations.

6.12.3.1 The Decoding Region (16S rRNA)

The decoding region consists of a segment of domain III of 16S rRNA that is intimately associated with codon–anticodon interaction (Figure 2(a)). The anticodon of tRNA in the P site can be cross-linked to the highly conserved C1400,[23] as can mRNA.[24] Chemical footprinting experiments with tRNA define a subset of nucleotides potentially interacting with the codon–anticodon complex.[25] Aminoglycoside antibiotics, which promote mistranslation, protect a similar set of nucleotides,[26] and resistance to these antibiotics occurs via mutations or base methylations in this region (reviewed in reference 27).

Aminoglycoside antibiotics bind specifically to oligonucleotide analogues of the decoding region.[28,29] The structure of a 27 nucleotide analogue corresponding to C1404–C1412 and G1488–G1497 of the decoding region complexed with the aminoglycoside antibiotic paromomycin, has been examined by NMR spectroscopy.[30] This RNA consists of a helical stem interrupted by an asymmetric internal loop (Figure 2(a)). The loop contains a closing U1406–U1495 base pair followed by a Watson–Crick C1407–G1494 pair. A1408 engages in a noncanonical pair via its N-6 by hydrogen bonding to the N7 of A1493. A1492 is unpaired and bulged. These noncanonical base pairing interactions and the conformation of A1408, A1492, and A1493 are essential for formation of the aminoglycoside binding site. A Watson–Crick base pair between C1409 and G1491 is also essential for aminoglycoside binding, and mutations disrupting this base pair confer aminoglycoside resistance (reviewed in reference 27). Mutations at this base pair also affect translational accuracy.[31] The ability of this RNA to bind drugs supports the contention that it represents the native structure of the corresponding region of 16S rRNA in the ribosome.

Binding of paromomycin occurs in the major groove within the internal loop. Ring II of paromomycin spans the U1406–U1495 and C1407–G1496 base pairs; amino groups in ring II make

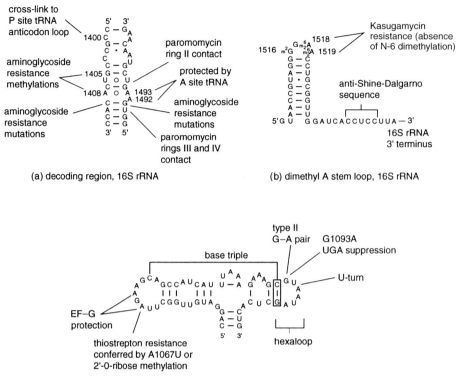

Figure 2   Secondary structures of functional sites characterized at high resolution by NMR spectroscopy. (a) The decoding region of 16S rRNA,[30] indicating sites of mutations, protections by tRNA, and sites of contact with the aminoglycoside antibiotic paromomycin detected by NMR spectroscopy. (b) The dimethyl A stem loop[39] and the adjacent 3′ terminus of 16S rRNA. The anti-Shine–Dalgarno sequence which base pairs with mRNA during translation initiation is indicated. (c) The GTPase center of 23S rRNA. The two loops which interact with ribosomal protein L11 and the antibiotic thiostrepton are brought into close proximity by a base triple between the G1092-A1099 base pair and C1072.[46] The structure of the 5′-GUAAUA-3′ hexaloop has been solved by NMR.[47]

hydrogen bonds to the O-4 of U1495 and N-7 of G1494. The A1408–A1493 pair, together with the bulged A1492, form a binding pocket for ring I of paromomycin. All eukaryal small subunit rRNAs have a G at position 1408 and bind aminoglycosides much less effectively than bacterial ribosomes. The C1409–G1491 base pair forms the "floor" of the drug-binding pocket.

Aminoglycoside-producing organisms typically avoid the inhibitory effects of the antibiotic through modification of specific rRNA residues. Methylation of A1408 prevents formation of the A1408–A1493 pair and prevents binding of a range of aminoglycosides. Resistance to kanamycin and gentamycin in the producing organisms is achieved through methylation of G1405; NMR studies have placed ring III of gentamicin close to G1405, and consequently, methylation may sterically hinder binding of these antibiotics. Analysis of RNA interactions with the two-, three-, and four-ring aminoglycosides (neamine, ribostamycin, and neomycin, respectively) indicate that rings I and II (found in all these aminoglycosides) interact with RNA in much the same way as paromomycin, while rings III and IV of the neomycin class of antibiotics make sequence-independent interactions with the phosphate backbone of the RNA, near the A1410–U1490 base pair. Comparison of the free and paromomycin-bound forms of the RNA indicates that in the absence of the antibiotic, A1492 and A1493 in the internal loop assume a dynamic, flexible conformation.[32] Binding of paromomycin in the major groove results in a shift of these residues towards the minor groove and a stabilization of their conformation. Antibiotic binding also altered the pattern of hydrogen bonding between A1408 and A1493; one hydrogen bond was found in the free form of the RNA while two occurred in the drug-bound form. Aminoglycoside antibiotics promote misreading of the genetic code and prevent translocation by increasing the nonspecific (codon-independent) affinity of tRNAs for the A site. Three bases implicated in aminoglycoside–ribosome interactions, A1408,

A1492, and A1493 are also protected from DMS modification by mRNA-dependent, A site tRNA binding.[25] These considerations have led to the suggestion that the universally conserved A1492 and A1493 may be involved in recognition of the correctly-formed codon–anticodon complex, and stabilization of their conformation that occurs upon drug binding may induce a high-affinity binding site for the codon–anticodon complex, with a concomitant loss in the specificity of decoding.[30,32] This study is notable as the first characterization at atomic resolution of the mode of interaction of an antibiotic and rRNA. Such studies are potentially of great importance for the rational design of novel antibiotic inhibitors of ribosome function.

6.12.3.2 The Dimethyl A Loop (16S rRNA)

One of the most conserved structural features of 16S-like rRNAs is the 3′ terminal dimethyl A loop (Figure 2(b)). This tetranucleotide loop has been cross-linked to domain IV of 23S rRNA[33] and the helix containing this loop has been cross-linked to initiation factor 3 (IF3),[34] suggesting it plays a role in subunit association during the initiation phase of translation. In prokaryotic ribosomes the loop has the sequence GGAA (positions 1516–1519, *E. coli* numbering), while its eukaryotic counterpart is UGAA. This loop is heavily modified, with 2-methylguanosine at the first position in prokaryotic 16S rRNAs, and *N*-6-dimethyladenosine at the third and fourth positions in all 16S-like rRNAs examined. Bacterial ribosomes lacking the *N*-6-dimethylations are resistant to the action of the antibiotic kasugamycin, an inhibitor of translation initiation,[35] and are less accurate.[36] While little or no growth defect in prokaryotes is associated with deficiency in this modification, and only mild effects on ribosome function are observed *in vitro*,[37] its absence is lethal in yeast.[38]

The structure of the fully modified loop was recently solved by NMR spectroscopy.[39] Despite having a canonical GNRA tetraloop consensus sequence, the dimethyl A loop differs substantially in its conformation from that of the corresponding unmodified loop.[21] While m^{2}G1516 is positioned as expected for a GNRA loop, *N*-6 dimethylation of A1519 prevents formation of a closing type II sheared G1516–A1519 pair characteristic of the GNRA family. Rather than being coplanar with m^{2}G1516, m$_2$^{6}A1519 is tilted upward toward the loop apex. The adjacent m$_2$^{6}A1518 stacks on m$_2$^{6}A1519. These methylations prohibit the extensive hydrogen bonding interactions found in GNRA loops, but only partially destabilize the loop.[39,40] The unusual conformation of the dimethyl-adenosines also changes the backbone conformation such that the turning phosphate is after A1519 rather than after G1517 (which also has a C2′ *endo* ribose pucker) and has the effect of positioning the bases in the major groove rather than in the minor groove as in GNRA tetraloops. While the NMR structure of the unmodified eukaryotic UGAA tetraloop has been solved,[41] it remains to be seen if the *N*-6-dimethylated UGAA loop adopts a conformation similar to that of the prokaryotic GGAA loop. Comparison of the methylated and unmodified structures illustrates the potential for modified nucleotides to affect a specific loop conformation.

6.12.3.3 The 1093–1098 5′-GUAAUA-3′ Hexaloop (23S rRNA)

The 1093–1098 5′-GUAAUA-3′ hexaloop, located in a three helix junction structure in domain II of 23S rRNA, comprises part of the binding sites for elongation factors G (EF-G) and Tu (EF-Tu) and ribosomal protein L11 (Figure 2(c)). Based on chemical modification experiments with ribosomal protein L11 and the antibiotic thiostrepton,[42,43] and footprinting and hydroxyl radical probing with EF-G,[44,45] the junction structure is thought to fold back on itself such that the two terminal loops, U1065 to A1073 and G1093 to A1098, are in close proximity. This is mediated in part by a base triple between the C1092–G1099 base pair and C1072.[46] Mutations and post-transcriptional modifications affect binding of the antibiotic thiostrepton, ribosomal protein L11, and EF-G.[42]

The structure of the G1093 to A1098 5′-GUAAUA-3′ hexaloop has been determined by NMR spectroscopy.[47] The loop contains a closing type II sheared G–A pair such as that seen in GNRA tetraloops. Interestingly, a G1093A mutation which confers a nonsense suppressor phenotype[31,48] could potentially form an analogous type II A–A pair. In addition, a U turn or uridine turn[49] occurs at residues U1094, A1095, and A1096, with a hydrogen bond between the O-2′ of U1094 and the N-7 of A1096 and a hydrogen bond between the N-3 of U1094 and the phosphate O-2 of U1097. Uridine turns are found in the anticodon and T loops of tRNA and in GNRA tetraloops.[50] The

5′-GUAAUA-3′ hexaloop provides an additional example of the existence of this secondary structure motif in terminal loops. U1097 is extended into solution, perhaps to engage in tertiary or RNA–protein interactions. Whether or not this structure is representative of a class of hexaloops remains to be determined.

6.12.3.4 The P Loop (23S rRNA)

The P loop, residues G2250 to C2254, located in domain V of 23S rRNA, forms a Watson–Crick base pair with the 3′ terminal CCA of P site tRNA,[51] and is an essential component of the peptidyl transferase center. Mutations in the loop affect translational fidelity[52] and cause severe defects in peptidyl transferase activity.[51,53]

The structure of an oligonucleotide analogue of the P loop was solved by NMR spectroscopy[54] (Figure 3). The most striking feature of the loop is the splayed out position of the loop bases, making them accessible for base pairing with the CCA end of tRNA. This arrangement is very different from that of other loops such as tetraloops and the anticodon loops of tRNAs, in which loop bases extend the stacking arrangement of the stem helix. G2252, which forms the canonical base pair with C74 of tRNA, and G2251 stack on one another facing the major groove of the helix. A reverse Watson–Crick base pair between G2251 and C75 of tRNA has been proposed.[54] Not surprisingly, mutations at G2251 or G2252 produce dominant lethal phenotypes and cause increased frameshifting and readthrough of nonsense codons. The G2253A and G2253U mutations have little or no detectable phenotype, in contrast to the G2253C mutation which produces a slow-growth, misreading phenotype.[52]

Interestingly, G2253, while protected from kethoxal modification by tRNA,[55,56] is located distant from G2251 and G2252 on the minor groove side of the stem. Protection of G2253 may result from a dynamic tertiary interaction induced by binding of tRNA in the P site.[54] G2250 and C2254 form an unusual noncanonical pair, with the N-4 of C2254 hydrogen bonding to the N-3 of G2250 and the N-2 of G2250 hydrogen bonding to the N-3 of C2254. However, C2254 is reactive to DMS in intact ribosomes but becomes unreactive upon binding of A site tRNA,[55] suggesting that the oligonucleotide analogue mimics the structure of the 2250 loop in ribosomes with a filled A site, but that these bases are unpaired with an empty A site. It seems likely, therefore, that the P loop is dynamic in structure, and that such conformational changes may be important for ribosome function.

6.12.3.5 The Sarcin–Ricin Loop (23S rRNA)

The NMR-derived structure of the eukaryotic sarcin–ricin loop is remarkable for the number of noncanonical pairing interactions forming a large internal loop[57] (Figure 4). The entire structure is essentially a composite of two secondary structure motifs observed elsewhere in ribosomal RNAs, a GNRA tetraloop and the E loop of eukaryotic 5S rRNA. The internal "E" loop consists of a type II sheared G–A pair, a U–A reverse Hoogsteen pair (N3U–N7A, O2U–N6A), a bulged G residue positioned in the major groove, and a symmetric A–A pair (N-6–N-7). The latter positions are an A–C juxtaposition in *E. coli* 23S rRNA. The backbone reverses direction at the symmetric A–A pair and is kinked at the following bulged G residue. A cross-strand stack occurs between the sheared G–A pair and the U–A reverse Hoogsteen pair. This internal loop is analogous in conformation to loop E of eukaryotic 5S rRNA.[12] The terminal GAGA tetraloop is separated from the internal loop by a Watson–Crick C–G pair.

The extensive interest in the sarcin–ricin loop stems from its interaction with elongation factors G and Tu,[44,58] and from the effects on protein synthesis of the cytotoxins α-sarcin and ricin.[59,60] In chemical footprinting experiments, EF–G protects the bulged G2655 and the tetraloop nucleotides A2660 and G2661, while EF–Tu protects G2655 and G2661 and to a weaker degree A2660 and A2665 (of the U–A reverse Hoogsteen pair).[44] Mutations in the sarcin–ricin loop affect the binding of the EF–Tu–tRNA–GTP ternary complex to the ribosome and perturb the accuracy of translation.[61–63] Hydroxyl radicals generated by Fe^{II}–EDTA tethered to amino acid 650 of EF–G produce cleavages in the sarcin–ricin loop in the GAGA tetraloop and in the 1093–1098 hexaloop in domain II,[45] placing these two structural elements in close proximity to EF–G and to one another.

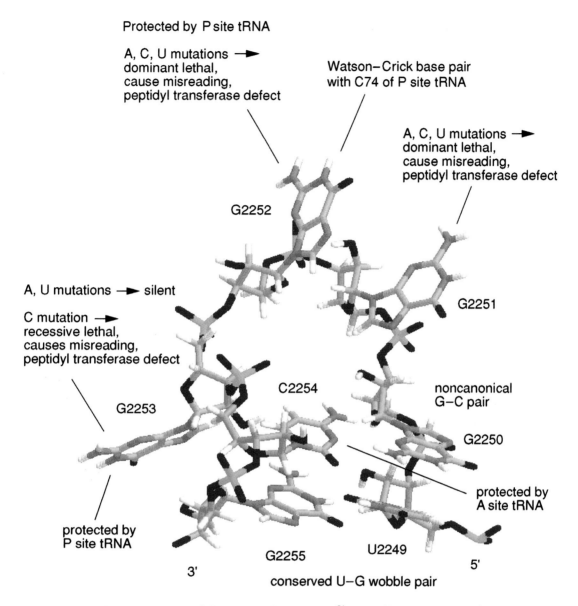

Protected by P site tRNA

A, C, U mutations →
dominant lethal,
cause misreading,
peptidyl transferase defect

Watson–Crick base pair
with C74 of P site tRNA

A, C, U mutations →
dominant lethal,
cause misreading,
peptidyl transferase defect

G2252

G2251

A, U mutations → silent

C mutation →
recessive lethal,
causes misreading,
peptidyl transferase defect

G2253

C2254

noncanonical
G–C pair

G2250

protected by
A site tRNA

protected by
P site tRNA

G2255

U2249

5'

3'

conserved U–G wobble pair

Figure 3 NMR-derived structure of the P loop of 23S rRNA.[54] Sites of interaction with the CCA terminus of P site tRNA are indicated, as are the phenotypes of mutations in the loop (pdb number 1VOP).

6.12.4 TERTIARY STRUCTURE OF rRNAs

Comparative sequence analysis continues to be the most effective method for revealing tertiary interactions such as pseudoknots, lone pairs, and base triples. While some of these interactions, such as pseudoknots, tend to be composed primarily of Watson–Crick pairings, others, such as base triples, are more difficult to recognize as they do not covary in a concerted fashion. Presumably other interactions have escaped detection for this same reason. Nevertheless, the number of tertiary interactions thus discovered can be expected to grow with the volume of the sequence database and with the continued development of algorithms for detecting nonconcerted covariations.[64]

Curiously, while many long-range tertiary interactions have been found in 23S rRNA by comparative sequence analysis, few have been found in 16S rRNA (see Figure 1). This is particularly striking considering that the 16S-like rRNA database is an order of magnitude larger than the 23S-like rRNA database.[1] Perhaps this is indicative of some fundamental difference in the principles guiding the global structure of these two molecules.

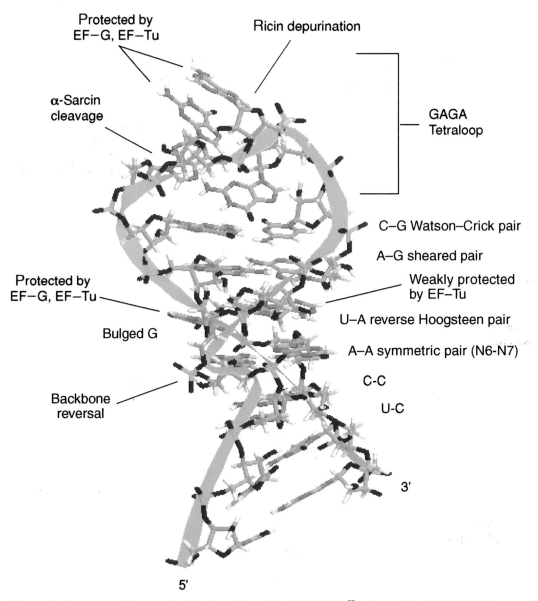

Figure 4 Structure of the sarcin–ricin loop from Rat 28S rRNA,[57] pdb number 1SCL. The loop consists of an E loop motif capped by a GAGA terminal loop. The noncanonical base pairing interactions are indicated, as are sites of protection from chemical probes by elongation factors EF–G and EF–Tu. Also shown are the sites of α-sarcin cleavage and ricin depurination. The path of the phosphodiester backbone is highlighted to emphasize backbone reversal at the bulged G residue and A–A symmetric pair.

Cross-linking studies have also indicated the proximity of sites distant from one another in the secondary structure models[4] and, in a number of instances, lend experimental support to phylogenetically-determined interactions. In addition to pseudoknots, lone pairs, and base triples, it is likely that there exists a number of tetraloop–tetraloop receptor interactions and loop–helix interactions of an unknown nature that are important for tertiary structure.

6.12.4.1 Pseudoknots, Lone Pairs, and Base Triples

A number of pseudoknot structures have been identified by comparative sequence analysis, and several have received experimental support. In 16S rRNA, three pseudoknots have been examined by mutagenesis (see Figure 1). One pseudoknot is formed between the triplets consisting of positions

17–19 and positions 916–918; canonical Watson–Crick pairing in this "central" pseudoknot is necessary for ribosome function.[65] A second pseudoknot in 16S rRNA occurs between the doublets 570/571 and 865/866. As demonstrated by site-directed mutagenesis experiments, the *E. coli* pseudoknot (5'-GU-3', positions 570 and 571, and 5'-AC-3', positions 865 and 866) is important for 30S subunit assembly and 70S ribosome formation. Either of the mutations U571A or A865U cause structural instability of the 30S subunits with severe consequences on growth rate. Restoration of Watson–Crick pairing potential by a double mutation, U571A/A865U, alleviates these defects.[66] Interestingly, both the central pseudoknot and the 571–865 pseudoknot are adjacent to the site of a conformational switch structure involving nucleotides 912–914 and 885–890.[67] Mutations perturbing the equilibrium of the switch structure also perturb the reactivity of bases in the 571–865 pseudoknot, suggesting a possible structural and functional relationship between these two regions.

A third pseudoknot found in 16S rRNA involves the 530 loop, an important functional site involved in the decoding of mRNA.[68] A pseudoknot forms between the loop positions 524–526 and positions 505–507 that are part of an asymmetric bulge. Some mutations disrupting these base pairing interactions have severe consequences on cell growth and ribosome function, while others confer resistance to the antibiotic streptomycin.[69]

In 23S rRNA a two base pair pseudoknot occurs in domain II between the doublets at positions 1005, 1006 and 1137, 1138. These nucleotides are located in two internal loops positioned near either end of a seven base pair Watson–Crick helix. Introduction of mismatches at either base pairing position impairs growth, while introduction of mismatches at both positions is lethal.[70] Nucleotides near the sites of these mutations also become more susceptible to nuclease cleavage and chemical modification. Consistent with the phylogenetic data, both growth and structural defects are rescued by compensatory changes that restore the potential for Watson–Crick pairing.

A series of long-range base pairing contacts in the core of 23S rRNA constrain the three-dimensional folding of rRNA.[71] This set of Watson–Crick base pairs, 1262–2017, 1269–2011, and 1270–2010, brings domains II, III, IV, and V closer together in the secondary structure model. Mutagenesis studies combined with enzymatic and chemical probing provide support for the proposed structure and for an extended helical structure including 1262–1270 and 2010–2017, and indicate the importance of this structure for ribosome function.

Base triples represent another class of tertiary interaction that have been observed directly in the crystal structures of tRNAs[49] and the *Tetrahymena* ribozyme.[2] Base triples have been proposed to play an equally important role in rRNA tertiary structure.[64] Some of the candidates identified in rRNAs are short range, whereas others cross secondary structure domains. These interactions are likely to exhibit a high degree of context dependence, and do not display the kind of one-to-one correlation observed for Watson–Crick base pairs.[64] A proposed base triple in 23S rRNA between U746 in domain II and the G2057–C2611 base pair in domain V is supported by a UV cross-link between these two regions[72] and footprinting of antibiotic inhibitors of peptidyl transferase at A2058 and in the 750 loop.[73] This base triple brings together two domains separated by a large distance in the primary and secondary structures of 23S rRNA. Recently, a base triple in the GTPase center in domain II of 23S rRNA, detected by comparative sequence analysis, was established by site-directed mutagenesis.[46]

6.12.4.2 Secondary Structure Motifs as Potential Sites of Tertiary Interaction

Secondary structure motifs, such as purine–purine tandems, are likely to be sites of tertiary interaction. Most conclusions about the importance of these motifs have been inferred from their conservation at specific sites in rRNA, and by analogy with smaller RNAs where their role in folding and tertiary structure is well established. In purine–purine mismatches, the Watson–Crick faces of both purines are exposed in the major and minor grooves, making them attractive as potential sites for tertiary interactions or as protein recognition sites. The extremely shallow minor groove at purine–purine tandems may facilitate contact between RNA helices.[74] These mismatches are also found often as part of terminal loops, where they may engage in tertiary interactions.[75]

Of the three main classes of tetraloops, the GNRA, UNCG, and UUCG tetraloops, the GNRA tetraloops are now known to engage in tertiary interactions with helices bearing specific tetraloop receptors.[2,17] The high frequency of GNRA loops in a variety of structural RNAs implies that tetraloop–tetraloop receptor interactions may be a common tertiary structure building block. However, GNRA tetraloop receptors have yet to be unambiguously identified in ribosomal RNAs, perhaps due to the lack of highly defined sequence parameters defining potential receptors. Selex experiments have indicated that the secondary structures of GAAA tetraloop receptors may be more variable than originally proposed,[76] making them potentially more difficult to recognize by sequence inspection. Other members of the GNRA class recognize simple motifs such as 5'-CC/3'-GG or 5'-CU/3'-AG tandems.[17] Our understanding of the contribution of tetraloops and mismatch tandems to rRNA tertiary structure could benefit from a comprehensive and systematic site-directed mutagenesis approach coupled with structure probing.

6.12.4.3 Evidence for Tertiary Structure from Functional Studies

Tertiary interactions serve to bring distant secondary structure elements together to form functional sites in the ribosome. Thus, close physical proximity and possibly direct contact can be inferred for structures which participate in a common ribosome function. Evidence for the involvement of specific sites of rRNA in individual steps of translation comes primarily from the experimental approaches of mutagenesis, chemical footprinting, and cross-linking of ribosome ligands (for a series of reviews, see references 4, 37, 68, 77–80).

The importance of functional data in the quest for tertiary structure information is particularly pertinent for regions that are highly conserved and not amenable to dissection by comparative sequence analysis. One such region is the peptidyl transferase center of 23S rRNA (for a review see reference 81). Peptidyl transferase activity is associated primarily with three highly conserved secondary structure elements of domain V: the central loop (a five-helix junction) and the A and P terminal loops. While these three structures are some distance from one another in secondary structure,[1] it is certain that, together with elements of domain IV and possibly domain II, they collectively form part of the active site of the peptidyl transferase. This conclusion is based in part on the large number of mutations in these three structures which produce defects in peptidyl transferase activity[53,68,82] and is supported by chemical footprinting of the CCA terminus of tRNAs in the A and P sites.[55] The challenge remains to determine precisely how these three elements are arranged in three-dimensional space to create the catalytic site of the ribosome.

6.12.4.4 Evidence for Tertiary Structure from Cross-linking, Chemical Footprinting, and Cleavage Experiments

A great deal of information regarding the overall arrangement of rRNAs in the ribosome has been obtained by a variety of cross-linking techniques (reviewed in references 4 and 83). In addition to direct rRNA–rRNA cross-links, constraints on rRNA folding can also be inferred from cross-links to tRNA, mRNA, or ribosomal proteins. Thus, two nucleotide positions which both cross-link to the same position on a ligand can be inferred to be in close proximity to one another. The limit of resolution of cross-linking methods depends in part upon the size and nature of the cross-linking agent. Site-directed cross-linking can be achieved by incorporation of photoreactive nucleotide analogues such as 4-thiouridine or azidoadenosines which are "zero length" cross-links. Zero length cross-links can also be obtained by utilizing the inherent propensities of some positions to form cross-links upon ultraviolet irradiation.

Chemical footprinting experiments have provided a wealth of information about global conformation. The protection of rRNA by the binding of ribosomal proteins, ligands such as tRNAs or translation factors, and antibiotics, contribute a significant number of constraints. This is particularly true for antibiotics which, by virtue of their small size, can make only a limited number of contacts with nucleotides that are in very close proximity to one another. For example, the antibiotic vernamycin B protects A752 in domain II, as well as A2062 and G2505 in domain V of 23S rRNA.[73] While A2062 and G2505 are both in the central loop of domain V, the proximity of A752 to these nucleotides has been confirmed by cross-linking of the 750 loop to the central loop.

Another powerful approach to defining distance constraints in the ribosome is the use of cleavage reagents tethered to specific sites on various ligands. These include 1, 10-*ortho*-phenanthroline-Cu[II84] and hydroxyl radicals generated by Fe[II]–EDTA.[68] These reagents catalyze cleavage of the RNA backbone independent of secondary structure. While effects on reactivity of base chemical probes can result from either proximity or long-range conformational effects, cleavage is completely dependent on proximity. rRNA residues neighboring specific nucleotides of tRNA or specific amino acid residues of a ribosomal protein or elongation factor can be identified in this way. These approaches are particularly informative if the crystal structure of the ligand is known, and if several different residues are used as sites of attachment of the cleavage reagent. Generally, results from cleavage experiments agree with those from chemical footprinting experiments and serve to complement the chemical probing approach.

6.12.4.5   Three-dimensional Reconstructions, X-ray Crystallography, Cryoelectron Microscopy

The accumulated data from cross-linking, mutagenesis, and footprinting experiments have been integrated in the construction of three-dimensional models of the large rRNAs.[3,85–87] One of the key elements of modeling is the placement of ribosomal proteins. Early neutron scattering data were used to position the mass centers of 30S subunit ribosomal proteins,[88] and, in conjunction with protein–RNA cross-linking data, to place constraints on 16S rRNA folding. These models are refined with the incorporation of data from the solution of individual ribosomal protein structures[89] in conjunction with data from experiments utilizing hydroxyl radicals generated from specific sites on ribosomal proteins of known structure,[68] and determination of the precise positions of the cross-linked nucleotide and amino acid.[90]

A 9 Å resolution X-ray crystallographic electron density map of *Haloarcula marismortui* 50S subunits was generated by molecular replacement using cryoelectron microscopic contours of *E. coli* 70S ribosomes.[6] This resolution is sufficient to discern the major and minor grooves of RNA helices. Establishing the path of 23S rRNA through the subunit will next require the identification of landmarks such as ribosomal protein binding sites. Models of 16S rRNA are also fitted to the cryoelectron microscopic contour.[85] It is expected that these approaches will ultimately lead to the derivation of an atomic-resolution structure of the intact ribosome.

6.12.5   CONFORMATIONAL SWITCHES IN rRNA

The realization of the central role of rRNA in ribosome function has led to the prediction that conformational changes in rRNAs occur during translation. This was inspired in part by the knowledge that conformational changes occur in protein enzymes during catalysis. However, direct evidence for specific changes in rRNA conformation and their association with specific ribosome functions has been difficult to obtain. For example, a reversible interconversion between functionally active and inactive forms of the 30S subunit, dependent upon heat, magnesium, and ammonium ion concentrations, has been reported,[91] and structure probing has provided clues to the nature of the transition.[92] However, its biological significance remains undetermined. Numerous proposals of alternate base pairing conformations during different stages of translation have been proposed,[67] but there is no direct evidence for the existence of any of these switch mechanisms or their association with specific ribosome functions. Currently, only one switch mechanism has been described which is supported by comparative sequence analysis, site-directed mutagenesis, and structure probing, although it is generally believed that other switch mechanisms will eventually be uncovered.

6.12.5.1   The 912-885/888 Conformational Switch in 16S rRNA

Site-directed mutagenesis and comparative sequence analysis, combined with structure probing of *E. coli* ribosomes, determined that two sets of pairing interactions are physiologically relevant structures and represent alternate conformations that occur during protein synthesis.[67,93] In the central region of *E. coli* 16S rRNA, three nucleotides, 5'-CUC-3' (912–910), can pair either with 5'-GAG-3' (888–890) (the "912-888" conformation), or with 5'-GGG-3' (885–887) (the "912-885" conformation) (see Figure 5). The dynamic change in rRNA structure involving these two alternate base pairing arrangements is apparently facilitated by ribosomal proteins S5 and S12, which interact with this region of 16S rRNA at the junction of the three major domains. Mutations in ribosomal proteins S5 and S12 have opposite effects on the fidelity of tRNA selection. Similarly, translational fidelity can be manipulated by site-directed mutations in 16S rRNA designed to perturb the equilibrium between the two rRNA conformations, interfering with base pairing in one or the other structure. 16S rRNA mutations favoring the 912-888 conformation increase translational fidelity and are incompatible with S12 hyperaccurate mutations but suppressed by an S5 error-prone mutation. In contrast, mutations predicted to shift the equilibrium toward the 912-885 conformation are error prone and compatible with hyperaccurate S12 mutations. By combining mutations at all three sites, 910–912, 882–885, and 888–890, a balance between the two proposed conformers is re-established and the phenotype reverted to that of the wild-type ribosome.

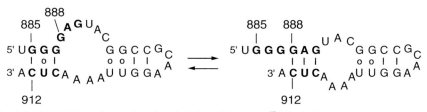

Figure 5 The 912–885/888 conformational switch in 16S rRNA.[67] The left structure represents the proposed base pairing interactions in the 912–885 conformation, the right structure the proposed 912–888 conformation. Mutations predicted to favor the 912–888 conformation are hyperaccurate, are incompatible with hyper-accurate mutations in ribosomal protein S12, and are suppressed by error-prone mutations in ribosomal protein S5. Mutations predicted to favor the alternative rRNA conformation are error prone.

While the precise function of this conformational switch has yet to be firmly established, ribosomes with mutations favoring the 912-885 conformation have greater affinity for tRNA. This, coupled with the effects on translational fidelity, immediately suggests a role in tRNA selection and proof-reading. However, chemical probing reveals significant changes in the rRNA structure at or near sites that have been cross-linked to mRNA, and implies that the switch could also be directly involved in translocation. It is noted that the simultaneous movement of three nucleotides during the conformational switch is consistent with the movement of mRNA by one codon.

The structures participating in the base pairing switch are conserved in all small subunit rRNAs, thus indicating that the switch mechanism is a fundamental process in all ribosomes. The apparent lack of long-range interactions between domains of 16S rRNA supports the idea that this molecule is more adapted to large transformational switches within and between domains than is the case for 23S rRNA.[68]

6.12.5.2 Potential Conformational Changes in the Ribosomal A Site to Accommodate Different Protein Factors

During translation, the ribosome must interact with many different tRNAs, as well as initiation, elongation, and termination factors. Recent high-resolution structures of some of these translation factors indicate however, that many of these diverse ligands share common structural features.[94–96] This has led to the development of the concept of molecular mimicry, where domains of protein factors or RNA molecules adopt similar conformations so as to utilize a single, common binding site on the ribosome. Consistent with these various proposals is the observation that the EF–Tu, EF–G, the IF1–IF2 (initiation factor) complex, and release factors all compete for binding to the A site of the ribosome. Nevertheless, the ribosome must distinguish among all these factors at the appropriate stage of translation. An intriguing possibility is that the binding site for these factors changes conformation to accommodate structural features unique to the individual proteins, shifting the binding equilibrium in favor of the appropriate factor. This hypothesis must await verification by high-resolution structural studies.

6.12.5.3 Other Potential Conformational Switches in rRNA

Additional switch structures are to be anticipated in functional centers of both 16S and 23S rRNA in light of the multiple hybrid transition states now known to occur during the various stages of translation.[97] For example, during the translation cycle, EF–Tu and EF–G alternate in binding to the ribosome and interacting with the sarcin–ricin loop in 23S rRNA. It has been proposed that the cytotoxins α-sarcin and ricin recognize differing conformations of the sarcin–ricin loop that are present during differing stages of the elongation cycle, and that the interconversion of these two conformations is promoted by (or at least associated with) the sequential action of EF–Tu and EF–G.[98]

Finally one must consider as a shift structure the intermolecular base pairing between the anti-Shine–Dalgarno (ASD) sequence (5′-CCUCC-3′) at the 3′ end of 16S rRNA and mRNA. During translation initiation, the ASD sequence base pairs with the Shine–Dalgarno sequence (5′-GGAGG-

3′) in mRNA upstream of the AUG initiation codon.[99,100] While this specific base pairing ceases once elongation begins, the ASD sequence continues to scan the translated codons and there is evidence that it can base pair with downstream sequences in mRNA.[101] Obviously, we are just beginning to appreciate the complex dynamic nature of rRNA during translation. Given the current rate of progress in the structural analysis of rRNA, a comprehensive understanding of its mechanism seems imminent.

6.12.6 REFERENCES

1. R. R. Gutell, in "Ribosomal RNA: Structure, Evolution, Processing, and Function in Protein Biosynthesis," eds. R. A. Zimmermann and A. E. Dahlberg, CRC Press, Boca Raton, FL, 1996, p. 111.
2. J. H. Cate, A. R. Gooding, E. Podell, K. Zhou, B. L. Golden, C. E. Kundrot, T. R. Cech, and J. A. Doudna, *Science*, 1996, **273**, 1678.
3. H. F. Noller, R. Green, G. Heilek, V. Hoffarth, A. Huttenhofer, S. Joseph, I. Lee, K. Lieberman, A. Mankin, C. Merryman, T. Powers, E. V. Puglisi, R. R. Samaha, and B. Weiser, *Biochem. Cell Biol.*, 1995, **73**, 997.
4. R. Brimacombe, P. Mitchell, and F. Muller, in "Ribosomal RNA: Structure, Evolution, Processing, and Function in Protein Biosynthesis," eds. R. A. Zimmermann and A. E. Dahlberg, CRC Press, Boca Raton, FL, 1996, p. 129.
5. F. Mueller, T. Doring, T. Erdemir, B. Greuer, N. Junke, M. Osswald, J. Rinke-Appel, K. Stade, S. Thamm, and R. Brimacombe, *Biochem. Cell Biol.*, 1995, **73**, 767.
6. N. Ban, B. Freeborn, P Nissen, P. Penczek, R. A. Grassucci, R. Sweet, J. Frank, P. B. Moore, and T. A. Steitz, *Cell*, 1998, **93**, 1105.
7. R. R. Samaha, B. O'Brien, T. W. O'Brien, and H. F. Noller, *Proc. Natl. Acad. Sci. USA*, 1994, **91**, 7884.
8. I. Nitta, Y. Kamada, H. Noda, T. Ueda, and K. Watanabe, *Science*, 1998, **281**, 666.
9. D. Gautheret, D. Konings, and R. R. Gutell, *J. Mol. Biol.*, 1994, **242**, 1.
10. J. SantaLucia, Jr. and D. H. Turner, *Biochemistry*, 1993, **32**, 12 612.
11. H. W. Pley, K. M. Flaherty, and D. B. McKay, *Nature*, 1994, **372**, 68.
12. B. Wimberly, G. Varani, and I. Tinoco, Jr., *Biochemistry*, 1993, **32**, 1078.
13. D. Gautheret, D. Konings, and R. R. Gutell, *RNA*, 1995, **1**, 807.
14. B. Wimberly, *Nature Struct. Biol.*, 1994, **1**, 820.
15. C. C. Correll, B. Freeborn, P. B. Moore, and T. A. Steitz, *Cell*, 1997, **91**, 705.
16. C. R. Woese, S. Winker, and R. R. Gutell, *Proc. Natl. Acad. Sci. USA*, 1990, **87**, 8467.
17. M. Costa, and F. Michel, *EMBO J.*, 1995, **14**, 1276.
18. Z. P. Cai, A. Gorin, R. Frederick, X. M. Ye, W. D. Hu, A. Majumdar, A. Kettani, and D. J. Patel, *Nature Struct. Biol.*, 1998, **5**, 203.
19. D. L. Abramovitz and A. M. Pyle, *J. Mol. Biol.*, 1997, **266**, 493.
20. C. Cheong, G. Varani, and I. Tinoco, Jr., *Nature*, 1990, **346**, 680.
21. H. A. Heus and A. Pardi, *Science*, 1991, **253**, 191.
22. F. M. Jucker and A. Pardi, *Biochemistry*, 1995, **34**, 14 416.
23. J. B. Prince, B. H. Taylor, D. L. Thurlow, J. Ofengand, and R. A. Zimmermann, *Proc. Natl. Acad. Sci. USA*, 1982, **79**, 5450.
24. J. Rinke-Appel, N. Jünke, R. Brimacombe, S. Dokudovskaya, O. Dontsova, and A. Bogdanov, *Nucleic Acids Res.*, 1993, **21**, 2853.
25. D. Moazed and H. F. Noller, *Cell*, 1986, **47**, 985.
26. D. Moazed and H. F. Noller, *Nature*, 1987, **327**, 389.
27. E. A. DeStasio and A. E. Dahlberg, *J. Mol. Biol.*, 1990, **212**, 127.
28. P. Purohit and S. Stern, *Nature*, 1994, **370**, 659.
29. H. Miyaguchi, H. Narita, K. Sakamoto, and S. Yokoyama, *Nucleic Acids Res.*, 1996, **24**, 3700.
30. D. Fourmy, M. I. Recht, S. C. Blanchard, and J. D. Puglisi, *Science*, 1996, **274**, 1367.
31. S. T. Gregory and A. E. Dahlberg, *Nucleic Acids Res.*, 1995, **23**, 4234.
32. D. Fourmy, S. Yoshizawa, and J. D. Puglisi, *J. Mol. Biol.*, 1998, **277**, 333.
33. P. Mitchell, M. Osswald, and R. Brimacombe, *Biochemistry*, 1992, **31**, 3004.
34. C. Ehresmann, H. Moine, M. Mougel, J. Dondon, M Grunberg-Manago, J.-P. Ebel, and B. Ehresmann, *Nucleic Acids Res.*, 1986, **14**, 4803.
35. T. L. Helser, J. E. Davies, and J. E. Dahlberg, *Nature New Biol.*, 1972, **235**, 6.
36. P. H. van Knippenberg, in "Structure, Function and Genetics of Ribosomes," eds. B. Hardesty and G. Kramer, Springer New York, 1986, p. 413.
37. P. R. Cunningham, C. J. Weitzmann, D. Negre, J. G. Sinning, V. Frick, K. Nurse, and J. Ofengand, in "The Ribosome: Structure, Function, and Evolution," eds. A. Dahlberg, W. E. Hill, A. Dahlberg, R. A. Garrett, P. B. Moore, D. Schlessinger, and J. R. Warner, ASM Press, Washington, DC, 1990, p. 243.
38. D. Lafontaine, J. Delcour, A. L. Glasser, J. Desgres, and J. Vandenhaute, *J. Mol. Biol.*, 1994, **241**, 492.
39. J. P. Rife and P. B. Moore, *Structure*, 1998, **6**, 747.
40. J. P. Rife, C. S. Cheng, P. B. Moore, and S. A. Strobel, *Nucleic Acids Res.*, 1998, **26**, 3640.
41. S. E. Butcher, T. Diekmann, and J. Feigon, *J. Mol. Biol.*, 1997, **268**, 348.
42. S. Douthwaite, B. Vester, C. Aagaard, and G. Rosendahl, in "The Translation Apparatus: Structure, Function, Regulation, Evolution," eds. K. H. Nierhaus, F. Franceschi, A. R. Subramanian, V. A. Erdmann, and B. Wittman-Liebold, Plenum, New York, 1993, p. 339.
43. G. Rosendahl and S. Douthwaite, *Nucleic Acids Res.*, 1994, **22**, 357.
44. D. Moazed, J. M. Robertson, and H. F. Noller, *Nature*, 1988, **334**, 362.
45. K. S. Wilson and H. F. Noller, *Cell*, 1998, **92**, 131.

46. G. L. Conn, R. R. Gutell, and D. E. Draper, *Biochemistry*, 1998, **37**, 11 980.
47. S. Huang, Y.-X. Wang, and D. E. Draper, *J. Mol. Biol.*, 1996, **258**, 308.
48. D. K. Jemiolo, F. T. Pagel, and E. J. Murgola, *Proc. Natl. Acad. Sci. USA*, 1995, **92**, 12 309.
49. W. Saenger, "Principles of Nucleic Acid Structure," Springer, New York, 1984, p. 338.
50. F. M. Jucker and A. Pardi, *RNA*, 1995, **1**, 219.
51. R. R. Samaha, R. Green, and H. F. Noller, *Nature*, 1995, **377**, 309.
52. S. T. Gregory, K. R. Lieberman, and A. E. Dahlberg, *Nucleic Acids Res.*, 1994, **22**, 279.
53. K. R. Lieberman and A. E. Dahlberg, *J. Biol. Chem.*, 1994, **269**, 16 163.
54. E. V. Puglisi, R. Green, H. F. Noller, and J. D. Puglisi, *Nature Struct. Biol.*, 1997, **4**, 775.
55. D. Moazed and H. F. Noller, *Cell*, 1989, **57**, 585.
56. D. Moazed and H. F. Noller, *Proc. Natl. Acad. Sci. USA*, 1991, **88**, 3725.
57. A. A. Szewczak, P. B. Moore, Y.-L. Chan, and I. G. Wool, *Proc. Natl. Acad. Sci. USA*, 1993, **90**, 9581.
58. A. Munishkin and I. G. Wool, *Proc. Natl. Acad. Sci. USA*, 1997, **94**, 12 280.
59. Y. Endo and I. G. Wool, *J. Biol. Chem.*, 1982, **257**, 9054.
60. Y. Endo, K. Mitsui, M. Motizuki, and K. Tsurugi, *J. Biol. Chem.*, 1987, **262**, 5908.
61. W. E. Tapprich and A. E. Dahlberg, *EMBO J.*, 1990, **9**, 2649.
62. P. Melancon, W. E. Tapprich, and L. Brakier-Gingras, *J. Bacteriol.*, 1992, **174**, 7896.
63. M. O'Connor and A. E. Dahlberg, *Nucleic Acids Res.*, 1996, **24**, 2701.
64. D. Gautheret, S. H. Damberger, and R. R. Gutell, *J. Mol. Biol.*, 1995, **248**, 27.
65. R. A. Poot, S. H. E. van den Worm, C. W. A. Pleij, and J. van Duin, *Nucleic Acids Res.*, 1998, **26**, 549.
66. A. Vila, J. Viril-Farley, and W. E. Tapprich, *Proc. Natl. Acad. Sci. USA*, 1994, **91**, 11 148.
67. J. S. Lodmell and A. E. Dahlberg, *Science*, 1997, **277**, 1262.
68. R. Green and H. F. Noller, *Ann. Rev. Biochem.*, 1997, **66**, 679.
69. T. Powers and H. F. Noller, *EMBO J.*, 1991, **10**, 2203.
70. G. Rosendahl, L. H. Hansen, and S. Douthwaite, *J. Mol. Biol.*, 1995, **249**, 59.
71. C. Aagaard and S. Douthwaite, *Proc. Natl. Acad. Sci. USA*, 1994, **91**, 2989.
72. P. Mitchell, M. Osswald, D. Schueler, and R. Brimacombe, *Nucleic Acids Res.*, 1990, **18**, 4325.
73. D. Moazed and H. F. Noller, *Biochimie*, 1987, **69**, 879.
74. S. A. Strobel, L. Ortoleva-Donnelly, S. P. Ryder, J. H. Cate, and E. Moncoeur, *Nature Struct. Biol.*, 1998, **5**, 60.
75. S. A. Strobel and J. A. Doudna, *Trends Biochem. Sci.*, 1997, **22**, 262.
76. M. Costa and F. Michel, *EMBO J.*, 1997, **16**, 3289.
77. L. Brakier-Gingras, R. Pinard, and F. Dragon, *Biochem. Cell Biol.*, 1995, **73**, 907.
78. M. O'Connor, C. A. Brunelli, M. A. Firpo, S. T. Gregory, K. R. Lieberman, J. S. Lodmell, H. Moine, D. I. Van Ryk, and A. E. Dahlberg, *Biochem. Cell Biol.*, 1995, **73**, 859.
79. R. A. Zimmermann, in "Ribosomal RNA: Structure, Evolution, Processing, and Function in Protein Biosynthesis," eds. R. A. Zimmermann and A. E. Dahlberg, CRC Press, Boca Raton, FL, 1996, p. 277.
80. R. A. Garrett, and C. Rodriguez-Fonseca, in "Ribosomal RNA: Structure, Evolution, Processing, and Function in Protein Biosynthesis," eds. R. A. Zimmermann and A. E. Dahlberg, CRC Press, Boca Raton, FL, 1996, p. 327.
81. K. R. Lieberman and A. E. Dahlberg, *Prog. Nucleic Acids Res.*, 1995, **50**, 1.
82. B. T. Porse and R. A. Garrett, *J. Mol. Biol.*, 1995, **249**, 1.
83. D. Mundus and P. Wollenzien, *Nucleic Acids Symp. Ser.*, 1997, **36**, 171.
84. W. E. Hill, D. J. Bucklin, J. M. Bullard, A. L. Galbraith, N. V. Jammi, C. C. Rettberg, B. S. Sawyer, and M. A. van Waes, *Biochem. Cell Biol.*, 1995, **73**, 1033.
85. F. Mueller and R. Brimacombe, *J. Mol. Biol.*, 1997, **271**, 524.
86. F. Mueller and R. Brimacombe, *J. Mol. Biol.*, 1997, **271**, 545.
87. F. Mueller, H. Stark, M. van Heel, J. Rinke-Appel, and R. Brimacombe, *J. Mol. Biol.*, 1997, **271**, 566.
88. M. S. Capel, D. M. Engelman, B. R. Freeborn, M. Kjeldgaard, J. A. Langer, V. Ramakrishnan, D. G. Schindler, D. K. Schneider, B. P. Schoenborn, I. Y. Sillers, *et al.*, *Science*, 1987, **238**, 1403.
89. V. Ramakrishnan and S. W. White, *Trends Biochem. Sci.*, 1998, **23**, 208.
90. B. Thiede, H. Urlaub, H. Neubauer, G. Grelle, and B. Wittman-Liebold, *Biochem. J.*, 1998, **334**, 39.
91. A. Zamir, R. Miskin, and D. Elson, *J. Mol. Biol.*, 1971, **60**, 347.
92. D. Moazed, B. J. Van Stolk, S. Douthwaite, and H. F. Noller, *J. Mol. Biol.*, 1986, **191**, 483.
93. J. S. Lodmell, R. R. Gutell, and A. E. Dahlberg, *Proc. Natl. Acad. Sci. USA*, 1995, **92**, 10 555.
94. P. Nissen, M. Kjeldgaard, S. Thirup, G. Polekhina, L. Reshetnikova, B. F. Clark, and J. Nyborg, *Science*, 1995, **270**, 1464.
95. K. Ito, K. Ebihara, M. Uno, and Y. Nakamura, *Proc. Natl. Acad. Sci. USA*, 1996, **93**, 5443.
96. S. Brock, K. Szkaradkiewicz, and M. Sprinzl, *Mol. Microbiol.*, 1998, **29**, 409.
97. D. Moazed and H. F. Noller, *Nature*, 1989, **342**, 142.
98. A. Gluck, Y. Endo, and I. G. Wool, *J. Mol. Biol.*, 1992, **226**, 411.
99. J. Shine and L. Dalgarno, *Proc. Natl. Acad. Sci. USA*, 1974, **71**, 1342.
100. J. A. Steitz and K. Jakes, *Proc. Natl. Acad. Sci. USA*, 1975, **72**, 4734.
101. R. B. Weiss, D. M. Dunn, A. E. Dahlberg, J. F. Atkins, and R. F. Gesteland, *EMBO J.*, 1988, **7**, 1503.

6.13
Turnover of mRNA in Eukaryotic Cells

SUNDARESAN THARUN and ROY PARKER
University of Arizona, Tucson, AZ, USA

6.13.1 INTRODUCTION

It has been clear for a number of years that the differential degradation of eukaryotic mRNAs plays an important role in the modulation of gene expression. The role of mRNA turnover in gene expression occurs in two general manners. First, individual mRNAs have different intrinsic rates of decay, which for some broadly expressed genes vary as much as differences in transcription rates.[1,2] In addition, there are a large number of examples where the regulation of gene expression occurs by modulation of the decay rate of an mRNA or class of mRNAs (for reviews, see Refs 3 and 4). Thus, in order to understand the modulation of gene expression, it is important to determine the mechanisms and control of mRNA degradation.

In the last few years several different and somewhat related mechanisms by which eukaryotic mRNAs are degraded have been defined. One mRNA decay pathway is initiated by shortening of the 3′ poly(A) tail (often referred to as deadenylation) followed by decapping and 5′ to 3′ exonucleolytic degradation of the transcript. Alternatively, transcripts can undergo 3′ to 5′ decay after poly(A) shortening. Decay of mRNAs can also be initiated independent of shortening of the poly(A) tail. For example, transcripts can be decapped in a deadenylation-independent manner leading to 5′ to 3′ degradation. Similarly, mRNA decay can also be initiated by endonucleolytic cleavages. In this chapter the authors discuss these pathways of degradation, the enzymes which catalyze the specific nucleolytic events, and how the activity of the nucleases may be modulated on individual transcripts to give rise to differential mRNA turnover.

6.13.2   ROLE OF DEADENYLATION IN THE DECAY OF EUKARYOTIC mRNAs

6.13.2.1   Deadenylation is a Key Step in mRNA Turnover

Following demonstration that the 3′ poly(A) tails of eukaryotic mRNAs are shortened in the cytoplasm,[5] several observations implicated removal of the poly(A) tail as an early step in the decay of many mRNAs. One set of observations came from analyzing the poly(A) shortening rates and the decay rates of different mRNAs (for a review see Ref. 6). For example, in the comparison of two adenovirus mRNAs, the unstable mRNA deadenylated significantly faster than the more stable transcript.[7] This observation suggested that the faster poly(A) shortening might be involved in faster mRNA decay. Correlations between the stability of a transcript and the metabolism of its poly(A) tail have also been observed by examining the poly(A) tail of an individual mRNA under different conditions that alter its decay rate. For example, when the transcripts for human growth hormone or avian liver apoVLDLII mRNA are stabilized by the presence of glucocorticoids and estrogen, respectively, the average poly(A) tail length is significantly longer.[8,9] Finally, additional evidence that poly(A) tail removal might be required for mRNA degradation came from examination of the stability of transcripts lacking poly(A) tails, which were either introduced into cells by microinjection or electroporation (reviewed in Ref. 10) or synthesized *in vivo* in the presence of cordecypin (3′ deoxyadenosine).[11] Since in these cases, unadenylated mRNAs were generally less stable than adenylated control mRNAs, the poly(A) tail was inferred to be required for maintaining the cytoplasmic stability of an mRNA.

Additional evidence that poly(A) tail shortening can be required for mRNA decay came from following the degradation of "pulses" of newly-synthesized transcripts produced by rapid induction and repression of regulatable promoters. In mammalian cells this approach has been performed by utilizing the c-fos promoter, which after serum stimulation becomes transiently active. By using constructs coding for c-fos and β-globin mRNAs driven by the c-fos promoter, the fate of these two mRNAs have been studied in a time dependent manner.[12,13] These experiments indicated that the unstable c-fos transcript deadenylates much faster than the stable β-globin mRNA and does not begin to be degraded until its poly(A) tail is shortened.

Further evidence that the c-fos mRNA requires poly(A) shortening before degradation came from the analysis of chimeric transcripts. Replacing the coding region of β-globin mRNA with that of c-fos mRNA results in both more rapid deadenylation and rapid decay of the chimeric mRNA as compared to the parental β-globin transcript. Moreover, a deletion in the c-fos coding region of this chimeric mRNA that leads to its stabilization also slows down its deadenylation rate.[12] Similar results have been seen when an AU-rich element, termed an ARE, within the c-fos 3′ UTR (untranslated region) is transferred to the 3′ UTR of the β-globin transcript. In this case, the resulting chimeric mRNA is highly unstable and has a rapid rate of deadenylation, suggesting that the c-fos ARE can promote decay, at least in part, by accelerating deadenylation.[12] Similar AU-rich sequences are found in the 3′ UTRs of several unstable mammalian oncogene and lymphokine mRNAs.[14] Evidence suggests that at least some of the other transcripts containing these elements are dependent on rapid deadenylation for their short half-lives.[15–17]

Similar "transcriptional pulse-chase" experiments have been performed on several yeast mRNAs. In this case, cells are grown in medium containing a "neutral" carbon source, such as raffinose, or a mixture of raffinose and sucrose, in which expression from the *GAL1* promoter is neither induced nor repressed.[18] Addition of galactose to these cultures causes a rapid induction of transcription from the *GAL1* promoter. After a brief induction, typically 10–15 min, transcription is quickly repressed by the addition of glucose. These yeast experiments have been particularly informative since the decay rate of the transcript, the poly(A) shortening rate, and the appearance of decay intermediates have all been examined. Examination of the stable *PGK1* and the unstable *MFA2* transcripts in this manner have led to the following observations.[19] (i) Following transcription repression there is a temporal lag before the decay of these mRNAs initiates. (ii) In each case, there is a good correlation between the length of this lag period and the time it takes for the poly(A) tail to be shortened from its initial length (from 50–80 residues long) to a short oligo (A) length (10–15 residues). Analysis of chimeric and mutant transcripts provide additional evidence that the lag period was required to allow deadenylation before decay of these transcripts could begin. For example, replacement of the 3′ UTR of the stable *PGK1* mRNA with the 3′ UTR of the unstable *MFA2* mRNA, which contains sequences capable of stimulating rapid deadenylation,[20] results in both an increased deadenylation rate and a corresponding decrease in the length of the lag period.[19] In addition, *cis*-acting mutations in *MFA2* mRNA that stabilize the transcript also change its deadenylation rate to an extent that partially or fully accounts for the increase in stability.[20]

Additional evidence that deadenylation precedes the decay of the *PGK1* and *MFA2* transcripts has come from the examination of mRNA decay products.[19,21] As described below, degradation of the transcript body in yeast is performed by the 5′ to 3′ *XRN1* gene exonuclease. Importantly, intermediates in the degradative process can be trapped by inserting strong RNA secondary structures, which block the exonuclease, into the mRNA.[19,21] Study of such decay intermediates (produced as a result of insertions) during a transcriptional pulse-chase experiment showed that they accumulated only after the poly(A) tails of the full-length mRNAs were shortened to an oligo(A) length.[19,22,23] Moreover, these intermediates have short poly(A) tails even when the full-length mRNA consists of a mixture of short and intermediate-length poly(A) tailed species.[19] Similar results are seen with decay intermediates trapped by deletion of the *XRN1* gene, where only deadenylated transcripts are seen to be decapped.[22,24] These results argue that decay products are generated only from full-length mRNAs that have been deadenylated.[19]

The data described above suggest that deadenylation is a prerequisite for degradation of the body of many eukaryotic mRNAs. An important issue is at what point during deadenylation the transcript becomes a substrate for subsequent decay events. The data from experiments in yeast argue that shortening to an oligo(A) tail is sufficient for later decay and that the poly(A) tail does not need to be fully removed. For example, degradation of at least three of the four yeast mRNAs examined is initiated once the poly(A) tail is shortened to the oligo(A) length (5–15 adenosine residues).[19] Moreover, the decay intermediates that are observed have oligo(A) tails.[19] Since *in vitro* studies have shown that the poly(A) binding protein (Pab1p) is unable to bind strongly to oligo(A) tracts shorter than 12 adenylate residues,[25] deadenylation to an oligo(A) tail could lead to the loss of the last Pab1p associated with the transcript and therefore may be functionally equivalent to complete removal of the poly(A) tail. Analysis of chimeric transcripts between the unstable c-fos and stable β-globin mRNAs suggests that deadenylation to a short poly(A) tail can be sufficient to initiate decay of mammalian mRNAs as well.[12] However, since the shortest poly(A) tails observed on these transcripts were approximately 25–60 A residues long, decay events following deadenylation may be triggered at a longer poly(A) tail length than in yeast.

Since the rate at which an mRNA is deadenylated contributes to its overall stability it is important to determine how different rates of deadenylation are controlled. In addition, some *cis*-acting sequences that stimulate mRNA turnover do so, at least in part, by accelerating mRNA deadenylation (see below). An important issue concerns how these sequences exert control over mRNA deadenylation, and what nucleases are the targets of this regulation.

6.13.2.2 Gene Products Involved in Deadenylation

One protein that affects deadenylation is the poly(A) binding protein, Pab1p, which binds to the 3′ poly(A) tail. The poly(A) binding protein has been cloned from several organisms and has a conserved organization consisting of four RNA binding domains and an extended C-terminus.[26–28] Experiments in yeast suggest that Pab1p is required for the proper rates of deadenylation. These experiments are slightly complicated since deletion of the *PAB1* gene is lethal.[26] Given this limitation, the role of Pab1p in deadenylation has been examined either in yeast strains that are conditional *pab1* mutants or in strains carrying a suppresser mutation that allows growth of *pab1*Δ strains. In such *pab1* mutant strains there is an increase in the length of bulk cellular poly(A).[29] Moreover, the *PGK1* and *MFA2* mRNAs are deadenylated three- and six-fold slower, respectively, in *pab1*Δ mutants than in *PAB1* strains.[30] Two general types of models can explain the reduced rates of deadenylation observed in *pab1* mutant yeast strains. In one view, proteins bound to the poly(A) tail in the absence of Pab1p could inhibit poly(A) shortening. Alternatively, Pab1p could directly influence deadenylation by interacting with the nuclease(s) responsible for shortening of the poly(A) tail (see below).

In contrast to the results above, examination of the role of Pab1p in cell-free extracts from human cells reveals that presence of a 3′ poly(A) tail can protect added transcripts from 3′ to 5′ exonucleolytic decay presumably when bound to the poly(A) binding protein.[31,32] At the present time the difference between these results and the yeast experiments remains unclear. Although unlikely, it is possible that the role of Pab1p in deadenylation differs between organisms. Alternatively, the *in vitro* system may be evaluating the function of an alternative 3′ to 5′ exonuclease that does not normally perform poly(A) shortening *in vivo*. This issue will be resolved as the poly(A) nucleases are identified and characterized in these different organisms.

The actual nuclease(s) involved in deadenylation is not yet known. Several nucleases have been identified which may be involved. For example, a 3′ to 5′ poly(A) nuclease has been purified from yeast extracts.[33] Consistent with a role for Pab1p in stimulating deadenylation in yeast, this activity, termed PAN, is largely Pab1p-dependent and exhibits little or no activity in the absence of Pab1p in crude extracts or following further purification.[33,34] Mutations in the two subunits of this enzyme, referred to as PAN2 and PAN3, do lead to an overall increase in the length of poly(A) tails *in vivo*, indicating that this activity does play a role somehow in deadenylation *in vivo*. Interestingly, further results suggest that this enzyme acts at a very early stage of deadenylation, perhaps while still in the nucleus.[35] However, since the actual rate of poly(A) shortening appears to be normal in pan2Δ and pan3Δ mutants, this enzyme does not appear to be the major enzyme responsible for cytoplasmic deadenylation.

Candidates for an enyzme involved in cytoplasmic deadenylation have also been described in extracts of mammalian cells. For example, several 3′ to 5′ exonucleases that degrade poly(A) with high preference have been either partially purified[36,37] or wholly purified to a 74 kDa protein from HeLa cell extracts.[38] Substrates containing nonpoly(A) sequences attached to the 3′ end of poly(A) were poorly degraded by this activity. Similarly, poly(A) molecules with a cordycepin moiety at their 3′ ends were also poor substrates. Future experiments with an anticipated clone of this protein should reveal the potential roles of this activity in deadenylation *in vivo*.

6.13.2.3   *cis*-Acting Sequences Affecting Deadenylation

Several *cis*-acting elements affecting mRNA deadenylation rate have been identified in yeast and more complex eukaryotes. These sequences are often localized to the 3′ untranslated regions, but can also be found within coding regions. For example, the 3′ UTR of the yeast *MFA2* mRNA is both necessary and sufficient for rapid deadenylation.[19,20] Sequences promoting deadenylation have also been localized to the 3′ untranslated regions of mammalian transcripts. One class of these elements is a set of AU-rich elements, termed AREs, found in the 3′ UTR of the transcripts for several oncogenes and lymphokines (for a review, see Ref. 17). These AREs are characterized by the presence of one or multiple copies of the pentanucleotide AUUUA and a high U content. By the analysis of chimeric transcripts, the AREs from many mRNAs, including the *c-fos*, GM-CSF, *c-myc*, and *jun*B mRNAs, have been shown to have the ability to accelerate deadenylation and decay rates of otherwise stable β-globin mRNA.[16,39]

In order to understand how these AREs function to promote mRNA decay, several groups have identified proteins that interact with the AREs of various transcripts and therefore might modulate the deadenylation rate.[40–45] However, only one of these proteins, termed AUBF, has been shown to stimulate decay in an *in vitro* system.[43] Progress in this area may be aided by the identification of proteins which bind to a small sequence UUAUUUA(U/A)(U/A) determined to be important for a functional ARE to stimulate mRNA decay.[46,47]

Examples of *cis*-acting sequences affecting deadenylation localized to the coding regions are found in the mammalian c-fos and yeast *MATα1* transcripts.[12,48] One of the striking similarities in these cases is that the function of both of these elements requires proper translation of the transcript. For example, in the yeast *MATα1* mRNA, the ability of a 32 nucleotide region to stimulate deadenylation and degradation is stimulated by a region of rare codons just 5′ to this 32 nucleotide element.[49,50] One interpretation of this observation is that the presence of a paused ribosome is required for this element to function.[50] Similarly, in order for the c-fos coding region to promote decay, the mRNA needs to be translated.[51] Since this coding region element is known to be recognized as RNA,[52] there may be a similar requirement for translating ribosomes to affect the function of this destabilizing element.

There are several additional cases wherein translation of the mRNA affects poly(A) shortening. For example, deadenylation, but not degradation following deadenylation, of the unstable mammalian c-myc mRNA is slowed when cells are treated with translation inhibitors.[53] In contrast, the insertion of a strong hairpin into the 5′ UTR of the yeast *PGK1* mRNA inhibits translation and leads to an increase in the rate of deadenylation.[23] Similar results where inhibition of in *cis* translation accelerates deadenylation have also been seen with mammalian transcripts.[54] Although how translation and control of deadenylation rate interact is unclear, one hypothesis is based on observations suggesting an important functional interaction between the 5′ and 3′ termini of eukaryotic transcripts. These observations include the fact that the poly(A) tail/Pab1p complex serves as both an enhancer of translation and an inhibitor of mRNA decapping[30] and that sequences in the 3′ UTR of the *MFA2* mRNA can influence mRNA decay following deadenylation.[19,20] In this model an mRNP (messenger ribonucleo-protein) structure involving the 5′ and 3′ ends of an mRNA would control rates of both translation and deadenylation. This mRNP structure would in turn be influenced by signals received from specific elements within an mRNA and respond by changing rates of deadenylation and/or translation. This model is attractive because it provides an explanation for how sequences located in the coding region of an mRNA are able to influence nucleolytic events at the 3′ ends of an mRNA since the element need only affect a central mRNP structure rather than the actual nucleolytic events themselves. Ultimately, identification of the specific *trans*-acting factors that mediate the effects of instability elements, and the determination of whether these factors directly or indirectly impact the deadenylation machinery, will facilitate an understanding of how poly(A) shortening is controlled.

6.13.3 DEADENYLATION TRIGGERS DECAPPING AND 5′ TO 3′ DECAY

One important issue is how deadenylation leads to mRNA degradation. In the absence of any strong experimental data, it had been assumed that loss of the poly(A) tail, and the associated poly(A) binding protein, would lead to degradation of a transcript by exposing the 3′ end to nonspecific nucleases. (e.g., Ref. 6). However, evidence has now accumulated in yeast indicating that the major effect of deadenylation is to lead to a decapping reaction that exposes the transcript to 5′ to 3′ exonucleolytic decay.[22,23] One of the key approaches used to determine the events following deadenylation was to insert strong RNA secondary structures, such as a poly(G) tract, which can form a very stable structure,[55] into the mRNA to block processive exonucleases.[19,21] In these experiments, mRNA fragments appear that are trimmed at their 5′ end to the site of the RNA secondary structure, even when the site of insertion is near the 5′ end of the transcript.[22,23] Since these fragments only appear after deadenylation, this observation suggests that poly(A) shortening triggers a cleavage event at the extreme 5′ end. Additional evidence for this conclusion came from the analysis of yeast strains lacking the major cytoplasmic 5′ to 3′ exonuclease, encoded by the *XRN1* gene.[56] In *xrn1*Δ strains, decapped forms of both the *MFA2* and *PGK1* transcripts accumulate following deadenylation.[22,23] Since many 5′ to 3′ exonucleases, including the product of the *XRN1* gene, are blocked by the cap structure located at the 5′ end of mRNAs,[56,57] cleavage of the cap linkage or one of the first few phosphodiester bonds would expose the transcript to 5′ to 3′ exonucleolytic degradation.

Experimental evidence suggests that the 5′ cleavage reaction is likely to be within the cap linkage itself and should be considered a decapping reaction. This possibility was first suggested by analysis of the decapped transcripts that accumulate in *xrn1*Δ strains. For example, the decapped *MFA2* transcripts that accumulate in *xrn1*Δ strains appear to be full length as judged by primer extension analysis.[22] Consistent with the reaction being cleavage of the cap linkage, an enzymatic activity capable of releasing m7GDP from the 5′ end of capped transcripts had been identified.[58] Evidence that this activity is actually required for mRNA decapping *in vivo* has come from the identification of a gene, termed *DCP1*, encoding the catalytic component of this decapping enzyme.[59,60] Strains deleted for the *DCP1* gene are unable to complete mRNA decapping *in vivo* resulting in increased stability of several mRNAs tested and show no decapping activity in cell-free extracts.[59] These results suggest that the nucleolytic event that exposes the mRNA to 5′ to 3′ decay is cleavage of the cap linkage releasing m7GDP and the mRNA body with a 5′ phosphate. Interestingly, the *PGK1* and *RP51A* transcripts accumulate in *xrn1*Δ strains as decapped species shortened at the 5′ end by a few nucleotides.[23,24] Although these species could arise by cleavage of one of the first few phosphodiester bonds, it is more likely that they arise by decapping followed by a 5′ to 3′ exonucleolytic nibbling that utilizes a nuclease different from the Xrn1p. This is supported by observations which suggest that there is a second less efficient 5′ to 3′ exonuclease in yeast[23] and that decapping of the *PGK1* mRNA requires the Dcp1p *in vivo* and occurs at the same site as the MFA2 transcript *in vitro*.[59,60]

These observations define a decay pathway common to stable and unstable mRNAs in yeast in which deadenylation triggers decapping, thereby leading to 5′ to 3′ decay of the transcript. The available evidence suggests that this pathway acts on most, if not all, yeast mRNAs. For example, it has been shown that the *MATα1* mRNA is also degraded by this same mechanism.[48] Moreover, in strains lacking the decapping enzyme (*dcp1*Δ), several yeast mRNAs including the unstable *MFA2, STE2, GAl10,* and *HIS3* mRNAs as well as the stable *ACT1* and *PGK1* transcripts are stabilized.[59] Similarly, in cells lacking Xrn1p, full-length deadenylated forms of the *Rp51A, CYC1, ACT1, MFα1,* and *MATα1* transcripts accumulate.[24,48] The deadenylated forms of several of these mRNAs (*ACT1*, *RP51A*, and *CYC1*) are degradable by purified Xrn1p suggesting that they lack a 5' cap.[24] Finally, Xrn1p is very abundant and primarily cytoplasmic.[61] These observations are consistent with the hypothesis that deadenylation triggers decapping of many yeast transcripts.

Since poly(A) tails and cap structures are common features of eukaryotic transcripts, an appealing model is that mRNA decay by deadenylation-dependent decapping and 5' to 3' digestion could be a conserved mechanism of mRNA turnover. Several observations suggest that decapping and 5' to 3' decay may occur in other eukaryotes. For example, mRNAs lacking the cap structure are rapidly degraded in many eukaryotic cells.[62] In addition, enzymes that could catalyze removal of the cap structure and subsequent 5' to 3' degradation of the transcript have been described in mammalian cells (e.g., Ref. 63). Decay intermediates consistent with mRNA decapping have been observed in several other eukaryotic cell types. For example, full-length mRNAs that are devoid of both the cap and most of the poly(A) tail have been detected from murine liver cells.[64] Similarly, mRNA decay intermediates that are shortened at their 5' ends have been identified in both plant and animal cells.[65,66] Furthermore, homologues of the yeast Xrn1p 5' to 3' exonuclease that degrade mRNA following decapping have been identified in several organisms.[61,67] Taken together, these observations suggest that deadenylation-dependent decapping followed by 5' to 3' exonucleolytic decay is a conserved mRNA decay mechanism in eukaryotic cells.

How would deadenylation stimulate decapping of the mRNA? One possibility is that decapping is an indirect result of deadenylation. For example, since the poly(A) tail of mRNAs can associate with the cytoskeleton,[68,69] poly(A) tail shortening could alter an mRNA's subcellular localization and thereby expose the mRNA to a decapping. An alternative model is that the poly(A) tail and the poly(A) binding protein might inhibit the decapping reaction directly by forming, or stabilizing, an mRNP structure involving the 5' and 3' termini of the mRNA. Such an mRNP structure has been suggested previously based on the stimulation of translational initiation by poly(A) tails,[29,70–72] and on the presence of circular polysome structures in electron micrographs.[73] Such an interaction is also supported by observations which suggest that base pairing between the 5' and 3' UTR sequences can occur *in vivo*.[74] Other results have provided insight into how such an interaction could form. This is based on the observation that the poly(A) binding protein, presumably bound to the 3' poly(A) tail, can associate with the eIF-4G component of the cap binding complex found at the 5' end of the transcript.[75] One simple model is that the same mRNP structure which stimulates translational initiation could inhibit decapping, perhaps by efficiently recruiting translation initiation factors to the 5' UTR and cap structure.

6.13.4 DEADENYLATION-INDEPENDENT DECAPPING OF mRNAs

In some cases mRNA decay is brought about by decapping and 5' to 3' decay of the transcript independent of poly(A) shortening. To date the only definitive example of this decay mechanism has come from the analysis of yeast transcripts that contain an early nonsense codon.[76] The degradation of mRNAs with nonsense codons is part of a conserved process, termed mRNA surveillance, which promotes the rapid turnover of aberrant mRNAs. These aberrant transcripts contain early nonsense codons,[77–80] unspliced introns,[81] or have extended 3' UTRs.[80] However, it should be expected that some other features of transcripts that promote rapid decay will do so by stimulating deadenylation-independent decapping.

mRNA surveillance may exist, in part, to increase the fidelity of gene expression by degrading aberrant mRNAs that if translated would produce truncated proteins. This process would be relevant since truncated proteins often have dominant negative phenotypes. A striking example of this phenomenon is seen in *C. elegans* where loss of function mutations in the *SMG* genes, which are required for mRNA surveillance, convert recessive nonsense mutations in the myosin gene *unc-54* into dominant negatives.[80] Since mRNA surveillance degrades unspliced and aberrantly processed transcripts, this decay mechanism might be most important in organisms with a large number of introns where processing errors might lead to aberrant transcripts giving rise to truncated proteins.

The evidence that premature nonsense codons trigger deadenylation-independent decapping consists of the following observations.[76] First, in transcriptional pulse-chase experiments in yeast, a *PGK1* transcript with an early nonsense codon starts decaying without any detectable temporal lag, in contrast to the wild-type transcript which only decays after a lag period during which deadenylation occurs.[19] Second, decay products trapped by the insertion of a poly(G) tract in the 3' UTR of the nonsense codon containing transcripts have long poly(A) tails. Third, full-length decapped *PGK1* transcripts containing nonsense codons are stabilized in *xrn1Δ* strains. Finally, in strains lacking the decapping enzyme encoded by the *DCP1* gene, transcripts with early nonsense codons are greatly stabilized.[59] These results suggest that the mechanism of degradation of these transcripts converges with the normal deadenylation-dependent decapping pathway at the nucleolytic step of decapping.

Two observations raise the possibility that nonsense codons might trigger deadenylation-independent decapping in mammalian cells. First, introduction of a nonsense mutation into the β-globin mRNA accelerates its decay rate more than 10-fold; the deadenylation rate almost remains unaffected. As a result, in transcriptional pulse-chase experiments, even at a time when 90% of the mutant mRNA is degraded the poly(A) tails are as long as > 50 nucleotides, therefore suggesting that decay is independent of poly(A) shortening.[12] In addition, examination of the turnover of a β-globin transcript with an early nonsense codon in murine erythroid tissue revealed the presence of capped decay products that were shortened at the 5′ end.[82] Although the mechanism by which these decay products receive a 5′ cap is unclear, the presence of such products imply that decay is proceeding in a 5′ to 3′ direction for these transcripts.

6.13.5 CONTROL OF mRNA DECAPPING

As described in previous sections, for the body of the mRNA to be degraded by the 5′ to 3′ exonucleolytic process, the mRNA needs to be decapped. Decapping appears to be a central point at which many of the inputs that control mRNA decay rates converge. For example, at least in yeast, the deadenylation rate of the mRNA affects overall transcript stability mainly by virtue of the ability of poly(A) tail to inhibit decapping.[22,23] In addition, specific sequences that modulate the mRNA decay rate can do so by affecting the rate of decapping.[20] Finally, as discussed above, the rapid decay of transcripts with early nonsense codons works by triggering extremely rapid mRNA decapping independently of poly(A) tail shortening.[76]

Results have demonstrated that in yeast cells a single decapping activity, dependent on, or encoded by, the *DCP1* gene, is responsible for the decapping of stable and unstable mRNAs, including those that are decapped independently of deadenylation. This observation suggests that features of mRNAs which alter the rates of decapping will do so by affecting the activity of the Dcp1p on individual transcripts. As discussed below, current evidence suggests that Dcp1p activity on a transcript will be modulated by proteins that recognize specific mRNA features, or sequences, and thereby affect the rate of Dcp1p cleavage.

The observation that decapping does not normally occur until the poly(A) tail is shortened to an oligo(A) length strongly suggests that the poly(A) tail inhibits this process. Inhibition of the decapping reaction by the poly(A) tail in yeast is mediated by the *PAB1* gene product. This conclusion is supported by the observation that in *pab1* mutant strains, mRNA molecules are decapped independently of deadenylation.[30] This suggests a model in which the shortening of the poly(A) tail to an oligo(A) length triggers decapping by promoting dissociation of the last Pab1p molecule.

In addition to the poly(A) tail, there are specific sequences within mRNAs that can affect the decapping rates of mRNA. For example, the unstable *MFA2* mRNA ($t_{1/2} = 4$ minutes) is rapidly decapped following deadenylation, while the stable *PGK1* transcript ($t_{1/2} = 35$ minutes) is decapped slowly after deadenylation.[19,22,23] These differences in decapping rates are at least partially determined by specific sequences within these mRNAs,[20,23] although these sequences are currently poorly defined. Similarly, as discussed above, the presence of an early nonsense codon can trigger rapid decapping.[76,83]

In each case where decapping rate responds to some mRNA feature, work has identified *trans*-acting factors that are required for modulation of decapping rate. For example, as discussed above, the inhibition of decapping by the poly(A) tail requires the associated poly(A) binding protein[30] to communicate the presence of a 3′ poly(A) tail. Similarly, the deadenylation-independent decapping caused by premature translational termination requires the *UPF1*, *UPF2*, and *UPF3* gene products to signal that termination is premature, thereby triggering rapid decapping.[78,84,85] In a similar manner, work has identified lesions in two genes, termed *MRT1* and *MRT3*, that appear to modulate the rates of decapping on substrates after deadenylation and therefore may function analogously to the *UPF* gene products.[86] Moreover, some dcp1 mutants have also been identified that are defective in deadenylation-dependent mRNA decapping *in vivo* but yield decapping enzymes exhibiting wild-type specific activity *in vitro*[87] which again supports the idea that decapping may be regulated *in vivo*.

There are two general types of hypotheses to explain how the decapping rate will be modulated on individual transcripts, either by the poly(A) tail or by specific sequences. As one possibility, the decapping rate would be modulated by affecting the interaction, or recruitment, of the decapping enzyme with each mRNA. The central feature of this model is that the decapping enzyme itself would be the site of regulatory inputs. An alternative is that the modulation of decapping activity on individual mRNAs occurs by alterations in mRNP structure that affect decapping rate. For example, all of the features that modulate the decapping rate could be envisioned to affect the interaction of proteins, such as translation initiation factors, with the 5' cap structure thereby sterically competing with the decapping activity. Such a model is appealing for two reasons. First, it would provide a common explanation for how the poly(A) tail could affect both translation initiation rates and decapping. In addition, such a model would provide a basis for the observation that the translation and turnover of eukaryotic mRNAs are often mechanistically coupled.

6.13.6 DEADENYLATION CAN ALSO LEAD TO 3' TO 5' DEGRADATION

Eukaryotic transcripts can also be degraded in a 3' to 5' direction following deadenylation. Evidence for this mechanism of degradation has primarily come from examining mRNA turnover in yeast when the 5' to 3' decay pathway is inhibited. For example, fragments of the yeast *PGK1* mRNA shortened at the 3' end are observed when the 5' to 3' decay pathway is inhibited by deletion of either the *XRN1* or *DCP1* genes.[23,59,76] Similarly, mRNA fragments with poly(G) tracts to block the 5' to 3' exonuclease are ultimately degraded 3' to 5', although at a relatively slow rate.[88] Interestingly, in wild-type yeast, the 3' trimmed fragments of the *PGK1* mRNA can also be observed at low levels. These observations argue that 3' to 5' degradation is generally a slower process than 5' to 3' degradation. Decay intermediates that are consistent with 3' to 5' decay following deadenylation have also been observed for the oat phytochrome A mRNA *in vivo*.[65] This observation, and the conservation of gene products required for this decay pathway (see below), suggest that 3' to 5' degradation of eukaryotic mRNAs is a conserved process.

In yeast, several observations now indicate that 3' to 5' nucleolytic degradation of the transcript body is a general mechanism of mRNA turnover capable of acting on many mRNAs. This evidence includes the analysis of mRNA decay when decapping and 5' to 3' degradation are blocked, which indicated that both the stable *PGK1* mRNA[23] and the unstable *MFA2* mRNA were degraded 3' to 5'.[88] Moreover, in strains defective for both 5' to 3' and 3' to 5' decay mechanisms, several mRNAs are extremely stable, indicating that these mRNAs are all substrates for 3' to 5' degradation.[88] These observations suggest that there are two general mechanisms for degrading the mRNA body following poly(A) shortening in yeast, decapping leading to 5' to 3' degradation, or 3' to 5' degradation. Moreover, since transcripts were extremely long-lived in the absence of these two mechanisms of degradation,[88] there are unlikely to be other *major* nucleolytic activities that can act to degrade mRNAs at a reasonable rate.

The generality of the 3' to 5' decay mechanism is strengthened by the observation that any combination of mutations which inactivate the 5' to 3' decay pathway with mutations that inactivate the 3' to 5' pathway leads to inviability.[88] This observation argues that efficient mRNA degradation, by either one of these pathways, is essential for viability. Thus, mRNA turnover is a redundant and essential process in yeast. Moreover, several of the proteins required for these mRNA decay mechanisms (see below), including *XRN1*, *SKI2*, *SKI6/RRP41*, and *RRP4*, have homologues in other eukaryotic cells, including mammals.[67,89-91] The existence of these homologues argues that these pathways of mRNA turnover occur in all eukaryotic cells and are likely to be the two general mechanisms of mRNA decay.

An interesting issue is the relationship between 5' to 3' and 3' to 5' degradation of mRNA and their respective roles in eukaryotic cells. Currently, the available evidence suggests that the major mechanism of mRNA decay in yeast is by decapping and 5' to 3' degradation.[19,22-24,59] However, the 3' to 5' mechanism of degradation is likely to have unique functions. For example, it is likely that particular mRNAs are preferentially degraded by the 3' to 5' mechanism even in wild-type cells. Similarly, since the pathways of mRNA degradation have only been examined under an extremely limited set of growth conditions, there may be specific conditions where the 3' to 5' mechanism is primary. In addition, the 3' to 5' mechanism of degradation may play an antiviral role by reducing expression from viral poly(A) transcripts.[92] Finally, it should be considered that in other eukaryotes the relative importance of these two mechanisms may be different. For example, in oat seedlings the phytochrome A mRNA is degraded by both 5′ to 3′ and 3′ to 5′ decay mechanisms at similar rates.[65]

Several proteins have been identified that are required for 3′ to 5′ degradation of mRNAs.[88] Additionally, the components of a multiprotein complex termed the exosome, which is known to be required for rRNA processing,[91] are also involved. Three of the proteins from this complex (Rrp4p, Ski6/Rrp41p, and Rrp44p) have been shown to have 3′ to 5′ exoribonuclease activity *in vitro*, and the remaining two (Rrp42 and Rrp43) have sequence similarity to bacterial 3′ to 5′ exonucleases.[91] While the complex is known to be required for proper 5.8 S rRNA processing in the nucleus in yeast, a homologous complex in HeLa cells is also found in the cytoplasm,[91] suggesting a potential role in 3′ to 5′ mRNA degradation. The essential observation suggesting that the exosome is involved in 3′ to 5′ degradation of mRNA is that mutations in components of the exosome lead to defects in 3′ to 5′ mRNA degradation. This observation has led to the working hypothesis that the exosome is the degradative activity that acts on mRNAs in a 3′ to 5′ direction.[88]

In addition to the exosome, three other factors are known to be required for 3′ to 5′ degradation; these are the products of the *SKI2*, *SKI3*, and *SKI8* genes.[88] Although the exact role of these factors in mRNA decay is not yet clear, mutations in these genes have other phenotypes that are consistent with a defect in 3′ to 5′ degradation of poly(A)⁻ mRNAs. For example, these mutants lead to an overexpression of the mRNAs from the double-stranded RNA killer virus, which could be due to a stabilization of the poly(A)⁻ viral mRNAs. In addition, poly(A)⁻ mRNAs introduced into yeast by electroporation showed a longer functional stability in *ski2Δ* and *ski8Δ* strains compared to wild-type strains.[92] Interestingly, the *ski2Δ* and *ski8Δ* strains also showed increased initial rates of protein production from electroporated poly(A)⁻ transcripts.[92] This observation was interpreted to indicate that these proteins function to repress translation of poly(A)⁻ mRNAs due to an alteration in the biogenesis of the 60S ribosomal subunit and that the longer functional mRNA stability was a consequence of differences in translation rates.[92] However, several observations now suggest that the *SKI2*, *SKI3*, and *SKI8* proteins affect 3′ to 5′ mRNA degradation more directly. First, polysome profiles in *ski2Δ*, *ski3Δ*, and *ski8Δ* mutants are identical to wild-type[92] and examination of 5.8S processing indicated that at least this aspect of rRNA processing was normal in these mutants.[88] Second, since the *ski2*, *ski3*, *ski6/rrp41*, *ski8*, and *rrp4* mutants affected the 3′ to 5′ degradation of an mRNA fragment, which was *not* being translated, it is unlikely that an increase in translation rate in the mutant strains could be indirectly protecting the RNA from 3′ to 5′ degradation. Given this, there are two possible explanations for the results with electroporated mRNAs. First, if there is a competition between 3′ to 5′ degradation and translation initiation for electroporated mRNAs, when the RNAs are first introduced into cells, more transcripts would be translated, but at the same initiation rate, in the mutant strains. Alternatively, the *SKI2*, *SKI3*, and *SKI8* proteins might function in remodeling mRNP structure, perhaps by promoting the disassociation of proteins from the 3′ UTR, which might decrease the translation rate and also make the 3′ end more accessible to the exosome.

A simpler model is that the *SKI2*, *SKI3* and *SKI8* proteins function to adapt, or recruit, the exosome to mRNA substrates. This is a particularly appealing model for the Ski2p, which is a member of the DEVH box family of proteins and thus a putative RNA helicase, because some 3′ to 5′ exonuclease complexes have been shown to have associated RNA helicases of this type.[93,94] In this view, other proteins would serve as "adaptors" for other exosome substrates, such as the 5.8S pre-rRNA. This hypothesis would explain why the *ski2Δ*, *ski3Δ*, and *ski8Δ* mutations do not affect 5.8S processing.[88] Strikingly, mutations in another DEVH protein closely related to Ski2p, Dob1p, show a defect in processing of the 5.8S pre-rRNA.[95] This observation suggests that the Dob1p might serve as the exosome "adaptor" for 5.8S pre-rRNA. This view make the testable predictions that Ski2p, and perhaps Ski3p and Ski8p, will show interactions with the exosome and will directly affect its ability to degrade mRNA substrates.

6.13.7 ROLE OF ENDONUCLEOLYTIC CLEAVAGES IN THE DECAY OF EUKARYOTIC mRNAS

There are now several examples of eukaryotic mRNAs whose degradation includes a sequence specific endonucleolytic cleavage. Evidence for this mechanism comes from the analysis of several mRNAs including the mammalian 9E3, IGFII, transferrin receptor (TfR), apoII and albumin mRNAs and Xenopus Xlhbox2B mRNA. In these cases, mRNA fragments are detected *in vivo* that correspond to the 5′ and 3′ portion of the transcript and are consistent with endonucleolytic cleavages.[96–101] In addition, endonucleolytic cleavages have been defined *in vitro* for the albumin mRNA[102] and in the coding region of the c-*myc* mRNA.[103] One interesting feature of several of these endonuclease cleavage sites is that there are multiple cleavages within a limited region of the transcript. For example, in the mouse albumin mRNA, Xenopus homeobox mRNA, and chicken liver apoII mRNA, several endonucleolytic cleavages have been observed in a small region and the various cleavage sites were found to have some sequence homology.[99–101,104] However, the cleavage site consensus sequences observed in each of these cases is different. This suggests that there may be a wide variety of endonucleases with different cleavage specificities. However, it is also possible that the specificity of a common endonuclease could be modified by association with different specificity determining factors.

Since sequence-specific endonuclease target sites are likely to be limited to individual mRNAs or classes of mRNAs, their presence allows for specific control of the decay rate of these transcripts. In some cases, the rate of endonucleolytic cleavage is modulated by the activity of protective factors that bind at, or near, the cleavage site and compete with the endonuclease.[98,103,105] For example, the binding of the iron response element binding protein in the TfR 3′ UTR in response to low intracellular iron concentrations inhibits endonucleolytic cleavage of this mRNA.[98] Therefore, some endonucleases may be constitutively active and the accessibility of the cleavage site is regulated.

In other cases, the endonuclease activity may be directly regulated. For example, the mammalian endonuclease RNase L is normally inactive and is only activated by oligomers of 2′, 5′ phosphodiester bonded adenylate residues, which are produced in response to the presence of double-stranded RNA.[106] Although RNase L is important in mediating interferon responses (e.g., Ref. 107), it has yet to be established whether this enzyme normally degrades any cellular mRNAs.

Given the role of deadenylation in mRNA turnover, it has been of interest to determine whether or not these endonuclease cleavage reactions are dependent on prior poly(A) shortening. At least in some cases it is clear that deadenylation to an oligo(A) tail length is not required for endonucleolytic cleavage since the 3′ fragment of the 9E3 and TfR mRNAs is polyadenylated.[96,98] In addition, cleavage of the Xlhbox2B mRNA is not affected by the adenylation state of the mRNA.[105] However, it is possible that endonucleolytic cleavage of some mRNAs could be dependent on the length of the poly(A) tail. For example, in the case of the mouse liver albumin mRNA, the downstream degradation intermediates resulting from *in vivo* endonucleolytic cleavages occurring in the middle region of mRNA are enriched in the poly(A)⁻ fraction but present in very low or undetectable levels in the poly(A)⁺ fraction.[101] Similarly, in the case of chicken apoII mRNA, the downstream fragments were also found to be poly(A)⁻.[100] These results raise the possibility that these cleavages could be deadenylation-dependent.

6.13.8 SUMMARY

It is now clear that there is a diversity of different pathways for mRNA turnover in eukaryotic cells. One simple integrated view is that all polyadenylated mRNAs would be degraded by the deadenylation-dependent pathway at some rate. For different transcripts, or potentially in different cell types, the relative importance of 5′ to 3′ and 3′ to 5′ decay may differ. In addition to this deadenylation-dependent "default" pathway, another layer of complexity would come from degradation mechanisms specific to individual mRNAs, or classes of mRNAs. Such mRNA-specific mechanisms would include sequence-specific endonuclease cleavage and deadenylation-independent decapping. Thus, the overall decay rate of an individual transcript will be a function of its susceptibility to these turnover pathways. In addition, *cis*-acting sequences that specify mRNA decay rate, and regulatory inputs that control mRNA turnover, are likely to affect all the steps of these decay pathways. Future work identifying and characterizing the gene products that catalyze and modulate the rates of these nucleolytic steps should provide a basis for the understanding of differential mRNA decay.

6.13.9 REFERENCES

1. C. V. Cabrera, J. J. Lee, J. W. Ellison, R. J. Britten, and E. C. Davidson, *J. Mol. Biol.*, 1984, **174**, 85.
2. M. Carneiro and U. Schibler, *J. Mol. Biol.*, 1984, **178**, 869.
3. J. Ross, *Microbiol. Rev.*, 1995, **59**, 423.
4. G. Caponigro and R. Parker, *Microbiol. Rev.*, 1996, **60**, 233.
5. D. Sheiness and J. E. Darnell, *Nature New Biol.*, 1973, **241**, 265.

6. S. W. Peltz, G. Brewer, P. Bernstein, P. Hart, and J. Ross, *Crit. Rev. Euk. Gene Exp.*, 1991, **1**, 99.
7. M. C. Wilson, S. G. Sawicki, P. A. White, and J. E. Darnell, Jr., *J. Mol. Biol.*, 1978, **126**, 23.
8. I. Paek and R. Axel, *Mol. Cell Biol.*, 1987, **7**, 1496.
9. A. W. Cochrane and R. G. Deeley, *J. Mol. Biol.*, 1988, **203**, 555.
10. P. Bernstein and J. Ross, *Trends Biochem. Sci.*, 1989, **14**, 373.
11. M. Zeevi, J. R. Nevins, and J. E. Darnell, *Mol. Cell Biol.*, 1982, **2**, 517.
12. A. B. Shyu, J. G. Belasco, and M. E. Greenberg, *Genes Dev.*, 1991, **5**, 221–234.
13. T. Wilson and R. Treisman, *EMBO J.*, 1988, **7**, 4193.
14. D. Caput, B. Beutler, K. Hartog, R. Thayer, S. Brown-Shimer, and A. Cerami, *Proc. Natl. Acad. Sci. USA*, 1986, **83**, 1670.
15. G. Brewer and J. Ross, *Mol. Cell Biol.*, 1988, **8**, 1697.
16. C.-Y. A. Chen, and A.-B. Shyu, *Mol. Cell Biol.*, 1994, **14**, 8471.
17. C.-Y. A. Chen and A.-B. Shyu, *Trends Biochem. Sci.*, 1995, **20**, 465.
18. M. Johnston, *Microbiol. Rev.*, 1987, **51**, 458.
19. C. J. Decker and R. Parker, *Genes Dev.*, 1993, **7**, 1632.
20. D. Muhlrad, and R. Parker, *Genes Dev.*, 1992, **6**, 2100.
21. P. Vreken and H. A. Raue, *Mol. Cell Biol.*, 1992, **12**, 2986.
22. D. Muhlrad C. J. Decker, and R. Parker, *Genes Dev.*, 1994, **8**, 855.
23. D. Muhlrad, C. J. Decker, and R. Parker, *Mol. Cell Biol.*, 1995, **15**, 2145.
24. C. L. Hsu and A. Stevens, *Mol. Cell Biol.*, 1993, **13**, 4826.
25. A. B. Sachs, R. W. Davis, and R. D. Kornberg, *Mol. Cell Biol.*, 1987, **7**, 3268.
26. A. B. Sachs, M. W. Bond, and R. D. Kornberg, *Cell*, 1986, **45**,827.
27. S. A. Adam, T. Nakagawa, M. S. Swanson, T. K. Woodruff, and G. Dreyfuss, *Mol. Cell Biol.*, 1986, **6**, 2932.
28. T. Grange, C. Martin de Sa, J. Oddos, and R. Pictet, *Nucleic Acids Res.*, 1987, **15**, 4771.
29. A. B. Sachs and R. W. Davis, *Cell*, 1989, **58**, 857.
30. G. Caponigro and R. Parker, *Genes Dev.*, 1995, **9**, 2421.
31. P. Bernstein, S. W. Peltz, and J. Ross *Mol. Cell Biol.*, 1989, **9**, 659.
32. L. P. Ford, P. S. Bagga, and J. Wilusz, *Mol. Cell Biol.*, 1997, **17**, 398.
33. A. B. Sachs and J. A. Deardorff, *Cell*, 1992, **70**, 961.
34. J. E. Lowell, D. Z. Rudner, and A. B. Sachs, *Genes Dev.*, 1992, **6**, 2088.
35. C. E. Brown and A. B. Sachs, *Mol. Cell. Biol.*, 1998, **18**, 6548.
36. J. Astrom, A. Astrom, and A. Virtanen, *J. Biol. Chem.*, 1992, **267**, 18 154.
37. J. Astrom, A. Astrom, and A. Virtanen, *EMBO J.*, 1991, **10**, 3067.
38. C. G. Korner and E. Wahle, *J. Biol. Chem.*, 1997, **272**, 10 448.
39. C.-Y. A. Chen, N. Xu, and A.-B. Shyu, *Mol. Cell Biol.*, 1995, **15**, 5777.
40. J. S. Malter, *Science*, 1989, **246**, 664.
41. P. R. Bohjanen, B. Petryniak, C. H. June, C. B. Thompson, and T. Lindsten, *Mol. Cell Biol.*, 1991, **11**, 3288.
42. D. A. Katz, N. G. Theodorakis, D. W. Cleveland, T. Lindsten, and C. B. Thompson, *Nucleic Acids Res.*, 1994, **22**, 238.
43. G. Brewer, *Mol. Cell Biol.*, 1991, **11**, 2460.
44. C.-Y. A. Chen, Y. You, and A.-B. Shyu, *Mol. Cell Biol.*, 1992, **12**, 5748.
45. E. Vakalopoulou, J. Schaack, and T. Shenk, *Mol. Cell Biol.*, 1991, **11**, 3355.
46. C. A. Lagnado, C. Y. Brown, and G. J. Goodall, *Mol. Cell Biol.*, 1994, **14**, 7984.
47. A. M. Zubiaga, J. G. Belasco, and M. E. Greenberg, *Mol. Cell Biol.*, 1995, **15**, 2219.
48. G. Caponigro and R. Parker, *Nucleic Acids Res.*, 1996, **24**, 4304.
49. R. Parker and A. Jacobson, *Proc. Natl. Acad. Sci. USA*, 1990, **87**, 2780.
50. G. Caponigro, D. Muhlrad, and R. Parker, *Mol. Cell Biol.*, 1993, **13**, 5141.
51. S. C. Schiavi, C. L. Wellington, A.-B. Shyu, C.-Y. A. Chen, M. E. Greenberg, and J. G. Belasco, *J. Biol. Chem.*, 1994, **269**, 3441.
52. C. L. Wellington, M. E. Greenberg, and J. G. Belasco, *Mol. Cell Biol.*, 1993, **13**, 5034.
53. I. A. Laird-Offringa, C. L. de Wit, P. Elfferich, and A. J. van der Eb, *Mol. Cell Biol.*, 1990, **10**, 6132.
54. M. Muckenthaler, N. Gunkel, R. Stripecke, and M. W. Hentze, *RNA*, 1997, **3**, 983.
55. J. R. Williamson, M. K. Raghuraman, and T. R. Cech, *Cell*, 1989, **59**, 871.
56. F. W. Larimer and A. Stevens, *Gene*, 1990, **95**, 85.
57. A. Stevens, *Biochem. Biophys. Res. Commun.*, 1978, **81**, 656.
58. A. Stevens, *Mol. Cell Biol.*, 1988, **8**, 2005.
59. C. A. Beelman, A. Stevens, G. Caponigro, T. E. La Grandeur, L. Hatfield, D. Fortner, and R. Parker, *Nature*, 1996, **382**, 642.
60. T. E. LaGrandeur and R. Parker, *EMBO J.*, 1998, **17**, 1487.
61. W. D. Heyer, A. W. Johnson, U. Reinhart, and R. D. Kolodner, *Mol. Cell Biol.*, 1995, **15**, 2728.
62. D. R. Drummond, J. Armstrong, and A. Colman, *Nucleic Acids Res.*, 1985, **13**, 7375.
63. M. Coutts and G. Brawerman, *Biochim. Biophys. Acta*, 1993, **1173**, 57.
64. P. Couttet, M. Fromont-Racine, D. Steel, R. Pictet, and T. Grange, *Proc. Natl. Acad. Sci. USA*, 1997, **94**, 5628.
65. D. C. Higgs and J. T. Colbert, *Plant Cell*, 1994, **6**, 1007.
66. S.-K. Lim and L. E. Maquat, *EMBO J.*, 1992, **11**, 3271.
67. V. I. Bashkirov, H. Scherthan, J. A. Solinger, J.-M. Buerstedde, and W.-D. Heyer, *J. Cell Biol.*, 1997, **136**, 761.
68. R. Lenk, L. Ransom, Y. Kaufmann, and S. Penman, *Cell*, 1977, **10**, 67.
69. K. L. Taneja, L. M. Lifshitz, F. S. Fay, and R. H. Singer, *J. Cell Biol.*, 1992, **119**, 1245.
70. M. T. Doel and N. H. Carey, *Cell*, 1976, **8**, 51.
71. G. Galili, E. E. Kawata, L. D. Smith, and B. A. Larkins, *J. Biol. Chem.*, 1988, **263**, 5764.
72. D. Munroe and A. Jacobson, *Mol. Cell Biol.*, 1990, **10**, 3441.
73. A. K. Christensen, L. E. Kahn, and C. M. Bourne, *Am. J. Anat.*, 1987, **178**, 1.
74. J. J. Van den Heuvel, R. J. Planta, and H. A. Raue, *Yeast*, 1990, **6**, 473.
75. S. Z. Tarun, Jr. and A. B. Sachs, *EMBO J.*, 1996, **15**, 7168.

76. D. Muhlrad and R. Parker, *Nature*, 1994, **370**, 578.
77. R. Losson and F. Lacroute, *Proc. Natl. Acad. Sci. USA*, 1979, **76**, 5134.
78. P. Leeds, S. W. Peltz, A. Jacobson, and M. R. Culbertson, *Genes Dev.*, 1991, **5**, 2303.
79. S. W. Peltz, A. H. Brown, and A. Jacobson, *Genes Dev.*, 1993, **7**, 1737.
80. R. Pulak and P. Anderson, *Genes Dev.*, 1993, **7**, 1885.
81. F. He, S. W. Peltz, J. L. Donahue, M. Rosbash, and A. Jacobson, *Proc. Natl. Acad. Sci. USA*, 1993, **90**, 7034.
82. S.-K. Lim, C. D. Sigmund, K. W. Gross, and L. E. Maquat, *Mol. Cell Biol.*, 1992, **12**, 1149.
83. K. W. Hagan, M. J. Ruiz-Echevarria, Y. Quan, and S. W. Peltz, *Mol. Cell Biol.*, 1995, **15**, 809.
84. B.-S. Lee and M. R. Culbertson, *Proc. Natl. Acad. Sci. USA*, 1995, **92**, 10 354.
85. P. Leeds, J. M. Wood, B.-S. Lee, and M. R. Culbertson, *Mol. Cell Biol.*, 1992, **12**, 2165.
86. L. K. Hatfield, C. A. Beelman, A. Stevens, and R. Parker, *Mol. Cell Biol.*, 1996, **16**, 5830.
87. S. Tharun and R. Parker, 1998, *Genetics*, in press.
88. J. S. J. Anderson and R. Parker, *EMBO J.*, 1998, **17**, 1497.
89. S. G. Lee, I. Lee, S. H. Park, C. W. Kang, and K. Y. Song, *Genomics*, 1995, **10**, 660.
90. A. W. Dangel, L. Shen, A. R. Mendoza, L. C. Wu, and C. Y. Yu, *Nucleic Acids Res.*, 1995, **23**, 2120.
91. P. Mitchell, E. Petfalski, A. Shevchenko, M. Mann, and D. Tollervey, *Cell*, 1997, **91**, 457.
92. D. C. Masison, A. Blanc, J. C. Ribas, K. Carroll, N. Sonenberg, and R. B. Wickner, *Mol. Cell Biol.*, 1995, **15**, 2763.
93. B. Py, C. F. Higgins, H. M. Krisch, and A. J. Carpousis, *Nature*, 1996, **381**, 169.
94. S. P. Margossian, H. Li, H. P. Zassenhaus, and R. A. Butow, *Cell*, 1996, **84**, 199.
95. J. de la Cruz, D. Kressler, D. Tollervey, and P. Linder, *EMBO J.*, 1998, **17**, 1128.
96. M. Y. Stoekle and H. Hanafusa, *Mol. Cell Biol.*, 1989, **9**, 4738.
97. F. C. Nielsen and J. Christiansen, *J. Biol. Chem.*, 1992, **267**, 19 404.
98. R. Binder, J. A. Horowitz, J. P. Basilion, D. M. Koeller, R. D. Klausner, and J. B. Harford, *EMBO J.*, 1994, **13**, 1969.
99. B. D. Brown and R. M. Harland, *Genes Dev.*, 1990, **4**, 1925.
100. A. Cochrane and R. G. Deeley, *J. Biol. Chem.*, 1989, **264**, 6495.
101. S. Tharun and R. Sirdeshmukh, *Nucleic Acids Res.*, 1995, **23**, 641.
102. R. L. Pastori, J. E. Moskaitis, and D. R. Schoenberg, *Biochemistry*, 1991, **30**, 10 490.
103. P. L. Bernstein, D. J. Herrick, R. D. Prokipcak, and J. Ross, *Genes Dev.*, 1992, **6**, 642.
104. R. Binder, S.-P. L. Hwang, R. Ratnasabapathy, and D. L. Williams, *J. Biol. Chem.*, 1989, **264**, 16 910.
105. B. D. Brown, I. D. Zipkin, and R. M. Harland, *Genes Dev.*, 1993, **7**, 1620.
106. R. H. Silverman, in "Ribonucleases: Structure and Function," eds. G. D'alessio and J. F. Riordan, Academic Press, New York, 1996.
107. B. A. Hassel, A. Zhou, C. Sotomayor, A. Maran, and R. H. Silverman, *EMBO J.*, 1993, **12**, 3297.

6.14
Ribonucleotide Analogues and Their Applications

SANDEEP VERMA, NARENDRA K. VAISH, and FRITZ ECKSTEIN

Max-Planck-Institut für Experimentelle Medizin, Göttingen, Germany

6.14.1 INTRODUCTION

Ribonucleotide analogues are beneficial for understanding catalytic RNA mechanisms and to study RNA–RNA, RNA–DNA, and RNA–protein interactions. Such studies often focus on the role of the phosphate group and thus employ a phosphate analogue, most often the phosphorothioate.

However, as the difference between the function of DNA and RNA might be located at the 2′ position, analogues which replace the 2′-OH group are often useful in understanding the role of this group in recognition processes or its involvement in reactions. Additionally, base analogues can be used for photoactivated cross-linking or to study the involvement of base exocyclic functional groups. This review will discuss some of the more commonly used analogues in the RNA field, concentrating on oligoribonucleotides and RNA, not considering nucleosides by themselves. A more comprehensive review of oligonucleotide analogues covering both oligodeoxynucleotides and oligoribonucleotides has appeared.[1] Another extensive review on nucleoside analogues and modified oligonucleotides is presented in Chapter 7.07, Volume 7.

6.14.2 INCORPORATION OF ANALOGUES INTO OLIGORIBONUCLEOTIDES

6.14.2.1 Chemical Synthesis

Automated, solid-phase DNA synthesizers can also be conveniently used for the synthesis of short oligoribonucleotides. The solid-phase synthesizers make use of fully protected ribonucleosides, anchored to a controlled pore glass support, at the 3′ end, via a chemically cleavable linker. A series of reactions couple ribonucleoside phosphoramidites sequentially to the support-bound nucleoside and after the last step, base hydrolysis is used to cleave the linker from the anchored nucleoside to yield a 2′-*O*-silyl protected oligomer. Desilylation is generally achieved by using tetrabutylammonium fluoride or similar reagents.

Development of effective 2′-OH protecting groups, use of efficient activators and oligomer purification methods have greatly improved RNA synthesis.[2-4] (see also Chapter 6.06). At the time of writing, synthesis of oligoribonucleotides using the solid-phase approach, is limited to a length of ~50 nucleotides. For the preparation of longer oligoribonucleotides, enzymatic ligation of chemically synthesized oligonucleotide fragments is preferred.

6.14.2.2 Enzymatic Ligation of Oligoribonucleotides

Enzymatic ligation is the method of choice for the synthesis of oligoribonucleotides that are too long to be obtained by chemical synthesis. Originally, RNA ligase was used for the ligation of the 5′ and 3′ termini of RNA fragments through the formation of a phosphodiester bond. This enzyme requires single-stranded ends for ligation and is thus restricted to the ligation of fragments in loop regions such as in the anticodon loop of tRNA.[5] An alternative method which employs T4 DNA ligase has been developed.[6] This enzyme requires the presence of the ends to be ligated on a duplex, which is achieved by annealing two oligoribonucleotide fragments to a complementary oligodeoxynucleotide bridge. This is a versatile method which permits efficient ligation and allows for the incorporation of nucleotide analogues. The methodology has been described in detail by Moore and Query.[7]

6.14.2.3 Enzymatic Incorporation by Polymerization

Template-directed synthesis of oligoribonucleotides and long transcripts is achieved by the use of T7 or SP6 RNA polymerases. This process is also well suited for the incorporation of modified nucleotides, provided their triphosphates are good substrates for these enzymes. This requirement, however, limits the number of modified ribonucleotides which can be used in transcription reactions and thus only a few modifications can be enzymatically introduced into oligoribonucleotides.

The phosphorothioate internucleotidic linkage is the most common enzymatically introduced modification as all of the four nucleoside α-thiotriphosphates are good substrates for RNA polymerases ((**1**)–(**4**)).[8] Substitution of one of the nonbridging oxygen atoms of a phosphate group by sulfur introduces chirality at the phosphorus center and consequently two diastereomers exist for the nucleoside α-thiotriphosphates. Only the Sp-diastereomers are good substrates for the polymerases. The enzymatic incorporation proceeds with inversion of configuration at phosphorus resulting in the formation of the Rp internucleotidic linkage. Thus, the enzymatic synthesis is limited to the incorporation of the Rp-phosphorothioate linkage.[8,9]

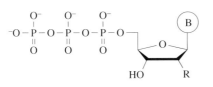

R = F, NH$_2$
(1) 2'-Modified nucleoside 5'-triphosphates

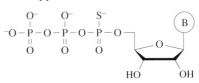

(2) Nucleoside 5'-α-thiotriphosphates

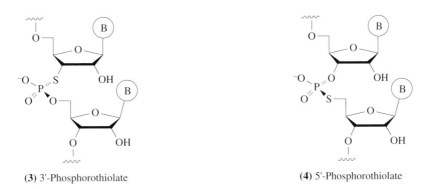

(3) 3'-Phosphorothiolate

(4) 5'-Phosphorothiolate

Of the ribose analogues, only 2'-amino- and 2'-fluoro-2'-deoxynucleoside triphosphates are substrates for the wild-type T7 RNA polymerase ((**1**)–(**4**)). In particular, 2'-amino nucleoside triphosphates are better substrates as observed from the synthesis of a luciferase transcript of 2500 nt in length.[10] In comparison, 2'-fluoro-2'-deoxy analogues are less readily incorporated. However, a mutant T7 RNA polymerase which incorporates 2'-deoxynucleotides better than the wild-type enzyme, also accepts 2'-fluoro analogues quite well.[11]

T7 RNA polymerase normally initiates transcription with a guanosine triphosphate, but it can also accept other guanosine derivatives such as guanosine 5'-monophosphorothioate[12] or guanosine 5'-γ-thiotriphosphate[13] for initiation. The applications of 5'-terminal phosphorothioate-containing constructs will be discussed in Section 6.14.3.3.

6.14.3 PHOSPHOROTHIOATE INTERNUCLEOTIDIC LINKAGES

The exchange of a nonbridging oxygen with a sulfur in the phosphorothioate is a minimal change since the negative charge of the phosphate group is retained and the van der Waals radius of sulfur (1.85 Å) is only 30% larger than that of oxygen (1.40 Å).[14] Sulfur is also more hydrophobic than oxygen, but the contribution of this factor to the difference in the properties of phosphates and phosphorothioates has not yet been determined. As mentioned before, the sulfur/oxygen exchange makes the phosphorus center chiral, resulting in two diastereomers of the phosphorothioate internucleotide linkage. The chirality of the phosphorothioate linkage permits the determination of the stereospecificity and of the stereochemical course of reactions occurring at phosphorus.[8] In addition, sulfur is a "soft" atom and it coordinates preferentially with "soft" metal ions, whereas oxygen is a "hard" atom and coordinates to "hard" metal ions.[15] Therefore, incorporation of phosphorothioates also allows for the identification of metal ion-binding phosphate group oxygen atoms in RNA or oligoribonucleotides.[16] The literature on nucleoside phosphorothioates, up to 1984, was reviewed by Eckstein.[8]

The oligonucleotide phosphorothioates can also be synthesized by chemical synthesis, using a sulfurizing reagent in the oxidation step. Chemical synthesis again produces a mixture of diastereomers, which in some cases can be separated, while the enzymatic incorporation results exclusively in the formation of the Rp-diastereomeric internucleotidic linkage.[8,9,17,18]

6.14.3.1 Phosphorothioates for Stereochemical Analysis

The chirality of the phosphorothioate group as an internucleotidic linkage in oligonucleotides permits analysis of the stereochemical course of transesterification reactions as long as the configurations of the starting material and the product are known. In most cases, the starting material for such studies is a transcript obtained by polymerization of a nucleoside α-thiotriphosphate with an RNA polymerase and therefore the configuration of the phosphorothioate internucleotidic linkage is known. Scheme 1 shows the stereospecificity of nucleoside phosphorothioate diastereomers with enzymes. (Literature may be consulted for individual reactions: polymerization;[8,19] reaction with RNase A;[20] reaction with RNase T₁;[21] reaction with SVPDE (snake venom phosphodiesterase);[19,22] reaction with nuclease Pl;[23] reaction with CNPase.[24]).

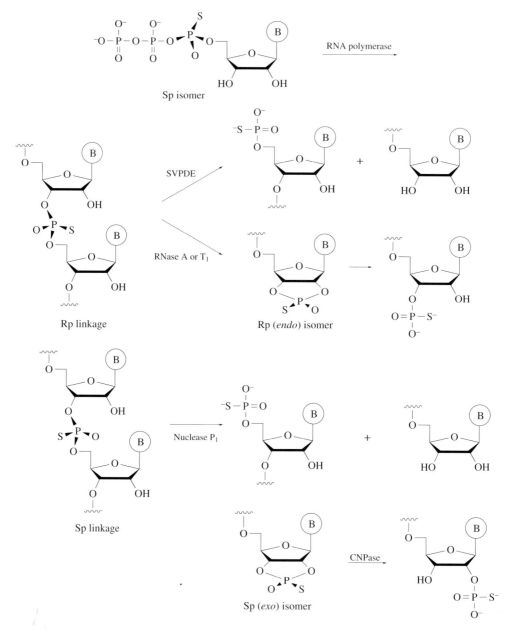

Scheme 1

The configuration of the phosphorothioate linkage in the product of a transesterification reaction can be determined in several ways depending on the type of product. If the product is a nucleoside 2′,3′-cyclic phosphorothioate, comparison can be made with an authentic sample or by using cyclic nucleotide phosphodiesterase which stereospecifically cleaves the Sp-isomer of cyclic phosphorothioates.[24] In cases where the product also contains a phosphorothioate internucleotidic linkage, the stereochemical preference of certain nucleases can be exploited (Scheme 1). Using such an approach, the stereochemical course of ribozyme-catalyzed reactions such as RNA cleavage by the hammerhead and the hairpin ribozymes have been determined.[25,26] These reactions proceed with inversion of configuration at phosphorus providing convincing evidence for an in-line mechanism in the absence of an intermediate. However, the ground state conformation of the hammerhead ribozyme, as established by X-ray crystallography,[27,28] is not compatible with such a mechanism and strongly suggests that a conformational change is needed to attain the transition state geometry.[29,30] The stereochemical course of the reactions catalyzed by the group I intron,[31,32] pre-mRNA splicing,[33,34] and group II intron self-splicing,[35,36] have been elucidated by stereopreferential degradation of the products. All these reactions also proceed with an inversion of configuration at phosphorus. Using phosphorothioates, a comparison of the stereochemical requirements of the individual steps has allowed similarities to be established between different splicing systems as determined for pre-mRNA splicing and group II splicing reaction.[37]

However, caution should be exercised in the mechanistic interpretation of rate differences observed between phosphate and phosphorothioate substrates in enzyme-catalyzed reactions.[38] From a purely chemical viewpoint, reactivity of the phosphorothioate group should only be marginally lower than that of the phosphate diester. Steric clashes in the transition state because of the larger van der Waals radius of sulfur, rather than the electronegativity difference, have been invoked to explain the reduction in reaction rate for the phosphorothioate.[39,40]

6.14.3.2 Phosphorothioate Interference

Phosphorothioates facilitate identification of the phosphate groups in an oligoribonucleotide involved in RNA–RNA or RNA–protein interactions. These studies exploit either the difference in size or a change in binding specificity when oxygen is substituted by sulfur. Binding of a metal ion to a specific phosphate group might be altered if it is replaced by a phosphorothioate, which makes the exact position of the phosphate group involved in metal ion binding easier to identify. Phosphorothioates are generally incorporated randomly by transcription for interference studies with oligoribonucleotides, so that on average only one phosphorothioate is present per molecule. The enzymatic incorporation results in the formation of an Rp-phosphorothioate linkage and thus an observed interference can be attributed to the pro-Rp oxygen of the replaced phosphate group. The success of this method depends on separation of the active species from the inactive species. Once separated, the precise position of the interfering phosphorothioate can be determined by iodine cleavage (*vide infra*).[41,42] Finally, a comparison of the phosphorothioate cleavage pattern of the active and inactive species allows identification of interfering positions in the reaction or interaction under investigation.

The phosphorothioate modification can also probe for phosphate–metal ion coordination in a reaction. This is based on the concept of "hard" and "soft" metal ions. Hard metal ions interact preferentially with the phosphorothioate oxygen atom, whereas soft ones interact with the phosphorothioate sulfur atom.[15] If a reaction is metal ion dependent and shows stereoselectivity for one diastereomer of the phosphorothioate in the presence of the hard metal ion Mg^{2+}, then use of a soft metal ion, such as Mn^{2+}, often reverses the diastereomer selectivity by making the other diastereomer the prefered substrate. Such a switch in stereospecificity indicates metal ion coordination to one of the oxygens of the phosphate group in the reaction. It is important to understand that this effect is only manifested if the phosphate–metal ion interaction is important in the rate-limiting step of the reaction. Also, the absence of such reversion does not necessarily indicate a lack of metal ion binding as pointed out for the 3′-5′-exonuclease reaction of the Klenow polymerase.[43]

6.14.3.2.1 Interference analysis in catalytic RNA

A study using phosphorothioate interference has identified 44 phosphate groups important for Mg^{2+} ion binding in the group I intron reaction.[44,45] For most of these positions, catalysis of the cleavage reaction could be restored with Mn^{2+} which indicates that the pro-Rp oxygen atoms are coordinated to the metal ion in the splicing reaction. The P4–P6 domain of this ribozyme has been crystallized and Mg^{2+} binding sites have been identified[46] (see also Chapter 6.04). In order to determine the importance of domains in RNA tertiary structure, they were individually replaced by the Rp-phosphorothioate and the folding was followed in the presence of Mg^{2+} or Mn^{2+}. This approach allowed for the identification of four phosphates critical for folding.

A development in the area of interference analysis is "nucleotide analogue interference mapping", which has been applied to identify the functional groups essential for group I intron catalysis[47] (see also Chapter 6.10). In this approach, a functional group modification in the nucleobase was coupled to the phosphorothioate linkage for easy identification. Modified group I intron constructs were prepared by substituting inosine α-thiotriphosphate in place of GTP, in the transcription reactions. Iodine-mediated phosphorothioate cleavage[41,42] was used to identify the positions of essential guanosine residues. In a parallel experiment, interference by the incorporation of GTPαS alone was used to separate the phosphorothioate effect from the inosine effect. This assay successfully mapped almost every guanosine residue essential for the catalytic activity of the group I intron ribozyme.

Four phosphate groups in the RNase P RNA were identified by phosphorothioate interference as important for an intramolecular, pre-tRNA cleavage reaction.[48] Beside these four positions, another study identified an additional twelve phosphate positions which were found to be important for the binding of the pre-tRNA.[49] Phosphorothioate substitution at the cleavage site of the pre-tRNA resulted in a 1000-fold reduction of the cleavage rate for both diastereomers, but it was found that the rate inhibition of the Rp-diastereomer could be largely restored by Cd^{2+}, a soft metal ion.[50] In yet another study, phosphorothioate substitutions of the pro-Rp oxygens in a pre-tRNA transcript revealed that direct coordination of Mg^{2+} ion to the pro-Rp oxygen of the scissile bond is required for RNase P catalysis.[51]

Rp-phosphorothioate interference also identified three phosphate groups, besides the one at the cleavage site, as important for catalysis by the hammerhead ribozyme.[52] The importance of one of these sites, where a metal ion is coordinated to the pro-Rp oxygen of the A9 phosphate, has been confirmed by X-ray structural analyses.[27,28] As suggested by the X-ray structure, a metal ion rescue could be observed with the Rp isomer at A9, which confirms the observation of metal ion coordination to this site. A Mg^{2+} ion has been found coordinated to the pro-Rp oxygen of the phosphate group at the cleavage site in a recent X-ray structure analysis.[30] These results are consistent with the cleavage of the Rp-phosphorothioate diastereomer in the presence of Mn^{2+} and that of the Sp-diastereomer in the presence of Mg^{2+} ion (Scheme 2).[26,53] However, the results of the phosphorothioate experiments have been challenged.[54] Zhou *et al.* observed a rate enhancement with Mn^{2+}, irrespective of the diastereomer used. Synthetic hammerhead ribozymes containing a mixture of phosphorothioate diastereomers have also been prepared.[55] The presence of the phosphorothioate Sp isomer at positions A6 and U16.1 severely reduced the rate of cleavage, indicating other important metal ion–phosphate interactions for this ribozyme.

Similar interference studies have identified important phosphate groups for the extended hammerhead ribozyme present in the satellite 2 transcripts[56] and for the hepatitis delta virus ribozyme.[57]

6.14.3.2.2   Interference analysis of phosphate–protein interaction

The phosphorothioate interference analysis and footprinting with iodine can also probe for the phosphate groups involved in RNA–protein interactions. This method was first applied to study the affinity of an oligoribonucleotide for the R17 coat protein.[58] It was extended to probe the interaction of *Escherichia coli* tRNA^{Ser} with its cognate aminoacyl synthetase.[42] The results suggested that contact with the tRNA variable loop might be located on the extended arm of the synthetase. All of these contacts, with one exception, were later confirmed by the X-ray crystal structure analysis of the related *Thermus thermophilus* complex.[59] Results obtained using the same approach with the yeast tRNA^{Asp} system were consistent with the interaction of the identity elements of the tRNA with the synthetase protein.[60] Phosphorothioate footprinting to investigate the interaction between tRNA^{Phe} transcripts from *T. thermophilus* and its synthetase revealed a complex and novel pattern of tRNA^{Phe}–synthetase interaction.[61] This strategy has also been used to study mitochondrial group I intron RNA splicing, which involves interaction with tyrosyl-tRNA synthetase.[62] These observations have led to the construction of a model which suggests that the recognition of the synthetase occurs by a tRNA-like structure in the group I intron.[63] The interaction of tRNA with the ribosome A and P sites was also investigated by phosphorothioate interference to map the essential "identity elements" involved in ribosome-tRNA recognition.[64]

6.14.3.2.3   Interference analysis for importance of 2′-hydroxy groups

A similar method was used to identify essential 2′-OH groups in the binding of tRNA to RNase P RNA.[65] dNTPαS was introduced into an RNase P RNA transcript by T7 RNA polymerase using

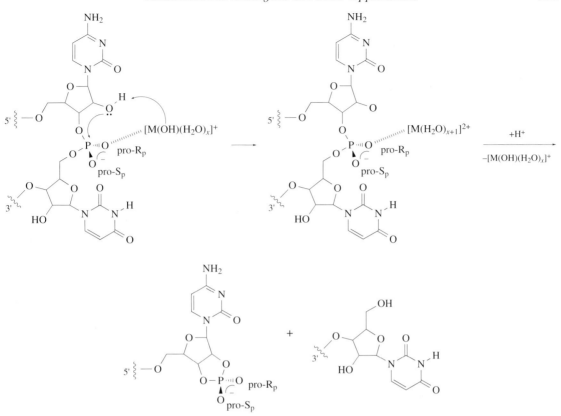

Scheme 2

a mixture of Mg^{2+} and Mn^{2+} in the reaction mixture to facilitate incorporation of the dNTPαS. Separation of the tRNA/RNA complex and comparison of the phosphorothioate iodine footprinting with the unbound RNA, revealed ∼20 positions for which 2′-deoxy substitution in the RNase P RNA impaired binding to the tRNA.

In the authors' laboratory, a similar study was undertaken to identify 2′-OH positions in the *E. coli* tRNA[Asp] responsible for charging by the cognate synthetase.[66] A T7 RNA polymerase mutant[67] was used for the random incorporation of dNTPαS into the tRNA transcript. In a separate experiment, NTPαS was incorporated to obtain tRNA transcripts to assess the effect of phosphorothioates alone on charging. Charged tRNAs were separated from the uncharged ones and iodine footprinting was performed on the two species. For several nucleotide positions, interference with charging could be attributed to the absence of the 2′-OH group. However, for some positions, the phosphorothioate effect could not be separated from the 2′-deoxy effect. Individual tRNAs, where the identified positions were singly replaced by the corresponding deoxynucleotide, were prepared and the kinetics of charging was determined. In general, the reduction in charging rate was between 2–10-fold.[66] This study differs from the one performed on yeast tRNA[Asp] using dNTP incorporation with a double mutant of the T7 RNA polymerase.[68] In the transcription reactions, one NTP was totally replaced by the corresponding dNTP. The tRNAs with all the uridines replaced by deoxyuridine and those with all guanosines replaced by deoxyguanosine, could not be charged. These results were interpreted on the basis of the X-ray structure of the native complex rather than evaluating tRNAs with individual deoxynucleotide replacements.

6.14.3.3 Terminal phosphorothioates for attachment

Oligonucleotides with a terminal phosphorothioate can be prepared by chemical synthesis,[69,70] by enzymatic 5′-thiophosphorylation using ATPγS,[13,71] by the attachment of pdCpS with RNA ligase to the 3′-end of oligoribonucleotides,[72] or by initiating transcription with guanosine derivatives such as GMPS or GTPγS.[12,13] Sodium periodate oxidation of a terminal ribonucleotide, linked to the oligonucleotide by a phosphorothioate internucleotidic linkage, followed by β-elimination is yet another alternative for postsynthetic labeling.[73] In oligonucleotides, terminal phosphorothioate groups have been used for a variety of purposes such as the separation of oligonucleotides or transcripts on mercury gels or columns,[13] reaction with haloacetyl derivatives of fluorescent dyes,[72] and with photoaffinity labels.[12]

6.14.4 S-BRIDGING PHOSPHOROTHIOLATES

Replacement of either the 3'- or the 5'-bridging oxygen of the phosphodiester internucleotidic linkage by sulfur, is another important modification used in oligoribonucleotides to probe mechanisms of ribozyme catalysis (Structures (1)–(4)). In the group I intron ribozyme-catalyzed trans-esterification reaction, cleavage of a substrate containing a 3'-S-phosphorothiolate internucleotidic substitution was about 1000-fold slower in the presence of Mg^{2+} when compared to an unmodified phosphodiester substrate.[74] However, in the presence of soft metal ions such as Mn^{2+} and Zn^{2+}, the cleavage activity was restored. This observation indicates a possible role of metal ions in stabilizing the negative charge on 3' oxygen in the transition state. The presence of a second metal ion involved in group I ribozyme catalysis was demonstrated using 3'-(thioinosinyl)-(3'-5')-uridine (IspU) dinucleotide as a substrate in the splicing reaction.[75] In a reaction mimicking the exon ligation step, IspU substrate required a second, thiophilic metal ion in optimum activity. Based on these observations, involvement of a second metal ion in activating the 3'-hydroxyl group of guanosine in the first step of splicing was proposed. The metal ion specificity switch from Mg^{2+} to Mn^{2+} using substrate analogues containing a 3'-S-phosphorothiolate was also determined for the first and second step of splicing of pre-mRNA. In contrast to the second step, the first step of splicing needs metal ion assistance.[76]

The mechanism of the hammerhead ribozyme-catalyzed reaction has also been probed using 5'-S-phosphorothiolates. This modification was introduced at the cleavage site in a DNA substrate.[77] However, an appreciable thio effect by changing the metal ion from Mg^{2+} to Mn^{2+} was not observed. There was concluded to be a lack of metal ion coordination with the 5'-oxygen of the leaving group in the transition state. Another group, which used an all-RNA substrate for similar experiments, came to a different conclusion.[78] They reasoned that attack at the 2'-OH is rate limiting for the modified substrate, whereas the departure of the leaving group is rate limiting for the unmodified one. Rate differences observed for the hydrolysis of 3'- and 5'-S-phosphorothiolate susbtrates with different metal ions should be interpreted with caution as even the metal ion-dependent hydrolysis of simple dinucleotides is not well understood.[79]

These examples elegantly document the power of sulfur substitution in the bridging oxygens of a phosphate internucleotidic linkage to study metal ion involvement in RNA-catalyzed reactions and complement the nonbridging oxygen substitution discussed earlier. However, it would be desirable to have more quantitative data on softness and hardness of the coordination partners in these systems which has been described only for ATPβS and related compounds.[15] Also, the absence of rescue effect with a thiophilic metal ion should not be interpreted as a lack of interaction in the native system, since metal ion rescue can be prevented or obscured by many factors.

6.14.5 MODIFICATION OF THE NUCLEOBASES

The heterocyclic ring of purine and pyrimidine bases provides the hydrogen bonding functional groups in nucleic acids. Thus base analogues, when introduced into oligonucleotides, can provide information on the importance of specific functional groups present in natural bases. In interpreting results obtained with base analogues, it should be realized that even a subtle change in the analogue can have dramatic effects due to the change in size, electronic distribution, nucleoside sugar conformation, tautomeric structure, or functional group pK_a values. The representative structures (5)–(19) of modified nucleosides discussed in subsequent sections are shown. Several synthetic approaches for the introduction of C-5 modifications in uridine and at C-8 of adenosine and guanosine have been reviewed.[80]

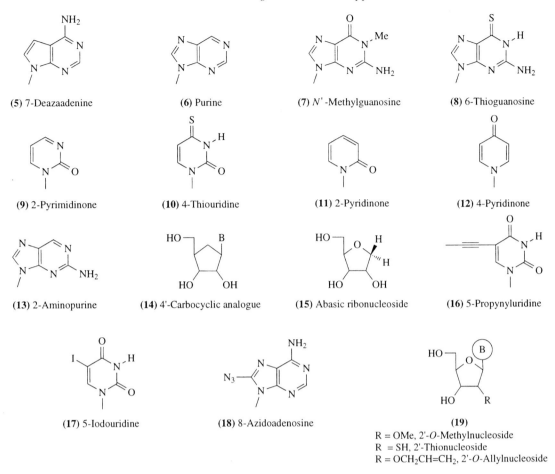

(5) 7-Deazaadenine **(6)** Purine **(7)** N'-Methylguanosine **(8)** 6-Thioguanosine

(9) 2-Pyrimidinone **(10)** 4-Thiouridine **(11)** 2-Pyridinone **(12)** 4-Pyridinone

(13) 2-Aminopurine **(14)** 4'-Carbocyclic analogue **(15)** Abasic ribonucleoside **(16)** 5-Propynyluridine

(17) 5-Iodouridine **(18)** 8-Azidoadenosine **(19)**

R = OMe, 2'-*O*-Methylnucleoside
R = SH, 2'-Thionucleoside
R = OCH₂CH=CH₂, 2'-*O*-Allylnucleoside

7-Deazaadenosine and purine ribonucleoside, lacking the N-7 nitrogen and the exocyclic amino group, respectively, have been incorporated into the branchpoint in a nuclear pre-mRNA and a group II intron to compare the effect in the two systems.[81] The changes in both systems were alike and further supported the functional similarity between pre-mRNA and group II intron splicing.

Another purine analogue, N'-methylguanosine (**7**), when substituted in the core region of the hammerhead ribozyme, completely inhibited its catalytic activity.[82] This analogue has also been incorporated into tRNA^{Asp}, where it abolished mischarging.[83]

2-Aminopurine (**13**) is often used for fluorescence measurements and will be discussed in Section 6.14.7. Other analogues such as purine (**6**), isoguanine, 2,6-diaminopurine, 6-thioguanine, and hypoxanthine have also helped to identify purine nucleosides essential for RNA catalysis.[84–87]

2-Pyrimidinone (**9**) and 4-thiouridine (**10**) ribonucleosides as well as 2- and 4-pyridinone ribonucleosides (**11**) and (**12**) have been used to assess the participation of conserved pyrimidine bases in making hydrogen bonding contacts in the catalytic core of the hammerhead ribozyme and to provide evidence for a Mg^{2+} ion-binding site in the catalytically active ribozyme.[88,89]

The hammerhead and the hairpin ribozymes remain the most studied of RNA molecules with base analogues. A detailed discussion of the results of these studies with the hammerhead ribozyme and their correlation with its X-ray structure has been published.[29] A summary of the studies with the hairpin ribozyme has also been discussed.[90]

A new area of application of base-modified ribonucleotide analogues involves *in vitro* selection of aptamers and ribozymes. Such analogues can provide new functional groups which might aid in the desired interaction or enhance catalytic properties of selected RNA molecules, by enlarging the functional space in the selection process. A precondition, of course, is that the corresponding analogue triphosphates should act as substrates for the polymerases employed in the selection procedure. The first report of such an application was the incorporation of 5-(propynyl)-2'-deoxy-uridine into a thrombin aptamer.[91] Another group has reported the incorporation of 5-pyridylmethyl- and 5-imidazolylmethyluridine during the selection processes designed to generate RNA sequences which can catalyze amide bond and carbon–carbon bond formation, respectively.[92,93]

Deletion of the purine or pyrimidine bases from the ribose sugar moiety gives rise to abasic nucleosides. These analogues preserve the sugar hydroxy groups and therefore any biochemical effect observed from their use results directly from the loss of the heterocyclic ring. Substitution of abasic ribonucleosides (15) in the catalytic core of the hammerhead ribozyme significantly impairs its catalytic activity.[94] However, it was possible to restore the activity of four abasic ribozymes by a simple addition of the missing base, indicating that the ribozyme structure can be altered to create a binding site for small ligands. Other studies using the ribo-abasic nucleosides include the recognition of branch-point adenosine in the group II intron,[87] the identification of essential base residues in loop B of the hairpin ribozyme,[95] and the importance of the stem-loop II and the conserved U residues in the catalytic core of the hammerhead ribozyme.[96,97]

6.14.6   RIBOSE MODIFICATION AT THE 2′-POSITION

Several sugar modifications, primarily at the 2′-position, have been employed to study the structure–function relationship in RNA. Such analogues can be used to investigate the conformational importance of the sugar pucker in RNA interactions. The furanose ring in oligoribonucleotides maintains a $C_{3'}$-*endo* sugar ring conformation.[98] In 2′-substituted nucleosides, the electronegativity of the substituent exerts a profound effect on the sugar conformation and as a result, the population of $C_{3'}$-*endo* conformer increases with an increase in the electronegativity. However, the conformational effect is often difficult to separate from the effect of the 2′-substituent on the hydrogen bonding at this position. For example, a 2′-fluoro-2′-deoxyribose sugar has a very high percentage of the $C_{3'}$-*endo* conformer, more than the deoxyribose, but is unable to support hydrogen bonding. Conversely, the 2′-amino derivative has an extremely low percentage of the $C_{3'}$-*endo* conformer compared to deoxyribose, but retains both hydrogen donating and accepting abilities similar to the 2′-OH group. Additionally, the 2′-amino group is an excellent nucleophile for substitution reactions and this property has been exploited in cross-linking experiments discussed in a subsequent section, and also in the attachment of cholesterol to the hammerhead ribozyme.[99] The synthesis of several 2′-modified nucleosides and their triphosphates has been reviewed.[80]

Sugar-modified analogues have been extensively used to study ribozymes and RNA aptamers to probe the importance of specific 2′-hydroxyl groups in catalysis and to provide resistance against nucleases. The 2′-hydroxyl modifications introduced to probe hammerhead ribozyme catalysis include fluoro and amino,[100,101] methoxy and allyloxy,[102] and 2′-C-allyl substituents.[103] 2′-O-methyl sugar modifications have been introduced to identify the positions of 2′-hydroxyl groups necessary for the hairpin ribozyme activity.[104] Interactions involving 2′-hydroxyl groups and the importance of having a ribose moiety at the cleavage site in the *Tetrahymena* ribozyme reaction, has been studied using 2′-deoxy- and 2′-fluoro-substituted nucleosides.[105–108] 2′-Deoxynucleosides have also been used to investigate the role of active site 2′-hydroxyl groups in group II intron molecular recognition and catalysis.[109]

A 2′-thio-2′-deoxynucleoside has been incorporated into a dinucleotide and into an oligodeoxynucleotide, but mechanistic studies using this modification have not been reported.[110,111] However, it is expected that they will also serve as interesting probes for establishing metal ion assistance for RNA in catalysis.

By employing a variety of selection protocols, high affinity nucleic acid aptamers that recognize a wide range of molecular targets such as antibiotics, amino acids, nucleoside triphosphates, cofactors, organic dyes, porphyrins, and proteins have been isolated.[112] In addition to specificity and affinity, improved chemical and enzymatic stability is also essential for the development and use of RNA-based aptamers in therapeutics and diagnostics. The substitution of 2′-fluoro- and 2′-amino-2′-deoxy pyrimidine nucleosides in RNA imparts resistance to nucleases[101] and moreover these modifications can be easily introduced during aptamer selection since the corresponding nucleotide triphosphates are good substrates for T7 RNA polymerase.[10,11] Using these analogues, aptamers have been isolated against basic fibroblast growth factor,[113] human thyroid stimulating hormone,[114] a monoclonal antibody which recognizes the main immunogenic region of human acetylcholine receptor,[115] and the keratinocyte growth factor.[116]

2′-O-methyl-modified, biotinylated oligoribonucleotides have been used as antisense probes to map U2 snRNP-pre-mRNA interactions and to study structure and function of U4/U6 snRNPs.[117,118] Modified nucleosides such as 2′-O-methyl have also been placed at either the 3′- or 5′-splice site of a nuclear pre-mRNA to understand chemical steps involved in the spliceosomal assembly.[6] *In vitro* studies have revealed that the presence of 2′-hydroxyl group is more important at the 3′-splice site for the second step in the splicing reaction which results in the ligation of two exons with a concomitant release of an intron.

Another sugar modification involves replacement of the *O*-4′ furanose ring oxygen by a methylene group to give carbocyclic analogues. Ribocarbocyclic analogues of cytidine and adenosine have been chemically incorporated into the hammerhead ribozyme and were found to confer RNase resistance.[119]

6.14.7 FLUORESCENT LABELS FOR OLIGORIBONUCLEOTIDES

2-Aminopurine (2AP), a fluorescent nucleoside analogue, has been extensively used to detect changes in oligonucleotide conformation and its applications have been reviewed.[120] It can substitute for adenosine in base pairing with uridine. The absorption and excitation maximum for the nucleoside is at 330 nm and has an emission at 380 nm. Importantly, the quantum yield of 2AP fluorescence, when substituted in oligonucleotides, depends on the degree of base stacking.[121] In one study, Mg^{2+} ion induced conformational perturbations in the tertiary structure of the hammerhead ribozyme were followed by monitoring the change in fluorescence of 2AP and the affinity constants for Mg^{2+} ion binding were determined.[122]

The guanosine-specific fluorescence quenching of 3′-fluorescein-labeled oligoribonucleotides was used to determine rate constants for substrate binding, cleavage, and dissociation of the hairpin ribozyme.[123] Two examples from the hammerhead ribozyme structural studies further emphasize the power and versatility of fluorescence resonance energy transfer (FRET) studies. In both the studies, helical termini of the ribozyme were labeled with a donor and an acceptor dye. In one study, FRET efficiency was used to determine the distances between these ends to propose a three-dimensional model for the ribozyme[124] and in another study, FRET as a function of Mg^{2+} concentration was used to follow the folding of the ribozyme.[125] There is considerable literature on fluorescent dyes attached at the termini of oligonucleotides for FRET measurements and their application in structural studies has been reviewed.[120]

6.14.8 NUCLEIC ACID CROSS-LINKING

The conformational flexibility of RNA–RNA duplexes or RNA–protein complexes can be restricted by site-specific cross-links using chemically modified oligonucleotides. Such cross-links can also determine the proximity of nucleotides or amino acids, since they would not be formed at distances greater than the length of the spacer. The reagents for cross-linking are usually attached to either the base or the sugar moiety and in most cases the sulfhydryl group is used for introducing cross-links.

6.14.8.1 Chemical Cross-links

A versatile strategy for the cross-linking of RNA oligomers results from the extension of the convertible nucleoside approach to oligoribonucleotides.[126] In this method, uridine and inosine derivatives that possess good leaving groups at the 4 or 6 position, respectively, are site-specifically incorporated in oligoribonucleotides (Figure 1). Subsequent reaction of the convertible nucleoside, at the oligonucleotide level, with a symmetrical ω,ω-dithiobis(alkylamine), results in the formation of cytidine and adenosine derivatives bearing an alkyl disulfide tether. Reduction of the disulfide linkage results in the formation of a free sulfhydryl group, which can be reoxidized to form a cross-link. Oligoribonucleotides can be cross-linked in this manner as exemplified by the formation of an intramolecular disulfide in an RNA ministem loop.[127] Thioethyl groups can also be attached to C-5 of pyrimidines and oxidized to the disulfides as described for the cross-linking of tRNA[Phe].[128]

However, there are systems where cross-linking via base modification is undesirable and, therefore, attachment of the cross-linking reagent via the sugar ring is preferred. This approach has been used to discriminate between the two models of the hammerhead ribozyme, one based on X-ray crystallographic analysis and the other based on FRET results.[129] In this method, the nucleotides to be linked were replaced by the corresponding 2′-amino nucleotides which were then reacted with 2-pyridyl 3-isothiocyanatobenzyl disulfide. After reduction, an intrastrand disulfide was introduced by oxidation and the catalytic activity of cross-linked ribozymes was determined. More reactive

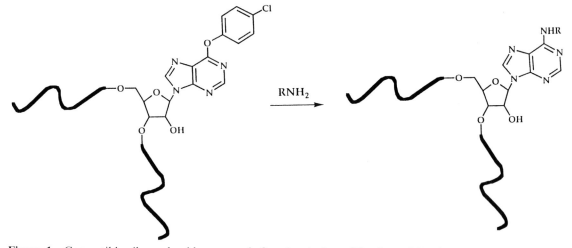

Figure 1 Convertible ribonucleoside approach for chemical modification of RNA. RNH_2 represents the attacking amine nucleophile which may contain a disulfide group for further cross-linking studies. Similarly, the C-4 position of uridine can also be activated. The ribbon structure represents a fully deprotected oligoribonucleotide, containing a convertible inosine derivative.

aliphatic isocyanates have been used for the interstrand cross-linking of the hairpin ribozyme as part of an effort to build a first three-dimensional model (Figure 2).[130] This method has been further extended to introduce interstrand disulfide cross-links in a group I ribozyme.[131] In this study, cross-links were formed by disulfide exchange reaction rather than oxidation as this former reaction occurs at a much faster rate (Figure 2). The cross-linked ribozymes were catalytically competent even though some cross-links bridged distances considerably farther apart than expected from the current model. These results reveal interesting dynamic properties of large catalytic RNAs. Alternatively, 2′-hydroxyl groups have also been derivatized with thioethyl groups to introduce intra- and interhelical disulfide cross-links in tRNA^{Phe}.[128]

6.14.8.2 Photocross-linking

6.14.8.2.1 Nucleobase cross-links

Natural nucleosides can be photoactivated to form new bonds within approachable distances. A good example is the formation of a cross-link between a cytidine and a uridine residue, which was observed upon irradiation of yeast tRNA^{Phe} and was found to be consistent with the X-ray structure.[132] Another example is described for the hairpin ribozyme, where a uridine and a guanosine in a loop were photochemically cross-linked in high yields.[133]

Usually, the cross-linking yields obtained with the natural nucleotides are low. Moreover, the wavelength required for activation is around 260 nm, which is not specific for a particular nucleotide. Thus, analogues which are more reactive and absorb at higher wavelength are usually preferred for photoactivated cross-linking. 4-Thiothymidine and 4-thiouridine (**10**), 5-bromouridine and 5-iodouridine (**17**) and 6-thioguanosine (**8**) meet this requirement and they also form stable Watson–Crick base pairs. The most frequently used analogues are 4-thiouridine (**10**) or 4-thiothymidine, which absorb at 331 nm and can be irradiated between 300 nm and 400 nm for activation.[134,135] Cross-linking can occur under anaerobic conditions by the addition at the 5, 6 double bond or at C-4 with loss of sulfur, by invoking a radical mechanism. Under aerobic conditions, sulfur becomes oxidized to sulfonic acid, which can subsequently be displaced by a nucleophile.

The first intrastrand RNA site-specific cross-link with 4-thiouridine was achieved upon irradiation of yeast tRNA^{Val}, which contains 4-thiouridine as a natural constituent.[136] 4-Thiouridine was also incorporated into oligoribonucleotides by enzymatic methods and depending on its proximal position, photocross-links were observed to ribosomal proteins and 16S RNA.[137]

4-Thiouridine has been incorporated at specific positions in an adenovirus pre-mRNA, using the Moore and Sharp ligation approach[6,7] to identify contacts to snRNAs during the splicing reaction.[138] Cross-links were observed with a loop sequence in U5 snRNA and with an invariant sequence of

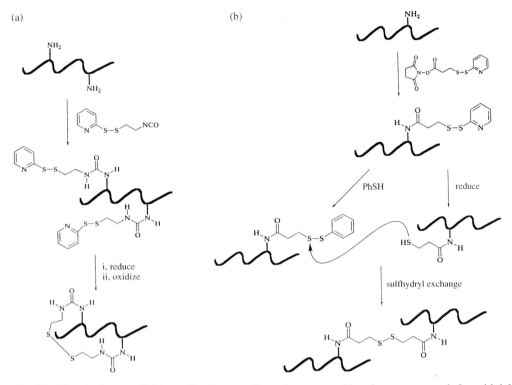

Figure 2 (a) Chemical cross-linking via the reaction of a sugar 2′-amino group and 2-pyridyl-3-iso-cyanatobenzyl disulfide. (b) Chemical crosslinking via the reaction of a 2′-amino group and *N*-hyd-roxysuccinimidyl-activated disulfide esters. The ribbon structure represents 2′-amino-modified RNA.

U6. Using a similar strategy, strand-specific cross-linking between 4-thiouridine-modified human tRNA[Lys] with HIV-1 reverse transcriptase subunits p66 and p51 have also been investigated.[139] 4-Thiouridine was used to cross-link small nuclear ribonucleoprotein particles U11 and U6 to the 5′-splice site of an AT-AC intron found in metazoan genes.[140]

6-Thioinosine and 4-thiouridine have been used as photoaffinity probes to study hammerhead and hairpin ribozyme conformations.[135] Multiple cross-links were found in both of the cases and the results were difficult to explain on the basis of models. It was concluded that these ribozymes can adopt multiple conformations making it difficult to identify catalytically competent conformations by this approach alone.

5-Bromouridine and 5-iodouridine (**17**), which can be activated at 308 nm and 325 nm, respectively, can also serve as convenient photoaffinity labels as exemplified by the cross-linking of an RNA transcript to the bacteriophage R17 coat protein.[141,142] The protein photodamage was considerably reduced with the iodo derivative.

The 5-position of uridine can be conveniently functionalized, without affecting the hydrogen bonding interaction, to attach various nonradioactive probes. In one of the first reports, biotin was covalently attached at the 5-position of uridine triphosphate for incorporation in RNA by transcription for a subsequent streptavidin-affinity isolation.[143] Many biotin phosphoramidites are now commercially available for solid-phase chemical coupling to oligonucleotides. A similar approach can also be used for the incorporation of noradioactive probes such as digoxigenin or fluorescein, instead of biotin.

In one study, multiple 5-methyleneaminouridine triphosphate residues were introduced in oligo-ribonucleotides by T7 RNA polymerase transcription of synthetic DNA templates.[144] The free amino group in the modified uridine was postsynthetically reacted with an azirinyl group-containing reagent. The photocross-linking results with the azirinyl–uridine derivative and 6-thioguanosine were used to study the binding of radiolabeled mRNA analogues to *E. coli* ribosomes. Precise cross-link positions were determined by a combination of ribonuclease H and T1 digestion and by primer extension analysis.

8-Azidoadenosine (**18**) and 8-azidoguanosine are frequently used for the photoaffinity labeling of enzymes.[145,146] Due to incompatibility of the azido group with phosphoramidite-based chemical synthesis, these analogues are not suitable for automated synthesis. However, enzymatic methods have been used to incorporate 8-azidoadenosine into the acceptor stem of tRNAs to probe its interaction with ribosomes.[147] 5-Azidouridine and 5-azidodeoxyuridine triphosphates are also suitable for photoactivated cross-linking and are substrates for RNA polymerases.[148] However, affinity labeling studies have not been reported for these analogues.

5-Sulfhydryluridine triphosphate has been incorporated enzymatically into RNA followed by subsequent derivatization with azidophenacyl bromide, for cross-linking to several RNA polymerases.[149]

6.14.8.2.2 *Terminal cross-linking*

Posttranscriptional reaction of a 5′-terminal guanosine phosphorothioate is an elegant method of attaching a photoactivatable azidophenacyl group to the transcripts. So far, it has mainly been used to investigate RNase P RNA interactions. Interestingly, the strategy involved transcription initiation at numerous positions and thus placed the terminal guanosine phosphorothioate along the RNase P RNA molecule. Upon irradiation of azidophenacyl-derivatized transcripts, intra- and intermolecular cross-links were obtained in the absence or presence of pre-tRNA.[150,151] The library of cross-links has facilitated significant refinement of the three-dimensional model of this ribozyme. A similar cross-linking study was also used to identify "neighbors" in the group I intron ribozyme.[152]

6.14.8.2.3 *Metal derivatives*

There have been few investigations regarding metal derivatives of oligoribonucleotides. In one study, 2′-*O*-methyl oligoribonucleotides were postsynthetically modified with transplatin to give a 1,3-intrastrand cross-linked oligomer.[153,154] When hybridized with complementary RNA strands, these intrastrand platinated oligomers triggered the formation of specific interstrand cross-links. The utility of transplatin-modified oligoribonucleotides was further demonstrated by introducing cross-links in Ha-ras mRNA and by concomitant inhibiton of cell proliferation.

6.14.9 CONCLUSIONS

The introduction of modified nucleotides into oligoribonucleotides or transcripts has greatly aided the understanding of mechanistic and structural aspects of many biochemical reactions involving RNA. Ribonucleotide analogues which can be enzymatically incorporated have found wider application because of their ease of introduction. Additionally, the analogues which confer resistance to modified RNA from nucleases are of particular interest for their possible use in the development of RNA-based therapeutics.

ACKNOWLEDGMENT

Work in this laboratory was supported by the Deutsche Forschungsgemeinschaft and the Fonds der Chemischen Industrie. A fellowship from the Alexander von Humboldt-Stiftung (N.K.V.) is gratefully acknowledged.

6.14.10 REFERENCES

1. S. Verma and F. Eckstein, *Ann. Rev. Biochem.*, 1998, **67**, 99.
2. M. J. Gait, C. Pritchard, and G. Slim, in "Oligonucleotides and Analogs—A Practical Approach," ed. F. Eckstein, IRL Press, Oxford, 1991, p. 25.
3. F. Wincott, A. Direnzo, C. Shaffer, S. Grimm, D. Tracz, C. Workman, D. Sweedler, C. Gonzalez, S. Scaringe, and N. Usman, *Nucleic Acids Res.*, 1995, **23**, 2677.
4. R. H. Davis, *Curr. Opin. Biotechnol.*, 1995, **6**, 213.
5. O. C. Uhlenbeck and R. I. Gumport, in "The Enzymes," 3rd edn., ed. P. D. Boyer, Academic Press, New York, 1982, vol. 15, p. 31.
6. M. J. Moore and P. A. Sharp, *Science*, 1992, **256**, 992.
7. M. J. Moore and C. C. Query, in "RNA–Protein Interactions: A Practical Approach," ed. C. Smith, Oxford University Press, 1998, p. 75.

8. F. Eckstein, *Annu. Rev. Biochem.*, 1985, **54**, 367.
9. A. D. Griffiths, B. V. L. Potter, and I. C. Eperon, *Nucleic Acids Res.*, 1987, **15**, 4145.
10. H. Aurup, D. M. Williams, and F. Eckstein, *Biochemistry*, 1992, **31**, 9636.
11. Y. Huang, F. Eckstein, R. Padilla, and R. Sousa, *Biochemistry*, 1997, **36**, 8231.
12. A. B. Burgin and N. R. Pace, *EMBO J.*, 1990, **9**, 4111.
13. G. L. Igloi, *Biochemistry*, 1988, **27**, 3842.
14. L. Pauling, "The Nature of the Chemical Bond," 3rd edn., Cornell University Press, Ithaca, NY, 1960.
15. R. G. Pearson, *Science*, 1966, **151**, 172.
16. V. L. Pecoraro, J. D. Hermes, and W. W. Cleland, *Biochemistry*, 1984, **23**, 5262.
17. G. Zon and W. J. Stec, in "Oligonucleotides and Analogs—A Practical Approach," ed. F. Eckstein, IRL Press, Oxford, 1991, p. 87.
18. W. J. Stec, A. Grajkowski, A. Kobylanska, B. Karwowski, M. Koziolkiewicz, K. Misiura, A. Okruszek, A. Wilk, P. Guga, and M. Boczkowska, *J. Am. Chem. Soc.*, 1995, **117**, 12019.
19. P. M. J. Burgers and F. Eckstein, *Proc. Natl. Acad. Sci. USA*, 1978, **75**, 4798.
20. P. M. J. Burgers and F. Eckstein, *Biochemistry*, 1979, **18**, 592.
21. F. Eckstein, H. H. Schulz, H. Rüterjans, W. Haar, and W. Maurer, *Biochemistry*, 1972, **11**, 3507.
22. F. R. Bryant and S. J. Benkovic, *Biochemistry*, 1979, **18**, 2825.
23. B. V. L. Potter, B. A. Connolly, and F. Eckstein, *Biochemistry*, 1983, **22**, 1369.
24. P. A. Heaton and F. Eckstein, *Nucleic Acids Res.*, 1996, **24**, 850.
25. H. van Tol, J. M. Buzayan, P. A. Feldstein, F. Eckstein, and G. Bruening, *Nucleic Acids Res.*, 1990, **18**, 1971.
26. G. Slim and M. J. Gait, *Nucleic Acids Res.*, 1991, **19**, 1183.
27. H. W. Pley, K. M. Flaherty, and D. B. McKay, *Nature*, 1994, **372**, 68.
28. W. G. Scott, J. T. Finch, and A. Klug, *Cell*, 1995, **81**, 991.
29. D. B. McKay, *RNA*, 1996, **2**, 395.
30. W. G. Scott, J. B. Murray, J. R. P. Arnold, B. L. Stoddard, and A. Klug, *Science*, 1996, **274**, 2065.
31. J. A. McSwiggen and T. R. Cech, *Science*, 1989, **244**, 679.
32. J. Rajagopal, J. A. Doudna, and J. W. Szostak, *Science*, 1989, **244**, 692.
33. M. J. Moore and P. A. Sharp, *Nature*, 1993, **365**, 364.
34. K. L. Maschhoff and R. A. Padgett, *Nucleic Acids Res.*, 1993, **21**, 5456.
35. G. Chanfreau and A. Jacquier, *Science*, 1994, **266**, 1383.
36. R. A. Padgett, M. Podar, S. C. Boulanger, and P. S. Perlman, *Science*, 1994, **266**, 1685.
37. P. A. Sharp, *Cell*, 1994, **77**, 805.
38. D. Herschlag, J. A. Piccirilli, and T. R. Cech, *Biochemistry*, 1991, **30**, 4844.
39. A. H. Polesky, M. E. Dahlberg, S. J. Benkovic, N. D. F. Grindley, and C. M. Joyce, *J. Biol. Chem.*, 1992, **267**, 8417.
40. J. P. Noel, H. E. Hamm, and P. B. Sigler, *Nature*, 1993, **366**, 654.
41. G. Gish and F. Eckstein, *Science*, 1988, **240**, 1520.
42. D. Schatz, R. Leberman, and F. Eckstein, *Proc. Natl. Acad. Sci. USA*, 1991, **88**, 6132.
43. C. A. Brautigam and T. A. Steitz, *J. Mol. Biol.*, 1998, **277**, 363.
44. R. B. Waring, *Nucleic Acids Res.*, 1989, **17**, 10281.
45. E. L. Christian and M. Yarus, *Biochemistry*, 1993, **32**, 4475.
46. J. H. Cate, R. L. Hanna, and J. A. Doudna, *Nature Struct. Biol.*, 1997, **4**, 553.
47. S. A. Strobel and K. Shetty, *Proc. Natl. Acad. Sci. USA*, 1997, **94**, 2903.
48. M. E. Harris and N. R. Pace, *RNA*, 1995, **1**, 210.
49. W. D. Hardt, J. M. Warnecke, V. A. Erdmann, and R. K. Hartmann, *EMBO J.*, 1995, **14**, 2935.
50. J. M. Warnecke, J. P. Fürste, W. D. Hardt, V. A. Erdmann, and R. K. Hartmann, *Proc. Natl. Acad. Soc. USA*, 1996, **93**, 8924.
51. Y. Chen, X. Q. Li, and P. Gegenheimer, *Biochemistry*, 1997, **36**, 2425.
52. D. E. Ruffner and O. C. Uhlenbeck, *Nucleic Acids Res.*, 1990, **18**, 6025.
53. S. C. Dahm and O. C. Uhlenbeck, *Biochemistry*, 1991, **30**, 9464.
54. D.-M. Zhou, P. K. R. Kumar, L.-H. Zhang, and K. Taira, *J. Am. Chem. Soc.*, 1996, **118**, 8969.
55. R. Knöll, R. Bald, and J. P. Fürste, *RNA*, 1997, **3**, 132.
56. O. Mitrasinovic and L. M. Epstein, *Nucleic Acids Res.*, 1997, **25**, 2189.
57. Y.-H. Jeoung, P. K. R. Kumar, Y.-A. Suh, K. Taira, and S. Nishikawa, *Nucleic Acids Res.*, 1994, **22**, 3722.
58. J. F. Milligan and O. C. Uhlenbeck, *Biochemistry*, 1989, **28**, 2849.
59. V. Biou, A. Yaremchuk, M. Tukalo, and S. Cusack, *Science*, 1994, **263**, 1404.
60. J. Rudinger, J. D. Puglisi, J. Pütz, D. Schatz, F. Eckstein, C. Florentz, and R. Giegé, *Proc. Natl. Acad. Soc. USA*, 1992, **89**, 5882.
61. R. Kreutzer, D. Kern, R. Giege, and J. Rudinger, *Nucleic Acids Res.*, 1995, **23**, 4598.
62. M. G. Caprara, G. Mohr, and A. M. Lambowitz, *J. Mol. Biol.*, 1996, **257**, 512.
63. M. G. Caprara, V. Lehnert, A. M. Lambowitz, and E. Westhof, *Cell*, 1996, **87**, 1135.
64. M. Dabrowski, C. M. T. Spahn, and K. H. Nierhaus, *EMBO J.*, 1995, **14**, 4872.
65. W. D. Hardt, V. A. Erdmann, and R. K. Hartmann, *RNA*, 1996, **2**, 1189.
66. T. Persson, C. S. Vörtler, O. Fedorova, and F. Eckstein, *RNA*, 1998, **4**, 1444.
67. R. Sousa and R. Padilla, *EMBO J.*, 1995, **14**, 4609.
68. R. Aphasizhev, A. Théobald-Dietrich, D. Kostyuk, S. N. Kochetkov, L. Kisselev, R. Giegé, and F. Fasiolo, *RNA*, 1997, **3**, 893.
69. J.-C. Francois, T. Saison-Behmoaras, C. Barbier, M. Chassignol, N. T. Thuong, and C. Hélène, *Proc. Natl. Acad. Sci. USA*, 1989, **86**, 9702.
70. S. M. Gryaznov and R. L. Letsinger, *Nucleic Acids Res.*, 1993, **21**, 1403.
71. B. C. F. Chu and L. E. Orgel, *DNA Cell Biol.*, 1990, **9**, 71.
72. R. Cosstick, L. W. McLaughlin, and F. Eckstein, *Nucleic Acids Res.*, 1984, **12**, 1791.
73. S. Alefelder, B. K. Patel, and F. Eckstein, *Nucleic Acids Res.*, 1998, **26**, 4983.
74. J. A. Piccirilli, J. S. Vyle, M. H. Caruthers, and T. R. Cech, *Nature*, 1993, **361**, 85.

75. L. B. Weinstein, B. C. N. M. Jones, R. Cosstick, and T. R. Cech, *Nature*, 1997, **388**, 805.
76. E. J. Sontheimer, S. G. Sun, and J. A. Piccirilli, *Nature*, 1997, **388**, 801.
77. R. G. Kuimelis and L. W. McLaughlin, *Biochemistry*, 1996, **35**, 5308.
78. D.-M. Zhou, N. Usman, F. E. Wincott, J. Matulic-Adamic, M. Orita, L.-H. Zhang, M. Komiyama, P. K. R. Kumar, and K. Taira, *J. Am. Chem. Soc.*, 1996, **118**, 5862.
79. J. B. Thomson, B. K. Patel, V. Jimenez, K. Eckart, and F. Eckstein, *J. Org. Chem.*, 1996, **61**, 6273.
80. B. E. Eaton and W. A. Pieken, *Annu. Rev. Biochem.*, 1995, **64**, 837.
81. R. K. Gaur, L. W. McLaughlin, and M. R. Green, *RNA*, 1997, **3**, 861.
82. S. Limauro, F. Benseler, and L. W. McLaughlin, *Bioorg. Med. Chem. Lett.*, 1994, **4**, 2189.
83. J. Pütz, C. Florentz, F. Benseler, and R. Giege, *Nature Struct. Biol.*, 1994, **1**, 580.
84. S. Bevers, G. B. Xiang, and L. W. McLaughlin, *Biochemistry*, 1996, **35**, 6483.
85. T. Tuschl, M. M. P. Ng, W. Pieken, F. Benseler, and F. Eckstein, *Biochemistry*, 1993, **32**, 11 658.
86. J. A. Grasby, K. Mersmann, M. Singh, and M. J. Gait, *Biochemistry*, 1995, **34**, 4068.
87. Q. L. Liu, J. B. Green, A. Khodadi, P. Haeberli, L. Beigelman, and A. M. Pyle, *J. Mol. Biol.*, 1997, **267**, 163.
88. J. B. Murray, C. J. Adams, J. R. P. Arnold, and P. G. Stockley, *Biochem. J.*, 1995, **311**, 487.
89. A. B. Burgin, Jr., C. Gonzalez, J. Matulic-Adamic, A. M. Karpeisky, N. Usman, J. A. McSwiggen, and L. Beigelman, *Biochemistry*, 1996, **35**, 14 090.
90. D. J. Earnshaw and M. J. Gait, *Antisense Nucleic Acid Drug Dev.*, 1997, **7**, 403.
91. J. A. Latham, R. Johnson, and J. J. Toole, *Nucleic Acids Res.*, 1994, **22**, 2817.
92. T. W. Wiegand, R. C. Janssen, and B. E. Eaton, *Chem. Biol.*, 1997, **4**, 675.
93. T. M. Tarasow, S. L. Tarasow, and B. E. Eaton, *Nature*, 1997, **389**, 54.
94. A. Peracchi, L. Beigelman, N. Usman, and D. Herschlag, *Proc. Natl. Acad. Sci. USA*, 1996, **93**, 11 522.
95. S. Schmidt, L. Beigelman, A. Karpeisky, N. Usman, U. S. Sorensen, and M. J. Gait, *Nucleic Acids Res.*, 1996, **24**, 573.
96. L. Beigelman, A. Karpeisky, and N. Usman, *Bioorg. Med. Chem. Lett.*, 1994, **4**, 1715.
97. L. Beigelman, A. Karpeisky, J. Matulic-Adamic, C. Gonzalez, and N. Usman, *Nucleosides Nucleotides*, 1995, **14**, 907.
98. W. Saenger, "Principles of Nucleic Acid Structure," Springer-Verlag, New York, 1984.
99. S. Alefelder, B. K. Patel, S. Th. Sigurdsson, and F. Eckstein, *Nucleic Acids Res.*, submitted.
100. D. B. Olsen, F. Benseler, H. Aurup, W. A. Pieken, and F. Eckstein, *Biochemistry*, 1991, **30**, 9735.
101. W. A. Pieken, D. B. Olsen, F. Benseler, H. Aurup, and F. Eckstein, *Science*, 1991, **253**, 314.
102. G. Paolella, B. S. Sproat, and A. I. Lamond, *EMBO J.*, 1992, **11**, 1913.
103. T. C. Jarvis, F. E. Wincott, L. J. Alby, J. A. McSwiggen, L. Beigelman, J. Gustofson, A. DiRenzo, K. Levy, M. Arthur, J. Matulic-Adamic, A. Karpeisky, C. Gonzalez, T. M. Woolf, N. Usman, and D. T. Stinchcomb, *J. Biol. Chem.*, 1996, **271**, 29 107.
104. B. M. Chowrira, A. Berzal-Herranz, C. F. Keller, and J. M. Burke, *J. Biol. Chem.*, 1993, **268**, 19 458.
105. A. M. Pyle and T. R. Cech, *Nature*, 1991, **350**, 628.
106. A. M. Pyle, F. L. Murphy, and T. R. Cech, *Nature*, 1992, **358**, 123.
107. D. Herschlag, F. Eckstein, and T. R. Cech, *Biochemistry*, 1993, **32**, 8299.
108. D. Herschlag, F. Eckstein, and T. R. Cech, *Biochemistry*, 1993, **32**, 8312.
109. D. L. Abramovitz, R. A. Friedman, and A. M. Pyle, *Science*, 1996, **271**, 1410.
110. C. L. Dantzman and L. L. Kiessling, *J. Am. Chem. Soc.*, 1996, **118**, 11 715.
111. M. L. Hamm and J. A. Piccirilli, *J. Org. Chem.*, 1997, **62**, 3415.
112. S. E. Osborne and A. D. Ellington, *Chem. Rev.*, 1997, **97**, 349.
113. D. Jellinek, L. S. Green, C. Bell, C. K. Lynott, N. Gill, C. Vargeese, G. Kirschenheuter, D. P. C. McGee, P. Abesinghe, W. A. Pieken, R. Shapiro, D. B. Rifkin, D. Moscatelli, and N. Janjic, *Biochemistry*, 1995, **34**, 11 363.
114. Y. Lin, D. Nieuwlandt, A. Magallanez, B. Feistner, and S. D. Jayasena, *Nucleic Acids Res.*, 1996, **24**, 3407.
115. S.-W. Lee and B. A. Sullenger, *Nature Biotech.*, 1997, **15**, 41.
116. N. C. Pagratis, C. Bell, Y.-F. Chang, S. Jennings, T. Fitzwater, D. Jellinek, and C. Dang, *Nature Biotech.*, 1997, **15**, 68.
117. S. M. L. Barabino, B. S. Sproat, U. Ryder, B. J. Blencowe, and A. I. Lamond, *EMBO J.*, 1989, **8**, 4171.
118. B. J. Blencowe, B. S. Sproat, U. Ryder, S. Barabino, and A. I. Lamond, *Cell*, 1989, **59**, 531.
119. F. Burlina, A. Favre, J.-L. Fourrey, and M. Thomas, *J. Chem. Soc., Chem. Commun.*, 1996, 1623.
120. D. P. Millar, *Curr. Opin. Struct. Biol.*, 1996, **6**, 322.
121. D. C. Ward, E. Reich, and L. Stryer, *J. Biol. Chem.*, 1969, **244**, 1228.
122. M. Menger, T. Tuschl, F. Eckstein, and D. Porschke, *Biochemistry*, 1996, **35**, 14 710.
123. N. G. Walter and J. M. Burke, *RNA*, 1997, **3**, 392.
124. T. Tuschl, C. Gohlke, T. M. Jovin, E. Westhof, and F. Eckstein, *Science*, 1994, **266**, 785.
125. G. S. Bassi, A. I. H. Murchie, F. Walter, R. M. Clegg, and D. M. J. Lilley, *EMBO J.*, 1997, **16**, 7481.
126. C. R. Allerson, S. L. Chen, and G. L. Verdine, *J. Am. Chem. Soc.*, 1997, **119**, 7423.
127. C. R. Allerson and G. L. Verdine, *Chem. Biol.*, 1995, **2**, 667.
128. J. T. Goodwin, S. E. Osborne, E. J. Scholle, and G. D. Glick, *J. Am. Chem. Soc.*, 1996, **118**, 5207.
129. S. T. Sigurdsson, T. Tuschl, and F. Eckstein, *RNA*, 1995, **1**, 575.
130. D. J. Earnshaw, B. Masquida, S. Müller, S. T. Sigurdsson, F. Eckstein, E. Westhof, and M. J. Gait, *J. Mol. Biol.*, 1997, **274**, 197.
131. S. B. Cohen and T. R. Cech, *J. Am. Chem. Soc.*, 1997, **119**, 6259.
132. L. S. Behlen, J. R. Sampson, and O. C. Uhlenbeck, *Nucleic Acids Res.*, 1992, **20**, 4055.
133. J. M. Burke, S. E. Butcher, and B. Sargueil, *Nucleic Acids & Molecular Biology*, eds. F. Eckstein and D. M. J. Lilley, Springer-Verlag, Berlin Heidelberg, 1996, vol. 10, p. 129.
134. T. T. Nikiforov and B. A. Connolly, *Nucleic Acids Res.*, 1992, **20**, 1209.
135. A. Favre and J.-L. Fourrey, *Acc. Chem. Res.*, 1995, **28**, 375.
136. A. Favre, A. M. Michelson, and M. Yaniv, *J. Mol. Biol.*, 1971, **58**, 367.
137. O. Dontsova, A. Kopylov, and R. Brimacombe, *EMBO J.*, 1991, **10**, 2613.
138. E. J. Sontheimer and J. A. Steitz, *Science*, 1993, **262**, 1989.
139. Y. Mishima and J. A. Steitz, *EMBO J.*, 1995, **14**, 2679.

140. Y.-T. Yu and J. A. Steitz, *Proc. Natl. Acad. Sci. USA*, 1997, **94**, 6030.
141. J. M. Gott, M. C. Willis, T. H. Koch, and O. C. Uhlenbeck, *Biochemistry*, 1991, **30**, 6290.
142. M. C. Willis, B. J. Hicke, O. C. Uhlenbeck, T. R. Cech, and T. H. Koch, *Science*, 1993, **262**, 1255.
143. P. R. Langer, A. A. Waldrop, and D. C. Ward, *Proc. Natl. Acad. Sci. USA*, 1981, **78**, 6633.
144. P. V. Sergiev, I. N. Lavrik, V. A. Wlasoff, S. S. Dokudovskaya, O. A. Dontsova, A. A. Bogadonov, and R. Brimacombe, *RNA*, 1997, **3**, 464.
145. B. Jayaram and B. E. Haley, *J. Biol. Chem.*, 1994, **269**, 3233.
146. A. J. Chavan, Y. Nemoto, S. Narumiya, S. Kozaki, and B. E. Haley, *J. Biol. Chem.*, 1992, **267**, 14 866.
147. J. Wower, S. S. Hixson, and R. A. Zimmermann, *Proc. Natl. Acad. Sci. USA*, 1989, **86**, 5232.
148. R. K. Evans and B. E. Haley, *Biochemistry*, 1987, **26**, 269.
149. B. K. He, D. L. Riggs, and M. M. Hanna, *Nucleic Acids Res.*, 1995, **23**, 1231.
150. M. E. Harris, J. M. Nolan, A. Malhotra, J. W. Brown, S. C. Harvey, and N. R. Pace, *EMBO J.*, 1994, **13**, 3953.
151. M. E. Harris, A. V. Kazantsev, J.-L. Chen, and N. R. Pace, *RNA*, 1997, **3**, 561.
152. J.-F. Wang, W. D. Downs, and T. R. Cech, *Science*, 1993, **260**, 504.
153. M. Boudvillain, M. Guerin, R. Dalbies, T. Saison-Behmoaras, and M. Leng, *Biochemistry*, 1997, **36**, 2925.
154. C. Colombier, M. Boudvillain, and M. Leng, *Antisense Nucleic Acid Drug Dev.*, 1997, **7**, 397.

6.15
Ribozyme Structure and Function

MASAKI WARASHINA, DE-MIN ZHOU, TOMOKO KUWABARA, and KAZUNARI TAIRA
University of Tsukuba, Japan

6.15.1 INTRODUCTION

Catalytic RNAs include hammerhead, hairpin, and hepatitis delta virus (HDV) ribozymes; group I and II introns; the RNA subunit of RNase P; and ribosomal RNA (Figure 1).[1–12] Among these catalytic RNAs, the first two ribozymes to be discovered, by Altman[2] and Cech,[1] respectively, were the RNA subunit of RNase P and a group I intron.[1,2] Within the following five years, small ribozymes, such as hammerhead, hairpin and HDV ribozymes, were discovered in studies of the replication, via a rolling-circle mechanism, of certain viroids, satellite RNAs and an RNA virus.[7–12] Of all these catalytic RNAs, the hammerhead ribozyme is the smallest.[13,14]

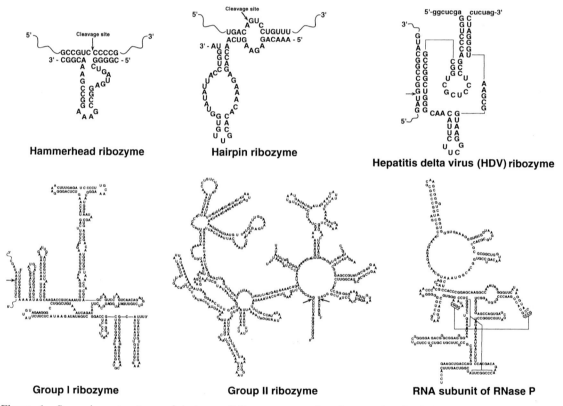

Figure 1 Secondary structures of six types of ribozyme, namely, a hammerhead, a hairpin, and hepatitis delta virus ribozymes; group I and group II introns; and the RNA subunit of RNase P. Each cleavage site is indicated by an arrow. The RNA subunit of RNase P cleaves precursors to tRNAs so no cleavage site is shown.

With respect to reaction mechanisms, large ribozymes, such as group I introns and the catalytic RNA subunit of RNase P, use external nucleophiles. By contrast, small ribozymes, such as hammerheads, hairpins and HDV ribozymes, use an internal nucleophile, namely, the 2′-oxygen at the cleavage site, with resultant formation of a cyclic phosphate. Since the large ribozymes do not require a 2′-OH group as a nucleophile at the cleavage site, the ribozymes of *Tetrahymena* and of RNase P can cleave DNA substrates in addition to RNA.[15–17] Over the past few years, ribozymes have been recognized as metalloenzymes.[7,9,18–33] In studies of the reaction mediated by the ribozyme from *Tetrahymena*, the existence was demonstrated, for the first time, of a metal ion catalyst that coordinates directly to and stabilizes the developing negative charge of the leaving 3′-oxygen, acting as a Lewis acid.[19] Moreover, there is now evidence to indicate that a metal ion activates the nucleophilic 3′-hydroxyl of guanosine in the same reaction, lending support to the proposed double-metal-ion mechanism of catalysis.[28] In the case of the reactions mediated by hammerhead ribozymes, base catalysis mediated by Mg^{2+}-hydroxide was first proposed on the basis of profiles of pH versus rate.[18] However, it was also pointed out that a general double-metal-ion mechanism, in which metal ions act as Lewis acids and coordinate directly to the 2′-hydroxyl and the leaving 5′-oxygen for activation of a nucleophile and for stabilization of a developing negative charge, respectively, might well explain reactions catalyzed by hammerhead ribozymes.[24,25,29–32] By contrast, the absence of metal-ion-mediated catalysis has been reported in the case of hairpin ribozymes.[34–36] Therefore, hairpin ribozymes can be classified as a distinct class of ribozymes that do not require metal ions as catalysts. It should also be pointed out that, under extreme conditions (in the presence of 1–4 M monovalent cations such as Li^+, Na^+, and NH_4^+), hammerhead ribozymes do not require divalent metal ions for catalysis.[37]

The extensive efforts that have been made over the 15 years since the discovery of ribozymes[1,2] have uncovered details of the mechanisms of ribozyme-mediated cleavage of RNA and studies of ribozymes have become very exciting. It is thought that ribozymes are fossil molecules that originated in the primitive RNA world and it is anticipated that the elucidation of their mechanisms of action will enhance our understanding of the life processes of primitive organisms. The rapidly developing field of RNA catalysis is of interest not only because of the intrinsic abilities of ribozymes but also because of their potential as therapeutic agents and specific regulators of gene expression.[9,38–50] The recent discovery of DNA enzymes, created by a selection procedure *in vitro*, has created even more excitement in the area of catalytic nucleic acids.[51–53]

6.15.2 HAMMERHEAD RIBOZYMES

6.15.2.1 General Features of Hammerhead Ribozymes

Hammerhead ribozymes are among the smallest catalytic RNAs. They are called "hammerheads" because of their two-dimensional structure.[8] The sequence motif, with three duplex stems and a conserved "core" of two non-helical segments that are responsible for the self-cleavage reaction (*cis* action), was first recognized in the satellite RNAs of certain viruses.[8] However, hammerhead ribozymes have been engineered in the laboratory to be able to act "in *trans*,"[13,14] and *trans*-acting hammerhead ribozymes, consisting of an antisense section (stem I and stem III) and a catalytic core with a flanking stem-loop II section, have been used as potential therapeutic agents and in mechanistic studies (Figure 2(a)).[14] Such RNAs can cleave oligoribonucleotides at specific sites (NUX, where N and X are A, G, C, U and A, C, U, respectively, with most efficient cleavage at GUC triplets).[63–68] In the case of *trans*-acting hammerhead ribozymes, most of the conserved nucleotides that are essential for the cleavage reaction are included in the catalytic core.[14] Therefore, RNA molecules consisting of only 30 or so nucleotides can be generated for use as artificial endonucleases that can cleave specific RNA molecules.

6.15.2.2 Sequence Requirements of Hammerhead Ribozymes

Phylogenetic analysis, extensive mutagenesis experiments and *in vitro* selection procedures have been used in attempts to probe the sequence requirements for the active structure of a hammerhead ribozyme.[7–9,53,65,69–75] In the catalytic core of the hammerhead ribozyme in Figure 2(a), replacement of certain nucleotides, with the exception of the residue U_7, reduced the cleavage activity dramatically. Selection *in vitro* for active hammerhead ribozymes revealed that active sequences corresponded broadly to the consensus core sequence, and no other sequences allowed efficient cleavage,[72,74] an indication that the consensus sequence derived from viruses and virusoids is probably the optimal sequence. Nonetheless, chemical modification at U_7 produced a non-natural ribozyme that had higher activity than the wild-type hammerhead ribozyme.[76] In stem I, any residues could be tolerated without substantial loss of cleavage activity. With the exception of the cleavage triplet, substitutions of nucleotides in stem III were also tolerated.

With respect to the important trinucleotide at the cleavage site, the NUX rule (where N can be A, U, G, or C, and X can be A, U, or C), which states that any oligonucleotide with a NUX triplex can be cleaved by hammerhead ribozymes, is generally accepted.[63–68] Among NUX triplets, GUC is cleaved most efficiently under k_{cat}/K_M conditions, with CUC and UUC being cleaved somewhat less efficiently.[68] Therefore, when a target site in a *trans*-acting system (an intermolecular reaction) is to be chosen, GUC or CUC may be preferable. However, in *cis*-acting systems (intramolecular

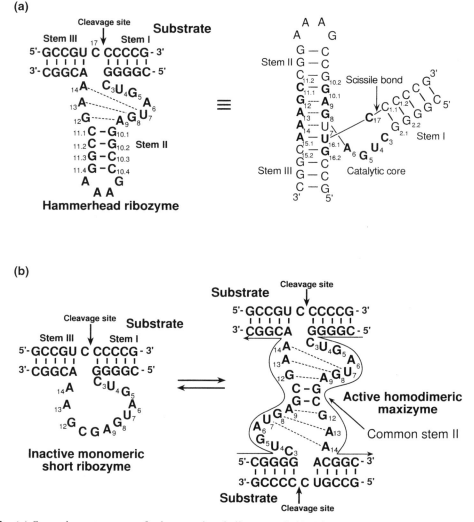

Figure 2 (a) Secondary structure of a hammerhead ribozyme (left). The structure on the right is that of the same hammerhead ribozyme, based on X-ray crystallographic data.[54-57] (b) A monomeric short ribozyme with low cleavage activity (left) and a homodimer of two maxizymes that has high activity (right).[58-62]

reactions), in which K_M values are irrelevant, other triplets, such as AUC, GUA, and AUA, may be chosen since these triplets are associated with high values of k_{cat}. In fact, the minus strand of the virusoid of Lucerne transient streak virus, (−)vLTSV, and the plus strand of the satellite RNA of barley yellow dwarf virus, (+)sBYDV, cleave the GUA triplet and the AUA triplet, respectively, in the hammerhead-catalyzed cleavage that occurs during their replication.[68] However, Eckstein and his co-workers have generated a hammerhead-like ribozyme that cleaved RNAs, after an AUG triplet, that were not cleaved by conventional hammerhead ribozymes. They used an *in vitro* selection procedure to identify this ribozyme and it had a different catalytic core sequence from those of the usual hammerhead ribozyme.[74,75]

In stem II, the $G_{10.1} \cdot C_{11.1}$ pair adjacent to the catalytic core is essential for stabilization of the catalytic core.[64,65,70,77] The nucleotides of the loop at the end of stem II can be varied, although, in many cases, a stable GAAA tetra-loop is used. Stem II is the only helix in the hammerhead ribozyme that is not involved in binding of the substrate and it can be shortened without complete loss of activity. The minimal length of helix II was found to be two base pairs and hammerhead ribozymes with a shorter stem II are designated minizymes. The activities of most such minizymes are very low.[69,77-83] However, it was found that some short ribozymes with short oligonucleotide linkers instead of the stem-loop II region can form homodimers or heterodimers that are very active (Figure 2(b)).[58-62] In order to distinguish monomeric forms of conventional minizymes that have extremely low activity from novel dimers with high-level activity, the latter very active short ribozymes capable of forming dimers are designated "maxizymes." Since the maxizymes tested to date have been more effective both *in vitro* and *in vivo* in reducing gene expression than "standard" hammerhead ribozymes, the novel homodimeric or heterodimeric maxizymes appear to have potential utility as gene-inactivating agents.[61,62,84]

6.15.2.3   Tertiary Structures of Hammerhead Ribozymes

Over the past few years, several attempts involving, for example, measurements of electrophoretic mobility,[85,86] nuclear magnetic resonance,[87–94] transient electric birefringence,[95,96] fluorescence resonance energy transfer (FRET),[97] and X-ray diffraction,[54–57,98,99] have been made to determine both the overall global structure and the detailed atomic structure of hammerhead ribozymes. The X-ray crystallographic structures determined by McKay's group and by Scott and Klug's group are nearly identical despite the difference in substrates used. McKay's group used an all-DNA substrate-analogue,[54] while Scott and Klug's group used an all-RNA substrate-analogue with a 2'-methoxy-2'-deoxyribose at the cleavage site.[55] Scott and Klug's group subsequently determined the structure of a freeze-trapped conformational intermediate of an unmodified all-RNA complex[56] by time-resolved crystallography. Changes were limited within the neighborhood of the cleavage site and no significant changes in the global structure of the freeze-trapped intermediate, as compared to the structure of the above-mentioned stable complex, were observed.

In all crystals, ribozymes had a γ-shaped configuration, in which stem I formed an acute angle with stem II and stems II and III were stacked coaxially to form a pseudo-A-form helix, in agreement with results inferred from fluorescence energy transfer,[97] electrophoretic,[83,86] and chemical cross-linking[100,101] studies (Figure 2(a), right).[54–57] There were two reversed-Hoogsteen $G \cdot A$ base-pairs between G_8–A_{13} and A_9–G_{12}, and a non-Watson–Crick $A_{14} \cdot U_7$ base-pair that consisted of one hydrogen bond. They were followed by stem II and were stacked coaxially onto the non-Watson–Crick $A_{15.1} \cdot U_{16.1}$ base-pair, with resultant formation of a pseudo-A-form helix by stems II and III. Four nucleotides ($C_3U_4G_5A_6$) formed a "uridine-turn" motif, allowing the phosphate backbone to turn and connect with stem I.

The tertiary structures of ribozymes can be stabilized by metal ions. The roles of metal ions in ribozyme-catalyzed reactions are of two distinct types. In one case, metal ions act as catalysts during the chemical cleavage step while, in the second case, they stabilize the active conformation of the ribozyme·substrate complex that is required for the reaction. Some metal ions may have dual functions. Lilley's group[85,86] has systematically examined the ion-dependent changes in conformation of a hammerhead ribozyme by monitoring shifts in electrophoretic mobility. Three discrete conformations were observed: in the absence of Mg^{2+} ions, the hammerhead ribozyme existed in an extended form in which the catalytic core appeared to be unstructured; at a low concentration of Mg^{2+} ions, stems II and III were aligned coaxially, forming a pseudo-continuous helix with stem I being adjacent to stem III; and at high concentrations of Mg^{2+} ions, stem I was reoriented and was adjacent to stem II, as in the crystal structure (Figure 2(a), right).[85,86] Such reorientation of the helical arms by divalent metal ions could also be achieved, to a lesser extent, by use of singly charged cations such as Na^+, although the Na^+-mediated folded structure was different from the Mg^{2+}-mediated structure. Using NMR spectroscopy, the authors investigated the effects of metal ions on the formation of an active complex between a hammerhead ribozyme and its substrate upon addition of Mg^{2+} ions.[92] In the absence of Mg^{2+} ions, no complex between the ribozyme and a substrate with deoxy-C_{17} was formed because the substrate-recognition regions of the ribozyme formed intramolecular base-pairs. In other words, the ribozyme remained in an inactive conformation in the absence of metal ions. Upon addition of Mg^{2+} ions to the mixture of ribozyme and substrate, the substrate-recognition regions of the ribozyme opened up and the ribozyme–substrate complex was formed. It is important to note that, in this system, monovalent Na^+ ions could not replace Mg^{2+} ions: the ribozyme remained in its inactive conformation in the absence of Mg^{2+} ions and in the presence of Na^+ ions.[92]

Metal ion-binding sites have been identified by capture of metal ions within the crystal structure.[55–57,98–99] Scott's group proposed various Mg^{2+}-binding sites, two of which appear to be important for catalysis.[55–57] At the first site, it was proposed that a Mg^{2+} ion bound to the *pro-R* oxygen of the 5'-phosphate of A_9 with further hydrogen bonding associated with N-7 of $G_{10.1}$, and this binding was suggested to have both structural and catalytic roles.[57,98] The second site seemed to be in the vicinity of the cleavage site. It was proposed that, at this site, a Mg^{2+} ion binds directly to the *pro-R* oxygen of the scissile phosphate in the freeze-trapped conformational intermediate.[56] The hydrated Mg^{2+} ion might participate directly in catalysis by acting as a base to facilitate the deprotonation of the 2'-OH of C_{17}, prior to nucleophilic attack at the scissile phosphate. However, crystal structures generally represent energy minima and do not provide direct and detailed structural information about transition states unless the structural data represent a deliberately designed analogue of a transition state. Because the conformation in all the available crystallographic structures does not allow in-line attack by the 2'-OH on the scissile phosphorus–oxygen bond, which is absolutely required for activity, none of the structures can represent the exact catalytic conformation.[54–57,98,99]

Indeed, further analysis by Scott's group suggested the probable invalidity of the earlier proposals[56] that the Mg^{2+} ion bound to the *pro-R* oxygen of the scissile phosphate in the ground state might move, together with the phosphate, into a conformation more suitable for in-line attack: Scott's group trapped an intermediate with an advanced change in conformation, in which the phosphate had moved considerably.[99] However, the metal ion (this time a Co^{2+} ion) remained associated with *N7* of $A_{1.1}$ and did not move with the *pro-R* oxygen.[99] This finding is in agreement with conclusions, based on kinetics, that the authors have drawn (see below).

6.15.2.4 Mechanisms of Action of Hammerhead Ribozymes and Catalytic Metal Ions

6.15.2.4.1 *Kinetic framework of hammerhead ribozymes*

Hammerhead ribozymes cleave RNAs after a NUX triplet to generate 2',3'-cyclic phosphate and 5'-hydroxyl termini. The minimum reaction scheme, consisting of at least three steps, is shown in Figure 3.[102–104] First, the substrate associates with the antisense arms of the hammerhead ribozyme through base pairs at stems I and III to form a Michaelis–Menten complex (k_{assoc}). Next, the phosphodiester bond at the cleavage site in the bound substrate is cleaved by the action of metal ions (k_{cleav}) to produce a 2',3'-cyclic phosphate and a 5'-hydroxyl group.[102,105,106] Finally, the cleaved fragments dissociate from the ribozyme (k_{diss}) and the liberated ribozyme is now available for a new series of catalytic events. The rate and equilibrium constants for a ribozyme with binding arms of 16 nucleotides in length were defined using a combination of steady-state kinetics, pre-steady-state kinetics, and equilibrium measurements by Uhlenbeck's group, as shown in Figure 4.[107–109]

The efficiency of binding of a hammerhead ribozyme to its substrate is influenced by the length

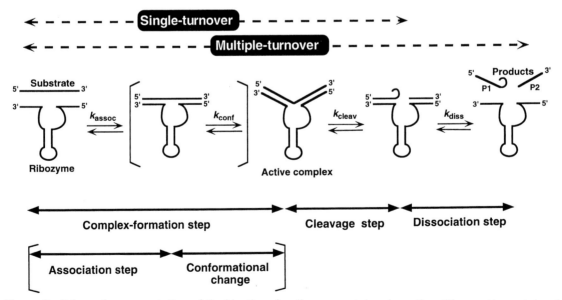

Figure 3 Schematic representation of the kinetics of a ribozyme-catalyzed reaction. The reaction catalyzed by a hammerhead ribozyme consists of at least three steps. The substrate first binds to the ribozyme (k_{assoc}). A conformational change may be required for formation of an active ribozyme·substrate complex (k_{conf}). The phosphodiester bond of the bound substrate is cleaved (k_{cleav}). The cleaved fragments dissociate from the ribozyme (k_{diss}), and the liberated ribozyme is now available for a new series of catalytic events.

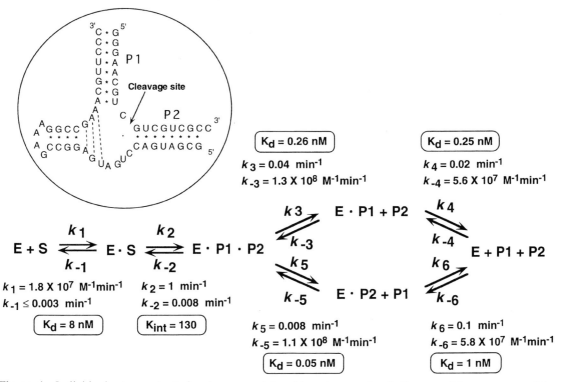

Figure 4 Individual rate constants for cleavage catalyzed by a hammerhead ribozyme. Measurements of the kinetic parameters were made by Uhlenbeck and his co-workers. The hammerhead ribozyme used in their study is shown within a circle.[107–109]

and sequence of the antisense arms of the ribozyme (stems I and III). Kinetic models of the action of hammerhead ribozymes and analysis of thermodynamic parameters predict that ribozymes with short antisense arms have a higher turnover rate than their counterparts with longer arms.[108,110–112] In general, ribozymes that have antisense arms with more than seven base pairs in one arm exhibit a decrease in the rate of dissociation of the product, so that the product-dissociation step is rate-limiting under multiple-turnover conditions. In contrast, when each of the antisense arms is shorter than five base-pairs, the rate of the reaction represents the rate of the chemical cleavage step because of higher product-dissociation rates. It is important to use ribozymes with short arms in functional analyses of kinetics *in vitro*, to ensure measurement of the rate of the chemical step, if reactions are to be carried out under multiple-turnover conditions.[9,103,104,113]

The dependence on temperature of the rate-limiting step was detected by analysis of an Arrhenius plot generated from results obtained with a ribozyme·substrate complex with 11 base pairs of the binding helices.[103,104] Distinct changes in the slope of the plot provided evidence for three different rate-limiting steps in the reaction. At mid-range temperatures of 25–50 °C, the chemical cleavage step (k_{cleav}) was the rate-limiting step, an indication that the cleaved fragments dissociated from the ribozyme at a higher rate than the rate of the chemical reaction ($k_{cleav} < k_{diss}$). At temperatures below 25 °C, the cleaved fragments adhered to the ribozyme more tightly and the product-dissociation step became the rate-limiting step ($k_{diss} < k_{cleav}$). Above 50 °C, the rate of the reaction decreased because, at such high temperatures, the formation of the Michaelis–Menten complex (formation of a duplex) was hampered by thermal melting.

The contributions to binding and catalysis of several individual base-pairs formed between a ribozyme and its substrate, have been investigated with substrates that were truncated at the 5′ or 3′ end.[114] Addition of residues close to the cleavage site contributed to the chemical step of the hammerhead-catalyzed reaction but not to the substrate-binding step, whereas base pairs distal to the cleavage site contributed exclusively to binding and had no effect on the chemical step. These results led to the proposal of a "fraying model," in which each ribozyme·substrate helix can exist in either an unpaired state (inactive "$(E·S)_{open}$" complex in Figure 5) or in a helical state (active "$(E·S)_{closed}$" complex in Figure 5), with the closed state being required for catalysis. According to this model, the cleavage rate depends on the concentration of the active $(E·S)_{closed}$ complex relative to that of the inactive $(E·S)_{open}$ complex, since both helices that involve the binding arms must be formed if the ribozyme is to cleave its substrate. This model predicts that, as far as the chemical step is concerned, the more stable the ribozyme–substrate helix that is formed, the higher will be the rate of the chemical cleavage step. This model effectively explains the results observed after mutagenesis of stems I and III.

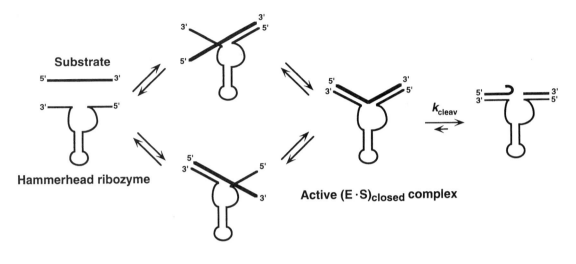

Figure 5   The fraying model with open and closed ribozyme·substrate helices that was proposed by Hertel *et al.*[114] Each ribozyme·substrate helix can exist in either an unpaired state [inactive "(E·S)$_{open}$" complex] or a helical state [active "(E·S)$_{closed}$" complex], with the closed state being required for catalysis. When either stem I or stem III (Figure 2) has sufficiently reduced stability, one or the other of the inactive (E·S)$_{open}$ complexes accumulates at equilibrium.

However, it must be added that not all results can be explained by the above-described model. In one case, the rate of cleavage by a DNA-armed ribozyme, which generally forms a weaker DNA·RNA duplex in the binding arms (stems I and III) than the corresponding RNA·RNA duplex, was higher than the rate of cleavage by the parental all-RNA ribozyme, despite the significantly lower concentration of the active (E·S)$_{closed}$ complex of the DNA-armed ribozyme as compared to that of the parent all-RNA ribozyme.[104,115–118] This finding argues against the proposal of an identical and unique structure for the active (E·S)$_{closed}$ complex for both natural all-RNA ribozymes and chimeric DNA-armed ribozymes, prior to the transition state. It is likely that the hybrid helices of the DNA-armed ribozyme–substrate complex form a slightly different (E·S)$_{closed}$ structure, resulting in the significantly higher activity of the DNA-armed ribozyme as compared to that of the all-RNA ribozyme (the reactivity difference between the two types of complex should be greater after correction for the different concentrations of the active (E·S)$_{closed}$ complexes). The authors emphasize here that the $\Delta H^{\neq}$ value for the chemical cleavage step (the enthalpy of activation) was identical for both all-RNA ribozymes and DNA-armed ribozymes and that the higher cleavage activity of the DNA-armed ribozyme originated exclusively from the greater value of $\Delta S^{\neq}$ (the entropically driven enhancement of cleavage).[104] Apparently, the DNA-armed ribozyme·substrate complex [the active (E·S)$_{closed}$ complex] is closer to the transition state structure because of the hybrid helices and, thus, a more limited reorganization (a conformational change of a smaller magnitude) will lead to the transition state structure (see below for details of the transition state structure).

6.15.2.4.2   *Energy diagram for the chemical cleavage step of hammerhead ribozyme-catalyzed reactions*

Hammerhead ribozymes catalyze an endonucleolytic transesterification reaction at the phosphodiester bond, using divalent metal ions as cofactors. The reaction is considered to be roughly equivalent to the nonenzymatic hydrolysis of RNA and the chemical cleavage requires two events.[31] In the first step, which yields transition state 1 (***TS1***), the 2′-hydroxyl group attacks the adjacent scissile phosphate as an internal nucleophile. In the second step, which yields transition state 2

(*TS2*), the 5′-oxygen of the next residue is separated to produce 2′,3′-cyclic phosphate and 5′-hydroxyl termini. The entire reaction is likely to be concerted in hammerhead ribozyme-catalyzed reactions (Figure 6, left).[31]

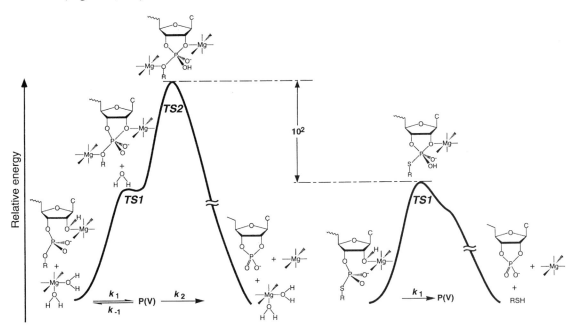

Figure 6 Energy diagrams for cleavage by a hammerhead ribozyme of normal and 5′-thio RNA substrates. The rate-limiting step of the reaction with the normal substrate is the cleavage of the P—(5′-O) bond (left). The rate-limiting step of the reaction with the 5′-thio substrate is shifted from the cleavage of the P—(5′-O) bond to the formation of the P—(2′-O) bond (right). The build-up of charge at the 5′-leaving group occurs after the rate-limiting step and, therefore, any metal ion bound at this position is kinetically insignificant in the case of the 5′-thio substrate.

In the case of nonenzymatic hydrolysis, the cleavage of the P—O(5′) bond (to yield *TS2*) was identified to be the overall rate-limiting step for natural RNAs, whereas the attack of the 2′-hydroxyl group on phosphorus (to yield *TS1*) was the rate-limiting step for 5′-thio RNA, in which the 5′-oxygen at the scissile phosphate had been replaced by sulfur.[26,30,31,119,120] In the latter case, *TS2* of 5′-thio RNA was significantly stabilized by the good thio leaving-group, as compared with the case of the natural RNA.[26,31] Following the same arguments as used in the analysis of the nonenzymatic hydrolysis of RNA, it is concluded that attack by the 2′-oxygen at C_{17} on the phosphorus (to yield *TS1*) must be the rate-limiting step for the 5′-thio RNA substrate (Figure 6, right) and the departure of the 5′-leaving group must be the rate-limiting step for the parental 5′-oxy RNA-substrate (Figure 6, left), in cleavage reactions catalyzed by a hammerhead ribozyme, because the 5′-thio RNA substrate was cleaved almost two orders of magnitude more rapidly than the parental 5′-oxy RNA substrate, at pH 6.0 in the presence of 0.3 mM Mg^{2+} ions.[26,30,31] Note that, if the rate-limiting step were the attack by the 2′-oxygen at C_{17} on the phosphorus (to yield *TS1*) in both cases, both substrates should have been hydrolyzed at the same rate since sulfur is expected to enhance both specifically and significantly the cleavage of the P—(5′-S) bond, because the pK_a of a thiol is more than 5 units lower than that of the corresponding alcohol.[121,122] Therefore, it can be concluded that the departure of the 5′-leaving oxygen is the overall rate-limiting step not only in the nonenzymatic hydrolysis of RNA but also in the hammerhead ribozyme-catalyzed reaction with a natural RNA substrate.

6.15.2.4.3 *Catalytic metal ions and the double-metal-ion mechanism of catalysis for hammerhead ribozyme-mediated reactions*

Since the departure of the 5′-leaving oxygen (via *TS2*) is the overall rate-limiting step in the nonenzymatic hydrolysis of RNA, any catalyst that accelerates the cleavage of RNA should lower the energy barrier for formation of *TS2*. The cleavage of RNA can, in general, be accelerated by a number of catalytic factors. One is a basic group that enhances the nucleophilicity of the 2′-oxygen by deprotonating the 2′-hydroxyl group and another is an acidic group that can neutralize and stabilize the leaving 5′-oxyanion. Furthermore, any environmental conditions, including electrophilic catalysis, that can stabilize the pentavalent phosphate, which is either a transition state or a short-lived intermediate, should enhance the cleavage of RNA.[123,124] Such catalytic factors include metal ions or protons, irrespective of whether cleavage is catalyzed by an enzymatic or a nonenzymatic system.

It is generally accepted that metal ions are required for the chemical cleavage step in reactions catalyzed by hammerhead ribozymes, which are metalloenzymes.[18–33] The possible roles of metal ions in ribozyme-catalyzed reactions are as follows. (i) A metal-coordinated hydroxyl group might act as a general base, abstracting the proton from the 2′-hydroxyl group (Figure 7(a)) or, alternatively, a metal ion might act as a Lewis acid to coordinate directly with the 2′-oxygen to accelerate its deprotonation (Figure 7(b)). (ii) The developing negative charge on the 5′-oxygen leaving group might be stabilized by direct coordination with a metal ion (Figure 7(c)) or by a proton that is provided by a solvent water or by a metal-bound water molecule (Figure 7(d)). (iii) Direct coordination of a metal ion to the nonbridging oxygen (Figure 7(e)) or hydrogen bonding between metal-bound water and the nonbridging oxygen might stabilize the trigonal-bipyramidal transition state or an intermediate by rendering the phosphorus center more susceptible to nucleophilic attach (electrophilic catalysis).

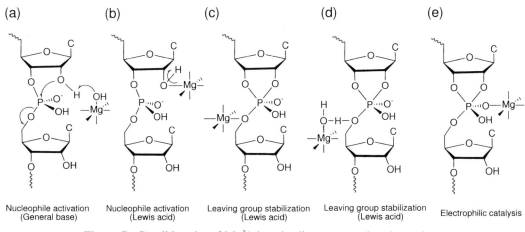

Figure 7   Possible roles of Mg^{2+} ions in ribozyme-catalyzed reactions.

In the case of the proteinaceous enzyme RNase A, which does not require metal cofactors, the acid/base catalysts are provided by two histidine residues within the catalytic pocket (Figure 8(a)). Since such acid/base functionality can, in principle, be replaced by Mg^{2+}-bound water moieties, it was proposed that Mg^{2+}-bound hydroxide acts as a general base to deprotonate the attacking 2′-OH and that a Mg^{2+}-bound water moiety acts as an acid to stabilize the developing negative charge of the leaving 5′-oxygen (Figure 8(b)).[7] An alternative, more generally accepted mechanism is a single-metal-ion mechanism of catalysis.[56,119,120] In this case, for stabilization of *TS2*, it is not a metal ion but a proton that is invoked (Figure 8(c)). This mechanism is supported by the finding that no metal ion was located close to the 5′-leaving oxygen in the X-ray crystal structure of a freeze-trapped conformation intermediate of a hammerhead ribozyme.[56] By contrast, from molecular orbital calculations,[23,24,125,126] we predicted that the direct coordination of Mg^{2+} ions with the attacking or the leaving oxygen might promote formation or cleavage of the P—O bond (Figure 8(d)).

The cleavage activity of ribozymes increases linearly with pH from pH 5.5 to pH 9.0, with a slope of about unity,[18,24,83,104] and this result is consistent with all the mechanisms outlined in Figure 8(b)–(d). An increase in the concentration of OH^- ions stimulates formation of possible active species in all cases. A higher concentration of OH^- ions increases the concentration of metal hydroxide and facilitates catalysis by deprotonation of the nucleophilic 2′-hydroxyl group, which leads to an increased concentration of the attacking 2′-oxyanion species. Similarly, a higher concentration of OH^- ions, which can deprotonate the 2′-hydroxyl group that is bound directly to the metal ion, generates more of the attacking 2′-alkoxide species. Thus, the profile of pH versus rate cannot be used to distinguish between the three possible mechanisms.

The mechanism that the authors propose from their analysis (Figure 8(d)) should be dis-

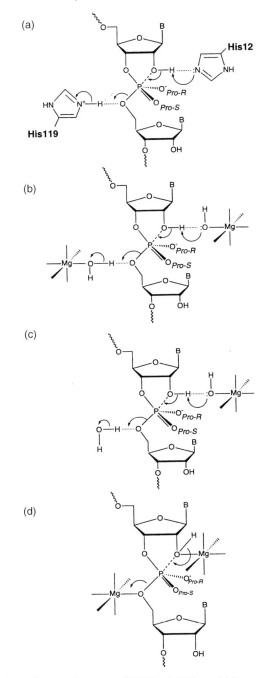

Figure 8 Proposed mechanisms for the cleavage of RNA. (a) The acid/base system, exemplified by the two histidine residues in RNase A. (b) Proposed mechanism for hammerhead ribozyme-catalyzed cleavage, in which Mg^{2+}-bound hydroxide and Mg^{2+}-bound water moieties act as base and acid catalysts, respectively. (c) Proposed single-metal-ion mechanism. (d) The mechanism proposed by the present authors and also by von Hippel's group, in which Mg^{2+} ions act as Lewis acids by directly coordinating to the attacking and the leaving oxygens.[24-32]

tinguishable from the mechanisms outlined in Figures 8(b) and 8(c) because, in the latter two cases, a proton-transfer process is involved in formation of the transition state whereas in this case (Figure 8(d)) it is not.[25,27] In order to examine whether a proton-transfer reaction occurs in the formation of the transition state, the authors investigated solvent isotope effects (k_{H_2O}/k_{D_2O}) for reactions catalyzed by hammerhead ribozymes.[25,27,31,104] It was found that, after correction for the difference in acidity of 2′-OH (or 2′-OD) in H_2O versus that in D_2O (ΔpK_a; see Refs. 27 and 31 for details), there was no actual kinetic isotope effect in the step that leads to cleavage of a phosphodiester bond by a ribozyme. This observation can be interpreted only in terms of a mechanism (Figure 8(d)) in which transfer of a proton does not occur in the transition state.

The observation used to support the popular single-metal-ion mechanism of catalysis (Figure 8(c)) is the inverse correlation between the pK_a of metal ions and their ability to promote cleavage of RNA.[18,127] The lower the pK_a, the higher the cleavage rate at a given concentration of metal ions. As pointed out by von Hippel[29,32] metal ions with lower values of pK_a (stronger acids) might be present at higher concentrations in the form of solvated metal hydroxides at a given pH. Such ions should, however, be correspondingly weaker bases and, therefore, they should be less able to remove the 2'-OH proton. As a final result, there should be virtually no correlation between the pK_a of metal ion and the cleavage activity if the mechanism shown in Figure 8(c) is the operative mechanism. In other words, the dependence on pK_a cannot be explained by the solvated metal hydroxide acting as a base. By contrast, deprotonation of the 2'-hydroxyl group can be greatly accelerated by the direct binding to it of a metal ion, in particular, a metal ion with a relatively low pK_a, because the pK_a of a 2'-hydroxyl group with a bound metal ion can be reduced by 3 to 8 units.[128] Both the divalent metal ion-dependent activity of ribozymes and the absence of a proton-transfer process in the ribozyme-mediated hydrolysis of RNA support the mechanism shown in Figure 8(d).[24-27,29-32,129]

Piccirilli and his co-workers demonstrated the direct coordination of a metal ion to the attacking 2'-hydroxyl group by using a 2'-thio RNA substrate in which the 2'-oxygen was replaced by sulfur.[129] Thio-substituted substrate analogues have frequently been used to probe the rate-limiting step of ribozyme-catalyzed reactions and to identify the site of coordination of a metal ion with a phosphate oxygen.[19,26-28,30,31,34-36,104,119,120,129-140] In attempts to characterize the rate-limiting step, the arguments are based on the "thio effect" (where the thio effect is defined as $k_{phosphate}/k_{phosphorothioate}$).[141] In the identification of metal-binding sites, the experimental approach relies on the difference in affinity between divalent ions. For ATPβS, the Mg^{2+} ion is coordinated 30 000-fold more strongly to oxygen than to sulfur while the strength of coordination of an Mn^{2+} ion to oxygen and to sulfur is more or less equal[142,143] Thus, the discrimination by Mn^{2+} ions between oxygen and sulfur atoms is poor, while the weak binding of a Mg^{2+} ion to sulfur results in a very large thio effect. Therefore, if direct coordination of a metal ion to a certain oxygen is involved in a ribozyme-catalyzed reaction, replacement of the oxygen by sulfur should reduce the rate of the reaction in the presence of Mg^{2+} ions while the rates should be basically similar in the presence of Mn^{2+} ions.[19,28,30,134-137,140] This effect of Mn^{2+} ion is generally referred to as the "rescue effect" (the rescue effect is defined as $k_{cleave(Mn)}/k_{cleav(Mg)}$). In the cleavage of a 2'-thio RNA substrate by a hammerhead ribozyme, the Mn^{2+}-rescue effect was observed when metal ions were changed from Mg^{2+} to Mn^{2+} ions and this result provided proof of the direct coordination of a Mg^{2+} ion to the attacking 2'-oxygen, lending further support to the mechanism shown in Figure 8(d).[129]

Mn^{2+}-rescue experiments were also used to investigate whether direct coordination of a metal ion is involved at the leaving 5'-oxygen in the cleavage reaction catalyzed by a hammerhead ribozyme. Two kinds of substrate, the parental 5'-oxy substrate and the corresponding 5'-thio substrate, in which the leaving 5'-oxygen at the scissile phosphate was replaced by sulfur, were subjected to ribozyme-mediated cleavage at pH 6.0 in the presence of either 0.3 mM Mg^{2+} ions or 0.3 mM Mn^{2+} ions.[30,31] As expected, both substrates were cleaved more rapidly in the presence of Mn^{2+} ions than in the presence of Mg^{2+} ions since metal ions with a lower pK_a produce higher concentrations of active species (M—O⁻—R; metal-bound-2'-alkoxides). The pK_a of a Mn^{2+}-bound water molecule is 10.6 and that of a Mg^{2+}-bound water molecule is 11.4.[128] Thus, at a given pH, the concentration of the active species (Mn^{2+}—O⁻—R) is roughly 6.3 times higher in Mn^{2+}-containing solutions than the concentration of the corresponding active species (Mg^{2+}—O⁻—R) in Mg^{2+}-containing solutions. Therefore, 6.3-fold is the theoretical maximum extent of the difference in the steady-state concentrations of the two active species, namely, the metal-bound 2'-alkoxides (M—O⁻—R), which are determined by the pK_a of each metal ion. However, ribozyme-mediated cleavage of a natural RNA substrate was 14 times more efficient in the presence of Mn^{2+} ions than in the presence of Mg^{2+} ions, a result that cannot be explained only by the difference in concentrations of active species (M—O⁻—R) since the difference should be 6.3-fold at most. This discrepancy suggests the involvement of more than one catalytic metal ion.[30,31]

In the ribozyme-mediated cleavage of the corresponding 5'-thio substrate, the reaction was only 2.2 times more efficient in the presence of Mn^{2+} ions than in the presence of Mg^{2+} ions. This value is significantly lower than the above-mentioned value for the parental RNA substrate of 14. This difference is understandable when the energy diagram in Figure 6 is considered, since the rate-limiting step for the cleavage of the 5'-thio substrate (Figure 6, right) has shifted from formation of **TS2** to that of **TS1** and since any direct coordination of a metal ion to the 5'-leaving group that occurs after the rate-limiting step is kinetically insignificant. Even in this case, the ratio of 2.2 for the cleavage of the 5'-thio substrate seems too small because the concentration of Mn^{2+}-bound nucleophile should have been 6.3 times higher than that of the Mg^{2+}-bound nucleophile. The value of 2.2 represents the lower nucleophilicity of the Mn^{2+}-bound 2'-alkoxide as compared to that of the Mg^{2+}-bound 2'-alkoxide. The nucleophile with a higher pK_a (Mg^{2+}-bound 2'-O$^-$) is expected to be less stable, more reactive, and a better nucleophile than the nucleophile with a lower pK_a (Mn^{2+}-bound 2'-O$^-$). Therefore, if there were no involvement of metal ions in formation of **TS2** during the cleavage of the normal RNA substrate (Figure 6, left), the overall value of $k_{cleav(mn)}/k_{cleav(Mg)}$ should have been lower than the theoretical maximum value of 6.3.

Another role of a metal ion, namely, as an electrophilic catalyst that coordinates directly with the *pro-R*p oxygen of the scissile phosphate has been proposed, in the case of hammerhead ribozymes (Figure 7(e)), on the basis of results of a similar "rescue" experiment with Mn^{2+} ions.[127,132,133] Thio substitution at the *pro-R*p oxygen at the cleavage site of a substrate (to yield **RpS**) for a hammerhead ribozyme resulted in a large thio-effect that was relieved by replacement of Mg^{2+} ions by Mn^{2+} ions, which have higher affinity for sulfur than do Mg^{2+} ions. This observation led to the general conclusion that a Mg^{2+} ion is directly coordinated with the *pro-R*p oxygen.[127,132,133] In this arrangement, the bound metal ion can act as an electrophilic catalyst and, thus, the proposed mechanism is very attractive as an explanation for the catalytic activity of metalloenzymes.[22] However, our re-examination of thio effects argues against the generally accepted mechanism of electrophilic catalysis, namely, the direct coordination of a metal ion with the *pro-R*p oxygen.[138] In an attempt to quantitate the rescue ability of Mn^{2+} ions (recall that the rescue effect is defined as $k_{cleav(Mn)}/k_{cleav(Mg)}$), our careful examination of the rescue ability of Mn^{2+} ions with these isomers demonstrated that Mn^{2+} ions could rescue the reaction not only with the **RpS** isomer but also with the **SpS** isomer in which the *pro-S*p oxygen at the cleavage site of a substrate was replaced by sulfur, and rescue occurred to a similar extent in both cases ($k_{cleav(Mn)}/k_{cleav(Mg)} = \sim 20$). Moreover, as described above, the rate of ribozyme-mediated hydrolysis of the unmodified, natural substrate was also 14-fold higher in the presence of Mn^{2+} ions than in the presence of Mg^{2+} ions. Since Mn^{2+}-mediated cleavage occurs about 20-fold more rapidly than Mg^{2+}-mediated cleavage with **RpS**, with **SpS** and, also, with the natural substrate, it seems unlikely that the previously observed "rescue" effects[127,132,133] support the proposed direct and specific coordination of a metal ion to the *pro-R*p oxygen in the transition state of the hammerhead ribozyme-catalyzed reaction. The previously observed "rescue" effects must have originated, at least in part, from the intrinsic properties of the metal ions, which include the specific values of pK_a of the different metal ions, as discussed above. The observed rescue effect of Mn^{2+} ions ($k_{cleav(Mn)}/k_{cleav(Mg)}$) of about 20 for these substrates is too large to be explained by a single-metal-ion model since the theoretical maximum value of 6.3 is the calculated ratio of catalytically active species that is based on the difference in pK_a of the two aqueous metal ions, Mg^{2+} and Mn^{2+}. Since anything above this value must be due to involvement of a second metal ion, the ribozyme-mediated cleavage of not only the natural substrate but also of **RpS** and **SpS** substrates should proceed via the double-metal-ion mechanism of catalysis (Figure 6, left, and Figure 8(d)), without the specific interaction of a metal ion with the *pro-R*p oxygen (as shown in Figure 7(e)) of the scissile phosphate in the transition state.

All things considered, it seems clear that the double-metal-ion mechanism of catalysis (Figure 6, left, and Figure 8(d)) is the operative mechanism in reactions catalyzed by hammerhead ribozymes, although no switch in metal ion specificity was observed for the 5'-thio substrate (because of a change in the rate-limiting step). The reaction scheme shown in Figure 6 (left), which is based on our kinetic measurements with natural and 5'-thio substrates,[30] is identical to the mechanism, recently proposed by von Hippel's group, that is based on results of competition experiments with La^{3+} ions.[32]

6.15.3 HAIRPIN RIBOZYMES

6.15.3.1 General Features of Hairpin Ribozymes

A hairpin ribozyme was first discovered in the (−)-strand of the satellite RNA associated with the tobacco ringspot virus (−)sTRSV.[10,105,144] This satellite RNA replicates via a rolling-circle mechanism. During the replication, concatinated strands of satellite RNAs are produced. The concatinated strands are self-cleaved to yield monomeric strands and then self-ligation occurs to generate monomeric circles. The hairpin ribozyme is responsible for the self-processing reaction of the (−)-strand.[145–147] Other naturally occurring hairpin ribozymes have also been discovered in the

$(-)$-strands of the satellite RNAs associated with chicory yellow mottle virus (sCYMV1) and arabis mosaic virus (sArMV).[148] Although the native sTRSV hairpin ribozyme is catalytically the most efficient, the other two ribozymes are also very active. The replication of the $(+)$-strand of these satellite RNAs is catalyzed by hammerhead ribozymes, but the ligation activity of hammerhead ribozymes is very much lower than that of hairpin ribozymes, indicating that the original (hammerhead) ribozyme-catalyzed ligation reactions might have been replaced by reactions catalyzed by proteinaceous enzymes.

6.15.3.2 Sequence Requirements of Hairpin Ribozymes

Within the 359-nucleotide-long, negative-strand satellite RNA of tobacco ringspot virus, the catalytically essential domain was identified as two minimal sequences by deletion analysis.[145] A 50-nucleotide catalytic portion can cleave a 10- to 14-nucleotide substrate portion at a phosphodiester bond located between residue A_{-1} and residue G_{+1} and can catalyze ligation of the cleaved products with a terminal 2′,3′-cyclic phosphate on the 5′-fragment and a 5′-OH group on the 3′-fragment.[105,145] Calculations of minimal energy for folding, mutagenesis experiments and limited phylogenetic comparisons have revealed a possible secondary structure of the hairpin ribozyme, and results of *in vitro* selection lend support to this model.[145–154] A model of the secondary structure of the hairpin ribozyme consists of four helices and two internal loops (Figure 9). In a *trans*-acting hairpin ribozyme, helices I and II are formed by interactions between the substrate and the ribozyme while helices III and IV are formed by sequences within the ribozyme itself. Internal loop A, which contains the cleavage and ligation site, is located between helices I and II. The other internal loop B is located between helices III and IV. Helix IV is capped by a hairpin loop that is not important for catalysis.[155] Most of the residues in loops A and B are essential for efficient cleavage. By contrast, all residues in helices I and IV, with the exception of the base pairs that flank loop A, can be varied provided that the base pairs within helices can be formed. Moreover, the lengths of helices can be increased without loss of activity.[150–153]

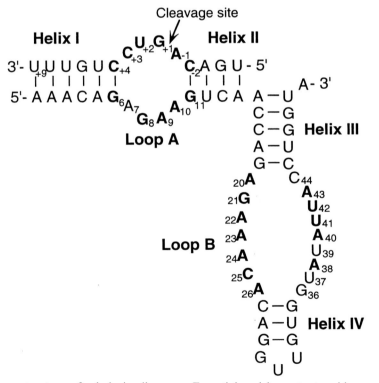

Figure 9 Secondary structure of a hairpin ribozyme. Essential and important residues are shown in bold letters.

One of the most important features of the hairpin ribozyme is a strong requirement for G_{+1} immediately 3′ of the cleavage site (Figure 9). Replacement of this residue by A, U, or C reduces the catalytic efficiency by a factor of at least 10^5 but does not affect the substrate-binding activity to the ribozyme.[150,152,153,156] These observations indicate that G_{+1} immediately 3′ of the cleavage site is necessary for a step after the substrate-binding step. Molecular modeling studies suggested that only a G residue could provide a chemically useful functional group, a 2-amino group in this case, close to the scissile phosphodiester bond. To confirm this hypothesis, G_{+1} was replaced by inosine or 2-aminopurine.[156] The 2-aminopurine-containing substrate was cleaved efficiently, whereas the inosine-containing substrate was not cleaved.[156] These results suggest strongly that the 2-amino group of G_{+1} plays an essential role in catalysis, while the keto group of G_{+1} might not be directly involved in the chemical reaction.

The functional groups of all the essential purine residues (G_{+1}, G_8, A_9, A_{10}, and G_{11} in loop A and G_{21}, A_{22}, A_{23}, A_{24}, A_{38}, A_{40}, and A_{43} in loop B) have also been investigated, in attempts to clarify the secondary structure and the overall conformation, by chemical modification.[157] Purine riboside (P) and inosine (I), for elimination of an exocyclic amino group; N^7-deazaadenosine (^{7}cA) and N^7-deazaguanosine (^{7}cG), for elimination of N^7-nitrogen; and O^6-methylguanosine (O^6Me-G) for elimination of the N^1-hydrogen of G, were used as analogues of adenosine and guanosine, respectively, and cleavage activities were compared to those of the unmodified ribozyme. Compared with alterations of a hammerhead ribozyme, in which elimination of a functional group from any of most of the conserved purine residues (with the exception of G_5 and A_{14}) has relatively small effects on ribozyme-mediated cleavage, alteration of a single functional group of an essential purine in the hairpin ribozyme caused substantial loss of cleavage activity in many cases. The decreases in cleavage activity were almost all found to have been due to changes in k_{cat} rather than in K_M. Therefore, most of the effects appeared to be due to the loss of hydrogen-bonding interactions that are essential for the formation of the transition state, such as inter- and intra-loop contacts and binding of Mg^{2+} ions, rather than for the binding of the substrate to the ribozyme in the ground state. Only changes of A_{24} to ^{7}cA, A_{40} to P, A_{40} to ^{7}cA and A_{43} to P in loop B were permissible, and the modified ribozymes retained 50–90% of wild-type activity.

Substitution by O^6-MeG of G_{+1} caused a dramatic loss of cleavage activity, as did introduction of inosine at this site, whereas 2-aminopurine had little effect.[156] However, in the case of O^6-MeG, it was assumed that the extra proximal methyl group sterically prevented proper location of a Mg^{2+} ion at the putative binding site at N-7 of this residue. Divalent cations (Mg^{2+} ions are preferred) are required for the cleavage-ligation reaction mediated by hairpin ribozymes.[145,158] Changes in Mg^{2+}-binding activity and in the cleavage rate of mutant ribozymes, depending on the concentration of Mg^{2+} ions, suggested that the N^7-nitrogen of G_{+1} might be the Mg^{2+}-binding site in the ground state and the N^7-nitrogen of A_9 might be the binding site in the transition state.[157,159]

The Mg^{2+}-binding sites have been investigated systematically by modification-interference analysis with thio substitutions at the *pro-R* oxygen of the phosphate backbone of the ribozyme and its substrate.[155] Thiophosphate substitutions at all the guanosine, cytidine and uridine residues of the ribozyme did not affect the ribozyme-mediated cleavage to any great extent, but the presence of multiple adenosine phosphorothioates in the ribozyme decreased its cleavage activity 25-fold. Experiments with ribozymes that included some adenosine phosphorothioate moieties suggested that this thio effect was mainly due to interference with the binding of a Mg^{2+} ion to A_7, A_9, and A_{10}. By contrast, phosphorothioate moieties in the substrate did not affect the efficiency of cleavage by the ribozyme.

Other Mg^{2+}-binding sites that performed the above-described modifications were also proposed by Burke's group.[160] Removal of the 2′-hydroxyl group of G_{11} or A_{24}, two essential residues, decreased the cleavage rate significantly but did not affect the substrate-binding activity. The reduction in cleavage activity of ribozymes with deoxy-G_{11} or deoxy-A_{24} was reversed by increased concentrations of Mg^{2+} ions. Therefore, two other Mg^{2+}-binding sites in the transition state were proposed to be the 2′-hydroxyl groups at G_{11} and A_{24}.[160] The most important hydroxyl group is that of A_{-1}, which acts as a nucleophile in the cleavage of the scissile phosphodiester bond.[161] Other 2′-hydroxyl groups that are important for tertiary interactions are reported to be those at A_{10}, C_{25}, U_{37}, A_{38}, and U_{41}.[159,160]

6.15.3.3 Tertiary Structures of Hairpin Ribozymes

From calculations of the minimal energy for folding of the negative-strand satellite RNA of tobacco ringspot virus, which has a large RNA duplex that connects the 5′ end of the substrate and 3′ end of the ribozyme,[145] it has been inferred that a bend exists between helices II and III. In order to prove this hypothesis, hairpin ribozymes were synthesized in which different numbers of cytidine linker units or 1,3-propanediol phosphate linker units were inserted at the junction between helices II and III.[162–164] For an efficient reaction, a linker of a certain length was required in both cases, suggesting a requirement for flexibility in formation of a complex in an active conformation. Separation of hairpin ribozymes at the hinge region had a minimal effect on the efficiency of catalysis but increased the K_M.[165,166] With respect to binding efficiency of a substrate, the coaxial stacking of helices II and III is energetically favorable.[167,168] However, the active complex of a hairpin ribozyme seems to have a sharp bend at the junction between these helices. The tertiary interactions between loop A and loop B probably provide a favorable environment for activation of 2′-hydroxyl group by a certain functional group within the ribozyme, bringing the distant functional group close to the attacking 2′-hydroxyl group. The putative bent structure is supported by results of other experiments that include mutagenesis experiments,[153,169] *in vitro* selection studies,[152] analyses of requirements for functional groups,[157,160] kinetic and thermodynamic analyses,[170] and chemical modification experiments.[171]

A solution structure of a partial hairpin ribozyme consisting of loop A, helices I and II, and including the cleavage site, was identified by NMR spectroscopy.[172] In the solution structure, G_{+1}, immediately 3′ of the cleavage site, formed a sheared $G \cdot A$ base pair with A_9, in which the 2-amino group of G_{+1} was involved in a hydrogen-bonding interaction, while the 2-amino group of G_{+1} had previously been considered to be essential for catalysis.

6.15.3.4 Mechanisms of Action of Hairpin Ribozymes and Roles of Metal Ions

6.15.3.4.1 *Kinetic framework of hairpin ribozymes*

Hairpin ribozymes cleave RNA to generate 2′,3′-cyclic phosphate and 5′-hydroxyl termini and they ligate these termini, in the reverse reaction, to form 3′,5′-phosphodiesters. The minimum reaction scheme for a hairpin ribozyme is basically the same as that for a hammerhead ribozyme (Figure 3). First, the substrate binds to the ribozyme, via the base pairs that form helix I and helix II, to yield a Michaelis–Menten complex (k_{assoc}). Then a specific phosphodiester bond, located between residue A_{-1} and residue G_{+1} in the bound substrate, is cleaved (k_{cleav}). Finally, the cleaved fragments dissociate from the ribozyme. The ligation reaction of hairpin ribozymes is the reverse of this scheme.

Rates and equilibrium constants for individual steps in the reaction scheme have been defined, as they have in the case of hammerhead ribozymes.[170,173] Substrates and products bind to hairpin ribozymes at rates nearly equal to those observed for other catalytic RNAs. The substrate dissociation rate, which is much lower than the rate of cleavage, allows almost every bound substrate to be cleaved before its dissociation.[170,173] *Trans*-acting hairpin ribozymes cleave their substrates with cleavage rate constants of 0.2 to 0.3 min^{-1}, and such rate constants are similar to the rate constants of about 1 min^{-1} of the chemical cleavage step in hammerhead ribozyme-mediated cleavage reactions under identical conditions.[107–109,170,173] These rates are about 10^5-fold higher than the cleavage rate in the nonenzymatic hydrolysis of RNA. The cleavage rate constant of the hairpin ribozyme may be affected by ligation of the product if cleaved products remain bound to the ribozyme long enough to undergo ligation. However, stabilization of the binding helices by extension of the lengths of helices I and II has little effect on the cleavage rate constants, an indication that, in general, as long as the substrate has 14 nucleotides or fewer, dissociation of one or both cleaved products is sufficiently rapid not to allow ligation.[170]

The minimal effects of pH and of substitution with sulfur of one of the non-bridging oxygens at the scissile phosphate on the rate constant for cleavage by the hairpin ribozyme suggested the possible existence of a rate-limiting step that corresponded to a change in conformation.[34–36,145,155,161] The minimal effects on the cleavage rate of destabilization of the ribozyme–substrate complex as a result of shortening of the binding helices[170] imply that the putative rate-limiting change in conformation is independent of the lengths of helices I and II but is a consequence of tertiary interactions. However it is still unclear whether or not the putative change in conformation can be the rate-limiting step. *Trans*-acting hairpin ribozymes catalyze intermolecular ligations at a rate of about 3 min^{-1} under saturating conditions and this rate is generally about 10- to 15-fold higher than the cleavage rate.[170] In addition, the rate of ligation by hairpin ribozymes is much higher than that by hammerhead ribozymes, which ligate their cleaved substrates with the rate of 0.008 min^{-1}.[108,170] Therefore, the hairpin ribozyme can be regarded as a single-stranded RNA ligase rather than as an endoribonuclease.

6.15.3.4.2 Roles of metal ions and the rate-limiting step in hairpin ribozyme-mediated cleavage·ligation reactions

Hairpin ribozymes cleave RNA molecules between the N_{-1} (usually an A residue) and G_{+1} residues. The mechanism of the endonucleolytic transesterification reaction catalyzed by hairpin ribozymes is generally similar to that of other reactions catalyzed by small catalytic RNAs, such as hammerhead ribozymes and the hepatitis delta virus ribozyme. First, the 2'-hydroxyl group of N_{-1} attacks the scissile phosphate as a nucleophile. Then the 5'-oxygen of the adjacent G_{+1} is separated to produce 2',3'-cyclic phosphate and 5'-hydroxyl termini. In the case of a hammerhead ribozyme, as described above, it is likely that divalent metal ions, such as Mg^{2+} ions, act as catalysts by co-ordinating directly to both the attacking 2'-oxygen and the leaving 5'-oxygen (Figure 8(d)).[25,27,29–32,104] The requirements for metal ions of reactions catalyzed by a hairpin ribozyme are similar to those of reactions catalyzed by hammerhead ribozymes, and increases in rate of cleavage or ligation that depend on concentrations of metal ions have been observed.[145,158] However, the cleavage reaction occurs only when Mg^{2+}, Sr^{2+}, or Ca^{2+} ions are used and no reaction is observed when Mn^{2+}, Co^{2+}, Cd^{2+}, Ni^{2+}, Ba^{2+} or monovalent cations are used.[158] Most of these ineffective divalent metal ions promote the cleavage reactions catalyzed by hammerhead ribozymes.[127] In particular, as described above, Mn^{2+} ions, with their lower pK_a, can promote the cleavage by hammerhead ribozymes more efficiently than Mg^{2+} ions, with their higher pK_a.[18,30,120,138] In addition, spermidine can induce catalytic cleavage by hairpin ribozymes in the absence of metal ions, albeit at a very low rate, even in the presence of chelating agents (EDTA and EGTA). Thus, we cannot ignore the possibility of metal-independent catalysis in the chemical cleavage step.[158]

Three independent groups have provided clear evidence for this hypothesis using cobalt hexa-amine, $Co(NH_3)_6^{3+}$, which is an inert transition metal complex (i.e., $Co(NH_3)_6^{3+}$ can play a structural role but cannot play a catalytic role in the chemical cleavage step), in place of divalent metal ions.[34–36] Even in the absence of other metal ions, $Co(NH_3)_6^{3+}$ promotes cleavage reactions mediated by hairpin ribozymes as effectively as Mg^{2+} ions and $Co(NH_3)_6^{3+}$-promoted reactions are not quenched by addition of a chelating agent (EDTA).[34–36] In addition, $Co(NH_3)_5Cl^{2+}$, which has a reduced charge and should have reduced ability to act as an electrostatic catalyst, had the same effect on the rate of hairpin ribozyme-mediated reactions as Mg^{2+} ions and $Co(NH_3)_6^{3+}$.[36] These observations provide clear evidence that, in the chemical cleavage step of hairpin ribozyme-catalyzed reactions, no metal ions act as Lewis acid catalysts to enhance the nucleophilicity of the 2'-oxygen and to stabilize the leaving 5'-oxyanion by coordinating directly to these oxygens (or, alternatively, no metal-bound hydroxide acts as a general base catalyst that abstracts the proton from the 2'-hydroxyl group), and no metal ion, as an electrophilic catalyst, coordinates directly to the *pro-R* oxygen at the scissile phosphate.

What might replace the roles of metal ions in the catalysis that is mediated by hairpin ribozymes? One potential base catalyst or nucleophile might be the 2-amino group of G_{+1} since substitution by inosine causes a dramatic loss of cleavage activity, while 2-aminopurine has little effect.[156] However, such a 2-amino group within the ribozyme would have a pK_a value that is too low for abstraction of the proton from the 2'-hydroxyl group. Such functional groups would have to be surrounded by networks, formed via tertiary interactions, that transiently stabilize the putative polar transition state. The possibility of such tertiary interactions was suggested previously by various groups of researchers.[157,159,160,162–165,169,171] However, the possibility of a 2-amino group acting as a catalyst was brought into question by a recent NMR study of loop A from the hairpin ribozyme.[36,172] In the NMR structure, G_{+1} was found to form a sheared G·A base pair with A_9 through hydrogen bonding between the 2-amino group and N^3 of G_{+1} and N^7 and the 6-amino group of A_9, regardless of the presence or absence of divalent metal ions.[172] It remains to be determined whether this sheared G·A base pair remains intact in the transition state in true hairpin ribozyme-mediated reactions.

Rates of hairpin ribozyme-mediated cleavage-ligation reactions in Mg^{2+}- or $Co(NH_3)_6^{3+}$-containing buffer are hardly affected by pH between pH 5 and pH 10,[35,145] while the rates of the cleavage reaction catalyzed by hammerhead ribozymes increase linearly with pH, reflecting the deprotonation of the 2'-hydroxyl group prior to the rate-limiting chemical cleavage step.[18,24,31,104] Since the former and unusual independence of pH was observed under conditions of excess ribozyme, where no product dissociation-step was involved,[35] it seems possible that the rate-limiting step of the hairpin ribozyme-catalyzed reaction might not be the chemical cleavage step. In other words, neither the deprotonation of the 2'-hydroxyl group nor the stabilization of the leaving 5'-oxygen by a proton appeared to influence the rate between pH 5 and pH 10. Therefore, the existence of an alternative rate-limiting step that involved a conformational change was suggested.[35] However, relatively small but unequivocal thio-effects were observed in the reaction mediated by a hairpin ribozyme under pH-independent conditions. Thio substitution at one of the nonbridging oxygens at the scissile phosphate of a substrate (*R*p*S* or *S*p*S* analogue) had a limited effect on the cleavage-ligation reactions mediated by hairpin ribozymes in the presence of Mg^{2+} ions and, also, in the presence of $Co(NH_3)_6^{3+}$.[34–36,155,161] This small thio-effect suggests the possibility that the chemical cleavage step is rate-limiting at these pH values. Therefore, the possibility cannot be excluded that the chemical cleavage step might be part of the rate-limiting step of the reaction catalyzed by hairpin ribozymes. It is to be noted that independence of pH can still be a reflection of the chemical cleavage step if the difference between pK_a values of acid and base catalysts is large, although, at the same time, another possibility cannot be excluded, namely, that the relatively small thio-effect might have originated from structural effects (e.g., via a change in the rate of conformational change prior to the cleavage step) of the thio substitution.

6.15.3.5 Metal Ion Requirements for Structural Purposes

Although metal ions are not involved in the chemical cleavage step of hairpin ribozyme-mediated reactions, a requirement for and dependence on metal ions are clearly apparent and potential Mg^{2+}-binding sites have been identified, as described above.[155,157,158,160] In contrast to the case of hammerhead ribozymes, in which metal ions are required for both the chemical reaction and structural folding, it seems likely that metal ions are required only for structural proposes, in particular for internal loop–loop interactions, in the case of hairpin ribozymes.[34–36,158] This structural role is supported by the fact that a low concentration of spermidine enhances both Mg^{2+}- and $Co(NH_3)_6^{3+}$-promoted reactions. However, a high concentration of spermidine inhibits the Mg^{2+}-promoted reaction, an indication that, at high concentrations, spermidine robs Mg^{2+} ions of their binding sites and causes a subsequent decrease in rate because of the inefficient manner in which spermidine participates in the proper folding of ribozymes, as compared with Mg^{2+} ions.[34,158] Furthermore, monovalent cations (as NaCl, KCl, and NaOAc) inhibit the Mg^{2+}-promoted reaction but enhance the $Co(NH_3)_6^{3+}$-promoted reaction.[34,158] Monovalent cations, which have relatively limited ability to stabilize the favorable ribozyme–substrate complex that is required for catalysis, probably compete with Mg^{2+} ions for binding sites and, upon binding, they decrease the rate of the reaction.[34,158] However, some of the most important functions of metal ions in RNA folding are considered to be hydrogen-bonding interactions, via metal-bound water molecules, for stabilization of the tertiary structure, as well as to overcome the negative charge repulsion of the phosphate backbone by direct and indirect binding with phosphate groups. Mg^{2+} ions are better suited for this purpose than $Co(NH_3)_6^{3+}$. Nonetheless, monovalent cations can enhance the reaction promoted by $Co(NH_3)_6^{3+}$, which is a relatively poorer stabilizer.[34]

All things considered, the mechanistic pathway of the reactions catalyzed by hairpin ribozymes seems to be quite different from that of reactions catalyzed by other small ribozymes, in spite of the fact that both hairpin and hammerhead ribozymes can be isolated from a single virus.

6.15.4 HEPATITIS DELTA VIRUS (HDV) RIBOZYME

6.15.4.1 General Features of the HDV Ribozyme

Hepatitis delta virus (HDV) is a small RNA satellite virus of hepatitis B virus (HBV). In humans, coinfection with HDV and HBV together, or the superinfection of HBV-infected individuals, exacerbates the severity of the disease.[174,175] HDV particles have a circular RNA genome that includes a self-cleaving domain.[176,177] The virus is an unusual animal virus in that its replication is thought to proceed by a double rolling circle mechanism, wherein both the genomic and anti-genomic sequences promote self cleavage.[11,12,178,179]

The two HDV ribozymes, a genomic and an anti-genomic ribozyme, are similar in size and sequence and they have very similar secondary structures.[180] The minimum length of the sequence of the natural HDV ribozyme that is required for efficient cleavage is 85 nucleotides.[181] Several models of the secondary structure of HDV ribozymes have been proposed.[182–191] The pseudoknot model[183] is most consistent with the results of mutagenesis experiments, chemical probing studies and a photo-crosslinking study,[181,183,185,187–199] and, in the model, the secondary structures of both ribozymes have four helical regions (Figure 10).

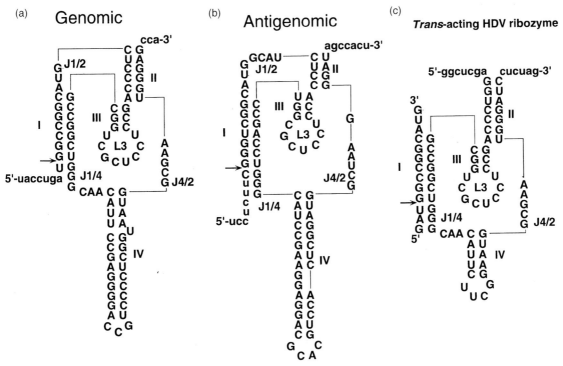

Figure 10 The pseudoknot model for the secondary structures of genomic (a) and antigenomic (b) hepatitis delta ribozymes. A *trans*-acting HDV ribozyme is also shown (c).

As is true for other ribozymes, HDV ribozymes can be designed to act "in *trans*," and several *trans*-acting ribozymes have been constructed for investigations of structure–function relationships.[181,182,185,188,200–202] A typical construct of this type is shown in Figure 10 (right). It was generated by removing the single-stranded region between helices I and II, to yield a long ribozyme and a short substrate, which was able to associate with the ribozyme sequence through formation of helix I.[185,200] In order to identify critical nucleotides and to improve the catalytic power, *in vitro* selection has also been performed using the *trans*-acting construct.[196,203,204]

6.15.4.2 Mechanistic Considerations of HDV Ribozyme-mediated Cleavage Reactions

With respect to requirements for metal ions, HDV ribozymes appear to resemble other catalytic RNAs. Thus, divalent cations are needed for efficient cleavage. However, there are slight differences from other ribozymes in terms of the kind and concentration of metal ions required for catalysis. HDV ribozyme-mediated reactions proceed efficiently in the presence of Ca^{2+} ions as well as of Mg^{2+} ions.[112,201,205] Moreover, Mn^{2+} and Sr^{2+} ions are also able to support efficient cleavage but Cd^{2+}, Ba^{2+}, Co^{2+}, Pb^{2+}, and Zn^{2+} ions are much less effective.[11,12,205,206] In general, the cleavage activity of HDV ribozymes increases with increasing concentrations of Mg^{2+} ions and then reaches a plateau value. For efficient cleavage, only 0.5 mM to 1 mM Mg^{2+} ions are sufficient, in contrast to the much higher concentrations of metal ions required for the optimal activity of hammerhead ribozymes.[139,186,205,207–209] However, a surplus of Mg^{2+} ions has an inhibitory effect.[139,210] Moreover, monovalent cations do not support cleavage at all but, as is the case for other ribozymes, spermidine can function to lower the required amount of metal ions.[205]

A remarkable property of HDV ribozymes is their tolerance to denaturing reagents, which, in some instances, even enhance cleavage activity.[210–212] For example, the antigenomic HDV ribozyme is fully active in the presence of 20 M formamide or 10 M urea.[186,211] A reasonable explanation for this effect is that it is a result of the correction of misfolding of RNA. Formamide and urea might accelerate the conversion from an inactive to an active conformation by lowering the energy barrier for the conformational change. It is likely that an equilibrium might exist between the active complex and the inactive complex and the change in conformation, with a high energy barrier, could become the rate-limiting step under single-turnover conditions (see below). Reactions catalyzed by HDV ribozymes are generally analyzed under single-turnover conditions because the cleaved products appear to stick tightly to the ribozyme at 37 °C and, thus, it is difficult to monitor kinetics under multiple-turnover conditions.[200]

As in the case of cleavage by other small ribozymes, the HDV-mediated cleavage of RNA produces products with a 5'-hydroxyl group and a 2',3'-cyclic phosphate, evidence for an intramolecular nucleophilic reaction.[11,12,213] Initial studies of self-cleaving HDV ribozymes, it was reported that the relative amount of product was unaffected by pH between pH 5 and pH 9.[12,207] This result was cited several times as an indication that the activity of HDV ribozymes is pH-independent (as is that of hairpin ribozymes). However, in the initial studies, the initial rates of the cleavage reaction at various pH values were not investigated. Therefore, the original results show only that the end point of the cleavage reaction is insensitive to pH. In fact, in the case of a more recently studied *trans*-acting HDV ribozyme, a bell-shaped profile of the reaction from pH 4 to pH 9 was obtained. The *trans*-acting HDV ribozyme that was used had high cleavage activity and had been improved by *in vitro* selection. The highest activity of the ribozyme was observed at pH 7.5.[139] A linear relationship between pH and the logarithm of the rate constant [$\log_{10}(k_{cat})$] from pH 4.0 to 6.0 with slope of unity indicated that the rate-limiting step of the reaction was the chemical cleavage step and reflected removal of a single proton from the 2'-hydroxyl group, as observed in the case of hammerhead ribozymes. It remains to be determined what mechanism generates a more nucleophilic 2'-oxyanion that attacks phosphorus and whether or not a metal ion plays a direct role. A *pro-R*p or *pro-S*p phosphorothioate substitution at the cleavage site did not interfere with cleavage. Moreover, neither a thio effect nor rescue by Mn^{2+} ions was observed.[139] Therefore, it is unlikely that a metal ion coordinates directly to the *pro-R*p oxygen or to the *pro-S*p oxygen to support electrophilic catalysis in HDV ribozyme-catalyzed reactions. Acidic and basic moieties with pK_a values around 7.5 remain to be identified.

6.15.5   GROUP I RIBOZYMES

6.15.5.1   General Features of Group I Ribozymes

RNA splicing is a fundamental feature of the processing of RNA in many organisms. The pre-rRNA of *Tetrahymena thermophila* was found to undergo "self-splicing" *in vitro* without the need for a protein catalyst and it was one of the first RNA molecules to be discovered to have enzymatic activity.[1] This and other "group I" self-splicing ribozymes promote two phosphoester-transfer reactions (Figure 11(a)) that result in the removal of an intervening sequence and the splicing of adjacent RNA domains.[3,6,46,53,214] Group I ribozymes were first defined on the basis of their common structural features and Figure 12 shows their general secondary structure, with lowercase letters indicating exons and uppercase letters indicating the intron, and with the cleavage and splicing sites indicated by thick arrows. The first and the second steps in RNA splicing by group I ribozymes and a detailed transition state structure for the first step are shown in Figures 11(a) and 11(b), respectively. The initial transesterification reaction promoted by this class of ribozymes uses the 3'-hydroxyl of guanosine or one of its 5'-phosphorylated derivatives as the nucleophile for S_N2 attack on the phosphate of the target internucleotide linkage (5' side exon/intron junction).[130,215–219] The second transesterification uses the newly formed 3'-hydroxyl moiety of the 5' exon as the nucleophile in the subsequent attack at the second splice-site junction (3' side exon/intron junction). Each reaction results in inversion at the phosphorus center.

RNA splicing by group I introns is extremely widespread and occurs in the generation of mature mRNAs, rRNAs, and tRNAs. Such introns have been found in mitochondrial, chloroplast and nuclear genomes of diverse eukaryotes, and they have also been found in prokaryotic and eubacterial genomes. However, their distribution is irregular. For example, some species of *Tetrahymena* have group I introns, whereas closely related species do not.

6.15.5.2   Secondary and Tertiary Structures of Group I Ribozymes

The common secondary structure of group I ribozymes was determined by comparative sequence analysis.[220,221] A common set of base-paired regions, named P1–P9, form the core of the structure

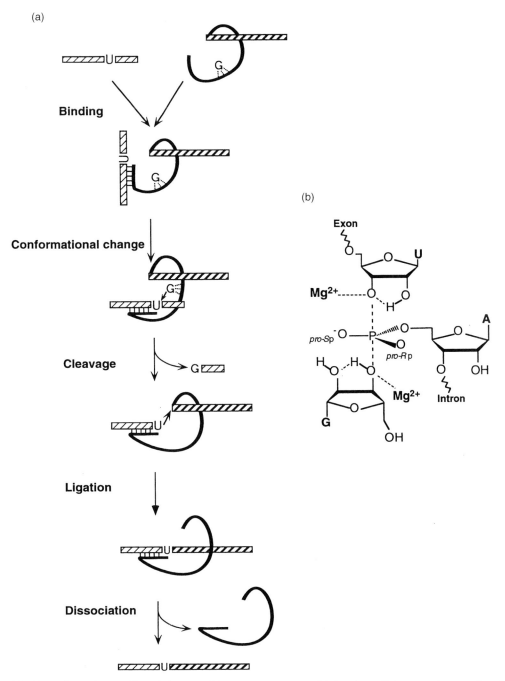

Figure 11 Reaction scheme (a) and the transition state structure (b) for the splicing reaction mediated by a group I intron.

(Figure 12). The base-paired regions P1, P3, P4, P6, P7, and P8 have been confirmed by site-directed mutagenesis and second-site mutational analysis. Mutation of a nucleotide involved in one of these core structures typically decreases the maximum velocity of splicing or increases the K_M for guanosine, while a second-site mutation can restore base-pairing and also restore splicing activity. These results show the primary importance of nucleotides in their contribution to the formation and stabilization of duplex regions. The guanosine-binding site is in the major groove of conserved duplex region P7 and the nucleophilic guanosine binds at this site via a hydrogen bond, using the same functional groups as used to form a G–C base pair.[222] The 5′-exon splicing site is in duplex region P1 and the nucleophilic attack of guanosine on a uridine residue leads to the first splicing reaction. The newly generated 3′-hydroxyl group of the 5′-exon fragment subsequently attacks the

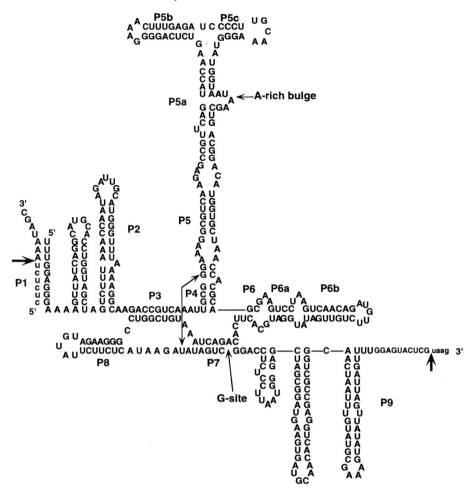

Group I ribozyme

Figure 12 Secondary structure of a group I intron. Lowercase and uppercase letters indicate exons and an intron, respectively. Splicing sites are indicated by thick arrows.

3′-exon splicing site in region P9, with resultant liberation of the internal guide sequence (IGS).[223] Extra sequences that are not shown in Figure 12, such as stem-loop structures, are peripheral to the core and are much more variable. In fact, they can be deleted completely without major loss of splicing function.[214]

A crystal structure of part of the *Tetrahymena* ribozyme, composed of the region P4–P6, has been described.[224,225] The X-ray crystal structure can be summarized as follows. The P4–P6 region consists of two helical regions that are packed side by side. Helices P6b, P6a, P6, P4, and P5 form a straight column on one side, and helices P5b and P5a are stacked on the other side. A bend of about 150° between helices P5 and P5a allows the P5abc extension to interact with one helical face of the conserved core region, with contact between the P4 helix and the A-rich bulge in the P5a region and contact between the tetraloop receptor (junction of P6b/P6a) and the GAAA tetraloop at the end of P5b. The P5c region protrudes from the plane of the helical stack of P5a and P5b.

The X-ray crystal structure of the P4–P6 region provides only limited information about the overall structure of group I ribozymes because details of the structure of other domains, such as the substrate-binding P1 domain, the P2 domain, and the guanosine-binding P7 domain, remain unknown. Site-directed mutagenesis and kinetic analysis of RNA splicing have been used to identify tertiary structures other than those in P4–P6.[226] A long-range interaction between the fifth base-pair of P4 and U within the junction of P8 and P7 has been found (indicated by a thin double-headed arrow in Figure 12), so that base triples bring the P3–P9 domains into proximity to the P4–P6 domains. Other information about the tertiary structure of group I introns remains to be determined.

6.15.5.3 Sequential Steps in Ribozyme-catalyzed Cleavage of RNA

In the self-splicing of RNA, the *Tetrahymena* intron lowers the energy barrier for two highly specific transesterification reactions. Identification of the elemental steps of these reactions and determinations of k_{cat} and K_M should provide a useful quantitative description of the efficiency with which the enzymatic reaction proceeds. Thus, for complete analysis, it was necessary to convert this self-splicing intron into a *trans*-acting enzyme that can catalyze multiple-turnover reactions. This conversion allowed detailed mechanistic studies of the transesterification reaction, as well as possibilities for exploiting the reaction in novel ways.[227–234]

The sequential steps of the cleavage of RNA by the *Tetrahymena* ribozyme, as determined by Cech's group, are shown in Figure 11(a). Prior to the chemical step, the free RNA substrate binds first to the ribozyme in two steps: the RNA substrate first anneals to the ribozyme's internal guide sequence (IGS) by complementary base-pairing to form a Michaelis–Menten complex. Then the helix formed between the ribozyme and substrate docks into the cleft that forms the active site[235–237] via multiple tertiary interactions with the backbone and the G·U pair at the cleavage site.[238–242] Once the nucleophilic G residue is bound to the intron and the ternary complex E·G·S is formed,[222,243–245] transesterification proceeds very rapidly with a high rate constant (~ 80 min^{-1} at 30 °C).[239,246,247] The E·P complex formed is more stable than the unreacted E·S complex because of destabilization by the substrate at the active site.[248] Release of the newly formed 5' fragment, which is retained on the ribozyme by base-pairing and tertiary interactions, is the rate-limiting step of the multiple-turnover reaction ($k_{cat} = k_{off} = 0.1$ min^{-1}).[246] Multiple turnover can be speeded up by increasing the rate of release of the product, for example, by the use of mismatched substrates or mutant substrates that have weaker tertiary interactions with the enzyme.

6.15.5.4 Chemistry and Roles of Metal Ions in Reactions Catalyzed by Group I Ribozymes

Self-splicing of group I introns proceeds by two consecutive transesterification reactions, as shown in Figure 11(a). The first of these reactions is initiated by guanosine, which attacks the phosphorus atom at the 5' splicing site and forms a new 3',5'-phosphodiester bond with the first nucleotide of the intron (Figure 11(b)). The 5' exon, now terminating in a free 3'-hydroxyl group, then attacks the phosphorus atom at the 3' splice site, with resultant ligation of the exons and excision of the intron. Both the G-addition reaction and the ligation reaction occur with inversion of the configuration at the phosphorus atom, and this inversion is consistent with an in-line, S_N2 mechanism.[214]

In contrast to proteinaceous enzymes, of which only a subset require a metal ion for catalytic function, all naturally occurring RNA enzymes require metal ions for stabilization of their structures and for catalytic action.[22] In the case of the *Tetrahymena* ribozyme, several divalent metal ions can serve in a structural capacity to promote appropriate folding, but only Mg^{2+} and Mn^{2+} ions can support splice-site cleavage and exon ligation to any significant extent.[249] A study of a ribozyme-catalyzed reaction analogous to the 5' splice-site cleavage was made with the 3'-oxygen of the leaving group replaced by sulfur. Evidence for participation of a metal ion as a Lewis acid, for stabilization of the leaving group, was first obtained with the *Tetrahymena* ribozyme by investigating Mn^{2+} rescue with a 3'-thio substrate.[19] The replacement of the 3'-leaving oxygen by sulfur reduced the activity of the *Tetrahymena* ribozyme in the presence of Mg^{2+} ions by more than 1000-fold. However, a change of metal ions from Mg^{2+} to Mn^{2+} ions reversed the effect, and the P—(3'-S) bond was cleaved in the presence of Mn^{2+} ions nearly as rapidly as the P—(3'-O) bond. These results indicate that a metal ion contributes directly to catalysis by coordination to the 3'-leaving oxygen in the transition state, stabilizing the developing negative charge on the leaving group (Figure 11(b)).[19] A more recent study of the reverse reaction of the first cleavage using a 3'-thio substrate demonstrated that thiophilic metal ions, such as Mn^{2+} or Cd^{2+} ions, were absolutely required for both the forward and the reverse reaction.[28] Therefore, the double-metal-ion mechanism[20] seems operative in this system too, with a metal ion activating the nucleophilic 3'-hydroxyl of guanosine and another metal ion neutralizing the developing negative charge on the 3'-leaving oxygen of the upstream exon by direct coordination of metal ions to both attacking and leaving oxygens (Figure 11(b)). In contrast to the first cleavage reaction at the 5' splice site, in the second ligation step, 3'-sulfur substitution at the 3' splice site provided no evidence for any interaction between a metal ion and the leaving group.[140] Such results suggest that the two steps of the splicing reaction proceed by different catalytic mechanisms.

6.15.6 GROUP II RIBOZYMES

6.15.6.1 General Features of Group II Ribozymes

The self-splicing of group II introns is required for the expression of essential genes in many organisms,[250] and the reaction may provide a simple model for catalysis by the eukaryotic spliceosome.[251] The secondary structure of a group II intron is typically divided into six domains (Figure 13), of which domain 1 and domain 5 are essential for catalysis.[250] Domain 5 is a short hairpin region of 34 nucleotides that contains most of the phylogenetically conserved nucleotides in the intron and is regarded as a central component of the active site of each group II intron.[252] Group II introns are spliced out from their primary transcripts by a two-step mechanism (Figure 14(a) and 14(b)).[136,253] In step 1, the 5′ splice site is attacked by the 2′-hydroxyl group of an adenosine residue near the 3′ end of the intron to generate a two-part intermediate that consists of the upstream exon and the intron-second-exon "lariat" RNA. In step 2, the 3′ end of the upstream exon attacks the 3′ splice site to produce the spliced-exon RNA and the released intron lariat.

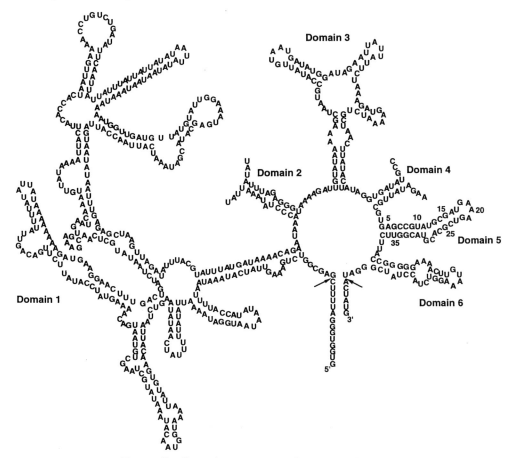

Figure 13 Secondary structure of a group II intron.

6.15.6.2 Tertiary Interactions in Group II Ribozymes

Domain 1 is strictly required for the catalytic activity of the group II ribozymes. Results of a stopped-flow fluorescence spectroscopic assay for RNA folding revealed that, in the presence of Mg^{2+} ions, the K_d for the substrate and domain 1 was very close to the K_d (k_{off}/k_{on}) calculated from an oligonucleotide-cleavage assay, indicating that domain 1 can fold independently and contains all the residues that are important for ground-state binding of the substrate and, moreover, that Mg^{2+} ions are required specifically for the formation of the distinct tertiary structure of group II introns.[254] Domain 2 was believed for a long time to be relatively unimportant for self-splicing. However, it was found recently that domain 2 can form tertiary interactions not only with domain

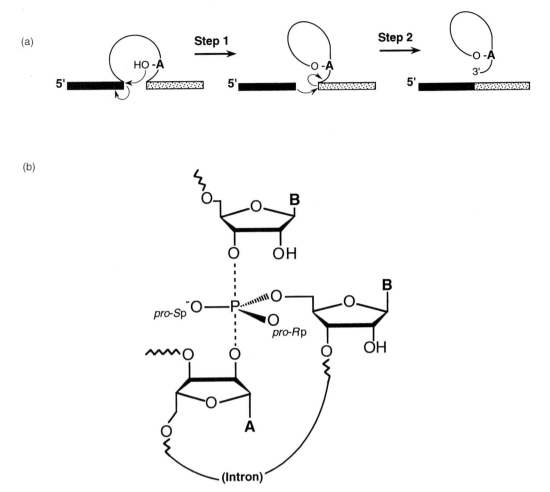

Figure 14 Reaction scheme (a) and a putative transition state (b) for the splicing reaction mediated by a group II intron.

1 but also with domain 6. These tertiary interactions not only stabilize the three-dimensional architecture of the RNA but are also important for dynamic interactions.[255]

Domain 5 is also an essential component of the active site of group II ribozymes. The role of the 2′-hydroxyl in the backbone was explored through the synthesis of a series of derivatives that contained deoxynucleotides at each position in domain 5. Kinetic screening showed that eight 2′-hydroxyl groups were critical for activity. The 2′-hydroxyl groups of nucleotides C_8, A_{20} and A_{21} appeared to play a critical role in molecular recognition because of the considerable increase in K_M when RNA was replaced by DNA, while the 2′-hydroxyl groups of nucleotides C_7, C_{26}, A_{27}, G_{29}, and G_{33} appeared to play a major role in catalysis since the K_M values after DNA substitution were almost the same as for the native RNA, with reductions in k_{cat} of about 10-fold.[256]

The 2′-hydroxyl group on the adenosine residue within the intron acts as the nucleophile during the first step of splicing by group II introns (Figure 14(a)). To understand how the ribozyme core recognizes this adenosine, the effects on catalytic activity of certain mutations were quantified. The results indicated that a low level of variability was tolerated at the branch-site without any apparent loss of fidelity. Moreover, analyses of mutants and modified nucleotides at the branch-site revealed that the recognition of the adenine moiety is different from the corresponding recognition during spliceosomal processing.[257]

6.15.7 RIBONUCLEASE P

6.15.7.1 General Features of the RNase P Ribozyme

The RNA subunit of ribonuclease P (the RNase P ribozyme) was one of the first two ribozymes to be discovered.[2,4] RNase P catalyzes the processing of the precursors to all kinds of tRNAs (pre-tRNAs). The enzyme cleaves pre-tRNAs, removing 5′ precursor-specific sequences and generating the mature 5′ ends of the tRNAs. The unique feature of the RNase P ribozyme is that it catalyzes intermolecular reactions in nature. Furthermore, RNase P has been found in all organisms in which it has been sought, such as eukaryotes, bacteria, and archaea bacteria. RNase P from all sources consists of an RNA subunit and a protein subunit. For example, RNase P from *Escherichia coli* consists of a large RNA component of about 400 nucleotides and a relatively small protein component of about 120 amino acids. In nature, the RNA subunit, the ribozyme, acts together with the protein component but the catalytic center is in the RNA subunit. The protein subunit of RNase P was initially thought to be responsible for the catalytic activity. However, the RNA subunit of bacterial RNase P was shown to be able to cleave pre-tRNA in the absence of the protein component at high ionic strength *in vitro*, and, thus, the RNA component was found to act as a ribozyme.[2,258] By contrast, none of the RNA subunits of archaea bacterial and eukaryotic RNase P has catalytic activity in the absence of the protein component. In all cases, the protein component is essential for catalytic activity *in vivo*. The substrate-binding domain is located in the RNA subunit.[259] RNase P recognizes its substrate via the helix formed by the coaxial stacking of the acceptor stem and the common (TΨC) arm and, thus, it can also cleave small RNAs that have structures similar to that of tRNAs (e.g., 4.5 S RNA of *E. coli*).[259–272] A putative model of the secondary structure of the RNase P ribozyme was deduced from phylogenetic comparative analysis and the results of cross-linking studies, and it is depicted in Figure 1. The minimum consensus sequence for catalysis by the RNase P ribozyme is shown in Figure 15.[5,273]

6.15.7.2 Roles of the Protein Component and Metal Ions

The role of the protein component of RNase P is still unclear but it is likely that it stabilizes the proper folding of the RNA. In the case of large RNA molecules, such as the RNase P ribozyme, negative charge repulsion of the phosphate backbone might prevent proper folding of the RNA. Thus, the positive charge of the basic protein might overcome the charge repulsion and stabilize the folded RNA.[2] For the reaction catalyzed by the RNase P ribozyme, both divalent cations and monovalent cations are required. Mg^{2+} or Mn^{2+} and K^+ or NH_4^+ are suitable as divalent cations and monovalent cations, respectively.[274–277] Monovalent cations are considered to be involved in stabilization of the structure, and high ionic strength allows catalytic activity in the absence of the protein component. By contrast, divalent cations are required for the chemical cleavage step rather than for structural purposes and binding of the substrate, although divalent metal ions can also stabilize the active conformation of the ribozyme–substrate complex.[17,275–280]

Unlike all other ribozymes, the RNase P ribozyme cleaves pre-tRNA via direct hydrolysis, rather than by transesterification, to generate a 5′-precursor-specific sequence with 3′-hydroxyl termini and a mature tRNA with a 5′-phosphate. Thus, a chimeric pre-tRNA substrate that contains a deoxyribonucleotide at the cleavage site is also cleavable.[17] However, details of the mechanism of the reaction are not fully understood. The cleavage rate increases linearly with increases in the concentration of hydroxide ions, indicating that a hydroxide ion rather than a neutral water molecule is the nucleophile.[17] Divalent cations, such as Mg^{2+} ions, have been proposed to activate a water molecule to form an attacking hydroxide ion in the chemical cleavage step. In this mechanism, it is essential that the attacking hydroxide be oriented apically to the 3′ oxygen so that it can attack the phosphate group through an in-line S_N2-like pathway. Mg^{2+} ions might be responsible for this orientation and proposed models for catalysis by RNase P are shown in Figure 16.[17,275,281] In an earlier model (Figure 16(a)), a catalytic Mg^{2+} ion coordinates directly to the *pro-S* oxygen at the scissile phosphate, stabilizing the transition state, and one of the metal-bound hydroxide moieties removes a proton from an outer-sphere water molecule to provide the attacking hydroxide ion at the apical position. Furthermore, the hydrogen-bonding interaction of the 2′-OH with the catalytically important metal-bound water molecule, which stabilizes the developing negative charge on the 3′-oxygen and/or acts as a proton donor, was shown to be important since substrates in which the 2′-OH adjacent to the scissile phosphate was removed or replaced by a 2′-*O*-methyl group were hydrolyzed at a very reduced rate.[17]

A modified model (Figure 16(b)) was proposed recently, based on Mn^{2+}- and Cd^{2+}-rescue experiments and the use of modified substrates with either an *R*p- or an *S*p-phosphorothioate at the cleavage site.[282,283] In this model, there is direct metal ion-coordination to the *pro-R* oxygen at the scissile phosphate and no interaction between a metal ion and the *pro-S* oxygen at the scissile phosphate. This model is consistent with the conventional double-metal-ion mechanism proposed

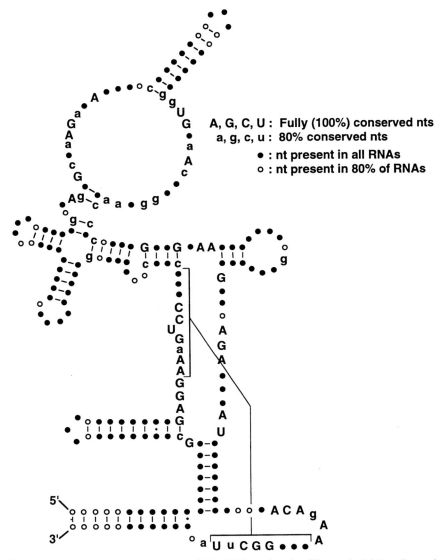

Figure 15 The minimum consensus sequence of the RNA subunit of bacteria RNase P, as deduced from phylogenetic comparisons.[5] Fully (100%) conserved nucleotides (nts) are indicated by uppercase letters and those that are at least 80% conserved but are not invariant are indicated by lowercase letters. Nucleotides that are not conserved in terms of specific identity but are present in all sequences are indicated by filled circles, while those that are absent in at least one case are indicated by open circles. The base pairs indicated by closed and open dots are, respectively, conserved non-canonical (G:G or A:C) interactions and a pairing that is frequently G:G rather than canonical.

by Steitz and Steitz.[20] In this model, two metal ions coordinate directly to the *pro-R* oxygen and one of them activates the attacking water molecule, while another metal ion stabilizes the developing negative charge of the leaving 3'-oxygen. It is possible that an indirect interaction occurs between the *pro-S* oxygen and metal-bound water molecules that are not shown in Figure 16(b).

The proposed functions of Mg^{2+} ions are qualitatively similar in both cases. The first is to activate a nucleophile (in this case, a water molecule), by acting as a base or by direct coordination to lower the pK_a of the attacking water molecule, in order to generate a more nucleophilic hydroxide group that can attack the phosphorus atom at the scissile phosphate in an S_N2-type reaction. The second function is to stabilize the transition state, with a Mg^{2+} ion acting as an electrophilic catalyst, by direct coordination of the metal ion to the non-bridging oxygen of the scissile phosphate. The third function is to stabilize the developing negative charge of the leaving 3'-oxygen either directly or via donation of a proton from the metal-bound water molecule. Ultimately, the RNase P ribozyme belongs to the class of metalloenzymes, even though it acts as a ribonucleoprotein in nature. The exact mechanism of the reaction remains to be determined.

(a)   (b)

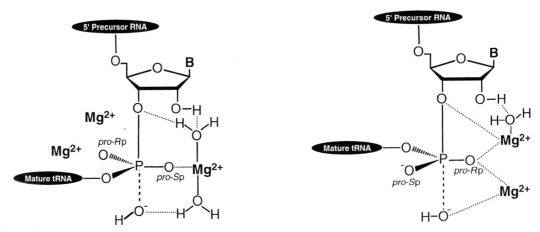

Figure 16 Proposed models of the transition state catalyzed by RNase P.[17,275,282] Two models are widely cited in the literature.

6.15.8 DNA ENZYMES

6.15.8.1 General Features of DNA Enzymes

Biological catalysis is dominated by protein enzymes, but it is clear that a variety of RNA molecules can also act as biological catalysts in certain cleavage and ligation reactions. DNA molecules have almost the same chemical composition as RNA molecules, differing in base composition, with thymidine in DNA versus uridine in RNA, and in sugar structure, with deoxyriboses in DNA versus riboses in RNA. The methyl group at the C-5 of thymidine has minimal impact on the structure of nucleic acids. Thus, a single-stranded DNA could, in principle, approximate the tertiary structure of a similar RNA.[284,285] Thus, it was postulated that DNA might also have enzymatic activity. A double helix of DNA molecules cannot create the structural and chemical environments necessary for catalysis of a wide range of reactions.[286] If the DNA could be configured in an appropriate single-stranded form, functional DNA molecules might be formed.

6.15.8.2 Artificial DNA Enzymes

Several attempts to generate DNA enzymes by *in vitro* selection have been successful and DNA molecules with enzymatic activity have been isolated, including a Pb^{2+}-dependent DNAzyme with RNA-cleavage activity by Joyce's group,[51] a DNAzyme with DNA ligase activity by Szostak's group,[287] and a DNAzyme with self-cleaving activity by Breaker's group.[288] A novel DNAzyme with activity similar to that of a hammerhead ribozyme was also found by Joyce's group by *in vitro* selection (Figure 17).[52] Compared with RNA molecules, DNA molecules have several beneficial features. For example, DNA is amenable to molecular design by rational and combinational methods; DNA can be synthesized with ease; and DNA and its analogues are more stable than their RNA counterparts. These intrinsic properties of DNA give it high potential value as a tool for therapeutic and industrial applications.

The DNAzyme, with activity similar to that of a hammerhead ribozyme is also a metalloenzyme. Almost all of the DNAzymes isolated to date by *in vitro* selection require metal cofactors, such as Pb^{2+}, Mg^{2+}, Zn^{2+}, Mn^{2+} or Ca^{2+} ions.[51,289,290] A study by *in vitro* selection suggests that a certain kind of phosphodiester-cleaving DNAzyme might act independently of divalent metal ions,[291] although the DNAzyme with such activity had a more complicated structure than that of the DNAzyme found by Joyce's group. The independence of metal ions exhibited by the DNAzyme demonstrates that nucleic acid components might play a substantial role, for example, in substrate positioning, stabilization of the transition state, or general acid/base catalysis. It is possible and even likely that artificial DNA enzymes will be engineered to operate inside cells as novel therapeutic agents.[292]

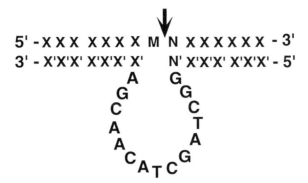

Figure 17 Secondary structure of DNA enzyme that was isolated by Joyce's group.[52] In the substrate, base X can form a base pair with base X′. M denotes nucleotide A or G, and N denotes nucleotide U or C. Base N′ can form a base pair with base N. Cleavage occurs at the site indicated by an arrow.

6.15.9 CONCLUSIONS

In this chapter, the structures and properties of various ribozymes have been described. Most, but not all, of these ribozymes are metalloenzymes. In terms of their mechanisms of action, hammerhead and *Tetrahymena* ribozymes are the ribozymes that have been most extensively studied. Despite differences in reaction mechanisms, the role of metal ions is practically identical in reactions catalyzed by both hammerhead and *Tetrahymena* ribozymes. In both cases, a double-metal-ion mechanism of catalysis has been suggested, with metal ions coordinating directly with the attacking and the leaving oxygens. The first metal ion, which coordinates directly to the nucleophilic oxygen, lowers the pK_a of the R—OH proton. The R—OH proton is then lost to the solvent. The resulting alkoxide serves as the attacking nucleophile. The second metal ion absorbs the developing negative charge on the leaving oxygen by coordinating directly to it. At one time, it was believed that metal ions play the same role in all reactions catalyzed by all kinds of ribozymes that are metalloenzymes. However, in the case of hairpin ribozymes, the role of metal ions appears to be strictly structural, and cleavage reactions can occur in the absence of catalytic metal ions and in the presence of EDTA and spermidine. Other ribozymes appear to require catalytic metal ions although the exact mechanism of each specific reaction remains to be determined. The mechanism of the metal-independent cleavage of RNA by hairpin ribozymes and by some DNAzymes will be of great interest. Elucidation of metal-dependent, as well as metal-independent, mechanisms of cleavage should help us to design better enzymes that might be useful as future therapeutic agents.

6.15.10 REFERENCES

1. T. R. Cech, A. J. Zaug, and P. J. Grabowski, *Cell*, 1981, **27**, 487.
2. C. Guerrier-Takada, K. Gardiner, T. Marsh, N. Pace, and S. Altman, *Cell*, 1983, **35**, 849.
3. R. F. Gesteland and J. F. Atkins (eds.), "The RNA World," Cold Spring Harbor Laboratory Press, New York, 1993.
4. S. Altman, *Adv. Enzymol.*, 1989, **62**, 1.
5. N. R. Pace and J. W. Brown, *J. Bacteriol.*, 1995, **177**, 1919.
6. T. R. Cech, *Angew. Chem., Int. Ed. Engl.*, 1990, **29**, 759.
7. J. Bratty, P. Chartrand, G. Ferbeyre, and R. Cedergren, *Biochim. Biophys. Acta*, 1993, **1216**, 345.
8. R. H. Symons, *Annu. Rev. Biochem.*, 1992, **61**, 641.
9. K. R. Birikh, P. A. Heaton, and F. Eckstein, *Eur. J. Biochem.*, 1997, **245**, 1.
10. J. M. Buzayan, W. L. Gerlach, and G. Bruening, *Nature*, 1986, **323**, 349.
11. L. Sharmeen, M. Y.-P. Kuo, G. Dinter-Gottlieb, and J. Taylor, *J. Virol.*, 1988, **62**, 2674.
12. H.-N. Wu, Y.-J. Lin, F.-P. Lin, S. Makino, M.-F. Chang, and M. M. C. Lai, *Proc. Natl. Acad. Sci. USA*, 1989, **86**, 1831.
13. O. C. Uhlenbeck, *Nature*, 1987, **328**, 596.
14. J. Haseloff and W. L. Gerlach, *Nature*, 1988, **334**, 585.
15. D. Herschlag and T. R. Cech, *Nature*, 1990, **344**, 405.
16. D. L. Robertson and G. F. Joyce, *Nature*, 1990, **344**, 467.
17. D. Smith and N. R. Pace, *Biochemistry*, 1993, **32**, 5273.
18. S. C. Dahm, W. B. Derrick, and O. C. Uhlenbeck, *Biochemistry*, 1993, **32**, 13 040.
19. J. A. Piccirilli, J. S. Vyle, M. H. Caruthers, and T. R. Cech, *Nature*, 1993, **361**, 85.
20. T. A. Steitz and J. A. Steitz, *Proc. Natl. Acad. Sci. USA*, 1993, **90**, 6498.
21. M. Yarus, *FASEB J.*, 1993, **7**, 31.
22. A. M. Pyle, *Science*, 1993, **261**, 709.
23. T. Uchimaru, M. Uebayasi, K. Tanabe, and K. Taira, *FASEB J.*, 1993, **7**, 137.

24. M. Uebayasi, T. Uchimaru, T. Koguma, S. Sawata, T. Shimayama, and K. Taira, *J. Org. Chem.*, 1994, **59**, 7414.
25. S. Sawata, M. Komiyama, and K. Taira, *J. Am. Chem. Soc.*, 1995, **117**, 2357.
26. D.-M. Zhou, N. Usman, F. E. Wincott, J. Matulic-Adamic, M. Orita, L.-H. Zhang, M. Komiyama, P. K. R. Kumar, and K. Taira, *J. Am. Chem. Soc.*, 1996, **118**, 5862.
27. P. K. R. Kumar, D. M. Zhou, K. Yoshinari, and K. Taira, *Nucleic Acids Mol. Biol.*, 1996, **10**, 217.
28. L. B. Weinstein, B. C. N. M. Jones, R. Cosstick, and T. R. Cech, *Nature*, 1997, **388**, 805.
29. B. W. Pontius, W. B. Lott, and P. H. von Hippel, *Proc. Natl. Acad. Sci. USA*, 1997, **94**, 2290.
30. D.-M. Zhou, L.-H. Zhang, and K. Taira, *Proc. Natl. Acad. Sci. USA*, 1997, **94**, 14343.
31. D.-M. Zhou and K. Taira, *Chem. Rev.*, 1998, **98**, 991.
32. B. W. Pontius and P. H. von Hippel, *Proc. Natl. Acad. Sci. USA*, 1998, **95**, 542.
33. S. Kuusela and H. Lönnberg, *Current Topics in Solution Chemistry*, 1997, **2**, 29.
34. A. Hampel and J. A. Cowan, *Chem. Biol.*, 1997, **4**, 513.
35. S. Nesbitt, L. A. Hegg, and M. J. Fedor, *Chem. Biol.*, 1997, **4**, 619.
36. K. J. Young, F. Gill, and J. A. Grasby, *Nucleic Acids Res.*, 1997, **25**, 3760.
37. J. B. Murray, A. A. Seyhan, N. G. Walter, J. M. Burke, and W. G. Scott, *Chem. Biol.*, 1998, **5**, 587.
38. N. Sarver, E. M. Cantin, P. S. Chang, J. A. Zaia, P. A. Ladne, D. A. Stephens, and J. J. Rossi, *Science*, 1990, **247**, 1222.
39. R. P. Erickson and J. Izant (eds.), "Gene Regulation: Biology of Antisense RNA and DNA," Raven Press, New York, 1992.
40. T. R. Cech, *Curr. Opin. Struct. Biol.*, 1992, **2**, 605.
41. S. Altman, *Proc. Natl. Acad. Sci. USA*, 1993, **90**, 10898.
42. J. Ohkawa, N. Yuyama, Y. Takebe, S. Nishikawa, and K. Taira, *Proc. Natl. Acad. Sci. USA*, 1993, **90**, 11302.
43. M. Yu, E. Poeschla, and F. Wong-Staal, *Gene Therapy*, 1994, **1**, 13.
44. J. T. Jones, S. W. Lee, and B. A. Sullenger, *Nature Med.*, 1996, **2**, 643.
45. H. Kawasaki, J. Ohkawa, N. Tanishige, K. Yoshinari, T. Murata, K. K. Yokoyama, and K. Taira, *Nucleic Acids Res.*, 1996, **24**, 3010.
46. F. Eckstein and D. M. J. Lilley, *Nucleic Acids Mol. Biol.*, 1996, **10**.
47. A. Wada, T. Shimayama, D.-M. Zhou, M. Warashina, M. Orita, T. Koguma, J. Ohkawa, and K. Taira, in "Controlled Drug Delivery, Challenges and Strategies," ed. K. Park, ACS Professional Reference Book, American Chemical Society, Washington, DC, 1997, p. 357.
48. J. Ohkawa, Y. Takebe, and K. Taira, *Methods Mol. Med.*, 1998, **11**, 83.
49. H. Kawasaki, J. Song, R. Eckner, H. Ugai, R. Chiu, K. Taira, Y. Shi, N. Jones, and K. K. Yokoyama, *Genes Dev.*, **12**, 233.
50. H. Kawasaki, J. Ohkawa, R. Eckner, T. P. Yao, K. Taira, R. Chiu, D. M. Livingston, and K. K. Yokoyama, *Nature*, 1998, **393**, 284.
51. R. R. Breaker and G. F. Joyce, *Chem. Biol.*, 1994, **1**, 223.
52. S. W. Santoro and G. F. Joyce, *Proc. Natl. Acad. Sci. USA*, 1997, **94**, 4262.
53. R. R. Breaker, *Chem. Rev.*, 1997, **97**, 371.
54. H. W. Pley, K. M. Flaherty, and D. B. McKay, *Nature*, 1994, **372**, 68.
55. W. G. Scott, J. T. Finch, and A. Klug, *Cell*, 1995, **81**, 991.
56. W. G. Scott, J. B. Murray, J. R. P. Arnold, B. L. Stoddard, and A. Klug, *Science*, 1996, **274**, 2065.
57. A. L. Feig, W. G. Scott, and O. C. Uhlenbeck, *Science*, 1998, **279**, 81.
58. S. V. Amontov and K. Taira, *J. Am. Chem. Soc.*, 1996, **118**, 1624.
59. T. Kuwabara, S. V. Amontov, M. Warashina, J. Ohkawa, and K. Taira, *Nucleic Acids Res.*, 1996, **24**, 2302.
60. S. V. Amontov, S. Nishikawa, and K. Taira, *FEBS Lett.*, 1996, **386**, 99.
61. T. Kuwabara, M. Warashina, T. Tanabe, K. Tani, S. Asano, and K. Taira, *Mol. Cell.*, 1998, **2**, 617.
62. T. Kuwabara, M. Warashina, M. Orita, S. Koseki, J. Ohkawa, and K. Taira, *Nature Biotechnology*, 1998, **16**, 961.
63. M. Koizumi, S. Iwai, and E. Ohtsuka, *FEBS Lett.*, 1988, **228**, 228.
64. C. C. Sheldon and R. H. Symons, *Nucleic Acids Res.*, 1989, **17**, 5679.
65. D. E. Ruffner, G. D. Stormo, and O. C. Uhlenbeck, *Biochemistry*, 1990, **29**, 10695.
66. R. Perriman, A. Delves, and W. L. Gerlach, *Gene*, 1992, **113**, 157.
67. M. Zoumadakis and M. Tabler, *Nucleic Acids Res.*, 1995, **23**, 1192.
68. T. Shimayama, S. Nishikawa, and K. Taira, *Biochemistry*, 1995, **34**, 3649.
69. D. M. Long and O. C. Uhlenbeck, *Proc. Natl. Acad. Sci. USA*, 1994, **91**, 6977.
70. K. L. Nakamaye and F. Eckstein, *Biochemistry*, 1994, **33**, 1271.
71. M. Ishizaka, Y. Ohshima, and T. Tani, *Biochem. Biophys. Res. Commun.*, 1995, **214**, 403.
72. J. B. Thomson, S. T. Sigurdsson, A. Zeuch, and F. Eckstein, *Nucleic Acids Res.*, 1996, **24**, 4401.
73. S. Fujita, T. Koguma, J. Ohkawa, K. Mori, T. Kohda, H. Kise, S. Nishikawa, M. Iwakura, and K. Taira, *Proc. Natl. Sci. Acad. USA*, 1997, **94**, 391.
74. N. K. Vaish, P. A. Heaton, and F. Eckstein, *Biochemistry*, 1997, **36**, 6495.
75. N. K. Vaish, P. A. Heaton, and F. Eckstein, *Proc. Natl. Acad. Sci. USA*, 1998, **95**, 2158.
76. A. B. Burgin, C. Gonzalez, J. Matulic-Adamic, A. M. Karpeisky, N. Usman, J. A. McSwiggen, and L. Beigelman, *Biochemistry*, 1996, **35**, 14090.
77. T. Tuschl and F. Eckstein, *Proc. Natl. Acad. Sci. USA*, 1993, **90**, 6991.
78. M. J. McCall, P. Hendry, and P. A. Jennings, *Proc. Natl. Acad. Sci. USA*, 1992, **89**, 5710.
79. D. J. Fu and L. W. Mclaughlin, *Proc. Natl. Acad. Sci. USA*, 1992, **89**, 3985.
80. F. Benseler, D. J. Fu, J. Ludwig, and L. W. Mclaughlin, *J. Am. Chem. Soc.*, 1993, **115**, 8483.
81. J. B. Thomson, T. Tuschl, and F. Eckstein, *Nucleic Acids Res.*, 1993, **21**, 5600.
82. P. Hendry, M. J. Moghaddam, M. J. McCall, P. A. Jennings, S. Ebel, and T. Brown, *Biochim. Biophys. Acta*, 1994, **1219**, 405.
83. P. Hendry, M. J. McCall, F. S. Santiago, and P. A. Jennings, *Nucleic Acid Res.*, 1995, **23**, 3922.
84. T. Kuwabara, M. Warashina, A. Nakayama, J. Ohkawa, and K. Taira, *Proc. Natl. Acad. Sci. USA*, 1999, in press.
85. G. S. Bassi, N. E. Möllegaard, A. I. H. Murchie, E. von Kitzing, and D. M. J. Lilley, *Nature Struct. Biol.*, 1995, **2**, 45.

86. G. S. Bassi, A. I. H. Murchie, and D. M. J. Lilley, *RNA*, 1996, **2**, 756.

87. A. C. Pease and D. E. Wemmer, *Biochemistry*, 1990, **29**, 9039.

88. H. A. Heus, O. C. Uhlenbeck, and A. Pardi, *Nucleic Acids Res.*, 1990, **18**, 1103.

89. O. Odai, H. Kodama, H. Hiroaki, T. Sakata, T. Tanaka, and S. Uesugi, *Nucleic Acids Res.*, 1990, **18**, 5955.

90. H. A. Heus and A. Pardi, *J. Mol. Biol.*, 1991, **217**, 113.

91. R. H. Sarma, M. H. Sarma, R. Rein, M. Shibata, R. S. Setlik, R. L. Ornstein, A. L. Kazim, A. Cairo, and T. B. Tomasi, *FEBS Lett.*, 1995, **357**, 317.

92. M. Orita, R. Vinayak, A. Andrus, M. Warashina, A. Chiba, H. Kaniwa, F. Nishikawa, S. Nishikawa, and K. Taira, *J. Biol. Chem.*, 1996, **271**, 9447.

93. J. P. Simorre, P. Legault, A. B. Hangar, P. Michiels, and A. Pardi, *Biochemistry*, 1997, **36**, 518.

94. R. P. Ojha, M. M. Dhingra, M. H. Sarma, Y. P. Myer, R. F. Setlik, M. Shibata, A. L. Kazim, R. L. Ornstein, R. Rein, C. J. Turner, and R. H. Sarma, *J. Biomol. Struct. Dyn.*, 1997, **15**, 185.

95. K. M. A. Amiri and P. J. Hagerman, *Biochemistry*, 1994, **33**, 13 172.

96. K. M. A. Amiri and P. J. Hagerman, *J. Mol. Biol.*, 1996, **261**, 125.

97. T. Tuschl, C. Gohlke, T. M. Jovin, E. Westhof, and F. Eckstein, *Science*, 1994, **266**, 785.

98. A. Peracchi, L. Beigelman, E. C. Scott, O. C. Uhlenbeck, and D. Herschlag, *J. Biol. Chem.*, 1997, **272**, 26 822.

99. J. B. Murray, D. P. Terwey, L. Maloney, A. Karpeisky, N. Usman, L. Beigelman, and W. G. Scott, *Cell*, 1998, **92**, 665.

100. A. Favre and J. L. Fourrey, *Acc. Chem. Res.*, 1995, **28**, 375.

101. S. T. Sigurdsson, T. Tuschl, and F. Eckstein, *RNA*, 1995, **1**, 575.

102. T. R. Cech and O. C. Uhlenbeck, *Nature*, 1994, **372**, 39.

103. Y. Takagi and K. Taira, *FEBS Lett.*, 1995, **361**, 273.

104. M. Warashina, Y. Takagi, S. Sawata, D.-M. Zhou, T. Kuwabara, and K. Taira, *J. Org. Chem.*, 1997, **62**, 9138.

105. J. M. Buzayan, A. Hampel and G. Bruening, *Nucleic Acids Res.*, 1986, **14**, 9729.

106. C. J. Hutchins, P. D. Rathjen, A. C. Forster, and R. H. Symons, *Nucleic Acids Res.*, 1986, **14**, 3627.

107. M. J. Fedor and O. C. Uhlenbeck, *Biochemistry*, 1992, **31**, 12 042.

108. K. J. Hertel, D. Herschlag, and O. C. Uhlenbeck, *Biochemistry*, 1994, **33**, 3374.

109. M. Werner and O. C. Uhlenbeck, *Nucleic Acids Res.*, 1995, **23**, 2092.

110. E. Bertrand, R. Pictet, and T. Grange, *Nucleic Acids Res.*, 1994, **22**, 293.

111. M. Homann, W. Nedbal, and G. Sczakiel, *Nucleic Acids Res.*, 1996, **24**, 4395.

112. R. Hormes, M. Homann, I. Oelze, P. Marschall, M. Tabler, F. Eckstein, and G. Sczakiel, *Nucleic Acids Res.*, 1997, **25**, 769.

113. J. Ohkawa, T. Koguma, T. Kohda, and K. Taira, *J. Biochem.*, 1995, **118**, 251.

114. K. J. Hertel, A. Peracchi, O. C. Uhlenbeck, and D. Herschlag, *Proc. Natl. Acad. Sci. USA*, 1997, **94**, 8497.

115. P. Hendry, M. J. McCall, F. S. Santiago, and P. A. Jennings, *Nucleic Acids Res.*, 1992, **20**, 5737.

116. S. Sawata, T. Shimayama, M. Komiyama, P. K. R. Kumar, S. Nishikawa, and K. Taira, *Nucleic Acids Res.*, 1993, **21**, 5656.

117. T. Shimayama, F. Nishikawa, S. Nishikawa, and K. Taira, *Nucleic Acids Res.*, 1993, **21**, 2605.

118. T. Shimayama, S. Nishikawa, and K. Taira, *FEBS Lett.*, 1995, **368**, 304.

119. R. G. Kuimelis and L. W. McLaughlin, *J. Am. Chem. Soc.*, 1995, **117**, 11 019.

120. R. G. Kuimelis and L. W. McLaughlin, *Biochemistry*, 1996, **35**, 5308.

121. T. B. Bruice, T. H. Fife, J. J. Bruno, and N. E. Brandon, *Biochemistry*, 1962, **1**, 7.

122. Z. Shaked, R. P. Szajewski, and G. M. Whitesides, *Biochemistry*, 1980, **19**, 4156.

123. W. P. Jencks, "Catalysis in Chemistry and Enzymology," McGraw-Hill, New York, 1969, p. 250.

124. J. K. Bashkin, and L. A. Jenkins, *Comments Inorg. Chem.*, 1994, **16**, 77.

125. K. Taira, M. Uebayasi, H. Maeda, and K. Furukawa, *Protein Eng.*, 1990, **3**, 691.

126. K. Taira, T. Uchimaru, J. W. Storer, A. Yliniemela, M. Uebayasi, and K. Tanabe, *J. Org. Chem.*, 1993, **58**, 3009.

127. S. C. Dahm and O. C. Uhlenbeck, *Biochemistry*, 1991, **30**, 9464.

128. T. Pan, D. M. Long, and O. C. Uhlenbeck, in "The RNA World," eds. R. F. Gesteland and J. F. Atkins, Cold Spring Harbor Laboratory Press, New York, 1993, p. 271.

129. J. A. Piccirilli, personal communication.

130. J. A. McSwiggen and T. R. Cech, *Science*, 1989, **244**, 679.

131. H. van Tol, J. M. Buzayan, P. A. Feldstein, F. Eckstein, and G. Bruening, *Nucleic Acids Res.*, 1990, **18**, 1971.

132. M. Koizumi and E. Ohtsuka, *Biochemistry*, 1991, **30**, 5145.

133. G. Slim and M. J. Gait, *Nucleic Acids Res.*, 1991, **19**, 1183.

134. D. Herschlag, J. A. Piccirilli, and T. R. Cech, *Biochemistry*, 1991, **30**, 4844.

135. R. A. Padgett, M. Podar, S. C. Boulanger, and P. S. Perlman, *Science*, 1994, **266**, 1685.

136. M. Podar, P. S. Perlman, and R. A. Padgett, *Mol. Cell Biol.*, 1995, **15**, 4466.

137. W. J. Michels and A. M. Pyle, *Biochemistry*, 1995, **34**, 2965.

138. D.-M. Zhou, P. K. R. Kumar, L.-H. Zhang, and K. Taira, *J. Am. Chem. Soc.*, 1996, **118**, 8969.

139. H. Fauzi, J. Kawakami, F. Nishikawa, and S. Nishikawa, *Nucleic Acids Res.*, 1997, **25**, 3124.

140. E. J. Sontheimer, S. G. Sun, and J. A. Piccirilli, *Nature*, 1997, **388**, 801.

141. S. J. Benkovic and K. J. Schray, in "The Enzymes," 3rd edn, ed. P. D. Boyer, Academic Press, New York, 1973, vol. 8, p. 201.

142. E. K. Jaffe and M. J. Cohn, *J. Biol. Chem.*, 1979, **254**, 10 839.

143. V. L. Pecoraro, J. D. Hermes, and W. W. Cleland, *Biochemistry*, 1984, **23**, 5262.

144. G. A. Prody, J. T. Bakos, J. M. Buzayan, I. R. Schneider, and G. Bruening, *Science*, 1986, **231**, 1577.

145. A. Hampel and R. Tritz, *Biochemistry*, 1989, **28**, 4929.

146. J. Haseloff and W. L. Gerlach, *Gene*, 1989, **82**, 43.

147. P. A. Feldstein, J. M. Buzayan, and G. Bruening, *Gene*, 1989, **82**, 53.

148. L. Rubino, M. E. Tousignant, G. Steger, and J. M. Kaper, *J. Gen. Virol.*, 1990, **71**, 1897.

149. A. Hampel, R. Tritz, M. Hicks, and P. Cruz, *Nucleic Acid Res.*, 1990, **18**, 299.

150. A. Berzal-Herranz, S. Joseph, and J. M. Burke, *Genes Dev.*, 1992, **6**, 129.

151. S. Joseph, A. Berzal-Herranz, B. M. Chowrira, S. E. Butcher, and J. M. Burke, *Genes Dev.*, 1993, **7**, 130.

152. A. Berzal-Herranz, S. Joseph, B. M. Chowrira, S. E. Butcher, and J. M. Burke, *The EMBO J.*, 1993, **12**, 2567.
153. P. Anderson, J. Monforte, R. Tritz, S. Nesbitt, J. Hearst, and A. Hampel, *Nucleic Acids Res.*, 1994, **22**, 1096.
154. M. B. De Young, A. M. Siwkowski, Y. Lian, and A. Hampel, *Biochemistry*, 1995, **34**, 15 785.
155. B. M. Chowrira and J. M. Burke, *Nucleic Acids Res.*, 1992, **20**, 2835.
156. B. M. Chowrira, A. Berzal-Herranz, and J. M. Burke, *Nature*, 1991, **354**, 320.
157. J. A. Grasby, K. Mersmann, M. Singh, and M. J. Gait, *Biochemistry*, 1995, **34**, 4068.
158. B. M. Chowrira, A. Berzal-Herranz, and J. M. Burke, *Biochemistry*, 1993, **32**, 1088.
159. S. Schmidt, L. Beigelman, A. Karpeisky, N. Usman, U. S. Sørensen, and M. J. Gait, *Nucleic Acids Res.*, 1996, **24**, 573.
160. B. M. Chowrira, A. Berzal-Herranz, C. F. Keller, and J. M. Burke, *J. Biol. Chem.*, 1993, **268**, 19 458.
161. B. M. Chowrira and J. M. Burke, *Biochemistry*, 1991, **30**, 8518.
162. P. A. Feldstein and G. Bruening, *Nucleic Acids Res.*, 1993, **21**, 1991.
163. Y. Komatsu, M. Koizumi, H. Nakamura, and E. Ohtsuka, *J. Am. Chem. Soc.*, 1994, **116**, 3692.
164. Y. Komatsu, I. Kanzaki, M. Koizumi, and E. Ohtsuka, *J. Mol. Biol.*, 1995, **252**, 296.
165. S. E. Butcher, J. E. Heckman, and J. M. Burke, *J. Biol. Chem.*, 1995, **270**, 29 648.
166. C. Shin, J. N. Choi, S. I. Son, J. T. Song, J. H. Ahn, J. S. Lee, and Y. D. Choi, *Nucleic Acids Res.*, 1996, **24**, 2685.
167. A. E. Walter, D. H. Turner, J. Kim, M. H. Lyttle, P. Müller, D. H. Mathews, and M. Zuker, *Proc. Natl. Acad. Sci. USA*, 1994, **91**, 9218.
168. A. E. Walter and D. H. Turner, *Biochemistry*, 1994, **33**, 12 715.
169. A. Siwkowski, R. Shippy, and A. Hampel, *Biochemistry*, 1997, **36**, 3930.
170. L. A. Hegg and M. J. Fedor, *Biochemistry*, 1995, **34**, 15 813.
171. S. E. Butcher and J. M. Burke, *J. Mol. Biol.*, 1994, **244**, 52.
172. Z. P. Cai and I. Tinoco, Jr., *Biochemistry*, 1996, **35**, 6026.
173. J. A. Esteban, A. R. Benerjee, and J. M. Burke, *J. Biol. Chem.*, 1997, **272**, 13 629.
174. M. Rizzetto, *Hepatology*, 1983, **3**, 729.
175. M. M. C. Lai, *Annu. Rev. Biochem.*, 1995, **64**, 259.
176. T. B. Macnaughton, Y.-J. Wang, and M. M. C. Lai, *J. Virol.*, 1993, **67**, 2228.
177. K.-S. Jeng, P.-Y. Su, and M. M. C. Lai, *J. Virol.*, 1996, **70**, 4205.
178. M. Y.-P. Kuo, J. Goldberg, L. Coates, W. Mason, J. Gerin, and J. Taylor, *J. Virol.*, 1988, **62**, 1855.
179. M. Y.-P. Kuo, L. Sharmeen, G. Dinter-Gottleib, and J. Taylor, *J. Virol.*, 1988, **62**, 4439.
180. S. P. Rosenstein and M. D. Been, *Nucleic Acids Res.*, 1991, **19**, 5409.
181. H.-N. Wu, Y.-J. Wang, C.-F. Hung, H.-J. Lee, and M. M. C. Lai, *J. Mol. Biol.*, 1992, **223**, 233.
182. A. D. Branch and H. D. Robertson, *Proc. Natl. Acad. Sci. USA*, 1991, **88**, 10 163.
183. A. T. Perrotta and M. D. Been, *Nature*, 1991, **350**, 434.
184. A. T. Perrotta, *Nucleic Acids Res.*, 1991, **19**, 1979.
185. M. D. Been, A. T. Perrotta, and S. P. Rosenstein, *Biochemistry*, 1992, **31**, 11 843.
186. J. B. Smith, P. A. Gottlieb, and G. Dinter-Gottlieb, *Biochemistry*, 1992, **31**, 9629.
187. H.-N. Wu and Z.-S. Huang, *Nucleic Acids Res.*, 1992, **20**, 5937.
188. A. T. Perrotta and M. D. Been, *Nucleic Acids Res.*, 1993, **21**, 3959.
189. G. Thill, M. Vasseur, and N. K. Tanner, *Biochemistry*, 1993, **32**, 4254.
190. H.-N. Wu, J.-Y. Lee, H.-W. Huang, Y.-S. Huang, and T.-G. Hsueh, *Nucleic Acids Res.*, 1993, **21**, 4193.
191. N. K. Tanner, S. Schaff, G. Thill, E. Petit-Koskas, A.-M. Crain-Denoyelle, and E. Westhof, *Curr. Biol.*, 1994, **4**, 488.
192. Y.-A. Suh, P. K. R. Kumar, F. Nishikawa, E. Kayano, S. Nakai, O. Odai, S. Uesugi, K. Taira, and S. Nishikawa, *Nucleic Acids Res.*, 1992, **20**, 747.
193. P. K. R. Kumar, Y.-A. Suh, H. Miyashiro, F. Nishikawa, J. Kawakami, K. Taira, and S. Nishikawa, *Nucleic Acids Res.*, 1992, **20**, 3919.
194. P. K. R. Kumar, Y.-A. Suh, K. Taira, and S. Nishikawa, *FASEB J.*, 1993, **7**, 124.
195. Y.-A. Suh, P. K. R. Kumar, J. Kawakami, F. Nishikawa, K. Taira, and S. Nishikawa, *FEBS Lett.*, 1993, **326**, 158.
196. J. Kawakami, P. K. R. Kumar, Y.-A. Suh, F. Nishikawa, K. Kawakami, K. Taira, E. Ohtsuka and S. Nishikawa, *Eur. J. Biochem.*, 1993, **217**, 29.
197. P. K. R. Kumar, K. Taira, and S. Nishikawa, *Biochemistry*, 1994, **33**, 583.
198. Y.-H. Jeoung, P. K. R. Kumar, Y.-A. Suh, H. K. Taira, and S. Nishikawa, *Nucleic Acids Res.*, 1994, **22**, 3722.
199. C. Bravo, F. Lescure, P. Laugâa, J.-L. Fourrey, and A. Favre, *Nucleic Acids Res.*, 1996, **24**, 1351.
200. A. T. Perrotta and M. D. Been, *Biochemistry*, 1992, **31**, 16.
201. M. Puttaraju, A. T. Perrotta and M. D. Been, *Nucleic Acids Res.*, 1993, **21**, 4253.
202. Y. C. Lai, J.-Y. Lee, H.-J. Liu, J.-Y. Lin, and H.-N. Wu, *Biochemistry*, 1996, **35**, 124.
203. F. Nishikawa, J. Kawakami, A. Chiba, M. Shirai, P. K. R. Kumar, and S. Nishikawa, *Eur. J. Biochem.*, 1996, **237**, 712.
204. F. Nishikawa, H. Fauzi, and S. Nishikawa, *Nucleic Acids Res.*, 1997, **25**, 1605.
205. Y.-A. Suh, P. K. R. Kumar, K. Taira, and S. Nishikawa, *Nucleic Acids Res.*, 1993, **21**, 3277.
206. H.-N. Wu and M. M. C. Lai, *Science*, 1989, **243**, 652.
207. H.-N. Wu and M. M. C. Lai, *Mol. Cell. Biol,.,*, 1990, **10**, 5575.
208. J. Rogers, A. H. Chang, U. von Ahsen, R. Schroeder, and J. Davies, *J. Mol. Biol.*, 1996, **259**, 916.
209. T. Sakamoto, Y. Tanaka, T. Kuwabara, M.-H. Kim, Y. Kurihara, M. Katahira, and S. Uesugi, *J. Biochem.*, 1997, **121**, 1123.
210. S. P. Rosenstein and M. D. Been, *Biochemistry*, 1990, **29**, 8011.
211. A. T. Perrotta and M. D. Been, *Nucleic Acids Res.*, 1990, **18**, 6821.
212. J. B. Smith and G. Dinter-Gottlieb, *Nucleic Acids Res.*, 1991, **19**, 1285.
213. M. D. Been and G. S. Wickham, *Eur. J. Biochem.*, 1997, **247**, 741.
214. T. R. Cech, *Annu. Rev. Biochem.*, 1990, **59**, 543.
215. G. Garriga and A. M. Lambowitz, *Cell*, 1984, **39**, 631.
216. G. van der Horst and H. F. Tabak, *Cell*, 1985, **40**, 759.
217. J. M. Gott, D. A. Shub, and M. Belfort, *Cell*, 1986, **47**, 81.
218. M. Q. Xu, S. D. Kathe, H. Goodrich-Blair, S. A. Nierzwicki-Bauer, and D. A. Shub, *Science*, 1990, **250**, 1566.

219. J. Rajagopal, J. A. Doudna, and J. W. Szostak, *Science*, 1989, **244**, 692.
220. F. Michel, A. Jacquier, and B. Dujon, *Biochimie*, 1982, **64**, 867.
221. R. W. Davies, R. B. Waring, J. A. Ray, T. A. Brown, and C. Scazzocchio, *Nature*, 1982, **300**, 719.
222. F. Michel, M. Hanna, R. Green, D. P. Bartel, and J. W. Szostak, *Nature*, 1989, **342**, 391.
223. T. R. Cech, *Science*, 1987, **236**, 1532.
224. J. H. Cate, A. R. Gooding, E. Podell, K. H. Zhou, B. L. Golden, C. E. Kundrot, T. R. Cech, and J. A. Doudna, *Science*, 1996, **273**, 1678.
225. B. L. Golden, E. Podell, A. R. Gooding, and T. R. Cech, *J. Mol. Biol.*, 1997, **270**, 711.
226. M. A. Tanner and T. R. Cech, *Science*, 1997, **275**, 847.
227. K. Kruger, P. J. Grabowski, A. J. Zaug, J. Sands, D. E. Gottschling, and T. R. Cech, *Cell*, 1982, **31**, 147.
228. A. J. Zaug and T. R. Cech, *Science*, 1986, **231**, 470.
229. A. J. Zaug, M. D. Been, and T. R. Cech, *Nature*, 1986, **324**, 429.
230. A. J. Zaug, C. A. Grosshans, and T. R. Cech, *Biochemistry*, 1988, **27**, 8924.
231. P. S. Kay and T. Inoue, *Nature*, 1987, **327**, 343.
232. J. A. Doudna and J. W. Szostak, *Nature*, 1989, **339**, 519.
233. A. A. Beaudry and G. F. Joyce, *Science*, 1992, **257**, 635.
234. R. Green and J. W. Szostak, *Science*, 1992, **258**, 1910.
235. P. C. Bevilacqua, R. Kierzek, K. A. Johnson, and D. H. Turner, *Science*, 1992, **258**, 1355.
236. D. Herschlag, *Biochemistry*, 1992, **31**, 1386.
237. T. B. Campbell and T. R. Cech, *Biochemistry*, 1996, **35**, 11 493.
238. P. C. Bevilacqua, K. A. Johnson, and D. H. Turner, *Proc. Natl. Acad. Sci. USA*, 1993, **90**, 8357.
239. T. S. McConnell, T. R. Cech, and D. Herschlag, *Proc. Natl. Acad. Sci. USA*, 1993, **90**, 8362.
240. A. M. Pyle, S. Moran, S. A. Strobel, T. Chapman, D. H. Turner, and T. R. Cech, *Biochemistry*, 1994, **33**, 13 856.
241. D. S. Knitt, G. J. Narlikar, and D. Herschlag, *Biochemistry*, 1994, **33**, 13 864.
242. S. A. Strobel and T. R. Cech, *Science*, 1995, **267**, 675.
243. B. L. Bass and T. R. Cech, *Nature*, 1984, **308**, 820.
244. M. Yarus, M. Illangesekare, and E. Christian, *J. Mol. Biol.*, 1991, **222**, 995.
245. D. Herschlag and T. R. Cech, *Biochemistry*, 1990, **29**, 10 172.
246. D. Herschlag and T. R. Cech, *Biochemistry*, 1990, **29**, 10 159.
247. D. Herschlag and M. Khosla, *Biochemistry*, 1994, **33**, 5291.
248. G. J. Narlikar, V. Gopalkrishan, T. S. McConnell, N. Usman, and D. Herschlag, *Proc. Natl. Acad. Sci. USA*, 1995, **92**, 3668.
249. D. W. Celander and T. R. Cech, *Science*, 1991, **251**, 401.
250. F. Michel and J. L. Ferat, *Annu. Rev. Biochem.*, 1995, **64**, 435.
251. H. D. Madhani and C. Guthrie, *Cell*, 1992, **71**, 803.
252. K. A. Jarrell, R. C. Dietrich, and P. S. Perman, *Mol. Cell Biol.*, 1988, **8**, 2361.
253. C. Guthrie, *Science*, 1991, **253**, 157.
254. P. Z. F. Qin and A. M. Pyle, *Biochemistry*, 1997, **36**, 4718.
255. M. Costa, E. Deme, A. Jacquier, and F. Michel, *J. Mol. Biol.*, 1997, **267**, 520.
256. D. L. Abramovitz, R. A. Friedman, and A. M. Pyle, *Science*, 1996, **271**, 1410.
257. Q. L. Liu, J. B. Green, A. Khodadadi, P. Haeberli, L. Beigelman, and A. M. Pyle, *J. Mol. Biol.*, 1997, **267**, 163.
258. C. Guerrier-Takada and S. Altman, *Science*, 1984, **223**, 285.
259. C. Reich, G. J. Olsen, B. Pace, and N. R. Pace, *Science*, 1988, **239**, 178.
260. W. H. McClain, C. Guerrier-Takada, and S. Altman, *Science*, 1987, **238**, 527.
261. A. K. Knap, D. Wesolowski, and S. Altman, *Biochimie*, 1990, **72**, 779.
262. D. Kahle, U. Wehmeyer, and G. Krupp, *EMBO J.*, 1990, **9**, 1929.
263. A. B. Burgin and N. R. Pace, *EMBO J.*, 1990, **9**, 4111.
264. D. L. Thurlow, D. Shilowski, and T. L. Marsh, *Nucleic Acids Res.*, 1991, **19**, 885.
265. K. A. Peck-Miller and S. Altman, *J. Mol. Biol.*, 1991, **221**, 1.
266. J. M. Nolan, D. H. Burke, and N. R. Pace, *Science*, 1993, **261**, 762.
267. W. D. Hardt, J. Schlegl, V. A. Erdmann, and R. K. Hartmann, *Biochemistry*, 1993, **32**, 13 046.
268. F. Y. Liu and S. Altman, *Cell*, 1994, **77**, 1093.
269. T. E. LaGrandeur, A. Hüttenhofer, H. F. Noller, and N. R. Pace, *EMBO J.*, 1994, **13**, 3945.
270. B.-K. Oh and N. R. Pace, *Nucleic Acids Res.*, 1994, **22**, 4087.
271. Y. Komine, M. Kitabatake, T. Yokogawa, T. Nishikawa, and H. Inokuchi, *Proc. Natl. Acad. Sci. USA*, 1994, **91**, 9223.
272. T. Pan, *Biochemistry*, 1995, **34**, 8458.
273. J. M. Nolan and N. R. Pace, *Nucleic Acids Mol. Biol.*, 1996, **10**, 109.
274. K. J. Gardiner, T. L. Marsh, and N. R. Pace, *J. Biol. Chem.*, 1985, **260**, 5415.
275. C. Guerrier-Takada, K. Haydock, L. Allen, and S. Altman, *Biochemistry*, 1986, **25**, 1509.
276. C. K. Surratt, B. J. Carter, R. C. Payne, and S. M. Hecht, *J. Biol. Chem.*, 1990, **265**, 22 513.
277. S. Kazakov and S. Altman, *Proc. Natl. Acad. Sci. USA*, 1991, **88**, 9193.
278. D. Smith, A. B. Burgin, E. S. Haas, and N. R. Pace, *J. Biol. Chem.*, 1992, **267**, 2429.
279. J. P. Perreault and S. Altman, *J. Mol. Biol.*, 1993, **230**, 750.
280. J. A. Beebe, J. C. Kurz, and C. A. Fierke, *Biochemistry*, 1996, **35**, 10 493.
281. K. Haydock and L. C. Allen, *Prog. Clin. Biol. Res.*, 1985, **172A**, 87.
282. J. M. Warnecke, J. P. Fürste, W.-D. Hardt, V. A. Erdmann, and R. K. Hartmann, *Proc. Natl. Acad. Sci. USA*, 1996, **93**, 8924.
283. Y. Chen, X. Li, and P. Gegenheimer, *Biochemistry*, 1997, **36**, 2425.
284. A. S. Khan and B. A. Roe, *Science*, 1988, **241**, 74.
285. J. P. Perreault, R. T. Pon, M. Y. Jiang, N. Usman, J. Pika, K. K. Ogilvie, and R. Cedergren, *Eur. J. Biochem.*, 1989, **186**, 87.
286. R. R. Breaker, *Nature Bio.*, 1997, **15**, 427.
287. B. Cuenoud and J. W. Szostak, *Nature*, 1995, **375**, 611.

288. N. Carmi, L. A. Shulz, and R. R. Breaker, *Chem. Biol.*, 1996, **3**, 1039.
289. R. R. Breaker and G. F. Joyce, *Chem. Biol.*, 1995, **2**, 655.
290. D. Faulhammer and M. Famulok, *Angew. Chem.*, *Int. Ed. Engl.*, 1997, **35**, 2837.
291. C. R. Geyer and D. Sen, *Chem. Biol.*, 1997, **4**, 579.
292. M. Warashina, T. Kuwabara, Y. Nakamatsu, and K. Taira, *Chem. Biol.*, 1999, in press.

Author Index

This Author Index comprises an alphabetical listing of the names of the authors cited in the text and the references listed at the end of each chapter in this volume.

Each entry consists of the author's name, followed by a list of numbers, for example

Templeton, J. L., 366, 385^{233} (350, 366), 387^{370} (363)

For each name, the page numbers for the citation in the reference list are given, followed by the reference number in superscript and the page number(s) in parentheses of where that reference is cited in the text. Where a name is referred to in text only, the page number of the citation appears with no superscript number. References cited in both the text and in the tables are included.

Although much effort has gone into eliminating inaccuracies resulting from the use of different combinations of initials by the same author, the use by some journals of only one initial, and different spellings of the same name as a result of the transliteration processes, the accuracy of some entries may have been affected by these factors.

(225), 264[76] (237), 267[263] (260), 267[278] (260)

Burgstaller, P., 20[65] (18), 78[43] (66), 147[65] (136)

Burkard, M. E., 37, 45[75] (25, 28, 29, 30, 31, 41, 42), 46[120] (31, 41, 43), 47[148] (37)

Burke, D. H., 267[266] (260)

Burke, J. M., 47[184] (43), 118, 120, 147[13] (118), 147[17] (119, 120), 148[86] (141), 166[51] (154), 232[104] (226), 232[123] (227), 249, 264[37] (237), 265[150] (248, 249), 265[151] (248), 266[152] (248, 249, 250), 266[155] (248, 249, 250, 252)

Burlina, F., 232[119] (227)

Burns, C. M., 108[52] (103, 105)

Burrows, C. J., 78[41] (66)

Butcher, S. E., 19[53] (16), 45[30] (22, 33), 147[18] (119, 120), 147[19] (119), 203[41] (195), 232[133] (228), 265[151] (248), 266[152] (248, 249, 250), 266[165] (250, 251)

Butow, R. A., 216[94] (213)

Buzayan, J. M., 60[43] (54), 166[12] (151), 187[53] (180, 183), 231[25] (221), 263[10] (247), 265[105] (240, 247, 248), 265[131] (246)

Byrne, E. M., 108[61] (105), 108[62] (105)

Cachia, C., 79[108] (72)

Cai, S.-J., 107[9] (98, 99)

Cai, Z. P., 203[18] (192), 266[172] (250, 251)

Cairo, A., 265[91] (239)

Calnan, B. J., 19[57] (18), 45[22] (22)

Campbell, T. B., 267[237] (257)

Cantin, E. M., 47[185] (43), 264[38] (237)

Canton, C., 108[52] (103, 105)

Cantor, C. R., 45[54] (23, 26, 27, 28, 29), 45[55] (23), 47[199] (43)

Capel, M. S., 204[88] (201)

Caponigro, G., 214[4] (205), 215[30] (207, 208, 211), 215[48] (208, 209)

Caprara, M. G., 231[62] (222), 231[63] (222)

Caput, D., 215[14] (206)

Carbon, P., 78[74] (65), 79[100] (69, 70)

Carey, N. H., 215[70] (210)

Carmi, N., 136, 147[73] (136, 144), 147[74] (137), 268[288] (262)

Carneiro, M., 214[2] (205)

Carpousis, A. J., 216[93] (213)

Carroll, K., 216[92] (212, 213)

Carter, B. J., 267[276] (260)

Carter, C., 108[25] (100)

Carter, Jr., C. W., 108[26] (100)

Carter, C. W. J., 60[12] (51), 60[13] (51)

Carter, K. C., 108[35] (101, 102)

Caruthers, H., 95[6] (82)

Caruthers, M. H., 45[49] (22, 28, 29), 45[74] (25, 29), 46[110] (30, 31), 167[101] (158), 231[74] (224), 263[19] (236, 244, 246, 257)

Casey, J. L., 108[50] (103), 108[51] (103)

Castanotto, D., 47[186] (43)

Cate, J. H., 16, 19[33] (13, 16), 45[18] (22, 33, 34, 36, 42), 60[64] (56), 61[70] (56, 57), 61[71] (57), 113[24] (110), 125, 147[31] (124, 125), 166[28] (151, 155, 156, 162), 166[61] (154, 155, 158), 167[93] (158), 203[2] (190, 192, 199), 204[74] (199), 231[46] (221), 267[224] (256)

Cavanagh, J., 19[5] (4)

Cavarelli, J., 79[134] (73, 74)

Cech, T. R., 14, 19[33] (13, 16), 19[39] (14), 44[2] (21), 44[3] (21), 45[18] (22, 33, 34, 36, 42), 46[108] (30, 43), 46[120] (31, 41, 43), 47[143] (36), 47[175] (43), 60[11] (51), 60[26] (52), 61[70] (56, 57), 78[36] (66), 96[54] (89), 113[1] (109, 112), 113[8] (109, 110, 111, 112, 113), 113[14] (109, 113), 129, 138, 139, 142, 146[1] (115), 147[31] (124, 125), 147[33] (125), 150, 203[2] (190, 192, 199), 215[55] (209), 231[31] (221), 231[38] (221), 232[75] (224), 232[105] (226), 233[142] (229), 233[152] (230), 236, 257

Cedergren, R., 79[149] (76), 263[7] (236, 237, 244), 267[285] (262)

Cedergren, R. J., 45[50] (22), 60[5] (50), 95[26] (86), 95[27] (86)

Celander, D. W., 78[31] (66), 166[17] (151), 166[20] (151, 152), 267[249] (257)

Cepanec, C., 79[107] (72)

Cerami, A., 215[14] (206)

Chaires, J. B., 46[79] (25)

Chamberlain, J. R., 113[20] (110, 111)

Chan, L., 107[10] (98, 99, 100), 107[13] (98), 108[27] (100)

Chan, Y.-L., 19[56] (16, 17), 204[57] (196, 198)

Chan, Y. L., 46[135] (34)

Chance, M. R., 78[39] (66, 77)

Chandler, A. J., 79[87] (68)

Chanfreau, G., 166[32] (152), 231[35] (221)

Chang, A. H., 266[208] (253)

Chang, B. H., 107[10] (98, 99, 100)

Chang, M.-F., 263[12] (252, 253, 254)

Chang, P. F., 47[185] (43)

Chang, P. S., 264[38] (237)

Chang, S. H., 78[62] (67)

Chang, Y.-F., 232[116] (226)

Chapman, K. B., 133, 147[54] (133, 144)

Chapman, T., 167[135] (162), 267[240] (257)

Charnsangavej, C., 108[27] (100)

Chartrand, P., 263[7] (236, 237, 244)

Chassignol, M., 231[69] (223)

Chattopadhyaya, J. B., 95[11] (83), 95[12] (83)

Chavan, A. J., 233[146] (229)

Chen, C. B., 79[132] (73)

Chen, C.-h. B., 79[97] (69)

Chen, C. X., 108[35] (101, 102)

Chen, C.-Y. A., 215[16] (206, 208), 215[17] (206, 208)

Chen, D.-S., 113[21] (110, 111)

Chen, J.-L., 166[41] (153), 233[151] (230)

Chen, L., 19[59] (18), 96[76] (93)

Chen, P.-J., 113[21] (110, 111)

Chen, S.-H., 107[9] (98, 99), 107[18] (99)

Chen, S. L., 96[81] (94), 232[126] (227)

Chen, X. Y., 45[32] (22, 38, 42)

Chen, Y., 231[51] (222), 267[283] (260)

Cheng, C. S., 203[40] (195)

Cheng, X. D., 108[41] (101)

Cheong, C., 19[22] (13), 45[19] (22, 34), 45[21] (22, 34), 203[20] (193)

Cheong, C. J., 12, 13, 19[20] (12, 13, 15)

Chevrier, B., 60[41] (53)

Chiba, A., 265[92] (239), 266[203] (253)

Chiu, R., 264[49] (237), 264[50] (237)

Choi, J. N., 266[166] (250)

Choi, Y. D., 266[166] (250)

Chow, C. S., 78[34] (66), 78[35] (66, 69)

Chow, W., 47[186] (43)

Chowrira, B. M., 147[18] (119, 120), 147[19] (119), 232[104] (226), 265[151] (248), 266[152] (248, 249, 250), 266[155] (248, 249, 250, 252)

Christensen, A. K., 215[73] (210)

Christian, A., 60[65] (56)

Christian, E., 167[144] (163), 267[244] (257)

Christian, E. L., 167[77] (157), 167[79] (157), 231[45] (221)

Christiansen, J., 216[97] (214)

Christodoulou, C., 95[16] (84), 96[82] (94)

Chu, B. C. F., 231[71] (223)

Chu, H., 108[52] (103, 105)

Church, G. H., 60[48] (55)

Church, G. M., 167[91] (158)

Ciafre, S., 147[64] (135)

Ciesiolka, J., 78[26] (66), 78[47] (66, 69), 167[88] (157)

Clark, B. F., 60[44] (54), 204[94] (202)

Clark, B. F. C., 44[13] (22, 40), 78[83] (67), 79[124] (73)

Clegg, R. M., 80[163] (76), 232[125] (227)

Cleland, W. W., 167[81] (157), 231[16] (219), 265[143] (246)

Cleveland, D. W., 215[42] (208)

Coates, L., 266[178] (252)

Cochrane, A. W., 215[9] (206), 216[100] (214)

Coetzee, T., 113[26] (110)

Cohen, M. A., 147[6] (115)

Cohen, S. B., 96[54] (89), 232[131] (228)

Cohn, M. J., 265[142] (246)

Colbert, J. T., 215[65] (210, 212, 213)

Cole, P. E., 45[37] (22)

Collins, R. A., 113[17] (110, 111), 166[58] (154), 166[59] (154)

Colman, A., 215[62] (210)

Comarand, M. B., 60[45] (54)

Conn, G. L., 204[46] (194, 195, 199)

Conn, M. M., 137, 144, 148[76] (137, 144)

Connell, G. J., 108[61] (105), 108[62] (105)

Connolly, B. A., 96[71] (93), 231[23] (220), 232[134] (228)

Cook, P. D., 96[42] (88)

Cormack, B. P., 167[133] (162)

Correll, C. C., 19[30] (13), 19[34] (13), 61[73] (59), 61[74] (59), 79[117] (72), 203[15] (192)

Cosstick, R., 147[33] (125), 167[102] (159), 231[72] (223), 232[75] (224), 264[28] (236, 244, 245, 246, 257)

Costa, M., 46[134] (34, 36), 60[66] (56), 124, 125, 140, 147[30] (124, 125), 147[32] (124, 140), 203[17] (192, 199, 200), 204[76] (200), 267[255] (259)

Coulson, A. R., 77[6] (64, 65)

Coutet, P., 215[64] (210)

Coutts, M., 215[63] (210)

Couture, S., 147[27] (124)

Cowan, J. A., 167[110] (159, 160), 264[34] (237, 246, 250, 251, 252)

Cox, C., 45[75] (25, 28, 29, 30, 31, 41, 42)

Crain, P. F., 80[164] (77), 107[4] (98), 107[5] (98)

Crain-Denoyelle, A.-M., 266[191] (252, 253)

Subject Index

PHILIP AND LESLEY ASLETT
Marlborough, Wiltshire, UK

Every effort has been made to index as comprehensively as possible, and to standardize the terms used in the index in line with the IUPAC Recommendations. In view of the diverse nature of the terminology employed by the different authors, the reader is advised to search for related entries under the appropriate headings.

The index entries are presented in letter-by-letter alphabetical sequence. Compounds are normally indexed under the parent compound name, with the substituent component separated by a comma of inversion. An entry with a prefix/locant is filed after the same entry without any attachments, and in alphanumerical sequence. For example, 'diazepines', '1,4-diazepines', and '2,3-dihydro-1,4-diazepines' will be filed as:-

    diazepines
    1,4-diazepines
    1,4-diazepines, 2,3-dihydro-

The Index is arranged in set-out style, with a maximum of three levels of heading. Location references refer to volume number (in bold) and page number (separated by a comma); major coverage of a subject is indicated by bold, elided page numbers; for example;

    triterpene cyclases, **299–320**
        amino acids, 315

See cross-references direct the user to the preferred term; for example,

    olefins *see* alkenes

See also cross-references provide the user with guideposts to terms of related interest, from the broader term to the narrower term, and appear at the end of the main heading to which they refer, for example,

    thiones
        see also thioketones